AF530721

Gene Editing:
Principles and Applications

B.D. Singh
N.S. Shekhawat

Published by

SCIENTIFIC PUBLISHERS (INDIA)
5 A, New Pali Road
P.O. Box 91
Jodhpur - 342 001 INDIA
E-mail: info@scientificpub.com
Website: http://www.scientificpub.com

ISBN : 978-93-89832-10-5 (HB)
eISBN : 978-93-89832-12-9

Printed in India

Respectfully dedicated to

Our Teachers

PREFACE

Organic evolution is fuelled by the genetic variation generated by spontaneous mutations that are fashioned by the forces of natural selection into adaptations and diversification of bioogical entities leading to progressive changes in their characteristics. Mutations are extremely useful for both basic and applied researches so that researchers have been in the constant search of tools and techniques for generating precise and predecided changes in the structures and/or functions of genes relevant to their objectives. This quest has resulted in the development of a suite of techniques ranging from chemical and physical mutagenesis, transgenic technology, *in vitro* site-directed mutagenesis, RNAi, antisense RNA and gene disruption technologies. These technologies have been very useful, but each of them offers some advantages and suffers from certain important limitations. For example, the transgenic technology enables utilization of any gene from the entire biological world so that the range, significance as well as novelty of its exploits are staggering. But the transgene integration is generally random and may lead to unpredictable consequences, which necessitates extensive evaluation for performance and biosafety assays. The regulatory approval of transgenic products is demanding and expensive, and their consumer acceptance is lukewarm. Similarly, the gene disruption technique has been used to create many extremely useful gene knockout animal models of human diseases. However, this technology is technically very demanding, involves sophisticated selection procedures, requires genuine ES cells, and is rather expensive. In addition, the frequency of gene integration by homologous recombination is very low.

It was observed during 1990s that the frequency of homology-directed gene integration can be greatly improved by deliberate induction of double-strand breaks (DSBs). These insights stimulated the development of reagents capable of cleavage at targeted genomic sites, which led to the creation of several site-directed nucleases like ZFNs, TALENs and CRISPR-Cas9. The CRISPR-Cas system is the easiest and the cheapest to redesign and implement for new target sites and is the most widely used. These SDNs, especially the CRISPR-Cas9, have revolutionized biological research since they allow precise targeting of desired genomic sites and introduction of all kinds of genetic changes the researchers would be interested in. The various types of genetic changes the researchers would like to induce may be classified into the following three broad groups: gene inactivation, mutation correction, and (iii) mutation induction leading to novel gene functions. Many genes may not be desirable or the researcher may be interested to analyse the functions of specific genes; in such cases, it would be desirable to specifically and reliably inactivate the desired genes. Many gene mutations lead to undesirable consequences, *e.g.*, human diseases; in such situations, it will be extremely useful to correct the changes in the mutant genes to restore their normal functions. In many other situations, it would be advantageous to modify an endogenous gene to enable it to perform a more useful function or to introduce a new gene into the genome of an organism to generate a useful novel phenotype that is not achievable with the endogenous genes.

SDNs induce DSBs at the targeted genomic sites and the genetic changes are produced during DSB repair by the normal cellular DNA repair pathways. These changes lead to gene inactivation due to indel formation due to nonhomologous end-joining. The correction of mutations in endogenous genes is achieved by sequence replacement due to homology-directed repair. This process can also lead to induction of the desired mutation in an endogenous gene or even integration of complete gene constructs at the specified genomic sites. Sequence replacement for mutation correction/induction or transgene integration requires a suitable donor DNA during exposure of the cells to the selected SDN system. In addition, the CRISPR-Cas9 system has been used to develop base editors for targeted chemical modification of C-to-T and A-to-G leading to single base substitutions at the specified genomic sites. It has also been used to develop epigenetic editors for histone modification and DNA methylation to activate or suppress the expressions of targeted genes. Finally, it has been used to create highly potent genetic devices called gene drives, which can propagate themselves and rapidly spread in crossbreeding experimental as well as natural populations. Gene drives can be used to eradicate harmful natural populations, rescue threatened populations or suitably alter their genetic makeup. But in the absence of appropriate guidelines, gene drives are yet to be field-tested. Finally, effective mechanisms need to be developed for preventing the misuse of gene editing and gene drive technologies by extremist groups.

Genome editing is being used to improve crop plants and other food sources, including animal species, and some of the products have started to reach the consumers. Therefore, policies and procedures for specific regulation of gene-edited plants and animals should be developed and the regulatory laws/ regulations of different countries need to be harmonized. Genome editing has been used to generate large animal models of human diseases that are useful for biomedical studies, including screening of drugs for disease management. Some of the gene-edited animals may possibly serve as donors of organs for humans. Human somatic cell genome editing has been used to develop some gene therapies that are already in clinical use. Human germline gene editing for clinical use is considered 'irresponsible' at present, but CRISPR-babies have already been borne in China. Thus, there is a need for an efficient and effective mechanisms for oversight and governance to prevent inappropriate use of the gene editing technology. The WHO of United Nations has constituted a committee that intends to develop a strong international regulatory framework for germline gene editing and 'a central registry on human genome editing research.'

The gene editing technology is a powerful tool that can be put to many extremely valuable uses, but it also opens the door to unexpected and unprecedented perils. For example, a gene drive targeting a female fertility gene is being field-tested by a British firm in Brazil for eradication of *Aedes aegypti* mosquitoes. This application has important benefits for human health, but gene drives can be designed to eliminate even useful species from the face of earth, which may have adverse effects on the ecosystem. Therefore, effective measures must be devised to prevent the misuse of gene editing in general and gene drives in particular by extremist groups.

The present book, Gene Editing: Principles and Applications, attempts to bring to readers the various developments in this rapidly evolving field in a concise, readable and easily comprehendible narrative. The book targets biologists in general and geneticists, biomedical researchers and plant breeders in particular. It is hoped that it will be useful to post-graduate students, research scholars and research workers concerned with analyses of biological phenomena and development of strains with novel and useful traits. The development and various features of chimeric oligonucleotides and the four major SDN systems, viz., meganucleases, ZFNs, TALENs and CRISPR-Cas9, are described in four chapters, one chapter deals with base editors, five chapters discuss the applications of these novel technologies in plants, animals, human health, gene regulation and natural populations and the last chapter discusses the ethical issues and regulations pertaining to the gene editing products.

We wish to emphasize that the field of gene editing is advancing at a rapid pace and it is difficult to keep abreast with the latest developments and include them all in a book like this. We have attempted to present as much and as up-to-date information about gene editing as it was practicable to help the readers appreciate the more recent developments in the field.

The authors appreciate the sincere efforts of Mr. Tanay Sharma and his production team, including reviewers, editors, graphics designers and others of Scientific Publishers (India), Jodhpur for their sincere and commendable efforts that have culminated in the speedy publication of this book in an excellent and attractive form. The authors are thankful to Dr Ramesh Solanki, Jodhpur for his help in correction of proofs.

In the end, we would like to express our sincere appreciation for the continued affectionate support from all our family members during our involvement in this project. This has been the prime source of our strength throughout our academic life particularly when there were activities like this that demanded the bulk of our time and attention

25.12.2020

B.D. Singh
N.S. Shekhawat

ABBREVIATIONS

5hmC, 5-hydroxymehtylcytosine

A1AD, alpha-1 antitrypsin deficiency

A3A, human cytidine deaminase APOBEC3A

AAV, Adeno-associated virus

ABE, adenine base editor

ADA, adenosine deaminase

ADAR1, adenosine deaminase acting on RNA 1

ADAR2, adenosine deaminase acting on RNA 2

agRNA, aptazyme-embedded sgRNA

AHAS, acetohydroxy acid synthase

AHC-HH, adrenal hypoplasia congenital and hypogonadotropic hypogonadism

AID*, hyperactive mutant of AID

AID*Δ, AID* with deleted NES

AID, activation-induced deaminase

AIDS, acquired immunodeficiency syndrome

AIDx, AID lacking NES

ALS, acetolactate synthase

APHIS. Animal and Plant Health Inspection Service

ART, antiretroviral therapy

ASFV, African swine fever virus

AteCas9, *Arabidopsis thaliana* codon-optimized Cas9

Aux/IAA9, AUXIN-INDUCED 9

AV vectors, Adenovirus vectors

BAT, brown adipose tissue

BE3, third generation base editor

BE4, fourth generation base editor

BER, base excision repair

BFP, blue fluorescent protein

BLESS, breaks labelling, enrichment on streptavidin, and sequencing

BLG, β-lactoglobulin

BSR, base-specifying residue (in case of RVDs)

CAC, Codex Alimentarius Commission

CAPS, cleaved amplified polymorphic sequences

CAR, chimeric antigen receptor

Cas9, CRISPR-associated 9

Cas9-HF, high-fidelity Cas9

Cas9p, plant codon-optimized Cas9

CCR5, C-C chemokine receptor type 5

CD4, cluster of differentiation 4

cDNA, copy-DNA or complementary DNA

CFTR, cystic fibrosis transmembrane conductor receptor

CoDA, context-dependent assembly

COs, chimeric oligonucleotides

CPB, Cartagena Protocol on Biosafety

CPPs, cell-penetrating peptides

CRISPR, clustered regularly interspaced short palindromic repeat

CRISPRa, CRISPR-activation

CRISPRi, CRISPR-interference

CRISPRm, multiplex CRISPR

crRNA, CRISPR RNA

CRY2, cryptochrome 2

CS domain, chromo-shadow domain of HP1a

CSA, carbon starved anther

CTLA-4, cytotoxic T lymphocyte–associated protein 4

CXCR4, C-X-C chemokine receptor type 4

D10A, Asp at position 10 replaced by Ala

DB-PACE, DNA-binding PACE

dCas9-SAM, dCas9-synergistic activation mediator

dCas9-VPR, dCas9-VP64-p65-Rta

DD, destabilizing domain; deaminase domain

DEP1, DENSE AND ERECT PANICLE 1

DETECTR, DNA endonuclease targeted CRISPR *trans*-reporter

DH, doubled haploid

Digenome-seq, *in vitro* nuclease-digested whole-genome sequencing
DMD, Duchenne muscular dystrophy
DNMT3A, human DNA methyltransferase
dPspCas13b, 'dead' PspCas13b
Drb2a, double-stranded RNA-binding protein 2a
DSB, double-strand break
dsODN, double-stranded oligodeoxynucleotide
DsRed, *Discosoma* red
dsRNA, double-stranded RNA
dU, deoxyuridine
eA3A, engineered A3A
EBV, Epstein–Barr Virus
ECJ, European Court of Justice
ecTadA, *E. coli* TadA
EFSA, European Food Safety Authority
eGFP, enhanced green fluorescent protein
EMA, European Medicines Agency
EMS, ethyl methane sulfonate
EPSPS, 5-enolpyruvylshikimate-3-phosphate synthase
ERACRs, elements to reverse the autocatalytic chain reaction
ESC, embryonic stem cell
eSpCas9, "enhanced specificity" SpCas9
EU, European Union
Fab, fragment antigen-binding
FACS, fluorescence-activated cell sorting
FAD2, Fatty acid desaturase 2
FAO, Food and Agriculture Organization
FDA, Food and Drug Administration
Fen-1, flap endonuclease 1
FKBP, FK506 binding protein 12
FLASH assembly, fast ligation-based automatable solid-phase high-throughput assembly
FoxP3, Forkhead box p3
FP, fluorescent protein
FRB domain, FKBP rapamycin binding domain
FRET, fluorescence resonance energy transfer
GATA-3, GATA binding protein 3
gDNA, guide DNA
GFP, green fluorescent protein
GI, Genetic interaction
GM, genetically modified
Gn1a, *GRAIN NUMBER 1a*
gRNA, guide RNA
GRON, gene repair oligonucleotide
GS3, *Grain Size* 3
GUIDE-seq, genome-wide, unbiased identification of DSBs evaluated by sequencing
hAID*Δ, engineered variant of hAID lacking the NLS sequence
hAID, human AID
HAR, hyperacute rejection
HATs, histone acetyltransferases
HBV, Hepatitis B Virus
HCV, Hepatitis C Virus
HDACs, histone deacetylases
HD-Ad vector, helper-dependent adenoviral vector
HDR, homology-directed repair
HDV, Hepatitis Delta Virus
HE, homing endonuclease
hESCs, human embryonic stem cells
HF-BE3, high fidelity BE3
hiPSCs, human induced pluripotent stem cells
HITI, homology-independent targeted integration
HIV-1, Human Immunodeficiency Virus-1
hLF, human lactoferrin
HMA, heteroduplex mobility assay
hPSCs, human pluripotent stem cells
HPV, Human Papillomavirus
HRMA, high-resolution melt analysis
HSA, human serum albumin
HSCs, hematopoietic stem cells
HSPCs, hematopoietic stem/progenitor cells
HTGTS, high-throughput, genome-wide translocation sequencing
HypaCas9, "hyper-accurate" variant of Cas9
IAA9, AUXIN-INDUCED 9
IBC, International Bioethics Committee
ICA, iterative capped assembly
IDLV capture, capture of integrase-deficient lentiviral vectors
IDLV, integrase-deficient lentivirus vector
IGD, integral gene drive

IL-2, interleukin 2
IL2rg, interleukin-2 receptor gamma chain
Indels, insertions/deletions
IPA1, *IDEAL PLANT ARCHITECTURE 1*
iPSCs, induced pluripotent stem cells
iSTOP, induction of stop codons
IVF, *in vitro* fertilization
JAK3, Janus kinase 3
kb, kilobase pair
KRAB, Kruppel-associated box
LAHEDES, LAGLIDADG homing endonuclease database and engineering server
LAM-PCR, linear amplification mediated PCR
LbCpf1, *Lachnospiraceae* bacterium ND2006 Cpf1
LDL, low-density lipoprotein
LHE, LAGLIDADG homing endonuclease
LIC, ligation-independent cloning
LMOs, living modified organisms
lncRNAs, long noncoding RNAs
LTR, long terminal repeat
LV, Lentiviruses
Mb, megabase pairs or megabases
MC, macrocalyx
MCPH1, *microcephalin 1*
MCR, mutagenic chain reaction
MCS, multiple cloning site
MEDEA. Maternal effect dominant embryonic arrest
MMEJ, microhomology-mediated end-joining
MNickases, meganickases
MNs, meganucleases
msgRNA, modified sgRNA
MSTN, myostatin
MTALs, MegaTALs
MTL, *MATRILINEAL*
NASEM, National Academies of Sciences, Engineering, and Medicine (USA)
ncRNAs, noncoding RNAs
NES, nuclear export signal
NGO, non-governmental organizations
NGS, next generation sequencing
NHEJ, nonhomologous end joining
NHPs, nonhuman primates
NIR, near infrared
NK58S, Nongken 58S
NLB, Northern leaf blight
NLS, nuclear localization signal
NmCas9, *Neisseria meningitidis* Cas9
Non-GM, non-genetically modified
NRAMP1, natural resistance-associated macrophage protein-1
nt, nucleotide(s)
ODM, oligonucleotide-directed mutagenesis
OPEN, oligomerized pool engineering
ORFs, open reading frames
OsmiR156, *Oryza sativa* microRNA 156
p65AD, activation domain of protein p65
PABE, plant adenine base editor
PACE, phage-assisted continuous evolution
PAGE, polyacrylamide gel electrophoresis
PAM, protospacer adjacent motif
PAMmers, PAM presenting oligonucleotides
PBMCs, peripheral blood mononuclear cells
PCR, polymerase chain reaction
PCSK9, Proprotein convertase subtilisin/kexin type 9
PD-1, programmed cell death protein 1
pDBD, programmable DNA-binding domain
PEG, polyethylene glycol
pegRNA, prime editing guide RNA
PEI, polyethylenimine
PEP, phosphoenolpyruvate
PERV, Porcine Endogenous Retrovirus
PGD, preimplantation genetic diagnosis
PGMS, photoperiod-sensitive genic male sterility
PITCh, precise integration into target chromosome
PNT, 'Plants with Novel Traits'
PRRS, porcine reproductive and respiratory syndrome
PSCs, pluripotent stem cells
PspCas13b, Cas13 from *Prevotella* sp.
QTLs, quantitative trait loci
QTN, quantitative trait nucleotide
rAPOBEC1, rat apolipoprotein B mRNA editing complex 1
rBE3, rice base editor 3
RCas9, RNA-targeted Cas9

RDOs, RNA-DNA oligonucleotides

REAL method, restriction enzyme and ligation method

REPAIR, RNA editing for programmable A to I replacement

REPAIR v1, REPAIR with hyperactive mutant of DD

REPAIR v2, REPAIR with hyperactive mutant of DD engineered for improved specificity

RE–PCR, restriction enzyme digestion suppressed PCR

re-TALE, recoded TALE

RFLP, restriction fragment length polymorphism

RFP, red fluorescent protein

Rin, ripening-inhibitor

RISC, RNA-induced silencing complex

RITS, RNA-induced transcriptional silencing

RNAi, RNA interference

rPGMS, reverse PGMS

RTDS™, Rapid Trait Development System™

RVD, repeat variable di-residue

SaCas9, *Staphylococcus aureus* Cas9

SAM, synergistic activation mediator

SCBD, Secretariat of the Convention on Biological Diversity

scFvs, single-chain antibodies

SCID, severe combined immunodeficiency

SCNT, somatic cell nuclear transfer

SDN, site-directed nuclease

SGN, structure-guided endonuclease

sgRNA, single guide RNA

sgSTOPs, sgRNAs iSTOP

SHERLOCK, Specific High-Sensitivity Enzymatic Reporter UnLOCKing

SID, mSin interaction domain

SlAGL6, Solanum lycopersicum AGAMOUS-LIKE 6

SlIAA9, Solanum lycopersicum IAA9,

SNP, single nucleotide polymorphism

SP5G, SELFPRUNNING 5G

SpCas9, *S. pyogenes* Cas9

SpCas9-HF, "high fidelity" SpCas9

SSB, single-strand break

ssDNA, single-stranded DNA

ssODN, single-stranded oligodeoxynucleotide

ssRNA, single-stranded RNA

STAR, simple TALE assembly reaction

swigRNA, switchable guide RNA

T7E1, T7 Endonuclease I

TadA*, TadA engineered to edit adenines in DNA

TadA, tRNA adenine deaminase

TALE, transcription activator-like effector

TALEa, TALE activator

TALE-CEs, TALE-chromatin editors

TALEN, transcription activator-like effector nuclease

TALEr, TALE repressor

TALER, TALE-recombinase

TALE-TFs, TALE-transcription factors

TAM, targeted AID-mediated mutagenesis

T-bet, T-box transcription factor TBX21

TCR, T cell receptor

TDG, thymine DNA glycosylase

T-DNA, transferred-DNA

TET, Ten-eleven translocation

TET1CD, TET dioxygenase1

Tev-ZFEs, I-TevI- zinc finger endonucleases

TGF-β, transforming growth factor β

TGMS, temperature-sensitive genic male sterility

Th1, Type 1 T helper

Th2, Type 2 T helper

TILLING, Targeting induced local lesion in genomes

tracrRNA, *trans*-activating chimeric RNA

Treg, regulatory T cell

trigDNA, trigger DNA

trigRNA, trigger RNA

TRV, Tobacco Rattle Virus

TSS, transcription start site

UDG, uracil DNA glycosylase

UGI, uridine glycosylase inhibitor

Ugp-1, UDP-glucose pyrophosphorylase1

ULtiMATE, USER-based ligation-mediated assembly of TAL effector

UNESCO, United Nations Educational, Scientific and Cultural Organization

UNO, United Nations Organization

USDA, United States Department of Agriculture

USER, uracil-specific excision reagent

UTR, untranslated region

V92A, Val at position 92 replaced by Ala

VPR, VP64-p65-Rta

WGS, whole genome sequence/sequencing

WHO, World Health Organization

Xoo, *Xanthomonas oryzae* pv. *oryzae*

XPC, Xeroderma pigmentosum complementation group C

X-SCID, X-linked severe combined immunodeficiency

zCas9, maize codon-optimized Cas9

ZF, zinc finger

ZFE, zinc finger endonuclease

ZFN, zinc-finger nuclease

ZFNickase, zinc-finger nickase

ZFP, zinc finger protein

γ-RV, Gamma-retroviral vectors

CONTENTS

1

DIRECTED MUTAGENESIS

The 2020 Nobel Prize in Chemistry was awarded to Emmauelle Charpentier of France and Jennifer Doudna of U.S.A. for the "development of a method of genome editing".

1.1. INTRODUCTION

The term mutation was first used by de Vries in 1900 to describe sudden changes in the characteristics of *Oenothera* sp. A simple definition of mutation would be a sudden heritable change in the characteristics of an organism. The basis of a mutational event may be a change in chromosome structure or even number, an altered base sequence of a gene/plasmagene or an epigenetic change, such as DNA methylation. Mutations occur spontaneously at low rates mainly due to errors during replication by DNA polymerase, activity of transposable elements, spontaneous deamination of 5-methylcytosine/cytosine, etc. The rates of spontaneous mutations vary depending on the organism, the gene in question, and several other factors. In general, mutant alleles produce deleterious effects and many incapacitating human diseases are caused by single mutant genes. Significantly, *mutations are the ultimate source of all genetic variation present in the biological world, and genetic variation is essential for organic evolution to proceed.* Thus, the genetic variation utilized for animal and crop improvement has originated from past mutations. There are several instances of spontaneous mutations being utilized for the establishment of new animal breeds/crop varieties. For example, the spontaneous semi-dwarf or reduced height mutants of wheat and rice provided the genetic foundation for the dramatic increase in global cereal production described by many as 'green revolution' (Singh, B.D., 2018, 2019).

Mutations are extremely important for a variety of biological investigations, including genetic analyses of phenotypes and dissection of biochemical pathways into individual reactions. However, spontaneous mutations are random, very low frequency (10^{-5} or lower per gene per generation) events. Therefore, researchers have long been trying to develop tools and techniques for inducing mutations in high frequencies and preferably in the desired genes. These persistent endeavours have been amply rewarded beginning with the discovery of physical and chemical mutagens (Section 1.2) and ending with the development of highly specific and efficient site-directed gene/genome editing tools that enable the desired modification in the targeted genes (Section 1.7).

The various types of genetic changes the researchers would like to induce may be classified into the following three broad groups: (i) gene inactivation, (ii) mutation correction, and (iii) mutation induction leading to novel gene functions. Many of the genes present in genomes of various organisms may not be desirable from one or the other perspective. In addition, the researcher may be interested to investigate the effects of inactivation of specific genes on phenotypes of the concerned organisms. In these and some other situations, it would be desirable to inactivate the relevant genes in a specific manner; several approaches can be used for achieving this objective (Sections 1.5-1.7). But when gene mutations lead to some undesirable consequences, *e.g.*, human diseases, it will be extremely useful to correct the changes in such mutant genes to restore their normal functions. This could be done either by providing a new normal copy of the concerned gene or a directed sequence replacement in the mutant gene (Sections 1.8.2, 1.8.3). In many situations, it would be advantageous to introduce a novel gene function into an organism to generate a useful phenotype that is not achievable with the endogenous genes of the species. In such cases, new genes are identified, isolated and cloned, and ultimately transferred and expressed into the concerned species using a suitable gene transfer technology (Section 1.3). The developments and applications of various modern strategies for genetic modification have been discussed in some detail by Lanigan et al. (2020).

1.2. PHYSICAL AND CHEMICAL MUTAGENESIS

The first notable success at deliberate induction of mutations was reported by Muller in 1927; he used X-ray treatment to induce mutations in *Drosophila.* One year later, in 1928, Stadler demonstrated the mutagenic action of X-rays in barley. These studies provided a tool for induction of heritable changes in genes and raised the possibility of controlled genetic manipulations in organisms. The significance of this discovery is reflected by the award of Nobel prize to Muller in 1946. These research findings stimulated the search for other physical and chemical agents capable of inducing mutations. As a result, several radiations including ultraviolet (UV) rays, and many chemicals like alkylating agents, base analogues, acridine dyes, nitrous acid, sodium azide, etc. were found to induce mutations. The physical and chemical mutagens induce mutations at frequencies that are several orders of magnitude higher than those of spontaneous mutations (Sturtevant, 1965).

The various mutagens have been used to induce mutations in many organisms and the recovered mutants have been extensively used for both basic and applied researches. For example, UV-induced mutants of *Neurospora crassa* were used for studies that led to formulation of the famous one-gene-one-enzyme hypothesis by Beadle and Tatum (1941). Mutagens have been widely used for improvement of agricultural and horticultural crops. As a result, 3,281 mutants belonging to over 200 different crop species have been released as varieties for commercial cultivation (IAEA Vienna Mutant Variety Database, https://mvd.iaea.org/ as on 30.10.2018). Mutagenesis has been successful in the induction of some useful new alleles that were not available/accessible to breeders. It has been particularly useful in correction of specific defects of otherwise well-adapted, high yielding and highly successful varieties. For example, peppermint (*Mentha piperita*) cultivation in U.S.A. is a multimillion-dollar industry and is based on a single peppermint clone. This clone became susceptible to *Verticillium* wilt and peppermint cultivation was threatened. A large-scale mutation breeding programme was initiated and a wilt resistant mutant with yield and quality comparable to those of the parental clone was isolated. This mutant clone saved the peppermint crop from being wiped out from U.S.A., and is now widely cultivated (Singh, 2018).

The use of physical and chemical mutagens is inexpensive, relatively simple and does not require technical expertise. But these agents generate a variety of adverse effects on the treated individuals and lead to considerable mortality. As a matter of fact, the optimum dose of a mutagen is generally regarded as the one, which causes nearly 50% mortality in the treated population. At the same time, the mutations induced by these mutagens are random as are spontaneous mutations. Mutagenic treatments generate very large numbers of genetic changes in the genomes of each individual in addition to the gene mutation responsible for its mutant phenotype. In any case, these mutagens can cause inactivation of endogenous genes and produce a low frequency (~0.1%) of mutant alleles with useful/beneficial effects. However, they cannot generate new gene functions that may be required in many situations; this objective can be achieved only by the transgenic technology.

1.3. TRANSGENIC TECHNOLOGY

Transgenic technology involves gene transfers into selected organisms using recombinant DNA technology or genetic engineering to achieve stable expression of these genes for generating the desired phenotypes. In simple terms, the recombinant DNA technology comprises production of genomic DNA fragments or cDNA (DNA copies of RNA, usually mRNA) molecules from an organism, integration of these DNA fragments into a suitable vector to produce recombinant DNAs and cloning them in a suitable host like *Escherichia coli*. The bacterial clone containing the desired DNA fragment or gene is identified, isolated and multiplied. The gene isolated from this clone is then used to create a gene construct comprising coding region of the gene and appropriate regulatory sequences needed for its expression in the organism, in which it is to be introduced. This gene construct is then transferred into the desired organism using a suitable gene transfer technology. The availability of restriction enzymes (Type II endonucleases) enabled the development of recombination DNA technology. Smith, Nathans and Arber were awarded the Nobel Prize for Physiology and Medicine in 1978 for their work on restriction enzymes (Old and Primrose, 2001).

Where required, chimeric genes can be created by combining coding and regulatory sequences from different genes to achieve the desired objectives. The transgenic technology can be used to transfer any gene of interest isolated from any organism or created by chemical synthesis into the organism of choice. The genes transferred by recombinant DNA technology are called transgenes and the individuals carrying and expressing them are known as transgenics. Generally, transgenes are integrated at random sites in the genomes of the recipient individuals, but in some organisms like yeast integration by homologous recombination seems to be rather frequent. The recombinant DNA technology has enabled the introduction of novel gene functions into various organism and the suppression of specified endogenous genes (Section 1.5). This technology has been widely used for basic investigations and for creation of transgenic organisms with enhanced usefulness to humans (Old and Primrose, 2001).

The recombinant DNA technology revolutionized our basic understanding of biology and has enabled molecular investigations with unprecedented precision, sensitivity and informativeness. For example, it allowed the development of first-generation DNA sequencing methods that provided us our first access to genome sequences and the development of genomics. In the applied arena, it enabled the production of human and other proteins of therapeutic significance in bacteria, yeast, animals and even plants. Transgenic crop varieties with very valuable features, such as resistance to stresses like insects, diseases, herbicides, drought, etc., improved quality traits, enhanced yield, etc. have been developed in many crops, and they are in cultivation since 1996. Some of these varieties are remarkable for their impact on the concerned crop. For example, a transgenic ringspot virus resistant papaya variety saved the papaya cultivation is Hawaii, U.S.A. Another remarkable development relates to the creation of transgenic drought tolerant varieties of maize and sugarcane; these varieties require up to 15% less irrigation water than their normal counterparts. The cultivation of transgenic crop varieties has expanded at a rapid rate from merely 1.7 million hectares in 1996 to 185.1 million hectares in 2016. The major part (54%) of this area is in developing countries.

The range, significance as well as novelty of the exploits of transgenic technology are staggering. But the transgene integration is generally random, which may create unpredictable and unintended effects. This necessitates extensive evaluation of transgenics for stable performance and absence of position effects. In addition, the biosafety assays required by regulatory authorities for approval of transgenic crop varieties for commercial cultivation require, on an average, about 5.5 years of time and investment of US $ 35 million. In any case, the transgenic or GM (genetically modified) food items are not readily accepted by consumers, particularly in European Union. Further, many countries like Russia and India either explicitly or effectively proscribe GM crop cultivation. Finally, the transgenic technology does not allow modification of specified endogenous genes in a predetermined manner, such as inactivation, desired mutation induction and mutation correction.

1.4. *IN VITRO* SITE-DIRECTED MUTAGENESIS

Site-directed mutagenesis or site-specific mutagenesis comprises introduction of specified changes at predetermined sites of the targeted genes and, thereby, their products. The various techniques used for site-directed mutagenesis are based on recombinant DNA technology. In simple terms, the targeted gene is first isolated and mutated *in vitro*. The mutant gene is then integrated, generally, at random sites in the genome of the selected organism; this gene would function in addition to the endogenous copy of the gene. The first report of site-directed mutagenesis appeared in 1978 and it used oligonucleotides as primers for extension by DNA polymerase. The oligonucleotides were complementary to the gene sequence around the intended mutation site and contained the desired change in the base sequence. Extension of this primer by DNA polymerase copied the rest of the gene sequence and thereby produced the mutant copy of the gene. The mutated copies of the gene were cloned in *E. coli* and sequenced to identify and isolate the copy with the desired mutation. These strategies are seldom used now since newer methods are faster, more convenient and some of them have high throughput.

Some other methods of *in vitro* site-directed mutagenesis are briefly described here. One of the methods uses a phagemid, *e.g.*, M13mp18/19, as vector to clone the targeted gene. Single-stranded version of the recombinant phagemid DNA is isolated and mixed with an oligonucleotide complementary to the specified site of the target gene. This oligonucleotide contains the pre-decided mutational change(s) preferably in its middle region. It will pair with the single-stranded gene in the targeted region and will be used as primer for replication of the M13 recombinant DNA. The nick remaining after replication is sealed and the recombinant phagemid DNA is introduced into *E. coli* cells, where replication will produce the mutated version of gene. Ultimately, the gene with the desired mutation is recovered from M13 recombinant DNA and used for achieving the desired objective.

Another strategy of *in vitro* site-directed mutagenesis is useful when the targeted site in the desired gene is flanked on either side by unique restriction sites. In such a case, the targeted region is excised by digesting it with the concerned restriction enzymes, and a synthetic oligonucleotide is integrated in its place. The synthetic oligonucleotide is homologous to the deleted sequence of the gene, except for the desired mutational changes deliberately included in its sequence. In this approach, several different versions of the gene, each containing a different mutation, can be simultaneously generated. This can be achieved by using a mixture of oligonucleotides, each containing a different mutational change, for integration into the gene in the place of the excised region. Suitable strategies have also been developed to introduce mutant oligonucleotides in such regions of the targeted genes that are not flanked by restriction sites. The above strategies (the phagemid- and restriction enzyme-based methods) allow base substitutions, base additions as well as deletions.

The cost of oligonucleotide synthesis is consistently decreasing so that chemical DNA synthesis is occasionally used to generate copies of complete genes with mutations at one or more specified sites of the genes. This approach allows even complete redesigning of genes to achieve specific objectives like codon optimization for expression in different host organisms. Some other methods of site-directed mutagenesis are whole plasmid mutagenesis and polymerase chain reaction (PCR)-based site-directed mutagenesis. PCR-based method uses a PCR primer having the desired mutational change so that all the copies of the gene amplified using this primer carry the desired mutation. Initially, the PCR procedure allowed mutational changes only in the end sequences of genes. But refinements in the PCR strategy now permit mutational changes anywhere in the targeted gene. This method is suitable for base substitutions. The PCR products must be cloned and sequenced to obtain the copy that has the desired mutation and is free from errors introduced during PCR.

In vitro site-directed mutagenesis is used for investigation of the effects of specific change in DNA sequences on the structures and functions of DNAs, RNAs and proteins, and for other basic investigations. It can also be used for protein and enzyme engineering, *i.e.*, production of proteins/enzymes with altered properties and/or functions. Obviously, these strategies do not allow a targeted *in vivo* modification of genes.

1.5. ENDOGENOUS GENE SUPPRESSION

Inactivation of specific genes can be desirable for both basic and application-oriented researches. In basic studies, inactivation of selected genes is useful for determination of their functions. In applied research, inactivation of a gene may be desirable when the concerned gene leads to some undesirable feature(s), or it may generate a more desirable phenotype. The various strategies used for endogenous gene inactivation/suppression include insertional mutagenesis, gene disruption, antisense RNA technology and RNA interference (RNAi). Insertional mutagenesis relies on integration of such sequences in genes that disrupt their functions. Examples of insertional mutagenesis include transposon mutagenesis, and viral vector and T-DNA (transferred-DNA) integrations generally within the coding regions of genes. Transposons are nucleic acid sequences that encode the enzymes needed for their movement from one location to the other within the genome. Several of these elements have been reconstructed for insertional mutagenesis in animal/plant species. Sleeping Beauty is one such transposon that has been used for random gene inactivation in several vertebrate species, including human cells in culture, and mouse and human embryonic stem cells. The integration of these DNA sequences occurs at random genomic sites and may inactivate or activate a gene located near the integration site (Chugunova et al., 2016). T-DNA is the part of Ti plasmid of *Agrobacterium* that is transferred into plant cells and is integrated at random sites in the plant genomes leading to similar consequences as transposon integration. These and other similar approaches have been widely used for developing mutant libraries used for gene function analyses, and isolation of genes associated with specific phenotypes. However, these methods simultaneously target multiple genes, cannot be used to inactivate single specific genes and their results cannot be reproduced.

Gene disruption involves integration by homologous recombination of a DNA sequence, the donor DNA sequence, within a gene; this integration prevents expression of the gene. The donor DNA sequence must be flanked by sequences homologous to the gene being disrupted to allow homologous recombination (HR) to occur between the gene and the donor DNA. In addition, appropriate marker genes are included in the donor DNA to allow selection for the products of HR events. The main limitation of this strategy is the very low frequency of HR and the need for highly sophisticated, tedious and time-consuming selection strategies. It has been reported that treatments with DNA damaging agents greatly increases the frequency of homologous recombination-based gene integration (Chugunova et al., 2016). Despite these difficulties, gene disruption has been used to develop gene 'knockout' mouse models of several human diseases that are extremely valuable for a variety of biomedical investigations.

The antisense RNA technology of endogenous gene suppression is based on antisense gene construct of the targeted gene, *i.e.*, the gene to be suppressed. In the antisense gene construct, the coding region of the targeted gene is inverted, *i.e.*, is rotated by 180^0, in relation to its promoter. As a result, the sense strand of this gene is transcribed to yield the antisense RNA. In contrast, the antisense strand of the native gene is transcribed, and this gives rise to the pre-mRNA transcript. When an antisense construct of a native gene is introduced into a cell, the cell will contain both the native gene as well as its antisense construct. The antisense construct will produce antisense RNA, which will base-pair with the pre-mRNA transcribed from the native gene. The resulting double-stranded RNA is processed by Dicer into small (~22 nucleotide-long) fragments that associate with either RISC (RNA-induced silencing complex) or RITS (RNA-induced transcriptional silencing) complex and ultimately lead to mRNA degradation or transcriptional silencing of the concerned gene.

RNAi is the strategy used by eukaryotes to combat invasion by RNA viruses. RNAi is caused by small double-stranded RNA (dsRNA) molecules that associate with RITS complex. The RITS complex degrades one of the strands of the dsRNA molecules and uses the other strand to target the mRNA having complementary sequence leading to its disintegration. Different sequences from a single gene differ in their effectiveness for causing RNAi. A highly effective sequence is selected from the target gene and supplied to the cells as dsRNA, which leads to transient RNAi that lasts until the exogenously supplied dsRNA is degraded. Alternatively, it can be used to design a gene construct, in which two inverted repeat copies of the sequence are separated by a small linker sequence. Expression of this gene in cells/plants yields a RNA transcript that folds to from a hairpin loop, which is processed to yield dsRNA. This approach generates long-lasting RNAi and has been used to produce transgenic plants with specific gene suppression. RNAi is a rapid, inexpensive and high-throughput technique for gene suppression especially when transient gene knockdown strategy is used. But the RNAi effect is incomplete, gives variable gene knockdown and produces unpredictable off-target effects. These features of RNAi limit its usefulness for associating many phenotypes with specific genes and restrict the practical application of the technology (Gaj et al., 2013).

1.6. GENE TARGETING BASED ON EMBRYONIC STEM CELLS

It was observed that when DNA is introduced into mammalian cells cultured *in vitro*, it can become incorporated, albeit in very low frequencies, into the cell genome at the homologous locus presumably by a process similar to HR. Ingenious strategies were devised to select the rare recombinant mouse embryonic stem (ES) cells having the introduced DNA/gene integrated into their genomes by homology-dependent recombination (HDR). The selected ES cells were injected into developing young embryos to recover whole mice carrying the introduced DNA/gene. The homologous gene integration was known to occur in yeast in much higher frequencies, and a method based on it was used to target specific genes in *Drosophila*. Subsequently, it was used to knockout the proto-oncogene *int-2* in mice (Mansour et al., 1988). The ES cell technology has since been used to produce a variety of 'knockout' mice that have been used as disease models. The knockout animals enable analyses of functions of specific genes in growth, development, health, reproduction, etc. These animals are also used as experimental models for various biomedical investigations, including the roles of specific genes in disease development and management (Carroll, 2014, 2016).

The chief limitation of this technology is that the frequency of homologous integration is very low as compared to random integration. This leads to difficulties in selection for the cells produced by homologous integration. In addition, genuine ES cell lines are not available for many animals so that whole animals carrying the disrupted genes cannot be recovered in these species. It was recognized that homology-dependent gene integration involved normal cellular processes of DNA repair. It was further discovered that HR during meiosis and gene conversion is initiated by DSBs in DNA; DSBs are lethal if they are left unrepaired (Youds and Boulton, 2011). It was observed that the frequency of homology-directed gene integration can be greatly improved by deliberate induction of DSBs. These insights stim-ulated the development of reagents capable of cleavage at targeted genomic sites (Section 1.7) to enhance homology-directed gene integration at specified genomic sites (Carroll, 2014, 2016).

In addition to promoting HDR, DSBs lead to inactivation of genes, in which they are located. The inactivation of genes occurs due to DSB repair by an error-prone DNA repair process called non-homologous end joining (NHEJ). NHEJ generally leads to either insertion or deletion of few to several bases at the DSB sites. In general, DSB repair by NHEJ is far more frequent than HDR. In addition, appropriate donor DNA containing the sequence that is to be integrated at the targeted genomic site must be provided for HDR (Carroll, 2016).

1.7. *IN VIVO* SITE-DIRECTED MUTAGENESIS BY SITE-DIRECTED NUCLEASES

The growing desire among scientists for developing tools and techniques that will enable them to dependably and easily modify or edit specific genes in the desired manner and at sufficiently high frequencies resulted in the creation of site-directed nucleases (SDNs). Targeted gene editing or genome editing comprises alteration in the base sequence of the specified genes as desired by the experimenter by targeted DSB induction using a SDN. It has been argued that the term 'genome editing' is preferable to 'gene editing' since the sequence alterations can involve genomic sequences other than genes. In this book, there is a noticeable bias for the term 'gene editing' since the induced genomic alterations are ultimately intended to modify gene function. Sometimes, the terms gene-targeting and genome engineering are also used to describe targeted changes in genes and genomes. It has been pointed out that all the above terms seem to suggest genomes to be 'static' like text pages that can be easily and precisely edited without any risk of unintended changes elsewhere. However, this image of gene/genome editing is far from real since SDNs induce, albeit in relatively much lower frequencies, off-target DSBs and indels (insertion/deletions) as well as unintended genomic changes at the on-target sites (Cribbs and Perera, 2017; NASEM, 2017).

Generally, genome editing/engineering involves either inactivation of the selected gene (this creates gene 'knockouts'), integration of the desired foreign gene or transgene (this yields gene 'knockins'), or modification of a specific gene by sequence replacement (this either restores function of a mutated gene or generates an allele with a novel function). Initially, a variety of different strategies were used to induce targeted cleavage of DNA. These early approaches included the use of oligonucleotides coupled to reagents like bleomycin that induced chemical cleavage or those that cause cross-linking in the DNA, peptide nucleic acids linked to DNA cleavage reagents, and self-splicing introns. The idea to use homing endonucleases (HEs) or meganucleases (MNs, nucleases encoded by certain introns) for targeted gene editing emerged out of the experiments with self-splicing introns.

The HEs induce DNA cleavage at specific sites in the genome and lead to integration of the concerned introns at the cleavage sites. The enzyme I-*Sce*I is an endonuclease encoded by group I introns of yeast mitochondria. It was used to introduce DSBs in synthetic target sites created in mammalian chromosomes, and thereby promote site-specific homologous recombination (Choulika et al., 1995). The meganuclease HO was also able to promote homologous recombination in the synthetic target sequence introduced in the target cell genome. Site-specific integration of a desired DNA sequence was achieved by using an HE in the following manner: the desired DNA sequence was first integrated into the intron encoding the HE. The HE encoded by this intron induced DSB at the genomic site recognized by the HE leading to integration of the intron along with the desired DNA sequence at the DSB site (Doudna and Charpetier, 2014). Continued research efforts led to the development of a group of diverse programmable nucleases capable of introducing DSBs at specified genomic sites. Kim and Chandrasegaran (1994) reported creation of the first chimeric nuclease, the zinc finger nuclease (ZFN), and its use for targeted DNA cleavage. Next, TALENs (transcription-activator-like effector nucleases) and CRISPR-Cas9 (clustered regularly interspaced short palindromic repeat-CRISPR-associated 9) were reported in 2011 and 2013, respectively (Table 1.1). The HEs, ZFNs, TALENs and CRISPR-Cas9 are called site-specific nucleases, programmable nucleases, engineered nucleases, "gene scissors", and SDNs.

Table 1.1. Salient features of site-directed gene editing tools, including epigenetic editors and chimeric oligonucleotides

Technology*	Constituents	Strategy	Remarks
Meganuclease (MN) (1994)*	Meganuclease engineered to bind and cleave the target site	MNs induce DSBs[@], which are repaired by NHEJ or HDR; mutation induced during DNA repair	Engineering of MNs for new targets is very difficult
Zinc-finger nuclease (ZFN) (2003)*	Zinc finger protein linked to FokI nuclease; two ZFNs used for each target site	ZFNs induce DSBs, which are repaired by NHEJ or HDR; mutation induced during DNA repair	Designing ZFNs for new targets is less challenging than that of MNs, but is far more challenging than that for TALENs
Transcription-activator like effector nuclease (TALEN) (2011)*	TALE protein linked to FokI nuclease; two TALENs used for each target site	TALENs induce site-specific DSBs, which are repaired by NHEJ or HDR; mutation induced during DNA repair	Designing TALENs for new targets is less challenging than that of ZFNs, but is far more challenging than that of CRISPR-Cas9
Clustered regularly interspaced short palindromic repeat (CRISPR)-Cas9 system (2013)*	sgRNA and Cas9 nuclease; sgRNA guides Cas9 to the target site and activates its nuclease activity	CRISPR-Cas9 induces site-specific DSBs, which are repaired by NHEJ or HDR; mutation induced during DNA repair	A single CRISPR-Cas9 is used for each targeted site; only a part of the sgRNA is to be redesigned for a new target site; the simplest to redesign and the cheapest SDN
Base editor	Single guide RNA, Cas9 nickase-cytidine or adenosine deaminase**	C deaminated to U (replaced by T during replication); A deaminated to I (replaced by G during replication)	Generates single base substitutions leading to reversion of mutant allele to wild type, generation of new mutant allele or gene inactivation by creation of a stop codon
Epigenetic editor	sgRNA, dead Cas9 linked to a histone or DNA modification enzyme	Catalyzes DNA methylation, histone acetylation or deacetylation, etc.	Leads to activation or repression of the target gene expression; targeted site in the promoter region of the gene
Oligonucleotide-directed mutagenesis (ODM) (1996)*	DNA:RNA hybrid chimeric oligonucleotide containing the desired alteration in base sequence	The altered base produces a mismatch, which is repaired by excision repair	Repair leads to altered base sequence; generally single base substitutions

[@] DSB, double-strand break; HDR, homology-directed repair; NHEJ, non-homologous end-joining; SDN, site-directed nuclease; sgRNA, single guide RNA.

* The year of first publication on gene editing in mammalian cells by the given site-directed nuclease.

** In case of C-to-T base editor, UGI (uridine glycosylase inhibitor) is also linked to nCas9 for inhibiting the excision of U by uridine glycosylase and, thereby, increasing the conversion to C-to-T.

Table 1.2. The various genetic applications of the SDNs and their modifications (based on Gaj et al., 2013 and other sources)

Genetic modification	Tool	Remarks
Gene inactivation (knockouts)	SDN[@] (ZFN, TALEN, CRISPR-Cas9), base editors	Targeted gene inactivated due to indels produced during NHEJ-mediated repair of the DSBs induced by SDNs or stop codons induced by base editors
Deletion of a gene (any genomic segment)	Two SDNs, each specific for a site at one end of the targeted region	Two DSBs induced (one at each of the two targeted sites); the sequence between the DSBs is deleted, but it may also be inverted, duplicated or translocated
Gene correction (reversion of mutant alleles of genes to their wild type alleles)	SDN with donor DNA* containing the wild type gene sequence	The mutated sequence is replaced with the donor DNA sequence during HDR of the DSB induced by the SDNs; restores normal gene function
Allele replacement (desired mutation induced in the targeted gene)	SDN with donor DNA containing the desired mutational change	The base sequence change included in donor DNA copied into the targeted gene by HDR of the DSB
Gene addition (integration of a specified gene in the genome)	SDN with donor DNA containing the complete gene construct in its middle region	The gene provided as donor; its terminal regions are homologous to the sequences flanking the DSB; integration of the gene at predecided genomic sites by HDR
Base substitution (base editing)	ZFPs, TALEs and CRISPR-dCas linked to cytidine or adenosine deaminase enzyme; some other proteins may also be included	The dCas lacks nuclease function; deamination leads to conversion of C to U (ultimately T) and of A to I (ultimately G); generates base transitions
DNA methylation or demethylation, histone methylation, acetylation or deacetylation; gene activation/repression	ZFPs, TALEs and CRISPR-dCas linked to the appropriate effector proteins needed for the desired histone and/or base modification	The dCas lacks nuclease function; DNA and/or histone modification leads to activation or suppression of gene expression
Gene drives** for genetic modification of natural populations	Cas9 and sgRNA gene constructs flanked by homology arms; where needed a cargo gene may also be included	CRISPR-Cas9 encoded by the gene drive cleaves the wild type chromosome at the site specified by the sgRNA; the gene drive is copied into the chromosome by HDR of the DSB; can eradicate, suppress, rescue or genetically modify crossbreeding natural populations

[@] Cas9, CRISPR-associated 9; CRISPR, clustered regularly interspaced short palindromic repeat; dCas, 'dead' Cas; DSB, double strand break; HDR, homology-directed repair; NHEJ, nonhomologous end joining; SDN, site-directed nuclease; sgRNA, single guide RNA; TALEN, transcription activator like effector nuclease; TALE, transcription activator like effector; ZFN, zinc finger nuclease; ZFP, zinc finger protein.

* The donor DNA is homologous to the sequence of genomic region, in which DSB is induced, except for the sequence change to be induced; the changed sequence is in the middle region of donor DNA. The changed sequence may be a complete gene construct.

** A gene drive is essentially a donor DNA having the Cas9 and sgRNA gene constructs; where needed, any desired gene (called cargo gene) may also be included.

SDNs are custom-designed to bind to the specified sequences of duplex DNA and to induce DSBs close to or within their binding sites. Since redesigning of HEs for new genomic targets is very difficult, their use is limited to few academic laboratories and some biotechnology companies (Chapter 2). One promising development seems to be the creation of megaTALs by fusing meganucleases with the DNA binding domains of TALENs. MegaTALs have very high cleavage activity and extremely high specificity, and they may emerge as potent therapeutic tools (Boissel et al., 2014; Wang, Y. et al., 2014). The ZFNs (Chapter 3) and TALENs (Chapter 4) are derived by fusing the DNA binding domains of transcription factors with the DNA cleavage domain of FokI, a Type IIS endonuclease. Each functional ZFN and TALEN SDN comprises a pair of ZFN and TALEN monomers, respectively, since the FokI nuclease domain must dimerize to effect DNA cleavage. Therefore, the two members of a ZFN or TALEN pair are designed to bind to neighboring genomic sites. But each functional CRISPR-Cas9 SDN comprises a single member, *i.e.*, a Cas9 protein and a sgRNA (single guide RNA). Both ZFNs and TALENs function as proteins, while the CRISPR-Cas9 functions as ribonucleoprotein. The Cas-9 nuclease must interact with the 3' region of sgRNA to become functional, while the 5' region of sgRNA pairs with the DNA duplex and positions Cas-9 at the targeted site (Chapter 5).

The ZFNs are difficult to redesign for new target sites and they are the costliest of SDNs. But some ZFNs are among the most specific SDNs. The TALENs are much simpler to redesign and substantially cheaper to implement than ZFNs. Therefore, TALENs rapidly gained popularity among biology researchers; they had just started to overtake ZFNs when the CRISPR-Cas9 system came on the scene. CRISPR-Cas9 is the easiest and the cheapest to design and use so that it has become the most popular SDN system for all biological systems for inducing a variety of targeted genetic manipulations (Table 1.2). Porter et al. (2019) provide an excellent overview of the various tools of genome editing in their concise practical guide for various applications of SDNs.

1.8. THE SDN TECHNOLOGY

The use of SDNs to induce targeted genetic changes is generally called SDN technology. This involves identification of the gene to be targeted and decision on the specific sequence alteration to be induced. A suitable SDN system is then selected, redesigned for the targeted genomic site and evaluated for its specificity and efficiency in inducing the intended genetic change at the targeted site. In case of CRISPR-Cas9, the 5' 20 nt (nucleotides) of the sgRNA are redesigned for each new genomic target, for which several software are available; the other components of the SDN remain unchanged. The SDNs can be delivered either as protein (ZFNs, TALENs), RNA, ribonucleoproteins (CRISPR-Cas9) or DNA (in a plasmid) directly into the animal cells or plant protoplasts. Alternatively, the SDN gene constructs can be introduced into cells by a suitable genetic transformation/transfection method. The expression vectors used for this purpose can be designed for either transient expression or stable transformation. When the SDN genes are stably integrated into the genome, segregation and selection can be used to remove them from the desirable mutant lines. In case of plants, plantlets are regenerated from the treated cells/tissues and mutant plants with heritable changes in their phenotype are selected. Mutant plants can also be selected using suitable molecular techniques, including PCR coupled with DNA sequencing. In case of animals, the SDNs can be delivered into zygotes or young embryos, which are then implanted into females to develop into young animals. There are basically three types of SDN technologies, *viz*., SDN-1, SDN-2, and SDN-3 technologies (Lusser et al. 2012, Podevin et al. 2013; Table 1.3).

1.8.1. SDN-1 technology

The SDN-1 technology uses only SDNs and relies on NHEJ-mediated repair of DSBs to inactivate the targeted genes. The SDNs induce DSBs at the targeted sites, which activate the cell's DNA repair mechanisms. The DSBs are predominantly repaired by NHEJ, which generally leads to deletion or insertion of one to few base pairs; sometimes there may be substitution of one or even few base pairs. These changes generally lead to inactivation of the targeted genes. Two pairs of specifically designed ZFNs or TALENs, or one pair of CRISPR-Cas9 can be used to produce two DSBs at nearby sites. This may lead to deletion, duplication, inversion or translocation of the DNA segment located between the two DSBs. This strategy can be used to delete complete endogenous genes. The SDN-1 technology will be useful for inactivation of genes associated with undesirable traits like susceptibility to diseases, production of antinutritional factors, etc. and for analysis of gene function.

Table 1.3. Salient features of the three types of site-directed nuclease (SDN) strategies for genetic alterations (Based on Lusser et al., 2012, Podevin et al., 2013)

SDN strategy	Salient features of strategy	Genetic change induced	Remarks
SDN-1	SDN introduced into cells	Repair of DSB[@] by NHEJ generally produces short indels; leads to gene inactivation	Low frequencies of off-target editing, and unintended on-target changes like large deletions, inversions, etc.
SDN-2	SDN along with a short donor DNA* introduced into cells; the donor DNA has the base sequence change desired at the target site	DSB repair uses donor DNA as template for HDR; produces the desired base sequence change at the DSB site	Can be used for correction of a mutation or induction of the desired mutation in the targeted gene
SDN-3	SDN along with a long donor DNA introduced into cells; the donor DNA has the complete transgene construct to be inserted at the target site	DSB repair uses the donor DNA for HDR leading to integration of the desired transgene construct at the DSB site	Transgenes integrated at predecided 'safe' genomic sites to avoid position effects; also enables transgene stacking, *i.e*., integration of many transgenes at a single locus

[@] DSB, double-strand break; HDR, homology-directed repair; NHEJ, non-homologous end joining

* The sequences at the two ends of donor DNA are homologous to the genomic sequences flanking the DSB, while its middle region has the complete transgene construct that is to be copied into the genome.

1.8.2. SDN-2 technology

Cells show extremely low frequency of integration of introduced DNA segments into their genomes by HR; this frequency is increased several thousand-fold by induction of DSBs. The SDN-2 and SDN-3 technologies exploit this feature of the cellular DNA repair systems. In case of SDN-2 technology, the selected SDN is used along with an appropriate donor DNA to achieve the pre-decided sequence replacement in the targeted gene. The donor DNA has the same base sequence as the targeted region of the gene, except for the desired mutational change located in its middle region. It serves as template for HDR of DSBs induced by the SDN. SDN-2 procedure can induce substitutions, insertions or deletions of up to few base pairs; these changes can inactivate a gene, correct a deleterious mutation in the targeted gene (targeted gene correction) or can induce a new mutation (gene/allele replacement; Lusser et al., 2012, Podevin et al., 2013).

1.8.3. SDN-3 technology

The SDN-3 technology is used for integrating a complete gene at the specified genomic site by HDR-mediated DSB repair. Therefore, the sequences flanking the ends of the transgene construct are homologous to the genomic sequences located on the two sides of the SDN-induced DSBs. This strategy allows additional copies of the desired endogenous genes or novel transgenes to be integrated at specified genomic sites. It can also be used for replacing the existing alleles of endogenous genes with new alleles, and for trait stacking. When genes for two or more traits are integrated at the same locus, it is called trait stacking. The targeting of desired genes to specific genomic sites would eliminate the risk of 'position effects' and unintended activation and/or inactivation of endogenous genes that may occur due to random transgene integration (Lusser et al., 2012, Podevin et al., 2013).

The SDN technologies can be used for improving important traits of organisms. The SDN-1 technology achieves gene inactivation, while the SDN-2 strategy generates pre-decided sequence changes in the targeted genes. The SDN-3 technology facilitates targeted integration of transgenes leading to more predictable and stable transgene expression. It also greatly reduces the risk of unintended changes in the expression of endogenous genes located at and around the transgene integration site. SDN-3 procedure can be used for stacking of desired endogenous genes or novel transgenes at the same locus or at different genomic sites. Further, the SDNs can be used to remove specific endogenous genes or transgenes like selectable markers by induction of two DSBs flanking the targeted genes/transgenes. The removal of unwanted marker genes after the recovery of desired transgenic individuals would have significant regulatory benefits. In addition, this strategy can be used to delete the selected genomic regions having highly repetitive DNA or some undesirable loci (Curtin et al., 2012).

In vitro site-directed mutagenesis involves cloning of the target genes and introduction of alterations in their sequences *in vitro* using synthetic oligonucleotides and, generally, PCR. It is used for modification of properties of proteins and enzymes. The target gene sequence must be known and prior knowledge of the relevant structural and functional features of the concerned protein will be helpful. The modified gene is ultimately introduced into the selected host where it is expressed. The gene editing technologies do not require cloning of the target genes and they introduce the desired sequence changes *in vivo*. The SDN-2 technology does use synthetic oligonucleotides as templates for HDR to bring about the desired sequence replacements. The changes in DNA sequences are brought about by cellular DNA repair enzymes. The chief objective of gene editing is to modify the characteristics of the concerned organisms by making suitable changes in their specific genes.

1.9. OLIGONUCLEOTIDE-DIRECTED MUTAGENESIS

Oligonucleotide-directed mutagenesis (ODM) uses chimeric oligonucleotides (COs) for induction of the desired mutations. The ODM procedure is also called oligonucleotide-directed gene targeting, chimeraplasty, etc. COs are double-stranded structures formed by internal base pairing in a single DNA strand that has two short RNA sequences at strategic locations and hairpin loops at its two ends (Section 2.11; Fig. 2.6). The COs are complementary to the targeted genomic sequence, except for one base that is to be mutated by ODM: this particular base of CO is changed to the base that is desired at the corresponding base position of the targeted sequence. ODM typically introduces substitution of one base at the targeted site of the gene. It can also generate substitutions for more than one base and addition/deletion for generally one base. The COs are delivered into cells by an efficient delivery technique. Inside cells, the COs unfold and move into the nuclei, where they base-pair with the homologous sequences of the targeted gene. This pairing will create a structural distortion due to the mismatched base deliberately included in the CO sequence (indicated as 'mutation' in Fig. 2.6). The distortion will be recognized and removed by cellular DNA repair systems by changing the base sequence of the targeted genomic region in at least some of the treated cells. This will generate the mutation specified by the CO at the targeted site of the gene. The COs can be designed to either induce the desired mutation or correct the existing mutation in the targeted gene. The COs are ultimately degraded by cellular enzymes within few hours following their introduction into cells (Komor et al., 2017a; Songstad et al., 2017).

1.10. BASE EDITING

Base editing directly and irreversibly converts one base into another at the targeted genomic site without induction of a DSB. It is achieved by base editors, which consist of the DNA binding module of a SDN and specific base deaminases that catalyse deamination of the targeted base. The zinc finger proteins from ZFNs, TALE proteins from TALENs, and dead or nickase version of Cas9 together with appropriate sgRNAs serve as the DNA binding modules of base editors. The base editors have either cytidine deaminase (for C-to-T conversion) or adenosine deaminase (for A-to-G transition). The adenosine deaminase used in base editors has been evolved from tRNA adenosine deaminase of *E. coli* by seven generations of genetic engineering. Some other proteins, *viz*., UGI (uridine glycosylase inhibitor) and Gam proteins are also used in the advanced versions of C-to-T base editors. Thus, the two groups of base editors together can produce all the four types of transitions, *viz*., C-to-T, T-to-C, A-to-G and G-to-A transitions. The C-to-T and A-to-G transitions are produced by direct base conversion, while the G-to-A and T-to-C transitions are achieved by editing their complementary bases A and C, respectively, located in the complementary strand of DNA duplex (Hess et al., 2017, Komor et al., 2017a,b,c).

The motivation for base editing was the realization that many human genetic diseases are generated due to a single T-to-C transitions in the concerned genes. Therefore, C-to-T base editors were the first to be developed to enable reversion of these transitions. The efficiency and specificity of base editors has been considerably improved, and base editing has been used for inducing targeted single base transitions in many different biological systems, including human cell cultures. The available base editors, however, are not truly single base editors as they edit other C or A residues present around the targeted base as well as at other genomic sites related to the targeted site. In addition, at least in some systems, they induce low frequencies of transversions and indels as well. Therefore, therapeutic applications of base editors must await further refinements in the technology, but base editing is expected to generate plant breeding products in the very near future (Chapter 6).

1.11. MODULATION OF GENE ACTION

Regulation of gene action in eukaryotes is quite complex since the genomic DNA is associated with proteins to form chromatin. The eukaryotic gene expression, therefore, can be regulated at several levels beginning with chromatin organization through transcription initiation to post-translational protein modification. The chief mode of gene regulation, however, is based on chromatin modification and/or transcription initiation. The first step in expression of a gene is conversion of the region of chromatin containing this gene into an 'active' state making the gene available for transcription initiation. The SDNs have been used to develop programmable transcription activators and repressors to regulate the expression of targeted genes. Since these activators/repressors catalyse epigenetic modifications, they are often called epigenome editors and the process of targeted gene regulation is termed as epigenome editing (Chapter 10).

The site-directed transcription activators/repressors are created by replacing the nuclease domain of TALENs with the desired effector domain. In case of CRISPR-Cas9, the effector domain is linked to either the C-terminus of dCas9 ('dead' Cas9, lacks nuclease activity) or to the sgRNA by interaction of an RNA hairpin with a fusion protein comprising a viral protein and the effector domain. In general, a single copy of an epigenetic effector induces only partial repression or activation. Therefore, either multiple copies of a single effector or several different effectors are deployed to achieve robust gene activation/repression. Suitable strategies for both these approaches have been devised and used. The CRISPR-dCas9 system has been modified for imaging targeted loci *in vivo* by either fusing a fluorescent protein to dCas9 or attaching it to sgRNA via RNA hairpin-viral protein interaction. For robust imaging, the fluorescence signal must be multiplied by a suitable strategy like dCas9-SAM or SunTag. The CRISPRrainbow method allows simultaneous imaging of multiple loci or of the same locus in different colours.

CRISPR-Cas9 system has been modified to permit several other applications. For example, the PAMmer technique allows Cas9 to bind to and cleave single-stranded RNA. It also enables imaging and tracking of single RNA molecules, and isolation of a specific mRNA. The epigenome editors have been used to activate expression of HIV1 genes in latently infected cells leading to death of these cells. It has also been used to develop a homing gRNA-based evolving DNA barcode that can be used to determine cell lineage relationships. Some other applications include identification and characterisation of *cis*-regulatory elements like promoters and enhancers, directed evolution of proteins, and specificity assay for programmable DNA binding systems. It is indeed difficult to predict the future uses of this very versatile tool, except for the general observation that the range of its various applications is likely to be awesome.

1.12. GENE DRIVES

Gene drives are SDN-based self-propagating genetic devices designed for introducing genetic alterations in crossbreeding natural populations. Gene drives based on CRISPR-Cas9 were designed and shown to work in laboratory populations in 2015, and they are now almost exclusively used. A simple gene drive has the expression cassettes for Cas9 and sgRNA targeting the desired genomic site; these cassettes are flanked by homology arms required for HDR. The base sequence of a homology arm is homologous to the sequence on one side of the DSB induced by the gene drive-encoded CRISPR-Cas9. Where needed, cargo and/or fluorescent marker genes may be included between the homology arms. The CRISPR-Cas9 induces DSB at the targeted site in the wild type chromosomes, and DSB repair by HDR leads to copying of the gene drive at the DSB site. Thus, in every gene drive heterozygote, the gene drive is copied into the wild type chromosome; this converts heterozygotes into homozygotes. This conversion leads to super-Mendelian inheritance pattern so that the gene drive rapidly spreads in the population till it reaches fixation. The rate of spread of gene drives is affected by several factors, including the species, and the strain/genotype of the target organism. Gene drives work effectively in laboratory populations, but their behaviour in natural populations is yet to be tested.

Gene drives can be used to control vectors of human/animal diseases like mosquitoes, promote agricultural production and sustainability, control invasive species, rescue endangered natural populations, develop desirable animal and crop strains/lines, etc. For example, gene drives have been developed for making mosquitoes incapable of transmitting malarial parasite and dengue-2 pathogen. The deployment of gene drives presents several risks, including off-target activity, spreading of gene drives into non-target populations and related species, negative effects on ecosystem, etc. The procedure for risk assessment for gene drives is yet to be developed as are the regulations and guidelines for their field testing and eventual deployment. Self-limiting gene drives, *e.g.*, daisy chain gene drives, have been developed to minimize the risk of gene drive persistence in target populations and their spread into non-target populations. The invasion of a population by an unwanted gene drive can be prevented by an immunizing gene drive or a precision gene drive. Unwanted gene drives existing in populations can be inactivated by gene 'brakes' or removed by excision drives. Reversal drives can undo the genetic changes induced by an earlier gene drive. Finally, split gene drives and synthetic site gene drives are designed to confine the gene drives to laboratory strains. In addition, several other gene drive designs have been proposed or tested for achieving specific objectives (Chapter 11).

1.13. ETHICAL ISSUES RELEVANT TO GENE EDITING

The direction and the pace of biotechnology innovations are affected by government policies, researches pursued in public institutions, and acceptance of the innovative products by general public. Genome editing of plants has been well-accepted and there have been little or no ethical concerns. The main concerns have been the safety issues like off-target action, and unintended on-target genomic rearrangements induced by the SDNs. The SDNs should preferably be delivered as proteins, mRNAs or ribonucleoproteins to eliminate the risk of integration of plasmid DNA into the cell genome. Further, the donor DNA plasmid delivered in SDN-2 and SDN-3 technologies may integrate at the target site. Comprehensive WGS (whole genome sequence/sequencing) analyses of the products should be done to eliminate the risk of foreign DNA integration into the genome. Similarly, gene editing of food animals raises few ethical questions. But gene editing of larger animals like nonhuman primates for use as human disease models and as organ donors does raise some ethical issues since they are closer to humans than other animals. This situation creates the following paradox: the closer an animal is to humans in evolutionary relationship, the better it would serve as a disease model, and the more objectionable would be its use as disease model. Another ethical objection is that such uses of animal species tend to reduce their status from living beings to mere commodities.

Human genome editing can be broadly divided into two categories, *viz,* somatic and germline gene editing. Further, both somatic and germline gene editing are of three types, *viz.*, gene editing for basic, preclinical and clinical researches. Somatic cell gene editing is generally accepted by the public; at least six gene therapies developed by this procedure are in clinical use and many others are in different stages of clinical trials. One of these therapies (Zolgensma, for treatment of spinal muscular atrophy) is the costliest drug on earth at 2.1 million US dollars for the full course of a single dose. But germline gene editing, especially heritable or clinical germline gene editing, is one of the most controversial subjects and it raises many ethical and regulatory issues. The ethical positions range from in-principle rejection to concerns about safety and societal inequalities, and potential exploitation of the would-be parents. It is generally agreed that it is 'irresponsible' to undertake heritable germline gene editing at present since more information on efficiency and safety aspects the gene editing platforms needs to be accumulated for reaching a rational decision in the matter.

Gene drives can be designed to eliminate a species from the face of earth, which may have adverse effects on the ecosystem. At the same time, effective measures must be devised to prevent the misuse of gene drives by extremists. These and some other issues have so far prevented their field tests, and the consensus seems to be to limit gene drive research to the laboratory till adequate safety measures have been developed.

1.14. REGULATION OF GENE EDITING

The regulation of biotechnology innovations ranges from voluntary self-regulation to legislations enacted by competent authorities. These regulations may be either product-based (tends to promote innovation) or process-based (discourages innovation). Similarly, the risk assessment approach may be either permissive (assumes a product to be safe until proved dangerous) or precautionary (considers a product dangerous until proved safe). The regulation in USA is product-based with permissive risk assessment; this regime promotes innovation and encourages consumer confidence. In contrast, the regulation in EU is process-based coupled with precautionary risk assessment; this tends to discourage innovations as well as consumer confidence. So far, no country has developed specific policies and procedures for regulation of gene-edited crop varieties or animal strains. But many gene-edited crop varieties have been approved for commercial cultivation in U.S.A., Canada and Brazil under the existing regulatory processes developed for GM crops. The countries following product-based approach of regulation tend to consider SDN-1 and some SDN-2 products as non-GM, while process-based regulatory systems treat them as GM. Thus, there is a need for international harmonization of the regulatory procedures for gene-edited crop varieties.

The regulation of gene edited food animals is beingly hotly debated since FDA (Food and Drug Administration), U.S.A. proposes to treat them as drugs, which seems unreasonable. Argentina and Brazil, on the other hand, would apply the same regulatory process to gene-edited animals that is applied to gene-edited plants. Many scientists have been demanding that authorities in U.S.A. follow the examples of Argentina and Brazil. There are no regulatory guidelines for deployment of gene drives, and it is not likely to be developed soon. Human somatic cell gene editing is used to develop gene therapies, many of which are either in clinical use or in different stages of clinical trials. In general, these gene therapies are regulated according to existing regulatory processes applicable to conventional gene therapies.

The regulation of human germline gene editing ranges from an enforceable and explicit ban on heritable germline editing through implicit but effective ban to explicit ban with doubtful enforceability. It is generally agreed that it is 'irresponsible' to undertake heritable germline gene editing at present. There is an urgent need for an international regulatory framework and WHO of the United Nations Organization has already initiated the process. It has appointed an Expert Advisory Committee on Developing Global Standards for Governance and Oversight of Human Genome Editing. At its first meeting (March 18-19, 2019), the committee agreed to develop 'a strong international governance framework' and establishment of 'a central registry on human genome editing research' by the WHO (WHO, 2019).

1.15. ACHIEVEMENTS

The genome editing, including base and epigenome editing, tools and techniques were initially developed for animal systems, especially, cultured human cells. They were subsequently successfully extended to many other organisms, including livestock and crop species. The SDN technology has been used for targeted gene inactivation, sequence replacement (to restore functions of mutated genes and to create new alleles of genes), integration of transgenes at specific genomic sites as well as trait stacking. The SDNs are being extensively used for studies like genome-wide screens for human gene functions, gene interaction analyses, and creation of disease models at cell, organoid and animal levels (Chapter 9) as well as to induce desired mutations in animals and plants leading to useful phenotypes. Many of the gene-edited lines of crop species have been approved for commercial cultivation. Some examples of gene-edited lines permitted by APHIS (Animal and Plant Health Inspection Service), USDA (United States Department of Agriculture) for cultivation are a tomato line bearing fruits with jointless pedicel, a *Camelina sativa* mutant line with enhanced oil yield and enhanced omega-3 fatty acid content, a maize line produced by HDR with resistance to Northern leaf blight, etc. But the first gene-edited plant to be declared as nonregulated article by APHIS, USDA was a nonbrowning mutant of button mushroom (*Agaricus bisporus*). ODM was used to develop SU Canola, a sulfonylurea resistant canola variety, being commercialized by Cibus. Similarly, several strains of gene-edited cattle, pig, goat and sheep have been developed for improved agricultural traits like milk/meat production, produce quality, disease resistance, etc. Some of the gene-edited animal strains are on the verge of commercialization (Chapter 8).

Gene editing has been successfully used for genetic modification of livestock species to achieve the following objectives: (i) creation of animal models of human diseases, (ii) development of livestock strains suitable as donors of organs for xenotransplantation in humans, (iii) generating animal bioreactors for the production of pharmaceutically valuable proteins, and (iv) improvement of livestock breeds for agricultural traits. Several animal models have been developed in NHPs, pigs and dogs, which are expected to be particularly useful for biomedical researches on complex human diseases concerned with brain and heart functions. Pigs appear to be the most promising as organ donors, and many of the genes involved in xenotransplant rejection have been modified in pigs leading to increased xenotransplant survival in NHPs. It is hoped that ultimately pigs may emerge as valuable organ donors for human patients suffering with terminal stage organ failure. Transgenic strains of cattle, goat and pig have been developed by SDN-3 technology to serve as bioreactors to produce human lactoferrin, human serum albumin (HSA), and lysostaphin.

The SDN system has been repurposed to develop base editors that irreversibly and directly modify either C-to-T or A-to-G. The direct base conversions can be used to indirectly convert T-to-C and G-to-A by editing C-to-T and A-to-G in the complementary strand. Base editing has been successfully applied to mammalian cells, many plants and other organisms as well. The SDNs also provided the basis for development of programmable transcription activators and repressors, including enzymes involved in chromatin modifications (the epigenome editors). The CRISPR-dCas9 has been used to devise tools for fluorescence imaging of chromosomes and other cell components including even single mRNA molecules and for tracking of these molecules. The CRISPR-Cas system is being widely used in biomedical research to understand the process of disease onset and development, to serve as an extremely sensitive diagnostic tool, to identify novel targets for drug development, and to explore the possibility of using gene editing for disease treatment.

Gene editing is being used to develop gene therapies for many diseases, some of which have been approved for clinical use, while several others are in different stages of clinical trials. A notable contribution of gene editing has been the development of 'universal' T cells, which are very less likely to cause graft-versus-host disease and can be used in any patient. The universal T cells have been approved for clinical use in U.S.A. The ZFNs, TALENs and CRISPR-Cas systems have been repurposed as nucleic acid (RNA and/or DNA) detection tools, *e.g.*, Cas12a-based DETECTR and Cas13a-based SHERLOCK. These tools have attomolar (10^{-18}) sensitivities and can be used in a variety of nucleic acid detection and/or quantification applications, including detection of specific strains of viruses, distinguishing pathogenic bacteria, etc. The CRISPR-Cas9 system has also been used to develop a variety of gene drives for genetic modification and control of natural populations. A gene drive targeting a female fertility gene is being field-tested by a British firm in Brazil for eradication of *Aedes aegypti* mosquitoes.

1.16. FUTURE PERSPECTIVES

The field of gene/genome editing is relatively young and a very active area of research. Some of the SDNs have been modified to develop programmable chimeric effectors that enable desired modulation of gene function, imaging of genomic regions/loci, etc. In addition, CRISPR-Cas9 has been used to create C-to-T and A-to-G base editors that convert C to T and A to G, respectively. These tools have greatly facilitated the analysis and unravelling of various biological processes including genome dynamics. Genome-wide gene inactivation would greatly facilitate reverse genetic analyses of biological processes and pathways. As our knowledge of the various aspects of genome organization and function, gene function and interaction in the development of various phenotypes, including human disease development increases, our ability to manipulate these features to our advantage would also increase. It may be expected that the applications of gene editing will rapidly expand as this powerful technology is further refined to satisfactorily resolve the issues related to efficient and safe delivery of SDNs particularly for therapeutic uses, generation of off-target effects and unintended on-target genetic alterations, and other relevant concerns. It has been suggested that the 'future applications of the nuclease technology may be limited only by the imagination and insight of researchers' (Carroll, 2015).

Genome editing is being vigorously applied to improve crop plants and other food sources, including animal species, and some of the products have started to reach consumers. However, not a single country has developed specific policies and procedures for regulation of gene-editing products, and they are regulated as per the regulation developed for GM crops/animals. Some of the products of gene editing have been exempted from regulation in USA and some other countries, but EU continues to treat them as GM. Similarly, gene-edited animals are proposed to be regulated as drugs by FDA, USA, while Argentina and Brazil intend to treat them at par with gene-edited plants. It has been suggested that policies and procedures for specific regulation of gene-edited plants and animals should be developed and that the regulatory laws/regulations of different countries need to be harmonized to foster research, development and innovations in this emerging technology.

This technology can generate strains of animals that would serve as useful models for the study of human diseases, drug screening, etc. and possibly serve as donors of organs for human patients. The use of large animals in biomedical research, including as disease models raises some ethical issues that need to be resolved through public engagement. Gene drives are potent genetic devices for the management of natural populations, but they remain confined to laboratory research. It is important that appropriate international guidelines are developed to facilitate their field-tests so that a rational decision can be taken about their deployment in the field. Finally, effective mechanisms need to be developed for preventing the misuse of gene editing and, especially, gene drive technologies by extremist groups.

Human genome editing for basic research and somatic cell gene editing for clinal applications is widely accepted and some gene therapies are already in clinical use. The most controversial issue concerns human germline gene editing. Germline editing for basic research is generally acceptable with some restrictions, but its clinical use is considered 'irresponsible' at present. The gene editing technologies need to be made more efficient and safer, and adequate data on these aspects need to be generated to reach a decision on use of heritable germline gene editing. Further, consensus needs to be developed on several important issues like the diseases/ conditions that would warrant germline gene editing, the levels of acceptable off-target and unintended on-target edits, clear separation of therapeutic and cosmetic or enhancement applications, etc. At the same time, efficient and effective mechanisms for oversight and governance need to be developed to prevent inappropriate uses of the technology. It is noteworthy that a strong international regulatory framework is being developed by the WHO of UNO. WHO also proposes to develop a 'a central registry on human genome editing research' that would serve as transparent database.

1.17. CONCLUSIONS

Researchers have been in search of tools and techniques for generating precise and predecided changes in the structures and/or functions of selected genes to facilitate both basic and applied researches. This quest has resulted in the development of chemical and physical mutagenesis, transgenic technology, *in vitro* site-directed mutagenesis, RNAi, antisense RNA and gene disruption technologies. These technologies have been very useful but each of them suffers from certain important limitations. For example, the gene disruption technique has been used to create many extremely useful gene knockout animal models of human diseases. However, this technology is technically very demanding, involves sophisticated selection procedures, requires genuine ES cells, and is rather expensive. In addition, the frequency of gene integration by homologous recombination is very low. The development of SDNs like ZFNs, TALENs and CRISPR-Cas9 has revolutionized biological research since they allow precise targeting of desired genomic sites. The CRISPR-Cas9, the latest SDN system, is the easiest and the cheapest to redesign and implement for new target sites and is the most widely used.

SDNs induce DSBs at the targeted genomic sites and the genetic changes are produced during DSB repair by the normal cellular pathways. These changes lead to gene inactivation due to indel formation during NHEJ-mediated DSB repair (SDN-1 technology), correction of existing mutations or induction of new mutations by sequence replacement due to HDR-based DSB repair (SDN-2 technology), and HDR-mediated integration of complete gene constructs at the specified genomic sites (SDN-3 technology). Mutations resulting from one to few base substitutions can also be produced by ODM based on chimeric oligonucleotides. The CRISPR-Cas9 system has been used to develop base editors for targeted chemical modification of C-to-T and A-to-G leading to single base substitutions at the specified genomic sites. It has been used to develop epigenetic editors for histone modification and DNA methylation to activate or suppress the expressions of targeted genes. It has also been used to create highly potent genetic devices called gene drives, which can propagate themselves and rapidly spread in crossbreeding natural populations. Gene drives can be used to eradicate harmful natural populations, rescue threatened populations or suitably alter their genetic makeup. But in the absence of appropriate guidelines, gene drives are yet to be field-tested. Finally, effective mechanisms need to be developed for preventing the misuse of gene editing and gene drive technologies by extremist groups.

Genome editing is being used to improve crop plants and other food sources, including animal species, and some of the products have started to reach the consumers. The crop varieties produced by SDN-1 and some SDN-2 methods have been exempted from regulation in USA and some other countries, but EU continues to treat them as GM plants. Similarly, gene-edited animals are proposed to be regulated as drugs by FDA, USA, while Argentina and Brazil intend to treat them at par with gene-edited plants. Therefore, policies and procedures for specific regulation of gene-edited plants and animals should be developed and the regulatory laws/ regulations of different countries need to be harmonized. This is important because the regulatory policies have a strong influence on the pace of research and development in emerging technologies. In general, product-based regulation with permissive risk assessment promotes innovation and consumer acceptance, while process-based regulation combined with precautionary risk assessment discourages innovations and consumer confidence.

Genome editing has been used to generate large animal models of human diseases that are useful for studies on onset and development of human diseases, screening of drug, etc. Some of the gene-edited animals may possibly serve as donors of organs for humans. The use of large animals, especially non-human primates, for biomedical research and as organ sources raises some ethical issues that need to be resolved through public engagement. Human somatic cell genome editing for basic research as well as clinal applications is widely accepted and some gene therapies are already in clinical use. Human germline gene editing for basic research is generally acceptable to public with some restrictions, but its clinical use is considered 'irresponsible' at present. It is agreed that the gene editing technologies need to be made more efficient and safer and adequate data need to be generated on their efficacy and safety to enable a decision on the future clinical use of germline gene editing. Further, consensus needs to be developed on several important issues like the diseases/conditions for which germline gene editing should be used, the levels of acceptable off-target and unintended on-target edits, clear delineation of cosmetic/ enhancement editing from clinical editing, etc. In addition, efficient and effective mechanisms for oversight and governance need to be developed to prevent inappropriate use of the technology. The WHO of UNO has constituted a committee that intends to develop a strong international regulatory framework for germline gene editing and 'a central registry on human genome editing research.'

■■■■

2

MEGANUCLEASES AND CHIMERIC OLIGONUCLEOTIDES

2.1. INTRODUCTION

Single gene mutations have been implicated in a large number of human genetic diseases, many of which cause considerable physical, emotional and financial hardships to the patients as well as their relatives. These diseases can be managed by repeated administration of appropriate therapeutic reagents, if available. Alternatively, normal alleles of the concerned mutant genes may be introduced into the patients to provide the missing gene functions; this constitutes gene therapy. Ideally, the introduced copy of the concerned gene should become incorporated in the genome at the same locus that is naturally occupied by the mutant gene and, preferably, replace the latter. However, the introduced genes generally integrate at random sites in the host cell genomes. But the introduced genes do lead to targeted correction of mutations in their endogenous counterparts present in the cell genomes albeit at very low frequencies (~0.1% of the cells). Most likely this mutation correction is mediated by a process similar to homologous recombination (Smithies et al., 1985; Thomas et al., 1986).

Thus, one of the major challenges in the development of safe and effective gene therapies was to achieve high frequency correction of defects in the disease-associated mutant genes. The processes of homologous recombination (HR) during meiosis and the events of gene conversion, including mating-type switch in yeast, are initiated by double-strand breaks (DSBs) in chromosomal DNA. Further, the frequency of homology-directed gene integration can be greatly improved by introduction of DSBs. These insights stimulated the search for and the development of reagents capable of targeted cleavage of DNA. In addition, the DSBs are generally repaired by nonhomologous end-joining (NHEJ) pathway, which typically induces short insertions or deletions (indels) at the break sites leading to inactivation of the targeted genes (Carroll, 2014, 2016).

In addition, the availability of genome sequences since 1990s has provided an opportunity to biologists to plan desired modifications in selected genes/genomic regions to answer specific questions or achieve desired genetic modifications. This generated a growing desire among scientists to develop techniques for specified genetic modifications by targeted induction of DSBs followed by their repair. The early approaches to induce DSBs included the use of oligonucleotides coupled to reagents like bleomycin for induction of chemical cleavage of DNA and self-splicing introns for targeted gene transfers. The experiments with self-splicing introns suggested the use of homing endonucleases (HEs) or meganucleases (MNs) for targeted gene editing. The HEs induce DNA cleavage at specific sites in the genome and lead to insertion of the concerned introns at cleavage sites (Doudna and Charpetier, 2014). The enzyme I-*Sce*I, an endonuclease encoded by group I introns of yeast mitochondria, was the first to be used for introduction of DSBs in synthetic target sites integrated into the mammalian chromosomes, which greatly promoted site-specific homologous recombination (Choulika et al., 1995). Subsequent researches have enabled the redesigning of meganucleases for desired genomic targets as well as the development of some other site-directed nucleases, *viz*., zinc finger nuclease (Chapter 3), transcription activator-like effector nuclease (Chapter 4) and CRISPR-Cas9 RNA-guided site-directed nuclease (SDN) systems (Chapter 5). A SDN or programmable nuclease can recognize the desired target site with high specificity, and it should not bind to other similar sites (off-sites) in the genome.

In this chapter, we will consider in some detail the development of a SDN system based on meganucleases that holds promise for therapeutic applications. There are some excellent reviews, including those by Stoddard et al. (2011, 2014) and Belfort and Bonocora (2014), that describe HEs and the genome editing tools derived from them. In addition, this chapter also deals with chimeric oligonucleotides (COs) that have been used for developing at least one crop variety that has been commercialized. The reviews dealing with chimeric oligonucleotides include those by Igoucheva and Yoon (2000), Liang et al. (2002), Lusser et al. (2012), Komor et al. (2017a), and Songstad et al. (2017).

2.2. HOMING ENDONUCLEASES

Homing endonucleases generate double-strand breaks (DSBs) at specific genomic sites and mediate the integration of a copy of their own reading frames at these sites. These proteins are relatively small, but they typically recognize long (20 to 30 base pairs) DNA sequences. These enzymes can mediate site-specific recombination leading to genetic changes, such as insertion, deletion, mutation, or correction of mutant DNA sequences. HEs are often associated with mobile introns and inteins in the genomes of a wide variety of microbial species. They are prevalent in microbes, their phages and viruses, and in the eukaryotic organelles, namely, mitochondria and chloroplasts. However, HE genes are not known to occur in the genomes of more complex organisms, including eukaryotes. The HE genes are selfish DNA sequences that have extraordinary ability to accumulate in microbial genomes. For example, ~11% of the total coding sequence in T4 phage genome consists of 15 different HE genes (Stoddard, 2011, 2014).

The first report on homing during early 1970s concerned the inheritance of dominant genetic marker 'Omega' in yeast. Homing involves integration of a copy of a self-splicing intron or intein in the allele of concerned gene that lacks the intron/intein. The process of homing is based on induction of a DSB in the recipient allele by the HE encoded by the self-splicing intron or intein. This is followed by homology-directed repair (HDR) of the DSB using the intron/intein as template (Fig. 2.1). Subsequently, the dominant marker 'Omega' was shown to be the gene for the site-specific endonuclease I-SceI. In general, the HEs are selfish elements that offer little benefit to and impose no significant cost on their hosts. However, some HEs do perform some functions that benefit their biological hosts. For example, some HEs increase the competitive ability of their host phage during mixed phage infections. At least one HE functions as transcriptional regulator, many HEs function as maturases (they participate in post-transcriptional splicing of the introns containing the HE sequences), while some HEs, *e.g*., HO endonuclease, are involved in gene conversion events in yeast. Some maturases lack endonuclease activity, while some others function as endonucleases as well. Finally, some of the HEs seem to be involved in protection and maintenance of the genomes of their hosts (Stoddard, 2011, 2014).

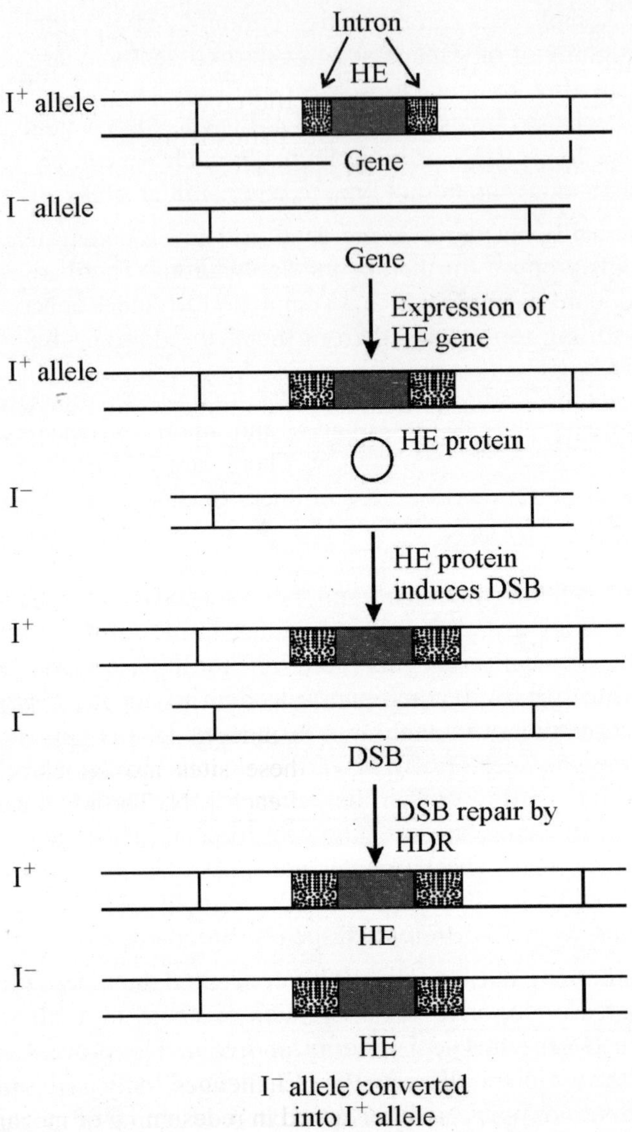

Fig. 2.1. A simplified representation of homing. The *HE* gene located within self-splicing intron of a gene encodes HE protein. The HE induces a DSB in the I^- allele of this gene. The DSB repair mediated by HDR using the I^+ allele as template leads to conversion of the I^- allele into I^+ allele and the intron along with the *HE* gene is copied into the intron-less I^- allele. HE, homing endonuclease; DSB, double-strand break; HDR, homology-directed repair; I^+, allele with self-splicing intron containing HE gene; I^-, allele lacking the self-splicing intron. The I^- allele has the intron insertion site where the HE protein induces DSB.

The HEs are divided into at least six different families. Each HE family is mainly confined within a distinct range of microbial hosts, has unique catalytic core, and appears to have evolved from a separate ancestral nuclease. There is some evidence that several HE families are yet to be discovered and characterised. The LAGLIDADG homing endonucleases (LHEs) are largely confined to species of archaea and organelles of algae and fungi. The LAGLIDADG domain of the LHEs contains their nuclease domains. The LHEs are one of the most sequence specific HEs. They make both sequence-specific and nonspecific contacts with bases in the major groove of DNA; the distribution of these bases in the target sites is uneven. The members of LHE family, also called meganucleases (MNs), are the main HEs that are used for site-directed genome modifications. The name 'meganuclease' reflects the name of the first discovered homing system Omega ('mega' derived from Omega) in yeast as well the large size of their recognition/target sequences (Stoddard, 2011).

2.3. MEGANUCLEASES

Meganucleases are encoded by mobile introns found in organellar (mitochondrial and chloroplast) genomes, and genomes of phages and microbial species, including eubacteria, archaeal species, algae, fungi and protists. There are more than 200 different natural meganucleases, which function as either homodimers or monomers. There seems to be a strong conservation of protein structures among MNs, but they do exhibit highly divergent target sequences. The homodimeric MNs are composed of two identical subunits of 160 to 200 amino acids. The monomeric MNs, on the other hand, comprise two LAGLIDADG domains joined in tandem by a variable linker peptide; each of these domains often consists of 100 to 120 amino acid residues. The monomeric MNs can recognize fully asymmetric target sites; the two LAGLIDADG domains of these enzymes cut the two strands of DNA duplexes at different rates so that nicked target sites are produced prior to induction of DSBs. The meganucleases have recognition sites of typically 20–30 bp, and these sites are largely divided into two parts recognized by the two LAGLIDADG domains of each MN. The MNs may bind many genomic sites with related DNA sequences with similar affinities, but they induce DSBs only at a small subset of these sequences. MNs cleave DNA around the center of the target site and typically generate 4-base 3' overhangs with free -OH groups (Fig. 2.2; Stoddard. 2011, 2014; Podevin et al., 2013).

The DNA binding and nuclease domains of meganucleases are interdependent and not clearly separated; this feature of MNs contrasts with those of ZFNs and TALENs. As a result, attempts to modify the DNA binding specificity of meganucleases present problems, and often reduce their nuclease activity. In view of these challenges, only a limited number of research workers and biotechnology companies are involved in redesigning of meganucleases (Curtin et al. 2012). The most widely used MNs for genome modification include I-SceI, I-CreI and I-DmoI. The DNA-binding domain of MNs has two components, each of which is located at one side of the DNA cleavage domain. Tailor-made MNs have been created in one of the following two ways: (1) modification of DNA binding specificity of the natural MNs by substituting a small number of amino acids in the DNA recognition domain, and (2) by combining the DNA binding domains from different MNs. The delivery of MNs into the cells is easy, and they have high specificity in DNA binding. But they seem to have limited flexibility in DNA binding and are difficult to engineer. It is encouraging that different groups of researchers, including those of at least three biotechnology companies, have succeeded in devising strategies that enable redesigning of MNs for recognition of and cleavage at many endogenous genomic sites (Podevin et al. 2013).

2.4. REDESIGNING MEGANUCLEASES FOR NOVEL GENOMIC SITES

Site-specific DNA cleavage by a meganuclease, I-SceI, from yeast was demonstrated in 1995 in cultured mouse cells. The target sequence for I-SceI had been integrated into the genomes of these cells prior to their exposure to the I-SceI. The MN I-SceI induced site-specific DSBs in the mouse genome, which promoted HDR using the donor DNA template and resulted in a functional *neo* marker gene. The I-SceI-sponsored HDR frequency was about 100-times higher than spontaneous homologous recombination (Choulika et al., 1995). This study prompted research efforts to engineer MNs with altered DNA binding and cleavage specificities so that MNs could be designed to target specific sites within the desired genes. The specificity of redesigned MNs has been assayed by different methods, including direct physical scrutiny of the protein-DNA interface, and repacking and optimization of the new protein-DNA contacts using computational algorithms. More direct methods of specificity assay involve elimination of function of reporter gene, inactivation of a toxin producing gene (bacterial selection strategies), restoration of function of reporter gene or a gene required for growth by HDR-mediated repair of the DSB induced by the engineered MNs (yeast-based recombination reporter assay), and *in vitro* compartmentation and PCR-based assay (Fig. 2.3; Stoddard. 2011).

The various approaches for creation of MNs with altered DNA recognition specificity include changing specific amino acids or groups of amino acids within the meganuclease DNA-binding surface as well as swapping of the whole LAGLIDADG domains or even the entire subunits from unrelated natural MNs. The latter approach established that the individual LAGLIDADG domains/subunits of various MNs typically function independent of each other. As a result, LAGLIDADG domains/subunits from two different MNs can be combined to create a functional chimeric MN that will recognize and bind to target sites comprising two half-sites characteristic for their component LAGLIDADG domains/subunits. Another approach uses closely related (with 40 to 70% sequence homology) MNs for systematic domain swapping combined with introduction of limited amino acid residue variation in the individual domains. This approach yielded about 75% catalytically active chimeric/hybrid MNs. The domain swapping approaches will generate a wide variety of novel chimeric MNs in addition to the naturally occurring enzymes. The amino acid substitutions in individual natural MNs have the potential to generate engineered MNs specific for the desired genomic target sites (Stoddard. 2011, 2014).

Thus, the target sequence for a MN can be divided into two half-sites, *viz.*, the left and the right half-sites, and MNs can be separately engineered to recognize and bind to each half-site. After the mutant MNs have been successfully created, the two mutated subunits can be combined to generate the MN specific for the desired target site. Several researchers have mutated the MNs, specifically I-CreI, to alter the recognition of individual bases in their target sequences. In addition, schemes for redesigning of MNs for altered specificity across adjacent groups or "pockets" of successive base-pairs were developed for a large-scale alteration in the target recognition by the mutated MNs. These experiments revealed that base recognition by the MNs was context dependent to a considerable extent. Therefore, such amino acid substitutions for recognition of individual bases that reduced MN activity or specificity sometimes had little adverse effect when they were part of some extensively modified groups of amino acid residues. At the same time, some alterations that had little adverse effect on MN specificity and activity, could not be combined with other alterations in the MN at adjacent positions (Arnould et al., 2006; Smith et al., 2006; Stoddard. 2011).

It is desirable to reduce the number of variants to be created and evaluated. A semi-rational approach achieves this by selecting the relevant amino acid residues on the basis of structural data. Another more elaborate method identifies the key residues with good accuracy using computational analysis of energy calculations. A computational approach for redesigning of MNs uses the Rosetta computational protein design algorithm; this method of redesigning yielded high frequency of MNs with altered specificities. Smith et al. (2006) used the semi-rational approach for generating I-CreI mutants and combined it with high throughput screening to identify hundreds of I-CreI variants with changed target sequence specificities. This procedure could identify I-CreI variants that were able to recognize and cleave targets differing from the wild type I-CreI target by up to 3 bp. It was realized that a large-scale application of this approach would not be able to generate I-CreI variants capable of recognizing and binding to target sequences differing from the wild type recognition sequence by several base pairs. It was clear that, in contrast to TALEs, there is no one-to-one relationship between the amino acid residues and the DNA nucleotides. This relationship is context dependent as is the case for zinc finger proteins; as a result, the same nucleotide can be correlated with different amino acid residues depending on the amino acid residues in the neighbourhood of the relevant residues.

Smith et al. (2006) devised a procedure for generation of efficient variants of the meganuclease I-CreI; this dimeric MN recognizes and cleaves a pseudopalindromic target sequence of 22 bp (Fig. 2.2). The two LAGLIDADG domains function independent of each other in target recognition. Therefore, they can be modified separately to generate I-CreI variants with novel specificities. Almost 25% of the amino acid residues of MNs are involved in sequence-specific and non-specific direct or indirect contacts in the major groove of DNA, the points of contact being unevenly distributed over the entire target sequence. The DNA bends in the middle of target site, which contributes to specificity of the MN. Evidence from crystal structure analyses of I-CreI and DNA complex indicated that nine amino acids of a single LAGLIDADG domain make direct contact with the target site. All possible substitutions at these nine positions would generate 20^9 I-CreI variants, screening of which was not possible by the available procedures. Therefore, two virtually independent subdomains, designated as 5 NNN (±3, ±4, ±5 nt) and 10 NNN (±8, ±9, ±10 nt), of a LAGLIDADG domain were identified. The ± sign signifies the nucleotide at the given position in both the left and the right half-sites of the target sequence (Fig. 2.2). The three amino acids at positions 44, 68 and 70 make direct contact with the nucleotides in the 5 NNN subdomain; they were substituted to create I-ScI variants. These variants were evaluated using a high throughput screening procedure to identify efficient variants specific for the 64 possible sequences of this trinucleotide subdomain.

Similarly, variants for the other subdomain, the 10 NNN subdomain, were created by amino acid substations at positions 28, 30, 33, 38 and 40 of I-ScI, and screened for selecting the efficient variants for the 64 possible trinucleotide sequences for this subdomain. The appropriate variants for the two subdomains were then selected based on the desired target sequence and combined to create a LAGLIDADG domain that would efficiently bind to this sequence. The LAGLIDADG domain that would bind to the other half of the pseudopalindromic target sequence was also modified according to the above general scheme. Finally, the two modified LAGLIDADG domains are combined to generate a dimeric I-ScI MN that would specifically bind to the desired target site. This scheme was used to create I-ScI variants that specifically cleaved the target site in human *RAG1* gene; this target site differs from that for the wild type I-CreI target by as many as 16 base pairs (Smith et al., 2006).

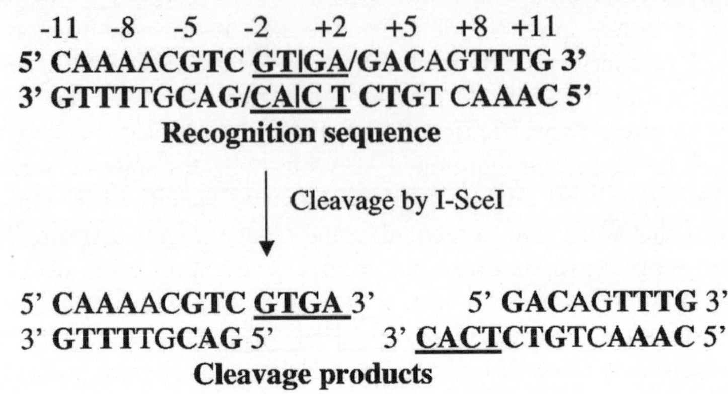

Fig. 2.2. The I-CreI recognition sequence is a pseudopalindrome. The base pairs with palindromic symmetry are depicted in bold letters. The 'central 4' bases are in bold letters as well as underlined. The vertical line demarcates the two half-sites of the recognition sequence, *i.e.*, the axis of symmetry, while the slashes depict the sites of cleavage that yield a 4-base 3' overhang with a free -OH group.

There is considerable flexibility in the MN redesigning: some redesigned efficient variants of MNs contain as many as 53 amino acid substitutions dispersed across their DNA binding surface. In addition, it is feasible to redesign MNs for target sites that differ for one or more bases from the authentic target site for the concerned MN; the variant bases can be located anywhere in the target site, except for the central 4 (-2,-1, +1, +2) base pairs of the target sequence (Fig. 2.2) since these are essential for the bending of target DNA during MN binding (Smith et al., 2006; Stoddard, 2011, 2014; Bogdanove et al., 2018).

Many extensively modified variants of the MN I-CreI have been developed and used for genome editing by two biotechnology companies. Their strategy, in simple words, comprises conversion of the naturally homodimeric I-CreI into a monomeric nuclease by joining the two subunits by a linker peptide. This is followed by separately modifying the two LAGLIDADG domains of this enzyme by amino acid substitutions at key positions to create domains specific for the two half-sites of the desired target sequence. The two modified LAGLIDADG domains are then combined to create a monomeric enzyme, which is further optimised for the desired target site. One company mainly used structural data to identify and combine individual amino acid substitutions to redesign the wild type LAGLIDADG domains. The second company, on the other hand, redesigned the MN using phenotypic screens for selection of specific and active variants from a semi-randomized library of I-SceI mutants created for the purpose. It is note-worthy that both these approaches resulted in replacement of largely the same amino acid side chains. Another biotechnology company has generated variants with novel DNA binding specificities of the monomeric MN I-OnuI. This company used a yeast surface-display and high throughput flow cytometry screen for evaluation of a semi-randomized library of I-OnuI variants. It may be pointed out that screening procedures used by the first two companies have relatively low throughput, while the screen used by the third company has high throughput (Stoddard, 2014).

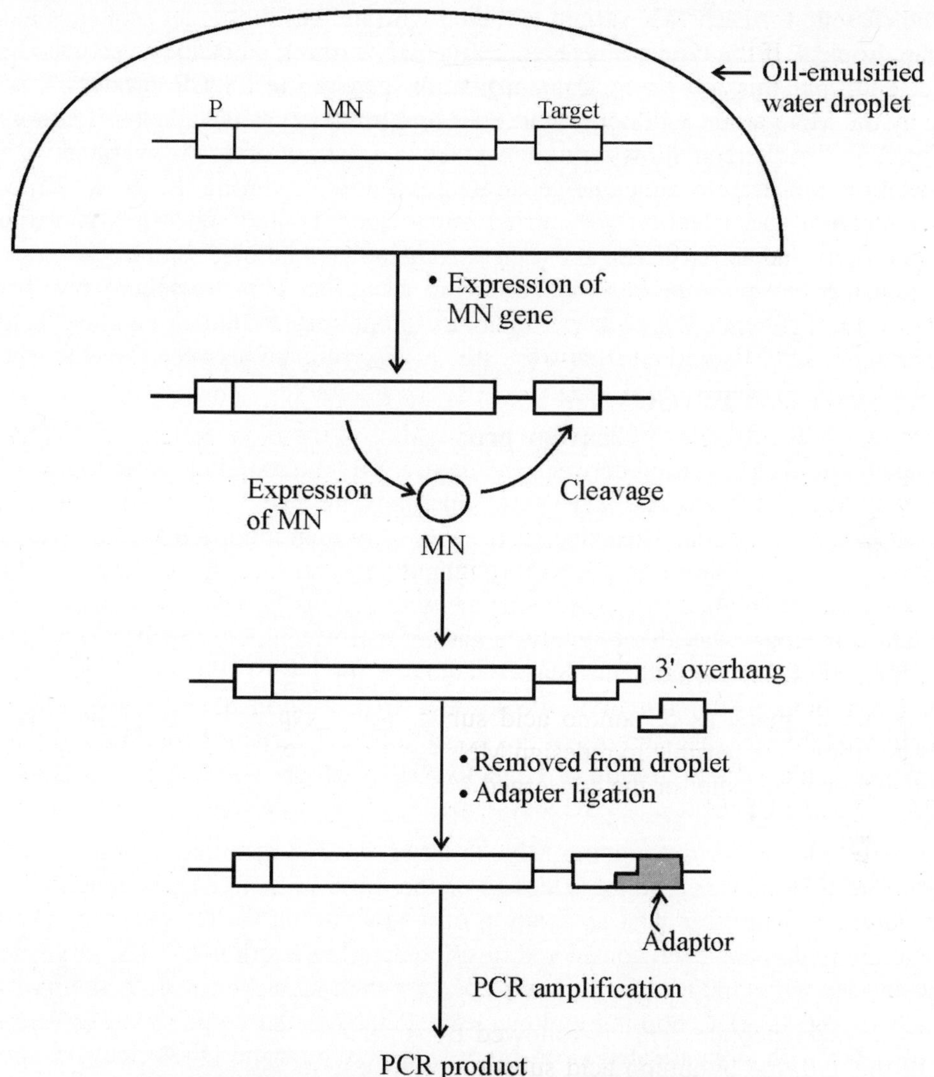

Fig. 2.3. A schematic representation of a simple, rapid and high throughput PCR-based screen for mutant libraries created during redesigning of MNs. The MN gene construct is linked to the desired target sequence and expressed within an oil-emulsified water droplet. Only when a mutant is specific for the target, it will cleave the target. The cleaved target will anneal to an adapter. PCR amplification will generate a detectable PCR product. The droplets are broken and the DNA segments are removed and used separately for annealing, PCR, etc.

A complementary strategy for engineering MNs for targeting novel genomic regions consists of computational screening of well-characterized wild-type MNs for identification of the most suitable MN for a given target site. The selected MN is then redesigned and selected for its activity assayed in compartmentalized oil-emulsified water droplets. This PCR-based screen is simple and fast as it does not require cell growth, which is the case with other assay methods. The expression cassette for each MN variant is linked with the desired target sequence and expressed in separate droplets. If the target is cleaved by the MN variant, an adapter becomes ligated to the cut target end, and this allows PCR amplification generating a PCR product. Lack of target cleavage by the MN variant will not permit PCR amplification as the adapter ligation cannot take place (Fig. 2.3). This screen allows simultaneous evaluation of up to 10^{10} variants; it also permits rigid control on temperature, time and substrate concentration during the assay. Repeated rounds of variant creation and selection are carried out to identify and select a MN variant with the desired specificity and activity. The above *in vitro* assay is combined with *in cellulo* (inside cells) evaluation using two-plasmid cleavage assay in bacteria. This procedure has been used to engineer several different MNs to target many different sites in human genome. It also allowed identification of such MN variants that were able to discriminate between target sites differing for a single nucleotide (Takeuchi et al., 2015).

Wang et al. (2014b) used progressive redesigning of the monomeric MN I-AniI to evolve a variant capable of discriminating between the mouse XID mutant allele from the wild type allele. The two domains of the enzyme were first separately modified by substitution of amino acid groups followed by selection. The selected variants were then modified at individual amino acids and selected for activity based on a high throughput yeast surface display method. The modified domains were then combined to yield an enzyme with activity comparable to that of the wild type enzyme. This procedure was able to evolve a variant that was highly specific for a genomic target differing for 9 bp from the authentic target sequence for I-AniI. Fusion of the selected variant with the DNA binding domain of a site-specific transcription activator-like effector (TALE) further enhanced the cleavage activity of the selected I-AniI variant. In addition, the I-AniI-TALE fusion enzyme achieved a high ratio of HDR to NHEJ, which is in contrast to the situation with ZFNs, TALENs and CRISPR-Cas9 SDN systems.

A general scheme for redesigning MNs to target desired genomic regions is outlined below (Fig. 2.4). The first step consists of selection of the most suitable target sequence for achieving the gene editing objective as well as creation of a MN variant for the purpose. The selection of target genomic region has been considered in some detail in Section 4.7. The target sequence and the most suitable wild type MN for this sequence are then identified with the help of LAHEDES. LAHEDES (LAGLIDADG homing endonuclease database and engineering server) provides an archive of LHEs with authenticated DNA-binding interactions and DNA cleavage specificities. It also provides a tool for identifying the sequences in the genomic regions of interest that can serve as targets for different wild type MNs. The LAHEDES is an open access webserver available at http://homingendonuclease.net (Taylor et al., 2012). The next step is the selection of the most suitable wild type MN for redesigning; this selection is based on several considerations, including the number of bases for which the genomic target sequence differs from the actual target of the wild type MN and the distribution of these bases in the target site, and the characteristics of the wild type MN, including its amenability to redesigning. In general, the target sequence should differ from the MN target sequence for as few bases as possible and most of the mismatches should preferably be in the 'outer 3' (-11, -10, -9, +9, +10, +11) base pairs. In addition, it is highly desirable that there is no mismatch in the 'central 4' (-2, -1, +1, +2) base-pairs (Fig. 2.2), but mismatch for one base-pair may be considered.

Selection of the genomic region to be edited by MN action

↓ LAHEDES Webserver

- Selection of the target sequence within the desired genomic region
- Identification of the most suitable wild type MN for the desired target sequence(s)

↓ LAHEDES Webserver

Identification of the amino acids to be substituted for creation of mutants of the selected MN that are expected be specific for the desired target

↓

- Specificificities and activities of the mutant libraries assayed using a suitable procedure (*e.g.*, Fig. 2.3)
  - The two LAGLIDADG domains/subunits modified independently

↓ Many rounds of mutation and selection

  - The two selected LAGLIDADG domains/subunits joined to yield a monomeric MN
  - The monomeric MN may need further optimization by mutation and selection

↓

A redesigned MN specific for the desired target sequence

Fig. 2.4. A simplified representation of a general scheme for redesigning MN for the desired genomic target sequence.

The next step is to create mutant libraries of the selected MN. This involves identification of the amino acid residues that are most responsible for sequence specificity for the bases at the mismatch sites in the target sequence; this can be achieved with the help of LAHEDES webserver. The mutant libraries of the MNs are then evaluated using a suitable assay procedure. Typically, several rounds of variant creation and selection are necessary for creating a MN variant with high specificity for the target sequence coupled with high activity. The two LAGLIDADG domains/subunits of the MN are redesigned separately, and the redesigned domains/subunits are then linked together to generate a monomeric MN. The monomeric form would need to be further optimized since its specificity and/or activity is likely to differ from those expected from its constituent domains/subunits that were separately redesigned and subjected to selection independent of each other (Anonymous, http://www. homingendonuclease.net/info/engineering_LHEs.pdf).

2.5. MEGANICKASES (MNICKASES)

Meganickases (MNickases) induce single strand breaks, *i.e.*, nicks, at their target sites. The nicks induced by MNickases are preferentially repaired by HDR. As a result, the proportion of HDR events is much higher than that of nonhomologous end-joining (NHEJ) events, which lead to mutations mainly due to indels. It may be pointed out that the efficiency of HDR is markedly reduced in comparison to that associated with double-strand breaks. However, the extent of reduction in the frequency of NHEJ events is far more than that in the HDR frequency. In addition, nickase action eliminates the risk of toxicity that is associated with DSB induction by MNs (Stoddard, 2011). MNickases are produced by inactivation of one of the nuclease sites of monomeric MNs. In case of homodimeric MNs like I-CreI, one mutated and one wild type monomer can be joined together by a linker to generate a monomeric nickase. The first nickase was produced by inactivation of one of the nuclease sites of the monomeric MN I-SceI. This was achieved by substitution of Lys residue in one of the two nuclease domain active sites of I-SceI. The nicked DNA strand depended on which of the two nuclease sites was inactivated (Niu et al., 2008). Subsequently, Smith et al. (2009) generated a nickase variant of the monomeric MN I-AniI by substitution of a single Lys residue at the border of the active site in the C-terminal domain of the enzyme. The nickase has the same specificity and binding affinity for the target DNA sequence as the wild type I-AniI nuclease. The MNickases may be more desirable than MNs for therapeutic applications involving sequence replacement for correction of disease-causing gene mutations.

2.6. OFF-TARGET ACTIVITY

There is little information about *in vivo* specificity and off-target activities of meganucleases. In general, the specificity assays like toxicity assay, binding-site selection *in vitro*, etc. are quite insensitive. In addition, it is not clear whether the findings from *in vitro* assays are a reliable indicators of *in vivo* specificities and off-target activities. The off-target cleavage by I-SceI, one of the most specific MNs, was assayed by monitoring the integration of adeno-associated virus (AAV) vector into the genomes of cultured human cells treated with I-SceI. About 61% of the integration events occurred at off-target sites. At one such site, the integration was more frequent than that at the authentic target site although this site differed from the authentic site for as many as four base pairs. It appears that the findings from *in silico* analyses and *in vitro* assays do not reliably predict the MN cleavage sites *in vivo*. Further, MN cleavage sites may differ among different cell types due to epigenetic modifications. Therefore, it is imperative that detailed investigation of the off-target activities of MN variants are carried out especially when they are being developed for therapeutic applications (Petek, 2010).

2.7. APPLICATIONS OF MEGANUCLEASES

Redesigned meganucleases have been used for unravelling the various aspects of cellular DNA strand break repair processes associated with gene conversion. They have been used for mutation correction by HDR, gene disruption/inactivation, insertion of defined sequences at targeted sites, targeted deletion of genomic segments and insertion of genes for multiple traits. They have been used in a variety of biological systems, including cultured human and rodent cells, and plant species like maize, *Arabidposis* and cotton. Engineered versions of the MN I-CreI have been used for HDR-mediated correction of mutations at a specific site of *RAG1* human gene; these mutations produce a subset of rare phenotypes of SCID (severe combined immunodeficiency) disease. The engineered MNs induced site-specific DSBs at the targeted site, which triggered HDR-mediated sequence replacement and mutation correction. Engineered variants of the I-CreI MN have been used to target other human genes associated with genetic diseases, *e.g.*, mutation correction in the genes *XPC* (Xeroderma pigmentosum complementation group C) and *RAG1*, and insertion of a missing exon in *dystrophin* or *DMD* gene (the mutant gene is associated with Duchenne muscular dystrophy disease).

A variant of the I-CreI MN has been used for targeted inactivation of maize genes *ligueleless-1* and *MS26* (cytochrome P-450 like gene for male fertility). Redesigned I-CreI has been used to induce deletions of targeted genomic regions in *Arabidopsis* and for insertion of genes for multiple traits in cotton. Engineered MNs have also been used as a component of a gene drive strategy for spreading deleterious mutations like inactivated fertility-related genes in the vector populations in order to control insect vectors for diseases. I-SceI exhibited high rates of homing in both male and female germlines of *Anopheles gambia* mosquitoes demonstrating that LHEs can be used to develop successful gene drives. Similar encouraging results have been obtained with other LHEs (Belfort and Bonocora, 2014; Stoddard. 2011, 2014).

It has been proposed that the use of MNs for gene editing would be facilitated by the following recent developments: (1) identification and characterization of a large number of wild type homing endonucleases, (2) increased chances of recombination involving the DSB sites due to generation of 4-bp 3' overhangs by MNs, (3) gene inactivation due to NHEJ when MNs are combined with 3' exonucleases, (4) creation of nickases, (5) creation of highly specific site-directed nucleases comprising small, single chain scaffolds that are easy to package and deliver into target cells, and (6) the development of very informative reporter systems and assays for reliable assessment of the efficiency of DSB repair and the repair pathway involved. These developments are expected to facilitate further refinement of the reagents and protocols leading to more extensive application of MNs for achieving appropriated genome editing objectives.

2.8. MEGATALs

MegaTALs (MTALs) are a type of hybrid site-specific nucleases produced by linking the DNA binding domain of TALENs with meganuclease redesigned for recognition of the desired target site. A hybrid SDN is created by linking the DNA binding domain of one SDN system with the nuclease domain from another SDN system. The first hybrid SDN was produced by Kleinstiver et al. (2012) by replacing the FokI nuclease domain of a ZFN with the nuclease domain of the homing endonuclease I-TevI; these fusion proteins are referred to as I-TevI-zinc finger endonucleases (Tev-ZFEs). The Tev-ZFEs are compact single chain SDNs that function as monomers; this eliminates the need for using a pair of ZFN monomers for gene editing and greatly facilitates the delivery of Tev-ZFEs.

The first megaTAL was constructed in 2014 by Boissel et al. (2014) by linking DNA-binding region of a TALEN to the N-terminus of meganuclease I-AniI via a short linker peptide (Fig. 2.5A). The motivation for creation of megaTAL was the general decline observed in cleavage activities of MNs redesigned for specificity to the targeted genomic sequences due to a reduction in their DNA binding affinity. It was reasoned that linking such MNs to DNA binding regions of TALENs designed to bind sequences few base-pairs (typically, 7 bp) removed from the MN-binding sites (Fig. 2.5B) may enhance the cleavage activities of the hybrid SDNs by increasing their DNA binding affinities. Indeed, the megaTAL cleavage activities were dramatically increased as compared to that of their isolated MN cleavage heads. It may be noted that the DNA binding domain of a TALEN comprises the N-terminal, TALE repeat and C-terminal domains of its TALE protein component. Further, the N-terminal and C-terminal domains of natural TALE proteins are, in general, suitably truncated for TALEN construction.

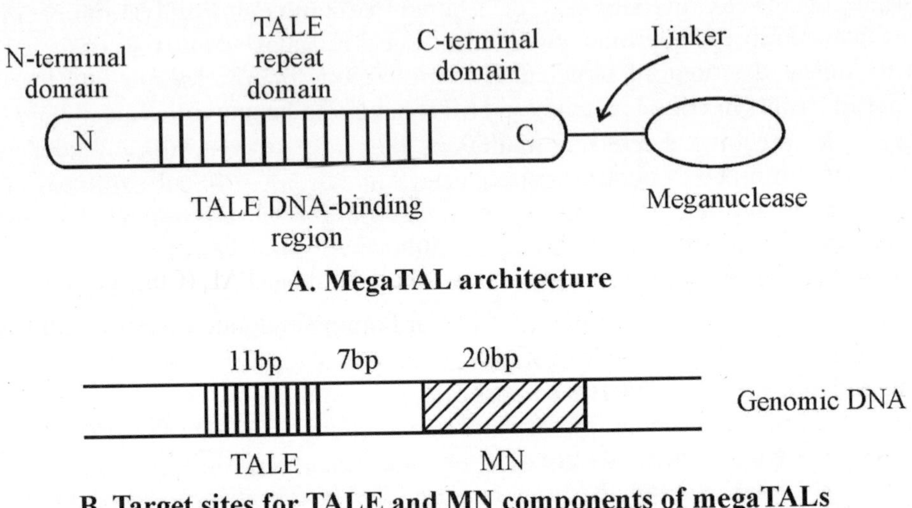

Fig. 2.5. Architecture and target sites for megaTALs. **A.** Schematic representation of MegaTAL architecture. The TALE DNA-bindng region is, in fact, the DNA-binding region of an efficient TALEN architecture. The N-terminal domain of TALE DNA-binding region comprises 154 amino acids, C-terminal domain has 63 amino acids, and the TALE repeat domain has 10.5 repeats, the target sequence being 11bp. The target sequence for I-AniI is 20 bp. **B.** The target sites for TAL and meganuclease components of megaTALs. TALE, transcription activator - like effector; TALEN, TALE nuclease; MN, meganuclease.

The various features of megaTALs make them highly suitable for therapeutic applications. A SDN for therapeutic application should have high on-target cleavage efficiency, be extremely specific for the target sequence so that it has little or, preferably, no off-target activity, and its delivery to the target cells should be simple, easy, efficient and inexpensive. MegaTALs have very high target site specificity due to the increased length of their target sequences comprising the target sequences for both their TALE and MN components. In addition, the cleavage heads of their MN components possess extremely high cleavage sequence specificity; these cleavage sequences are located in the central regions of their target sequences. It appears that the megaTAL architecture does not affect the specificity of MN cleavage heads, but it increases their on-target cleavage activity. Thus, megaTALs are extremely efficient and hyper-specific nucleases; they are compact in size, function as single chain monomers, and are suitable for viral and nonviral delivery methods in current use. In addition, their TALE components can be easily redesigned for any target DNA sequence. The 3' overhangs generated by megaTALs are efficiently processed by Trex2 exonuclease. The co-expression of megaTALs with Trex2 could achieve targeted gene inactivation in >70% of human primary cells. The gene inactivation frequency increased several-fold by combining Trex2 and a MN like I-CreI (Boissel et al., 2014; Wang et al., 2014b).

The very high on-target activity of megaTALs would allow the desired levels of on-target genetic modifications to be achieved by lower levels or reduced periods of expression of the megaTALs in the targeted cells; this would further reduce the risk of off-target action. It has been speculated that future efforts may focus on developing low affinity and high specificity mutants of LHEs and combine such mutants with the DNA-binding domains of TALENs to generate high activity megaTALs with very high specificity; such MTALs would be highly desirable for therapeutic applications. Wang, Y. et al. (2014b) noted that MTALs led to a high ratio of HDR to NHEJ repair of the DSBs. This is a very useful feature and supports the general idea that LHE-mediated gene editing favours HDR. MegaTALs have been developed for specific targeting of human gene *CCR5* involved in HIV-1 infection. Gene editing with this megaTAL was able to achieve 5-times higher HDR than the high activity TALEN SDNs specific for the identical targeted region of *CCR5*. But in the absence of donor DNA, the TALEN generated substantially higher frequencies of NHEJ-mediated DSB repair than the megaTAL (Curinga et al., 2015).

Homing endonucleases can be fused with other homing endonucleases or with Cas9 nuclease to construct SDNs that will induce two DSBs at genomic sites located at a short distance from each other. As a result, these reagents will reliably induce short deletions and thereby ensure inactivation of the targeted genes. For example, the nuclease domain of I-TevI has been fused with a meganuclease and Cas9 nuclease to generate MegaTev and TevCas9 SDN, respectively. The MegaTev and TevCas9 induce two DSBs at short distance from each other; one DSB is induced by the Tev nuclease domain, while the other DSB is due to the MN or Cas9 nuclease domain. The length of the deleted sequence is determined by the distance between the cleavage sites of I-TevI and meganuclease/Cas9 nuclease (Guha and Edgell, 2017).

2.9. MEGANUCLEASES, ZFNs AND TALENs

Meganucleases, ZFNs, and TALENs have been utilized extensively for genome editing in a variety of different cell types and organisms. MNs are engineered versions of naturally occurring restriction enzymes that typically have extended (mostly, 20-30 bp) recognition sequences. Engineering of MNs has been challenging for most academic researchers because their DNA recognition and cleavage functions are intertwined in a single domain and target base recognition is context dependent. The ZFNs are derived from ZFPs, the eukaryotic transcription factors having distinct DNA binding and transcription activation domains. The ZFNs are created by replacing the ZFP activation domain with the nuclease domain of FokI Type IIS restriction enzyme. The redesigning of ZFNs for novel target sequences has proved challenging since their base recognition function is context dependent similar to that for MNs. TALENs are also derived from eukaryotic transcription factors called TALEs that are produced by certain plant pathogenic bacteria. The TALENs are constructed by joining the DNA-binding domains of TALE proteins with the FokI nuclease domain. The redesigning of TALENs for new target DNA sequences is the easiest of the three SDN systems as the base recognition is essentially context-independent and there is a simple relationship between the RVDs (repeat variable diresidues) of the TALE repeat units and the bases recognized by them.

The length of target sequence is flexible in case of ZFNs and TALENs, but it is fixed for the MNs although different MNs may have targets of different lengths. The TALENs can access almost any genomic target sequence as a result of the simple one-to-one relationship between their RVDs and the target DNA nucleotides. But ZFNs recognize target sequences in triplets, as a result of which not all specific bases in the target sequence are recognized. The target access is the most restricted is case of MNs since the extensive redesigning efforts limit the choice to such target sites that differ for the least number of bases from their authentic target sequences. In addition, ZFNs and TALENs do not require specific sequences at cleavage sites, while MNs have a rigid requirement for specific cleavage sequences (the 'central 4' base pairs). In general, cleavage specificity is the highest for MNs, followed by TALENs and ZFNs. Therefore, off-target activity is expected to be higher for ZFNs and TALENs than that for MNs, which are believed to be the most specific SDNs. The ZFNs and TALENs function as dimers, while MNs function as monomers or dimers depending on the particular natural MN that was used to develop the meganuclease-based SDN. The DSBs induced by ZFNs and TALENs have 5' overhangs, but those generated by MNs have 3' extensions. The MNs have a compact organization, while ZFNs and TALENs have modular architecture. As a result, the delivery of MNs is easier than that of ZFNs, while TALEN delivery using some viral vectors may be problematic due to the nearly identical repeat units present in their DNA binding domains (Belfort and Bonocora, 2014).

2.10. ADVANTAGES AND LIMITATIONS OF MEGANUCLEASES

The chief advantage of MNs is their outstanding specificity, which is extremely important for therapeutic applications. The MNs are compact in size and they are either monomeric or monomeric versions can be created for the naturally homodimeric MNs; these features greatly facilitate their delivery into the target cells. They can be delivered into many cell types and more than one variant of a MN can be delivered into each cell to simultaneously target more than one genomic site in the cell. But their major limitation is the difficulty in redesigning MNs for the desired target sequences. This limitation is due to the lack of simple relationship between the amino acid sequences of DNA binding domains of MNs and the bases in the target sequences recognized by them since DNA recognition is context dependent. Several schemes for redesigning MNs have been devised but all of them involve considerable amount of protein engineering and several rounds of mutant creation and selection for the identification of MN variant specific for the desired target sequence. In addition, the redesigned MNs have reduced DNA binding affinity, which results in lower DNA cleavage activity. This difficulty can be ameliorated by creating megaTALs by fusing MNs to the DNA binding domains of TALENs. The megaTALs may emerge as useful therapeutic agents in view of their extremely high specificity and very high efficiency. There may be some limitation on the availability of suitable target sites in the genome due to the rigid requirement of the 'central four' base sequence in the targets for cleavage by MNs (Smith et al., 2006; Stoddard 2014; Boissel and Scharenberg, 2015).

Thus, in contrast to TALENs and CRISPR-Cas9 systems, LAGLIDADG homing endonucleases are more challenging to engineer for new target sequences, but they exhibit superior physical attributes, and DNA recognition and cleavage properties. Recent advances have overcome many of the hurdles in MN engineering, but their application for routine biological investigations and even in plant breeding activities is not practicable due to the cost and the effort involved in their redesigning. But it may be a viable option to develop customized MNs for specific therapeutic applications since once a successful MN variant has been created it can be used for therapy of all patients requiring the given treatment. This will more than justify the cost and effort involved in the development of the MN-based therapeutic agent.

2.11. CHIMERIC OLIGONUCLEOTIDES

Chimeric oligonucleotides, also called RNA-DNA oligonucleotides (RDOs), are double-stranded chimeric oligonucleotides comprising a single DNA strand that has two short (10 nucleotides long) RNA sequences at strategic locations (Fig. 2.6). The 3' and 5' end regions of this typically 68 nucleotides (nt) long chimeric single-strand fold back to yield the double-stranded structure with two hairpin loops formed by T residues; this effectively prevents concatamerization of the COs. In addition, the 5' and 3' ends of COs are appropriately modified to prevent their circularization as well as ligation with other DNA molecules; this precludes the integration of COs into the host cell genomes. A CO has a 25 bp long segment that is complementary to the target region of the gene of interest. One strand of this region is chimeric in that it has five deoxyribonucleotides flanked on either side by 10 ribonucleotide residues. The ribonucleotides are modified by 2'-O-methylation of their ribose residues to make the RNA sequence resistant to RNase H digestion (Igoucheva and Yoon, 2000; Komor et al., 2017a).

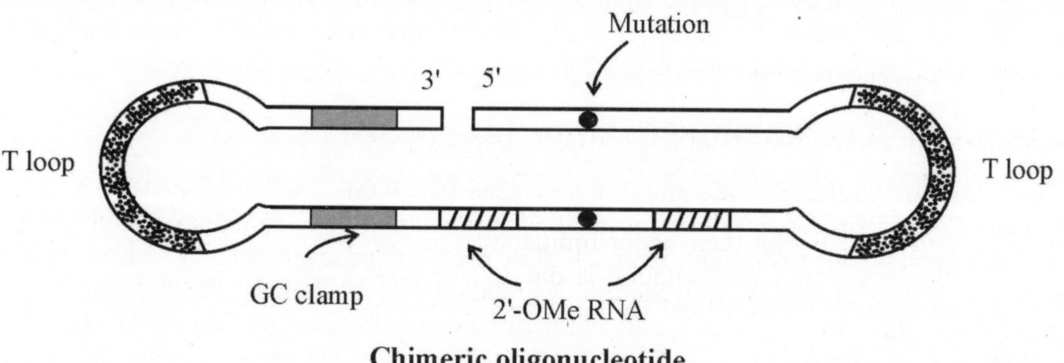

Chimeric oligonucleotide

Fig. 2.6. Schematic representation of the organization of a chimeric digonucleotide. GC clamp, sequence composed of G and C nucleotidis; mutation, changed base sequence to be introduced in the targeted gene; 2'-OMe RNA, RNA sequence with 2'-O-methylated ribose residues; T loop, a run of T residues to generate hairpin loops.

The CO strand containing RNA segments is likely to promote pairing with the target sequence and stabilize the labile intermediate generated during the process. The homologous region of CO comprising only DNA nucleotides seems to be mainly responsible for the sequence alterations due to DNA repair activities of the cells. Since annealing is the primary determinant of efficiency of mutation induction by COs, their chimeric state improves the efficiency by promoting and stabilizing their annealing to the targeted sequences. One to few nucleotides of the five deoxyribonucleotides located between the two RNA segments are different from those present in the targeted gene sequence. The bases that differ from the targeted sequence represent the substitutions intended to be introduced in the targeted gene. A 5 bp long segment called GC clamp is present towards the 3' T loop; the GC clamp increases the resistance of COs to digestion by cellular nucleases (Igoucheva and Yoon, 2000; Lusser et al. 2012; Komor et al., 2017a; Songstad et al., 2017).

It may be noted that the chimeric strand of COs has much lower efficiency in sequence correction than the complementary strand made up exclusively of DNA nucleotides. When a CO contains the mismatched nucleotide only in the strand composed exclusively of DNA nucleotides, it exhibits about two-fold higher mutation correction efficiency than the COs with mismatches in both the strands (Fig. 2.6). In addition, COs with continuous RNA sequence in one strand in the place of typical RNA-DNA-RNA sequences are 2-3-fold more efficient than the latter in mutation correctio/induction. Further, COs designed for the sense strand of the targeted sequences exhibit much higher efficiencies than those designed for the antisense strand (Gamper et al., 2000; Liang et al., 2002).

In addition, modified single-stranded all-DNA oligonucleotides of 25 nt can be used in place of COs for induction of specified sequence changes. These modified all-DNA oligonucleotides can be 3-4-times as active as the corresponding COs. Phosphorothioate-modified oligonucleotides are the most commonly used single-stranded all-DNA oligonucleotides, but they are toxic to cells when used *in vivo*. Another form of oligonucleotides that can be used for gene editing is the triplex-forming oligonucleotides, which bind to the major groove of DNA and form triplex DNA structures. However, these oligonucleotides have high affinity for polypurine/polypyrimidine sequences and are limited to induction/correction of mutations near homopurine/homopyrimidine genomic sequences (Gamper et al., 2000)

2.12. OLIGONUCLEOTIDE-DIRECTED MUTAGENESIS

The technique of oligonucleotide-directed mutagenesis (ODM) was developed by Yoon et al. (1996) and Cole-Strauss et al. (1996) for targeted correction of gene mutations in cultured human cells. Subsequently, in 1999, this approach was used to induce herbicide resistant plants in tobacco and maize. The ODM procedure is also called oligonucleotide-directed gene targeting, chimeraplasty, etc. This technique uses chimeric oligonucleotides (Section 2.11; Fig. 2.6) to typically introduce substitutions of one base at specific sites of the targeted genes; it can also be used to generate substitutions for more than one base and even addition/deletion for typically one base or even few bases (Lusser et al. 2012; Songstad et al., 2017). The COs are delivered into the target cells by an efficient delivery technique appropriate for the targeted cells/tissues. Once inside the cell, the COs unfold and move into the nucleus. Here, the CO segments will base-pair with the homologous sequences in the two DNA strands of the targeted genomic sequence; this will create a structure like that produced during DNA replication (Fig. 2.7). However, structural distortion will be created due to the mismatched base(s) that were deliberately included in the CO sequence (indicated as 'mutation' in Fig. 2.6). This distortion will be recognized by the cellular DNA repair system, and the mismatch repair pathway will remove this distortion by changing the base sequence of either the CO or the targeted genomic region. When the base sequence of the targeted genomic region is altered, the new base(s) in this sequence will be the ones specified by the DNA nucleotides included in the CO (Fig. 2.7). The mechanism of CO action does not depend on transcriptional state of cells; it also does not appear to require DNA replication as mutation correction has been obtained in G_0 cells. The COs are ultimately degraded by the cellular enzymes within few hours of their introduction into the cells (Igoucheva and Yoon, 2000; Lusser et al. 2012; Komor et al., 2017a; Songstad et al., 2017).

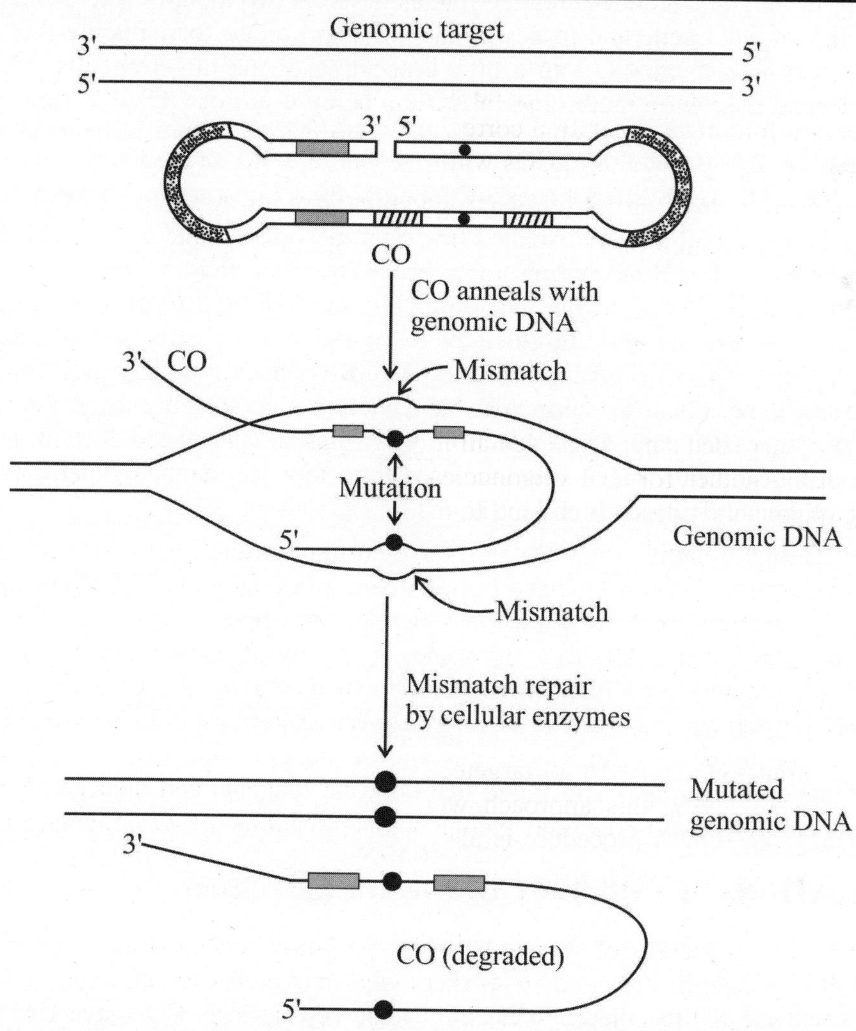

Fig. 2.7. Schematic representation of mutation induction by chimeric oligonucleotides. CO, chimeric oligonucleiolitide (details given in Fig. 2.6). CO is degraded by cellular nucleases.

The ODM enables induction of specified SNPs (single nucleotide polymorphisms) at the targeted sites of the selected genes. It can also induce directed sequence changes for two to five nucleotides as well. The CO-induced sequence changes can be designed to either induce desired mutations or correct the existing mutations in the targeted genes. It is important that the method of delivery of COs into the targeted cells should have high efficiency. A delivery method having low efficiency (delivery into 10-15% of the treated cells) may not induce mutations at frequencies that are significantly different from the spontaneous mutation rates (Songstad et al., 2017).

2.13. A GENERALIZED PROCEDURE FOR OLIGONUCLEOTIDE-DIRECTED MUTAGENESIS

The first step in oligonucleotide-directed mutagenesis is to identify the base/bases to be substituted in the targeted gene and to design the CO appropriate for inducing this substitution. The next step is to deliver the CO into a high proportion of the targeted cells. The method of delivery will depend mainly on the biological system being used for ODM. In case of plants, the COs can be delivered into cultured cells or tissues by particle gun bombardment. The bombarded cells are transferred onto a suitable selection medium favoring the mutant cells. The COs can also be delivered into protoplasts either by electroporation or with the help of PEG (polyethylene glycol); generally, the protoplasts are isolated from leaf mesophyll cells or from cell cultures. The protoplasts are then cultured according to a procedure that favors cell survival, cell wall regeneration and cell proliferation. The resulting calli are subjected to the appropriate selection agent. The selection procedure is designed to favor the mutant cells and thereby ultimately enhance the recovery of mutant plants. Since the ODM has been typically used for induction of herbicide resistance, the concerned herbicides have been used as selection agents in these studies. Generally, more than one round of selection was applied and the selection pressure, *i.e.*, concentration of the selection agent, was increased in a stepwise manner. Finally, plantlets were regenerated from the surviving cells and evaluated for the mutant trait.

In case of animal systems, the COs are delivered into cultured cells by microinjection and particle gun bombardment. The COs have been delivered *in vivo* with the help of liposomes and the polymer polyethylenimine (PEI). Liposomes have lower efficiency than PEI and both delivery systems are not tissue-specific. Modified liposomes, *e.g.*, polysaccharide modified liposomes, and modified PEI, *e.g.*, lactosylated PEI, have been shown to deliver the COs in a tissue-specific manner. It may be pointed out that the *in vivo* delivery systems should be nontoxic, have low immunogenicity, be tissue-specific, ensure passage of the COs into cell nuclei and, above all, have high efficiency. It has been concluded that both the reagents and their delivery procedures need considerable refinement to facilitate therapeutic applications of COs (Liang et al., 2002).

2.14. APPLICATIONS OF CHIMERIC OLIGONUCLEOTIDES

The use of COs for induction of targeted base substitution for mutation correction was first reported in 1996 by Cole-Strauss and co-workers who delivered COs into cultured human cells and successfully corrected the mutation causing sickle cell disease. COs have also been used in plants mainly to produce herbicide resistant mutants. ODM was first used to induce site-directed single base substitution in the *acetolactate synthase* (*ALS*) gene to generate resistance to the herbicide chlorsulfuron (Glean from DuPont). The nucleotide C in the codon CCA (encoding Pro-196) was substituted by A converting it into ACA (encoding Thr-196). The COs were delivered into cultured tobacco cells by particle gun bombardment and the treated cells were selected for survival on medium containing chlorsulfuron. The selected cells were tolerant to the herbicide. The CO was also able to correct a single base mutation in the *green fluorescent protein* (*GPF*) gene, and the corrected gene produced functional GFP (Beetham et al., 1999).

The ODM procedure was used to induce mutations in the *ALS/AHAS* (*acetohydroxy acid synthase*) gene to generate chlorsulfuron tolerant mutants in maize and *Brassica napus* as well. Subsequently, Pro-196 was substituted with the amino acids Ala, Thr, Gln, or Ser using ODM, and mutant tobacco plantlets were regenerated; these plants were highly resistant to the herbicide. In addition, plantlets with the substitution of Trp-573 with Leu-573 were also resistant to chlorsulfuron. The herbicide resistance was transmitted to sexual progeny of the mutant plants (Songstad et al., 2017). In the animal systems, COs have been typically used for mutation correction both in cultured mammalian cells and in whole animals. For example, COs were introduced into HeLa cells using PEI to successfully correct a single base mutation in the β-*globin* gene. Similarly, a mutation in the α-*globin* gene was corrected by application of COs. The CO designed for correction of rat *factor IX* gene mutation was introduced in mice by tail injection, and substitution of A with C in the Ser-365 codon was achieved. The mutation in the *tyrosinase* gene was corrected using COs in albino rats, and the treated rats had patches of black hairs attesting to the effectiveness of ODM (Igoucheva and Yoon, 2000; Liang et al., 2002).

Specific base substitutions in the *AHAS* gene have yielded imidazolinone (sulfonylurea) tolerant maize and canola plants (Table 2.1). The sulfonylurea tolerant canola has been commercialized as SU Canola by Cibus, U.S.A. In addition, COs have been used in combination with TALEN or CRISPR-Cas9 SDNs to induce specific base substitutions in *EPSPS I* and *AHAS* genes to produce Glyphosate resistant plants of flax and chlorsulfuron resistant plants of maize and rice, respectively. It is noteworthy that the use of SDNs (TALEN and CRISPR-Cas9) in combination with the COs greatly increased the frequency of recovery of the targeted gene mutations. However, the combined treatment may also induce some unintended deletions in addition to the planned nucleotide substitutions (Lusser et al. 2012; Wang et al., 2015; Songstad et al., 2017).

Table 2.1. Some selected examples of gene editing by redesigned meganucleases and chimeric oligonucleotides (based on Songstad et al., 2017)

Organism	Gene	Modification	Remarks
		Meganucleases (redesigned)	
Human	*RAG1*	Mutation correction by HDR	Mutation involved in a rare form of SCID (severe combined immunodeficiency) disease
	XPC	Mutation correction by HDR	Xeroderma pigmentosum is a genetic skin disorder; premature skin ageing; extreme sensitivity to sunlight/UV rays, which may result in cancer; due to defective DNA repair system
	DMD	Insertion of a missing intron in the mutant gene	Encodes dystrophin protein; mutations cause Duchenne muscular dystrophy disease
Anopheles gambia mosquitoes	Meganucleases	Insertion within selected genes, *e.g.*, fertility-related genes	A gene drive device used as a part of strategy for control of insect vectors of diseases
Maize	*Liguleless-1*	Inactivation	Presence of ligules at the junction of leaf sheath and leaf blade
	MS2	Inactivation	Cytochrome P-450 like gene for male fertility; mutant plants male sterile
Arabidopsis	Selected genomic regions	Deletions	–

Organism	Gene	Modification	Remarks
Cotton	Multiple genes	Insertion into cotton genome	–
		Chimeric oligonucleotides	
Tobacco	*ALS**	Single base substitution	Acetoacetate synthase; substitution of Pro-196 (CCA) with Thr (ACA), Ala (GCA), Gln (CAA), Ser (UCA), or Trp-573 (UGG) with Leu (UUG); resistance to herbicide chlorsulfuron
Maize	*AHAS**	Single base substitution	Substitution of Pro-165 with Ala (corresponds to tobacco Pro-196 with Ala); chlorsulfuron resistance
	*AHAS**	Single base substitution	Substitution of Ser-621 (AGT) with Asn (AAT); tolerance to imidazolinone (sulfonylurea) herbicide
	GFP	Single base substitution	Green fluorescent protein; premature stop codon TAG changed to TAC; GFP function restored
Rice	*ALS**	Single base substitution	Trp-548 substituted with Leu; herbicide chlorsulfuron tolerant callus lines and plants
Canola	*AHAS*	Single base substitution	Substitution of Ser-653 with Asn or Trp-574 with Leu; imidazolinone (sulfonylurea) tolerant plants; commercialized by Cibus as SU Canola, a non-GMO herbicide resistant crop variety
Arabidopsis	*BFP*	Single base substitution	Blue fluorescent protein; single base substitution converts BFP into GFP
Arabidopsis and *Linum usitatis-simum* (Flax)	*BFP*	Single base substi-tution in association with DSB induction	DSB (double-strand break) induced by phleomycin, TALEN or CRISPR-Cas9; greatly improved frequency of conversion of BFP to GFP
Flax	*EPSPS*	-	CRISPR-Cas9 combined with oligon-ucleotide with the desired mutated sequence; Glyphosate resistant plants
Maize	*AHAS**	-	CRISPR-Cas9 combined with oligonucleotide with the desired mutation (Pro-165 to Ser); chlorsulfuron resistant plants

* ALS is also known as AHAS (acetohydroxy acid synthase)

2.15. ADVANTAGES AND LIMITATIONS OF CHIMERIC OLIGONUCLEOTIDES

The ODM is a powerful tool for induction of desired single base substitutions in the targeted genes without induction of DSBs in the genome and the risk of toxicity caused by DSBs. ODM can induce even such mutations due to amino acid substitutions that were not available either in the gene pool or induced by many years of mutation breeding. For example, the mutation in canola *AHAS* gene leading to substitution of Ser-563 with Asn was induced by ODM, but this mutation could not be induced by several decades of chemical mutagenesis. The varieties generated by ODM are considered non-GM (non-genetically modified) and equivalent to those developed by conventional plant breeding. Therefore, they are exempt from the regulatory requirements mandatory for GM crop varieties (Komor et al., 2017a; Songstad et al., 2017).

One of the main limitations of ODM in the animal systems is that its efficiency is unpredictable and shows large variation (*e.g.*, 1-40%) depending on factors like target cell type, DNA sequence at the targeted site, transcription and DNA replication states of the cells, and availability of proteins Rad51 and MSH2. This feature makes ODM unsuitable for some applications like gene therapy. Storici et al. (2001) devised an elaborate procedure called "delitto perfetto" to overcome this limitation. In this procedure, an expression cassette containing a reporter gene and a marker gene enabling counter selection is first incorporated by homologous recombination into the targeted genomic locus of the cells. The cells with this cassette are isolated based on the reporter gene. They are then subjected to ODM leading to incorporation of the oligonucleotide into the genome by HR; in this process, the expression cassette integrated into the genome is excised out of the genome. The gene-edited cells are selected using the counter-selection procedure. The "delitto perfetto" strategy has been used in yeast and mammalian cells. However, it is complicated, has inherently low efficiency, and depends on two HDR events, a counter-selection procedure and the availability of protein Rad51 in the targeted cells (Komor et al., 2017a).

An efficient method of delivery of COs in the animal cells is not available, and there is urgent need for considerable refinements in the existing delivery procedures. At the same time, a reliable and sensitive procedure for determination of the desired gene correction events *in vivo* is not available. As a result, it is quite difficult the quantify the sequence changes induced by COs. In addition, there is almost complete lack of information about the off-target effects generated by COs. It has been argued that even whole genome sequencing may not provide reliable data since ODM generates SNPs, and many SNPs may arise spontaneously. In any case, therapeutic use of COs will require considerable refinements in the procedures and possibly the CO design to achieve high frequency mutation correction (Igoucheva and Yoon, 2000).

The genome editing procedures based on ODM have been considerably refined and applied in crop plants. ODM forms an important component of the Rapid Trait Development System (RTDS™) gene editing technology suite adopted and refined by Cibus, U.S.A. for development of completely transgene-free crop varieties. Cibus has already commercialized SU Canola, a canola variety developed by the ODM procedure (Songstad et al., 2017).

2.16. REGULATION OF ODM PRODUCTS

It has been argued with full justification that the genetic changes produced by ODM are essentially those generated by conventional mutagenesis. In addition, it does not lead to integration of any foreign DNA sequence in the cellular genome. It was, therefore, suggested that the products of ODM should be considered as non-GMO and should be exempt from the regulatory requirements for GM crop varieties. In fact, the ODM uses nucleic acids as mutagenic agents, while traditional mutagens comprise a variety of chemical and physical agents. In addition, the ODM has site-specific and user-defined mutagenic action, while these features are uncontrolled in traditional mutagenesis. Therefore, ODM is not only comparable but superior to the conventional mutagenesis procedures.

The Animal and Plant Health Inspection Service (APHIS), U.S. Department of Agriculture has ruled in the case of sulfonylurea (SU) tolerant mutant of canola developed by Cibus, U.S.A. that this canola line is exempt from the regulation applicable to GM crops. The PNT ('Plants with Novel Traits') procedure of Canada also considers varieties developed through ODM at par with those evolved by other plant breeding methods, including mutation breeding. The New Plant Breeding Techniques working group (set up by the European Commission in 2007) has concluded that ODM is a technique for mutagenesis and that it should be exempt from the European Union regulations applicable to GMOs. Thus, the general consensus seems to be to treat the products of ODM as non-GM and exempt from the regulatory requirements for GM crops (Songstad et al., 2017).

2.17. FUTURE PERSPECTIVES

The homing endonucleases were the first to be used to demonstrate the feasibility of DSB induction at targeted genomic sites to promote homologous recombination. The efforts to redesign meganucleases for novel target sequences have been successful, but they are hampered by context dependent DNA recognition and localization of DNA binding and cleavage activities in the same domain of MNs. In general, redesigned MNs show reduced DNA-binding affinity and, as a result, cleavage efficiency. One way to overcome this limitation is to create megaTALs, which have enhanced activity as well as target specificity. Thus, future efforts may be focused on generation of MN variants with high specificity for the targeted genomic sequence. The lower binding affinities of these MN variants may be compensated for by using them to create megaTALs; this approach would be particularly desirable for therapeutic applications of MNs.

It is generally expected that MNs and COs would be highly specific as they have rather long recognition sequences. However, all other site-directed nucleases do show off-target activities. It has been argued that since the DNA recognition components of these SDNs have evolved to tolerate some mismatches, some off-target activity may be intrinsic to them. This argument applies equally well to MNs as these endonucleases must coevolve with the host genome to survive over time. Thus far, there is little information on the off-target activities of both MNs and COs. It is imperative that this information is generated to allow decision on their suitability especially for therapeutic applications.

The methods of delivery of COs need to be considerably refined to achieve efficient delivery to the targeted cells. Further, easy, reliable and sensitive methods for determining the frequency of mutation induction/correction by COs, particularly under *in vivo* conditions, need to be developed. In addition, the present reagents and protocols for CO-based gene editing, especially in the animal systems. suffer from highly variable gene editing frequencies depending on a variety of factors, including cell type, targeted DNA sequence, the transcription and DNA replication states of the cells, etc. Therefore, the ODM procedures and reagents must be suitably refined to achieve predictable high frequency mutation induction/correction.

2.18. CONCLUSIONS

The discovery of HEs began with the apparently inconsequential yeast genetic marker Omega and ended up providing the initial tools and techniques for targeted genetic modification. The LAGLIDADG homing endonucleases or meganucleases are either dimeric or monomeric. The dimeric LHEs function as homodimers, and each monomeric subunit has one LAGLIDADG domain. The monomeric LHEs comprise a single subunit that has two LAGLIDADG domains. MNs are the most specific of HEs and they are perhaps the most useful for gene editing applications. The efforts directed at redesigning MNs for novel genomic targets have been quite successful, and it is now possible to develop MN variants specific for any genomic target provided the central four bases of the target sequence match those found in the authentic target site of the concerned MN. Each of the two LAGLIDADG domains of a MN recognizes half of the target site and functions independent of the other domain. There is no simple code governing the relationships between the amino acid residues and the DNA bases they recognize, and DNA binding is largely context dependent. Therefore, large numbers of MN variants are created and their binding to the desired target sequence is assessed by a suitable, preferably, high throughput assay procedure. The variants are generated by either changing amino acids contacting individual DNA bases or by varying the groups of amino acids that contact specific clusters of DNA bases in the target sequence. Generally, an amino acid is substituted by all the other 19 amino acids or a computational analysis may be used to reduce the numbers of substitutions.

The MN variants developed for specificity to new target sequences generally show reduced DNA binding affinities, which leads to lower cleavage activities as compared to the wild type MNs. This difficulty can be overcome by fusing the DNA-binding motifs from TALENs with the MN variants to create hybrid nucleases called megaTALs. The TAL component of megaTALs provides increased DNA binding affinity, which increases the cleavage activity of the MN component. It also enhances the target specificity of the megaTAL complex by including the additional TAL target sequence. It has been suggested that megaTALs may find therapeutic applications in view of their compact size, monomeric nature and very high activity coupled with extremely high specificity.

Chimeric oligonucleotides, also termed as gene repair oligonucleotide (GRON), are single-stranded sequences that comprise RNA and DNA segments and fold due to internal base-pairing to form double-stranded structures with two hairpin loops. A 25 bp sequence of COs is homologous to and contains the mutation to be induced in the target sequence. The COs unfold and pair with the target genomic region generating mismatches due to their mutant sequences. These mismatches are corrected by cellular DNA repair pathways, which copy the mutations from COs into the genomes of a proportion of the treated cells. COs generally induce substitutions of one to few bases, but they can also generate insertions and deletions of one to few bases.

The MNs have been used in a variety of organisms to produce targeted gene knockouts, gene knock-ins, sequence replacement for mutation correction or mutation induction through HDR, and production of short deletions in the genomic sequences. The COs have been used in microbial, animal as well as plant systems to induce or correct targeted mutations due to SNPs, and even to induce insertion or deletions of one to few bases. A herbicide tolerant canola mutant developed by ODM has been commercialized by Cibus, U.S.A., as SU Canola.

■■■■

3

ZINC FINGER NUCLEASES

3.1. INTRODUCTION

A site-directed nuclease (SDN) or programmable nuclease should be able to recognize the desired target site with high specificity, and it should not bind to other similar sites (off-target sites) in the genome. Thus, such a nuclease is expected to perform two functions, namely, binding to a specified DNA sequence and inducing cleavage at this or a neighboring site. A nuclease suitable for double-strand break (DSB) induction should have recognition sequence between 16 and 18 bp so that it would induce a single DSB at a unique target site in a large genome like human genome (size, 3×10^9 bp). The naturally occurring restriction enzymes (Type II endonucleases) are not suitable for inducing double-stranded breaks (DSBs) at targeted genomic sites because their recognition sequences are short (usually of 4 to 8 bp) palindromes, which would occur at several different locations in a genome. In addition, their DNA binding and nuclease domains overlap each other so that it is not feasible to modify their DNA binding domain in such a way that they bind to longer recognition sequences. Another group of naturally occurring endonucleases, the meganucleases, have comparatively much longer (typically of 20–30 bp) recognition sequences, but it has proved challenging to redesign these enzymes for cleavage at specified target sites (Chapter 2). Therefore, attention was focused on developing hybrid or chimeric endonucleases that could be used as SDNs.

The Type IIS restriction enzymes cleave DNA at some distance from their binding sites. This suggested that the DNA binding and DNA cleavage activities of these enzymes are performed by separate domains of the enzymes. This prompted the exploration of FokI, a Type IIS restriction enzyme, for creation of a chimeric programable endonuclease. These researches led to the development of zinc finger nucleases (ZFNs) as the first SDN. The ZFNs, in turn, stimulated the development of second generation SDNs, *viz.*, transcription activator like endonucleases (TALENs) and clustered regularly interspaced short palindromic repeat-CRISPR-associated 9 (CRISPR-Cas9), that have been relatively more successful than and virtually replaced the ZFNs at least from routine laboratory applications (Chandrasegaran and Carroll, 2016; Carroll, 2014, 2015, 2016).

3.2. FokI ENDONUCLEASE

FokI endonuclease has 66-kDa mass and recognizes a 5 bp (base pairs) long non-palindromic sequence, *viz.*, 5′-GGATG-3′. This endonuclease cleaves the DNA duplex 9 or 13 nucleotides (nt) downstream of the recognition sequence. Proteolytic cleavage revealed that FokI has separate DNA recognition (located at the N-terminus) and DNA cleavage (present at the C-terminus; mass

25 kDa) domains (Li et al., 1992). This nuclease induces non-specific cleavage, *i.e.*, the site of cleavage is not dependent on base sequence of the site. The crystal structure of FokI confirmed the above findings, and further revealed that the catalytic domain of the enzyme is sequestered by its DNA binding domain (Wah et al., 1998). It was also shown that the site of cleavage could be modified to 10 or 14 bases downstream of the binding site by adding four additional amino acids to the α-helix that links the two functional domains of the enzyme.

Thus, FokI seemed to be a likely candidate for the development of a programmable nuclease. One way this could be done was to swap its DNA recognition domain with those of other naturally occurring DNA-binding proteins capable of recognizing longer DNA sequences. Kim and Chandrasegaran (1994) engineered the first chimeric endonuclease by combining the homeodomain DNA binding domain of *Drosophila* Ubx protein with the cleavage domain of FokI. Subsequently, Chandrasegaran and co-workers created site-specific chimeric endonucleases by joining the zinc finger proteins (ZFPs) and the N-terminal domain (DNA binding domain) of yeast Gal4 protein with the FokI cleavage domain. These findings established that the cleavage domain of FokI endonuclease can be linked to appropriate DNA binding domains to enable it to cleave specified DNA sequences. Subsequently, the cleavage domain of FokI was modified to generate variants with altered dimerization properties (modifications in the dimerization interface) and even DNA cleavage activity. A hyperactive FokI variant, Sharkey, with >15-fold DNA cleavage activity has been isolated and used for ZFN construction (Gaj et al. 2013; Carroll, 2014; Chandrasegaran and Carroll, 2016).

3.3. ZINC FINGER PROTEINS

The zinc finger (ZF) motif was first described in the transcription factor IIIA present in *Xenopus* oocytes. The Cys2-His2 ZF domain is one of the most common eukaryotic DNA-binding motifs and the second most frequent protein domain encoded in the human genome. Each ZF motif has about 30 amino acids containing the invariant Cys2-His2 residues that together bind one zinc atom (Fig. 3.1A). A ZF has a unique structure consisting of two β sheets and one α helix that are stabilized by the zinc atom (Fig. 3.1B). The ZFs in a ZFP are linked with each other by a highly conserved linker sequence. The ZFs in a ZFP bind DNA in the following sequence: the N-terminal zinc finger binds to the triplet at the 3' end of the binding sequence in the contact strand, followed by the middle finger, and the C-terminal finger binds to the triplet at the 5' end of the sequence (Carroll, 2014; Chandrasegaran and Carroll, 2016).

Each ZF motif inserts into the major groove of DNA duplex and recognizes 3 consecutive bases in the contact or primary strand of the duplex (Fig. 3.1C). Amino acids at positions –1, +3, and +6 (in relation to the start of α helix) of a ZF make the chief contacts with DNA duplex, but the amino acid residue at position +2 is also involved in some contexts. A ZF having aspartic acid at the +2 position in its α-helix recognize 4 DNA bases since the aspartic acid residue recognizes A or C in the complementary non-contact strand located next to the 3 bases recognized by the amino acids in the regular positions. This interaction not only extends the contact site from 3 to 4 bp, but also affects the DNA recognition specificity of the neighboring ZFs in the concerned ZFPs/ZFNs. This situation complicates the creation of ZFPs capable of binding to the desired DNA sequences (Carroll, 2014; Chandrasegaran and Carroll, 2016).

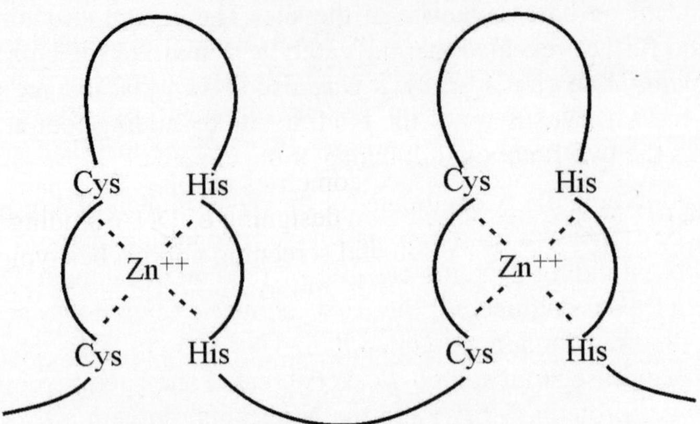

A. Two zinc fingers (Cys_2 His_2 type)

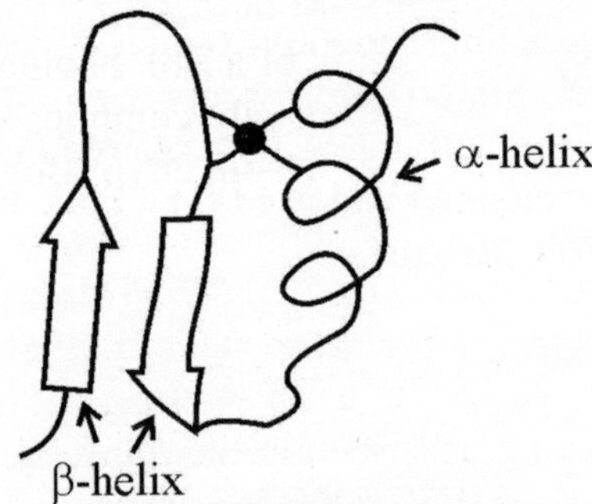

B. Structure of a single ZF

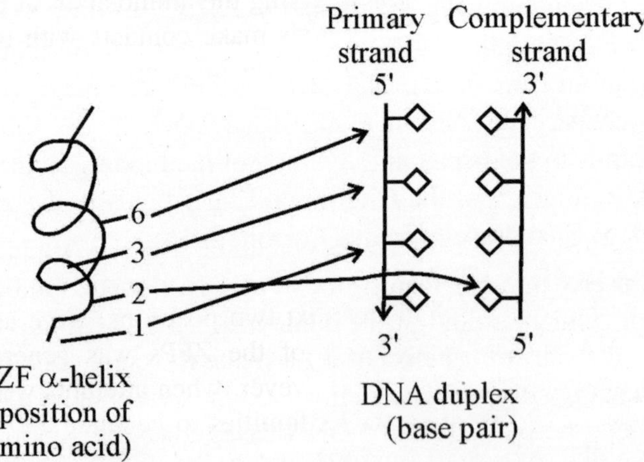

C. DNA sequence recognition by ZF

Fig. 3.1. Zinc finger structure and DNA sequence recognition. **A.** Simplified schematic representation of two zinc fingers. **B.** Structure of a single ZF. **C.** Scheme of DNA base recognition by ZFs.

3.4. CREATION OF NOVEL ZINC FINGER PROTEINS

Despite extensive research efforts, designing of DNA-binding proteins for recognition of and binding to specific DNA sequences remains a challenging area of research and development. The possible reasons for this situation may be the involvement of several different types of forces/factors in interactions between DNA and proteins, and the inability to predict the changes in the conformation of DNA backbone and the geometries of DNA base pairs induced by binding of proteins. As a result, the procedures for custom designing of DNA-binding proteins for binding to specified DNA targets have to use selection and screening approaches typically in combination with structure-based analyses and guidance. Four amino acid residues of the α-helix of each ZF make base-specific contacts with DNA, and they together recognize three base pairs. But base recognition by a ZF is also affected by three other amino acids interspersed between the four base contacting amino acids. In addition, base recognition by a ZF is context-dependent, *i.e.*, an individual ZF does not function independent of other ZFs in the ZFP. This context-dependence is thought to be the consequence of sequence-dependent interactions among the base triplets in the binding site, which influence the dispositions of the relevant base pairs. As a result, the designing of ZFs that are appropriate for the construction of a ZFP having high specificity and high binding affinity to the given DNA sequence is a technically complicated, laborious and time-consuming exercise (Carroll, 2014; Chandrasegaran and Carroll, 2016; Bogdanove et al., 2018).

Generally, 3 to 6 ZFs are joined together one after the other to yield a ZFP that will bind to a sequence of 9 to 18 bp. It was once hoped that the scientists would be able to design unique ZFs specific for each of the 64 triplets of DNA bases. It would then be a simple matter to assemble a ZFP specific for any given target sequence just by linking the appropriate ZFs selected from this set of 64 unique ZFs in proper order. Initially, novel ZFPs were constructed from a small collection of alternative amino acids that contacted DNA bases; this panel was developed by bioinformatic analysis of natural ZFPs. This approach met with limited success. Further, variants of three ZF containing ZFPs were created by varying the amino acids at positions −1, +1, +2, +3, +5 and +6 in the α-helix since these amino acids make contacts with the DNA duplex. In one scheme, these amino acids are changed in the middle ZF, while the amino acid sequences of the first and the third ZFs are kept invariant. Phage display is used to select ZFP variants that bind strongly to the desired 9-bp DNA targets. This approach was successful to some extent in identifying ZFs that apparently were specific to the triplets 5′-GNN-3′, 5′-ANN-3′, 5′-CNN-3′ and 5′-TNN-3′ (where N is any of the four deoxyribonucleotides or nucleotides). Similarly, ZFs were identified for binding to the 5′-ANN-3′, 5′-CNN-3′ and 5′-TNN-3′ triplets. Further, when the 16 different ZFs belonging to the 5’-GNN-3’ family (all ZFs specific for targets with G at their 5’ position and for any of the four bases at the next two positions) were assembled to yield ZFPs with three ZFs, the DNA binding specificity of the ZFPs was generally as predicted. This suggested the ZFPs to be modular in design. However, when attempts were made to combine the 5’-GNN-3’ ZFs with the ZFs from other triplet families to generate ZFPs, a large proportion of these ZFPs did not exhibit the DNA binding specificity predicted by their component ZFs (Chandrasegaran and Carroll, 2016; Bogdanove et al., 2018).

It is not feasible to generate and evaluate a library of all possible variant ZFPs that can be produced by simultaneous variation of the relevant amino acids in the three ZFs of the ZFPs since the number of such variants will be extremely large. Therefore, alternative strategies like sequential selection and bipartite selection were devised to select ZFPs with high DNA binding specificity. In sequential selection, a ZF is selected in the context of its neighboring ZFs. As a result, for each new ZFP multiple libraries of ZFPs are created and subjected to selection. The bipartite selection uses two ZFP libraries; in one of these libraries, the amino acid sequence of the N-terminal half of the ZFP is varied, while the C-terminal half is varied in the second library. Selection in the two libraries is performed simultaneously for binding to the 5' and 3' halves, respectively, of the target DNA sequence. The ZFPs selected from the two libraries are then recombined and evaluated for their binding specificity for the full target sequence. Both sequential and bipartite selection schemes enable creation of ZFPs with high specificity. However, they are too cumbersome, laborious and time-consuming for routine application. Further, they become increasingly demanding as the number of constituent ZFs increases in a given ZFP (Durai et al., 2005; Carroll, 2014).

The use of strings of two finger units is reported to improve the DNA-binding specificity of poly-zinc finger peptides. DNA binding specificity is also greater when three domains of two ZFs each are linked than when two domains of three ZFs each are joined together. The selection strategies based on phage display necessitate several rounds of ZFP selection in an *in vitro* setting. In contrast, cell-based selection strategies, *e.g.*, the bacterial one- and two-hybrid systems, and the yeast two-hybrid system, allow high specificity ZFP selection in one round in an *in vivo* setting. The selection schemes based in bacteria largely select for binding affinity rather than specificity in binding to the target sequence. In these selection systems, the DNA binding specificity of a ZFP is linked to the transcription of a reporter gene that can be readily assayed. A detailed discussion of the various cell-based ZFP selection strategies can be found in Durai et al. (2005). A hybrid selection approach that combines bacterial selection with modular assembly and empiric changes has been implemented by Sangamo Biosciences as a component of their proprietary platform for designing, evaluation and development of ZFNs (Porteus, 2016).

Many researchers have designed, selected, and characterized a large number of ZFs. The data on individual ZFs and ZF combinations specific for a variety of target sequences are stored in extensive catalogues. However, a new ZFP comprising three or more fingers has to be tested and validated since the fingers that are highly specific in one context do not necessarily function as expected in another context. It may be pointed out that the ZFs in ZFPs are not in perfect register with (alignment in relation to the bases in) the B-form DNA. The relative positions of the ZFs and DNA duplex bases can be improved by adding one-amino acid as a linker between two zinc-fingers. In addition, bases between successive triplets in the target sequence can be skipped by inserting longer linkers between ZF units (Carroll, 2014, 2016; Chandrasegaran and Carroll, 2016).

ZFPs can be linked to any effector domain to generate chimeric proteins with varied capabilities. They can be linked with transcription activator and repressor domains from transcription regulator proteins to create chimeric transcription activators and repressors, respectively, that act on the promoters specified by their DNA binding domains, *i.e.*, the ZFPs. Similarly, different ZFPs can be fused with a CpG doublet-specific DNA methyltransferase to create hybrid methylase enzymes, called zinc finger methylases, that would methylate cytosine residues at the targeted sites in the genome. Similarly, ZFPs can be joined to transposases, recombinases, histone methyltransferases and acetyltransferases (Gaj et al., 2013; Chandrasegaran and Carroll, 2016). They can be even linked to enzymes involved in base conversion leading to base substitutions as is described in some detail in Chapter 6. It may be emphasized that these chimeric proteins will carry out the functions provided by their effector domains in a genomic region/site specified by their DNA binding domains. In addition, the TALE proteins and the Cas-9 nuclease of the CRISPR-Cas9 system can also be used to create such chimeric proteins. It is noteworthy that in such applications, the nuclease function of Cas-9 protein is inactivated, *i.e.*, Cas-9 is 'dead' (dCas-9). In fact, the CRISPR-dCas9 system is currently almost exclusively used to create programmable chimeric effectors.

3.5. HIGH THROUGHPUT CREATION OF ZFPs

ZFs specific for several triplets, *e.g.*, 5'-GNN-3', 5'-ANN-3', 5'-CNN-3', etc., had been successfully designed. There are several procedures for designing of the appropriate ZFNs using these ZFs. The cell-based screening procedures have been further refined to allow high-throughput creation of new ZFNs. One approach is to generate a library of ZFPs that have been earlier tested and shown to bind to their target sequences with high specificity and affinity. The desired ZFN can be created by selecting a ZFP from this library that binds to a sequence that matches the desired target sequence. The functional ZFPs are chosen from a database using an approach like oligomerized pool engineering (OPEN) or context-dependent assembly (CoDA). The OPEN scheme uses a collection of ZF pools, in which each pool contains up to 95 divergent ZFs that bind to a given 3 bp sequence. For a ZFP with three fingers, the ZFs in the three pools relevant for the target 9-bp sequence are combined to generate a library of 3-finger ZFPs. The resulting 3-finger ZFPs are evaluated for their binding efficiency to the target sequence using a suitable cell-based selection strategy like the bacterial two-hybrid system, and the ZFPs with high binding specificity are identified. The OPEN approach identifies multi-ZF ZFPs that bind to specified target sites with high affinity and specificity. Thus, for every new target sequence, a library of multi-finger ZFPs must be created and evaluated. This is a cumbersome and time-consuming process, which limits the usefulness of the OPEN procedure (Durai et al., 2005; Chandrasegaran and Carroll, 2016).

The CoDA scheme is essentially an extension of the OPEN approach. This procedure uses the large number of functional 3-finger ZFPs produced, evaluated and stored in the OPEN database to facilitate the development of a 3-finger ZFP specific for a new target sequence. In a 3-finger ZFP, symbols F1, F2 and F3 denote the N-terminal, middle and C-terminal ZFs, respectively. Suppose, a ZFP specific for the target sequence 3' GGT AGC CTG 5' is to be designed. The CoDA procedure uses two sets of 3-finger ZFPs. The ZFPs in the two sets have the same central finger F2 that is specific for the triplet AGC. Further, the ZFPs in one of the sets have different N-terminal F1 ZFs that are specific for the triplet GGT, while the F3 specificity is irrelevant. Similarly, in the second set, different C-terminal F3 ZFs specific for the tripled CTG are included, while the F1 specificity is irrelevant. The ZFPs with the highest efficiencies from the two sets are used to construct a 3-finger ZFP specific for the target sequence 3' GGT AGC CTG 5' by joining the F1 from the first set with the common F2, which is further linked to the F3 from the second set of ZFPs (Fig. 3.2). Since the F1 and F3 ZFs in the two sets of ZFPs have been evaluated and selected using a bacterial two-hybrid DNA-binding assay and found to function well in the context of F2, the new ZFP derived from them is expected to function efficiently. Finally, the new ZFPs created in this way are evaluated for binding to the desired 9 bp target site. This procedure is reported to successfully produce high specificity ZFPs with a success rate of about 75%, which is comparable to the selection-based methods like OPEN. The CoDA method is simple to practice, uses a large library of F1 (319 ZFs), F2 (18 ZFs) and F3 (344 ZFs) units, and the reagents and software are publicly available. It enables assembly of dozens of 3-finger ZFPs in 1 to 2 weeks and there is no need for selection before applying the ZFPs to cells (Sander et al., 2011).

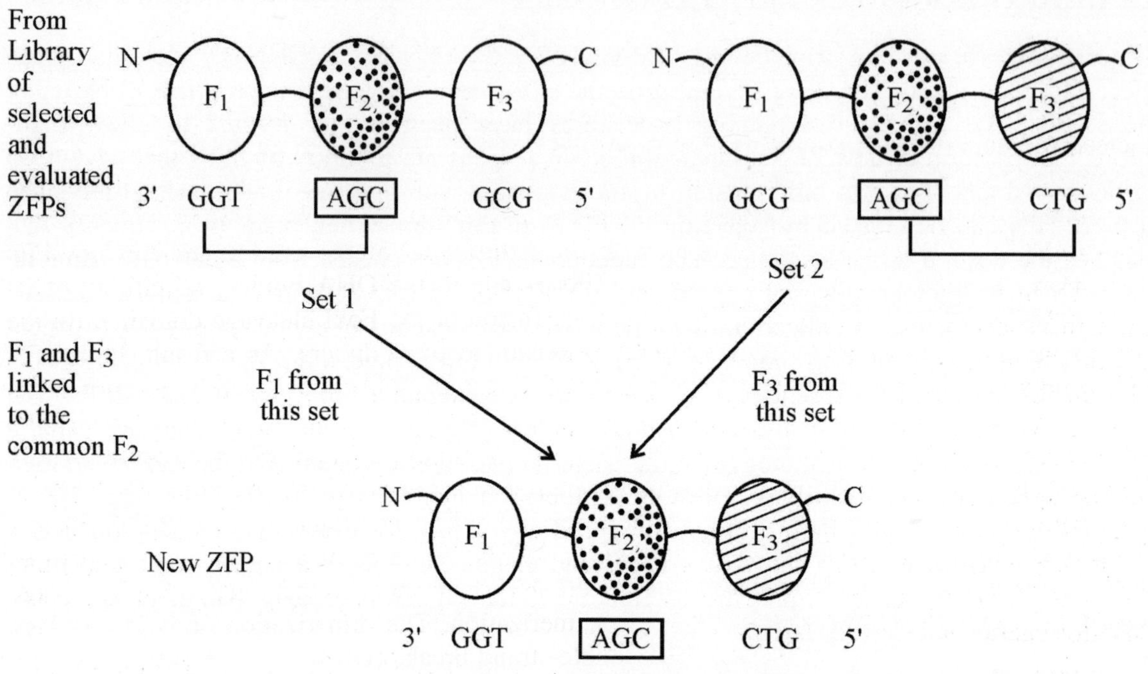

Fig. 3.2. Schematic representation of the context-dependent assembly (CoDA) of zinc finger proteins specific for new target sequences. Set 1 ZFPs have their N-terminal and middle ZFs specific for the first two triplets (from 3' end) of DNA bases, while the second set has ZFPs with their middle and C-terminal ZFs specific for the last two triplets. The middle ZF is the same in both the sets of ZFPs and specific for the middle triplet of the target sequence. The new ZFP specific to the target sequence is constructed by joining the N-terminal ZF from Set 1 and C-terminal ZF from set 2 with the common middle finger (based on Sander et al., 2011).

The selection-based engineering procedures like OPEN consider the effects of other ZFs of the ZFN on the DNA binding specificity of a given ZF. Therefore, these approaches can design ZFNs with greater DNA binding specificity than the other procedures. It takes about two to three months for a highly trained person to construct a specified ZFN using the OPEN procedure. In comparison, the CoDA procedure allows synthesis of ZFNs in less than a week, but the ZFNs designed by this procedure recognize and produce DSBs at about 50% of the targeted sites (Curtin et al. 2012). In any case, the construction and selection of ZF proteins with the given DNA specificity are highly demanding and sophisticated procedures.

Sangamo Biosciences, U.S.A. have created a proprietary database of functional ZFPs and ZF pairs using a variety of strategies. It appears that they are able to identify from this database such ZFPs that can be readily refined to develop ZFPs that are highly specific for a given binding site (Chandrasegaran and Carroll, 2016). Custom designed, tested and validated ZFNs for the desired target sites are marketed by Sigma-Aldrich (http://www.sigmaaldrich.com/life-science/zincfinger-nuclease-technology.html). Sigma uses the extensive library of tested single ZFs and ZF pairs maintained by Sangamo Biosciences, Inc., U.S.A. The price of custom designed ZFNs has substantially declined over time, but they remain quite expensive for most laboratories (Carroll, 2016).

3.6. DESIGNING ZINC-FINGER NUCLEASES

Zinc-finger nucleases (ZFNs) are man-made proteins comprising 3-4 zinc fingers linked to the nuclease domain of FokI; they were first produced in 1996 by Kim et al. (1996). Thus, a ZFN is custom designed to cleave DNA at the specified sequence, and contains two distinct domains, *viz.*, a ZF domain (for DNA sequence recognition) and a nuclease domain (for DNA cleavage). The cleavage domain is typically fused with the C-terminus of the zinc-finger domain (Fig. 3.3A). The specificity of DNA cleavage by ZFN is due to the DNA binding specificity of its constituent ZFs (Fig. 3.3B; 3.4). In the beginning, fusion of the FokI cleavage domain with the ZFs somehow reduced the ability of cleavage domain to form dimers. As a result, high ZFN concentrations were required to achieve cleavage of the target DNA. Subsequently, pairs of ZFNs binding to target sites located close to each other were used to achieve efficient cleavage. The dimerization interface of the cleavage domain was later modified to prevent homodimer (dimer formed by a single ZFN) formation; this modification also improved the cleavage efficiency of the ZFNs. Thus, a functional ZFN comprises a pair of ZFNs: each member of this pair has 3 or 4 ZFs linked to the FokI cleavage domain. The two ZFN members of a pair bind to sequences located close to and oriented in the opposite directions of each other so that the two FokI cleavage domains are brought together to enable their dimerization. The dimerization of FokI cleavage domain is needed for efficient induction of double-strand breaks (DSBs) in the target DNA (Fig. 3.3C). These ZFNs, called second generation ZFNs, are currently in common use. The optimum size of linker joining the ZFs with the cleavage domain is four amino acids; this linker separates the binding and the cleavage sites by 6 bp (Carroll, 2016).

Initially, ZFNs were designed by combining the existing ZF proteins with the cleavage domain of FokI. Experiments with these ZFNs suggested that ZFNs functioned as dimers to generate DSBs. They also revealed that fusion of FokI cleavage domain did not affect the binding affinity and the specificity of the constituent ZFPs. Further, these ZFNs did not cleave at the predicted consensus binding sites, and the cut sites appeared to be close to the binding sites of the constituent ZFPs. Subsequent studies confirmed that FokI dimerization is necessary for DNA cleavage. It was also found that a single ZFN required a pair of inverted binding sites located close to each other for efficient induction of DSBs in the target DNA. This suggested that the FokI cleavage domain of ZFNs must dimerize for efficient DSB production in the target DNA. Therefore, a pair of ZFNs is used for each target site; if each ZFN comprised three ZFs, the ZFN pair will have an effective recognition sequence of 18 bp. This will greatly enhance the specificity of the ZFN and reduce the chances of off-target cleavage of DNA (Chandrasegaran and Carroll, 2016). This principle has been extended to design heterodimeric ZF methylases to catalyze *in vivo* methylation of targeted genomic sequences.

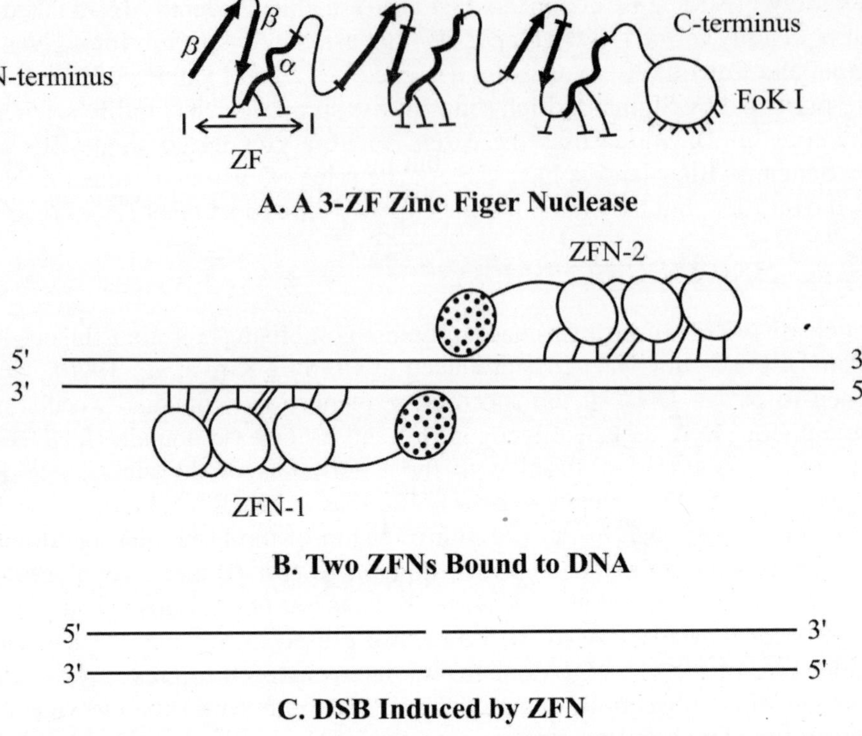

Fig. 3.3. DSB (double-strand break) induction by ZFN (zinc finger nuclease). **A.** A highly simplified representation of a ZFN. **B.** A pair of ZFNs bound to target DNA. **C.** DSB induced by ZFN in the target DNA. β, beta-sheets; α, alpha-helix; ZF, zinc finger; the three lines extended from each α-helix represent the side chains that contact DNA bases.

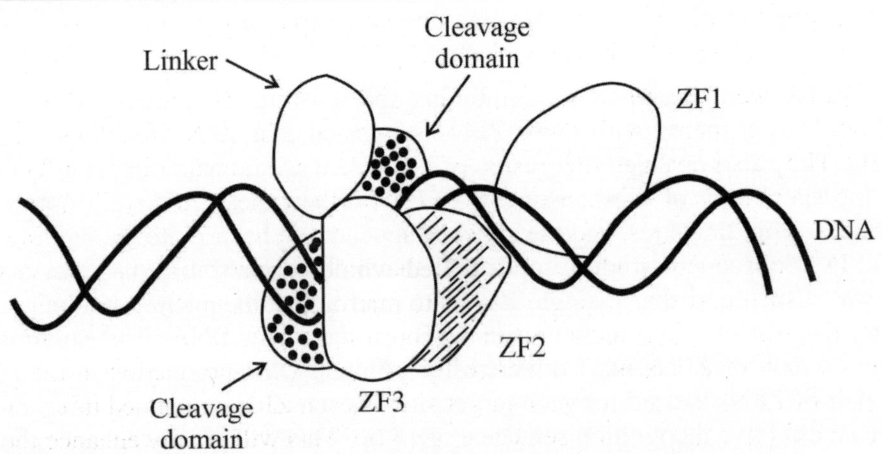

Fig. 3.4. A simplified schematic representation of a ZFN bound to DNA duplex. The linker joins the FokI cleavage domain with the zinc finger domain of ZFN. ZF1, ZF2 and ZF3, the three zinc fingers of ZFN.

3.7. ZINC FINGER NICKASES (ZFNICKASES)

A nickase is an endonuclease that produces nicks, *i.e.*, single-strand breaks (SSBs), in the substrate DNA. Zinc finger nickases (ZFNickases) were the first nickases to be designed and used; they provided the information needed for the development of nickases based on TALENs and CRISPR-Cas9 SDN systems that were subsequently created. They are custom designed and produced by genetic engineering just like ZFNs, but they cleave only one specified strand of the target DNA. ZFNickases are produced by inactivating the DNA cleavage activity of FokI nuclease domain of one of the two ZFN monomers designed for cleaving the targeted site (Ramirez et al., 2012). ZFNickases induce highly strand-specific nicks in the target DNA and stimulate homology-directed repair (HDR) since nicks do not activate nonhomologous end-joining (NHEJ) pathway of DNA repair. As a result, the use of ZFNickases greatly reduces the frequency of gene inactivation due to insertions/deletions (indels) generated by nonhomologous end-joining (NHEJ). However, the frequency of homology-directed repair following nicks induced by ZFNickases is lower than that obtained following induction of DSBs by regular ZFNs. The ZFNickases may be expected to show greatly reduced off-target activities since the nicks would be precisely repaired by base excision repair pathway of DNA repair (Carroll, 2016; Lee et al., 2016).

3.8. DELIVERY OF ZFNS AND DONOR DNAs INTO CELLS

The method of delivery of ZFNs and donor DNAs (if used) needs to be carefully considered for each new organism/cell type and application; this is true for the other SDN systems as well. The delivery methods will largely depend on the organism and the cell type, into which the ZFNs are to be introduced. In any case, the protein components of SDNs should contain signal peptides for nuclear localization. In case of cultured animal cells, the ZFNs can be placed in plasmid or viral vectors and delivered through direct DNA uptake, electroporation or liposome-based transfection. But electroporation or cationic lipid-based transfection of DNA/mRNA into animal cells can be toxic and is applicable to only certain cell types. They can also be introduced into the cells by ZFN encoding mRNA transfection. Further, the ZFNs in protein form can be added to the culture medium to enable their uptake by the cells since ZFNs possess high inherent positive charge; this approach is applicable to several different cell types. The delivery in RNA and protein form would minimise off-target activity, eliminate requirement of specific cell type and toxicity and/or immunogenicity of viral/nonviral delivery methods, and avoid several regulatory burdens associated with their delivery in form of DNA. Plasmid or viral vectors are used for the delivery of long, double-stranded donor DNAs. But short, single-stranded donor DNAs are generally added to the culture medium (Gaj et al., 2013; Carroll, 2014, 2016; Chugunova et al., 2016).

In case the objective is to produce gene-edited whole animals, it is necessary to express ZFNs in germ line cells. One way to achieve this is to microinject the mRNAs encoding the ZFNs along with the donor DNA (where required) into zygotes/very early stage embryos. This approach has been successful in several animal species, including insects, fish, frogs, mice, rats and rabbits. Another approach is to carry out gene editing *in vitro* using cultured cells, and then fuse the edited cells with enucleated eggs (somatic cell nuclear transfer). This method was used to produce gene edited pigs and cows. Where necessary, *e.g.*, in mice, rats, rabbits, pigs, cows, etc., the microinjected embryos are transplanted into the uteri of receptive females to allow them to complete their development. This approach obviates the necessity of producing transgenic lines and can induce relatively high frequencies of both NHEJ and HDR. This strategy has been used to produce knockins and knockouts in rats. Since rats do not have any useful embryonic stem cells, the knockin and knockout rats cannot be produced by the transgenic technology (Cui et al., 2011, Carroll, 2014, 2016).

The ZFNs can also be introduced into animal and plant cells as gene constructs by transfection/genetic transformation to produce stable transgenic lines capable of expressing ZFNs. The pair of ZFNs required for cleavage can be encoded by a single gene (Fig. 3.5). In this gene, the part of the gene construct encoding the ZF domain comprising three zinc fingers of the first ZFN is followed by the sequence encoding its nuclease domain. The subsequent segments of the gene construct encode a spacer, and the nuclease and ZF domains of the second ZFN member of the pair. In this gene, the two ZFNs are linked in tail-to-tail orientation, which facilitates dimer formation at the target site (Durai et al., 2005). In case of plants, the gene constructs can be delivered by *Agrobacterium*-mediated transformation, or DNA delivery by particle gun or by electroporation into protoplasts. It may be pointed out that only transient expression of the ZFNs is required or even desirable since continued SDN expression might increase the risk of off-target cleavage (Section 3.12). One way to achieve this is to regulate the expression of ZFN gene constructs by deploying a suitable inducible promoter like heat shock protein gene promoter to drive the ZFN gene constructs (Carroll, 2014, 2016).

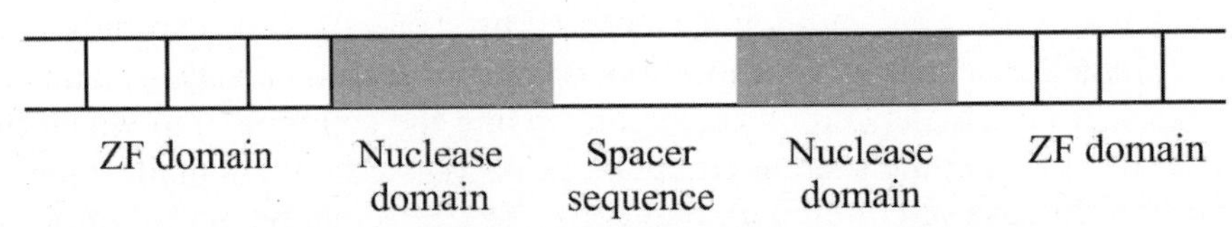

Fig. 3.5. A schemaitc representation of a gene encoding a pair of zinc finger nucleases (ZFNs), each ZFN comprising three zinc fingers (ZFs). The two ZFNs are joined in a tail-to-tail orientation by a spacer, which ensures dimer formation at the target site. Generally, the ZFs in the two ZFNs are specific to different target sequences located close to each other in complementary strands of the DNA duplex and oriented opposite to each other. The nuclease domain is the FokI cleavage domain; ZF, zinc finger; ZFN, zinc finger nuclease (Based on Durai et al., 2005).

3.9. *IN VIVO* DNA CLEAVAGE BY ZINC FINGER NUCLEASES

The DSB production stimulates DNA repair pathways and promotes HDR. The organization of target DNAs in form of chromatin does not affect either the cleavage activity of ZFNs or the repair of DSBs. However, it is still not clear as to how the ZFNs are able to recognize their target sequences sequestered in the chromatin structure. The first report of successful gene editing in a mammalian genome was published in 2003 (Porteus and Baltimore, 2003). They edited the gene encoding GFP (green fluorescent protein) stably integrated in the genomes of cultured human cells. This gene was interrupted by a sequence that served as the target for a ZFN of known specificity. Introduction of the appropriate ZFN along with the donor DNA suitable for correcting the interruption in the GFP gene resulted in expression of the GFP in a substantial proportion of the cells. The GFP expression resulted from removal of the interrupting sequence from the GFP encoding gene due to HDR using the donor DNA as template. It was noted that there was 2,000-fold increase in the frequency of HDR as a result of the DSBs introduced by ZFNs. The frequency of successful gene editing ranged between 3 and 5 per cent of the cells. The gene correction frequency could be improved by increasing the amount of donor DNA, enhancing the degree of homology between the GFP gene and the donor sequence, and by inducing DSBs at a higher frequency. However, there was evidence for off-target DSB induction by the ZFNs as the frequency of GFP-positive cells declined with time.

The first report of successful gene editing of a human gene came in 2005 from scientists working at the Sangamo Biosciences, U.S.A. Urnov et al. (2005) targeted the human gene encoding IL-2Rγ using a pair of ZFNs comprising four zinc fingers each. Mutations in the gene encoding IL-2Rγ are known to be associated with some cases of X-linked severe combined immunodeficiency (SCID). The ZFNs designed for this study were highly specific and very effective in inducing mutations in this gene by both HDR based on a donor DNA and NHEJ in the absence of a donor sequence. The frequency of repair of the IL-2Rγ encoding gene approached 20%. It was also reported that improved designing of the ZFNs to enhance specificity resulted in increased gene editing efficiency.

The ZFNs were shown to be effective in plants like *Arabidopsis* and cultured tobacco cells in inducing DSBs, gene inactivation by NHEJ as well as gene repair by HDR. In case of *Arabidopsis*, the ZFN encoding gene constructs and the target DNA for the ZFNs were introduced by *Agrobacterium*-mediated genetic transformation (Lloyd et al., 2005). But the ZFNs and the donor DNA were delivered by electroporation into tobacco protoplasts. As high as 10% of the transformed tobacco protoplasts showed HDR of reporter genes used in the study (Wright et al., 2005).

3.10. REPAIR OF DOUBLE-STRAND BREAKS

Gene editing using SDNs is dependent on induction of DSBs and their repair by the DNA repair pathways operating in the treated cells. Double-strand breaks in DNA are lethal and must be effectively and rapidly repaired by the affected cells. There are two main pathways for the repair of DSBs, *viz.*, nonhomologous end-joining and homology-directed repair or homologous recombination (HR)-based repair. In most organisms, DSBs are mainly repaired by NHEJ and HDR, but several other repair pathways also operate in cells. Under most experimental conditions, NHEJ predominates but excess of HDR over NHEJ repair has also been reported (Porteus, 2016). NHEJ pathway is operative in both mitotic and postmitotic cells, while HDR is confined to dividing cells. It may be pointed out that the repair of DSBs using sister chromatids or homologous chromosomes as template is not detectable since the repair products are identical to the original chromosomes. Similar is the case for products of DSB repair involving a precise religation of the two ends of the DSBs. The repair products can be detected only when there is sequence change at the DSB site either due to NHEJ or HDR (Carroll, 2015). NHEJ predominance is just fine when the gene editing aims to inactivate the target gene. But when it is desired to replace a specific sequence of the target gene with a predetermined sequence contributed in form of donor DNA, it will be desirable to have a high proportion of DSBs repaired by HDR.

3.10.1. Non-homologous end-joining (NHEJ)

DSBs are generally repaired by the NHEJ pathway in far higher frequencies than HDR-mediated repair even when a donor DNA is available for use as template. The NHEJ repair pathways are mainly of two types, *viz.*, Ku-dependent and Ku-independent pathways. In the Ku-dependent pathway, the first step consists of sensing of DSBs by Ku complex of two proteins, namely, Ku70 and Ku80. The Ku protein complex binds to the two broken ends of DNA at the DSB (Fig. 3.6, 3.7). It is then joined by the MRX complex (MRN complex in mammals), which acts as a bridge to bring and keep the two ends together. One of the proteins that is activated during the repair process is Artemis; it has both exonuclease and endonuclease activities. Artemis can trim any single-stranded protrusions at the ends generated by the DSB. Finally, DNA ligase IV in conjunction with protein XRCC4 joins the two ends to complete the NHEJ repair. In the Ku-independent pathway, the Ku proteins do not bind the ends at the DSB. This pathway has several variants, and it leads to longer deletions and insertions than the Ku-dependent pathway. The NHEJ process is frequently associated with indels of one to few nucleotides. These indels can add or delete one or few amino acids (when the number of added or deleted bases is a multiple of three) from the protein encoded by the concerned gene, disrupt the reading frame of the mRNA (if the number of added or deleted bases is not a multiple of three) encoded by the coding sequence, or disrupt the promoter/enhancer sequences that serve as binding sites for *trans*-acting factors. Therefore, the error-prone NHEJ pathway generally leads to inactivation of the concerned genes (Krebs et al., 2014; Chugunova et al., 2016).

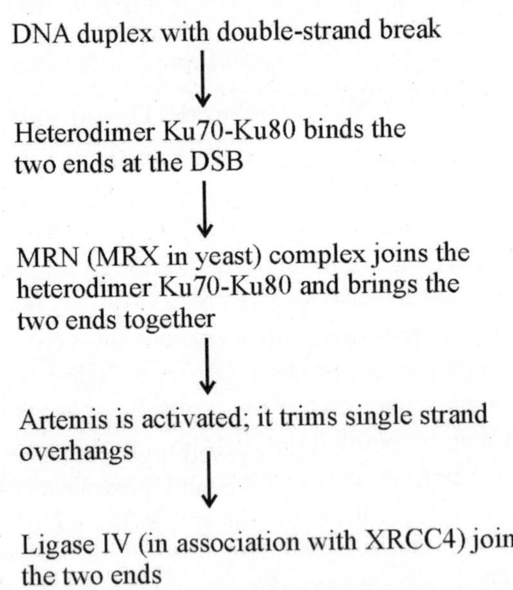

Fig. 3.6. A simplified description of the events during non-homologous end-joining repair of DNA double-strand breaks. Generally, there is either deletion or insertion of some base pairs at the site of break. DSB, double-strand break.

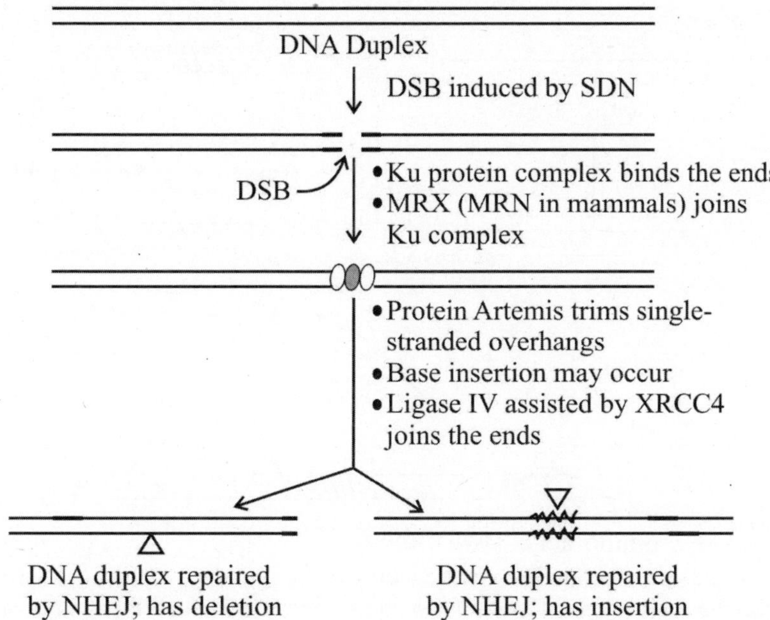

Fig. 3.7. Repair of double-strand break by non-homologous end-joining. During repair, some base pairs may be deleted or additional base-pairs because inserted; these generate indels that inactivate the concerned genes. DSB, double-strand break; NHEJ, non-humologous end joining; Δ, denotes deletion; ∇, indicates insertion.

3.10.2. Homology-independent targeted integration (HITI)

The homology-independent targeted integration (HITI) of transgenes and DNA sequences involves NHEJ mode of repair of the DSBs. NHEJ is intrinsically a precision repair system and the indels associated with NHEJ in studies with SDNs are believed to be the result of repeated action of SDNs until the target site sequence is changed. It was reasoned that if the genomic target site and the donor DNA plasmid were cleaved at the same time by, say, CRISPR-Cas9, NHEJ might lead to integration of the donor plasmid at the genomic target site. This was corroborated by Auer et al. (2014) in zebrafish by integrating *Gal4* gene into the *eGFP* gene already present in the zebrafish genome. They used CRISPR-Cas9 for DSB induction; therefore, the target sequence including the PAM (protospacer adjacent motif) for the CRISPR-Cas9 was inserted at the appropriate location (the 5' end of the transgene) into the donor plasmid. Suzuki et al. (2016) refined this technique further and achieved high frequency (over 50% of the transfected cells) targeted gene integration in both mitotic and postmitotic human cells. In its simplest application, HITI is used to insert a transgene within an endogenous gene at a downstream location of the gene (Fig. 3.8). Therefore, in-frame insertion of the transgene would generate a fusion protein comprising the amino acid sequences encoded by the endogenous gene and the integrated transgene. Appropriate strategies may be employed to enable the production of transgene encoded proteins free from the endogenous protein sequences.

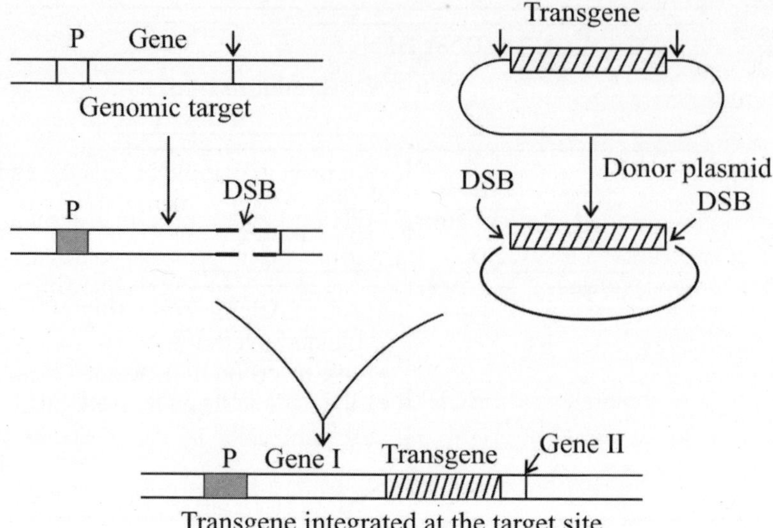

Fig. 3.8. A simplified schematic representation of homology-independent targeted integration (HITI) of transgenes. P, promoter; Gene, endogenous gene; DSB, double-strand break; Gene-I and Gene-II, sequences of the endogenous genes on the upstream and downstream sides of the integration site. Vertical arrows indicate the CRISPR-Cas9 target sequences (based on Suzuki et al., 2016).

The HTIT method can be used for addition of a novel transgene into the genome, insertion of the deleted exon(s) of an endogenous gene to restore gene function, and to replace an entire endogenous gene. For whole gene replacement, the endogenous gene will have to be deleted by two strategically placed DSBs, the donor DNA will be similarly cleaved free from the donor plasmid and two sgRNAs (single guide RNAs) will need to be employed for the purpose. This approach may also be used for the replacement of a part of an endogenous gene leading to sequence replacement and mutation correction (Auer et al., 2014; Suzuki et al. 2016).

3.10.3. Homology-directed repair (HDR)

In case of homology-dependent repair, DSB repair uses as template a DNA segment that is homologous to the genomic region surrounding the DSB; this template DNA is called donor DNA. HDR can mediate replacement of a genomic sequence by homologous sequence from the donor DNA, insertion of a novel DNA segment like transgene into the genome, or deletion of a genomic region. The insertion of novel segment occurs due to homologous recombination involving the genomic sequences flanking the DSB and the homologous sequences located on either side of the novel segment or donor DNA. The HDR ensures that there is no loss of critical genetic information from the genomic region involved in DSB. It may be pointed out that homologous recombination during meiosis in initiated by a DSB in one of the two homologous chromosomes; therefore, this type of DSB repair is often referred to as homologous recombination (HR) repair. Many of the proteins involved in meiotic recombination also participate in HDR. The HDR is carried out by the Rad52 group of proteins that were identified in yeast in relation to radiation damage; the homologues of these proteins are found in other eukaryotes, including mammals. In HDR, generally the sister chromatid or the homologous chromosome serves as the template. HDR typically occurs when the chromosomes are either nearly or fully replicated (late S and G2 phases of the cell cycle) so that a sister chromatid is available to serve as template (Sun and Zhao, 2013; Krebs et al., 2014).

A group of Rad52 proteins (MRX complex) binds to the two free ends at the DSBs and may keep the ends together (Fig. 3.9, 3.10). This complex, in association with exonucleases and helicases, generates single-stranded regions with 3'-OH ends. The protein Rad51 associates with the single-stranded regions and in association with some other proteins mediates invasion of a homologous DNA by these single-strands. Two other proteins (Rad54 and Rdh54) assist in the search of homologous donor DNA as well as its invasion by the single strands. The Rad54 and Rdh54 proteins are members of chromatin remodelling enzymes and may be involved in chromatin reorganisation associated with HDR. The HDR results in a structure that is like the Holliday junction produced during meiotic recombination; this structure is resolved by the same process that resolves the Holliday junction (Krebs et al., 2014).

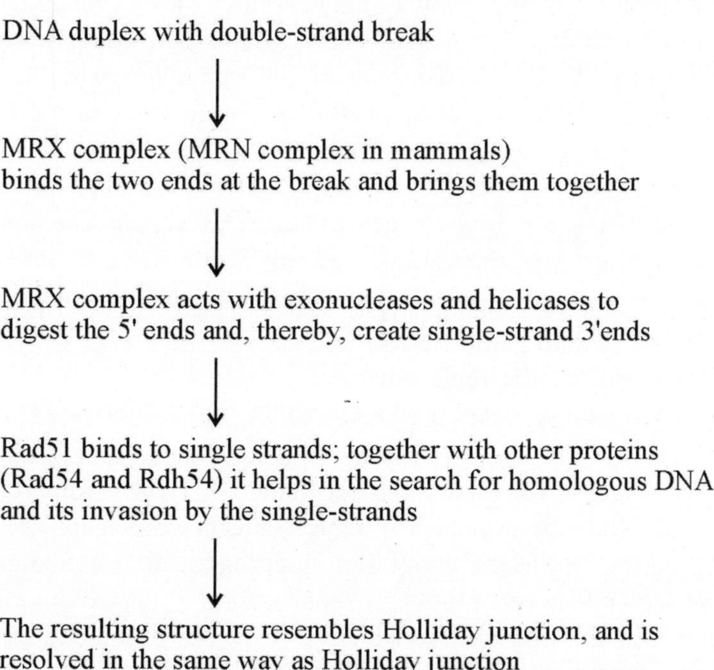

Fig. 3.9. A simplified description of the homology-directed repair of DNA double-strand breaks. This repair uses of homologous DNA sequence as template, and the process is similar to meiotic recombination. This repair pathway is based on the products of *RAD52* group of genes. Proteins Rad54 and Rdh54 most likely remodel the chromatin to facilitate HDR.

The donor DNA can be supplied as long double-stranded DNA or as short oligonucleotides. The donor DNA must be sufficiently homologous to the targeted region. In case of long, double-stranded donors, the homologous region ranges between 1,000 and 4,000 bp; this region is approximately equally distributed at the two ends of the donor DNA and is often called homology arms. The frequency of HDR declines significantly when the lengths of homology arms are 200 bp or less. The maximal HDR efficiencies are achieved when the arms of homology are centred at the DSBs. The desired sequence change located between the homology arms could be as small as a single nucleotide or it may be as large as 10 kb (kilobases), and this variation in length does not seem to affect the editing efficiency. It is noteworthy that the homology between the genomic sequences and homology arms need not be a perfect one. But the HDR frequency seems to be directly proportional to the amount of donor DNA delivered into the cells (Carroll, 2014, 2015, 2016; Porteus, 2016).

Alternatively, 30 to 40 nt long oligonucleotides can be successfully used as donor DNAs. The oligonucleotides (oligos) must be homologous to the sequences flanking the DSB and equally distributed on the two sides of the DSB. Both double-stranded and single-stranded oligos have been used as donors. Generally, oligo donors can sponsor only short sequence replacements at the cleavage site, but they can be designed to introduce deletions of several kilobases. The donor oligo in the latter case has one terminal region homologous to the target region on one side of the DSB, while its other terminal regions is homologous to the genomic region some distance away on the other side of DSB. HDR repair is up to 100-fold higher with linear donor DNA than that with circular donor provided in a plasmid. But in certain situations, circular donor may be preferable to linear donors. In case of primary human cells, the delivery of donor DNA in a form that avoids interferon response (evoked by naked DNA) is desirable; this can be achieved by delivery in an integrase deficient lentivirus vector or an AAV vector. The donor can be delivered *in vitro* by transfection or electroporation, and by hydrodynamic injection *in vivo*. The mechanism of integration of oligonucleotide donors seems to be different from normal HDR as it does not require Rad51, which is essential for HDR (Bibikova et al., 2003; Carroll, 2014, 2015, 2016; Porteus, 2016).

The sequence incorporated from the donor DNA into the targeted gene due to HDR corresponds to the conversion tract, which is the sequence of a gene involved in a gene conversion event. The length of conversion tract generated by HDR is important for the following reason. An effective SDN can be expected to induce sequence replacements in genomic regions that, on an average, correspond to the conversion tract length. In case the conversion tracts are long, the location of the DSB in relation to the sequence to be replaced in the target region may need not be very precise and it may vary considerably depending on the range of conversion tract lengths. But if the conversion tracts were short, the DSB site must be precisely located close to the site targeted for sequence replacement. The conversion tracts generated in mammalian cells following DSB induction by ZFNs tend to be rather short (merely 100–200 bp). In contrast, the conversion tracts obtained after ZFN cleavage are somewhat longer than 200 bp in the germ line cells of *Drosophila* larvae, but they are of many kilobases in embryonic cells. It appears that the length of conversion tract is not only affected by the species but also the cell type of the same species. It would be quite helpful in gene targeting if the factors affecting the lengths of conversion tracts were well known (Carroll, 2014, 2015).

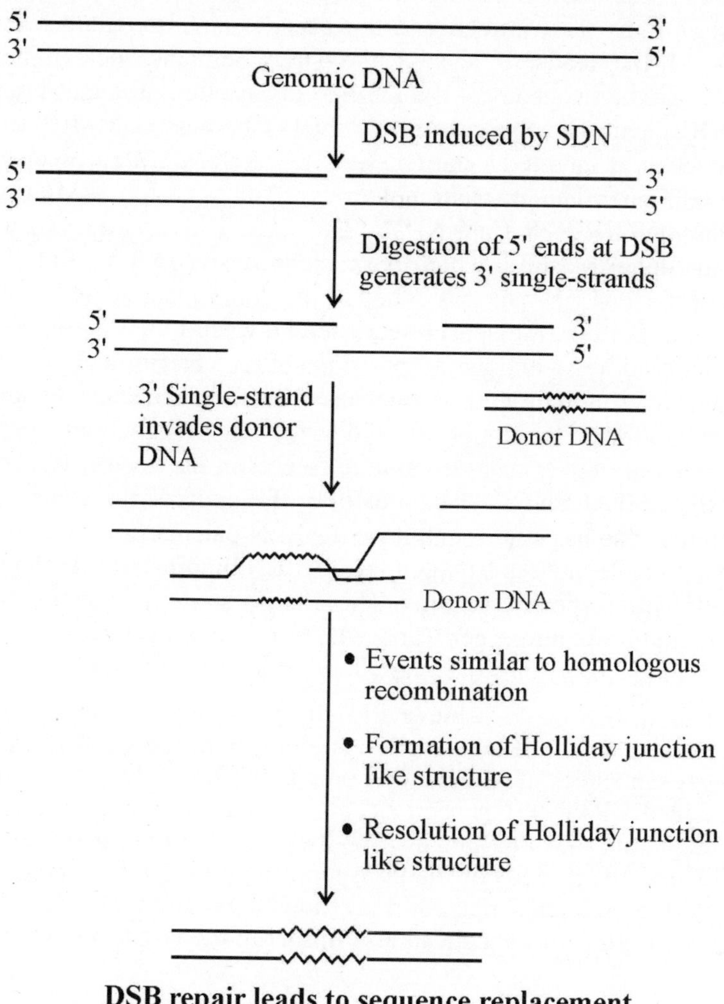

Fig. 3.10. Homology-directed DSB repair. The 5' ends at DSB are digested. The resulting 3' single-strands invade the homologous donor DNA and initiate homologous recombination like events. The Holliday junction like structure formed in the process is resolved leading to replacement of a sequence in the genomic DNA by that of donor DNA. The same strategy is used to integrate transgenes at targeted genomic sites.

3.10.4. Microhomology-mediated end-joining (MMEJ)

The microhomology-mediated end-joining (MMEJ) is a type of NHEJ DSB repair mechanism that is not based on Ku proteins. This NHEJ pathway is active during G1 and early S phases of the cell cycle. In contrast, HDR is active during late S and G2 phase cells. It uses short homologous sequences of 5-25 bp (microhomologous sequences) for error-prone joining of the broken ends. The 5' ends at the DSBs are digested till 3' overhangs with short homologous regions are generated. These short homologous sequences anneal, the unpaired 3' ends of the overhangs remaining as flaps are trimmed and the gaps, if any, are filled before the nicks are ligated. As a result, MMEJ leads to loss of the DNA segment located between the short homologous sequences on the two sides of the DSBs; thus, it often produces larger deletions than the Ku-dependent NHEJ pathway. Nakade et al. (2014) harnessed MMEJ to integrate donor vectors/transgene cassettes at targeted genomic sites by strategy called precise integration into target chromosome (PITCh). The microhomologous sequences for MMEJ were created by cleavage with TALENs and CRISPR-Cas9 SDNs. The same TALEN pair was used to cleave the genomic target and the donor plasmid, which is specially designed TAL-PITCh vector. For this reason, a target site for the TALEN pair was added in the donor plasmid at the 5' or upstream end of the transgene cassette. Further, the spacer sequence between the TALEN binding sites in the donor plasmid was modified by switching the positions of the anterior and posterior halves of the spacer to generate the microhomologous sequences following cleavage by the TALEN. This scheme enabled to integration of the donor plasmid at the targeted genomic site. The scheme can be modified to integrate only the transgene cassette free from the vector backbone by using two pairs of suitably designed TALENs to cleave at both the ends of the transgene cassette. This would require insertion of binding sites for the two TALEN pairs at the two ends of the transgene cassette. Further, the spacer sequences between the TALEN binding sites at the two ends must be suitably modified following a specific scheme. This PITCh strategy is applicable to a variety of biological systems, including cultured cells and whole animals. Further, this strategy would be more useful in those biological systems that show low frequencies of HDR.

CRISPR-Cas9 was also used to achieve PITCh in a gene in cultured human cells. This strategy requires the use of three different sgRNAs along with the Cas9 nuclease. Further, two different sgRNA target sequences must be added on either side of the transgene cassette present in a specifically designed CRIS-PITCh vector. This method was used to demonstrate targeted transgene cassette integration into the human *FBL* locus. In the PITCh strategy based on either TALEN or CRISPR-Cas9, the 5' ends of the transgene cassettes tend to show high frequency of precise MEMJ, while their 3' ends may or may not be repaired by MEMJ. It is, however, significant that none of the 10 off-target sites most similar to the target site were mutated (Nakade et al., 2014).

3.10.5. Enhancing the frequency of HDR

The frequency of HDR products may be as high as 20–30% under favourable conditions. But in many organisms, HDR frequencies are quite low and, in some cases, even <1%. The HDR products can be identified and isolated by using molecular screening procedures (Carroll, 2014, 2016). The proportion of HDR products can be increased by providing the donor DNA in linear form; HDR is up to 100-fold higher with linear donor DNA than that with circular donor provided in a plasmid (Bibikova et al., 2003). In addition, the design of donor DNA can affect the HDR frequency. It has been concluded that in the case of CRISPR-Cas9 system, the use of a ssDNA donor that is able to pair with the target DNA strand, which is the first to be released by Cas9 following cleavage can improve the rates of HDR-mediated genome-editing. HDR can be encouraged by inactivating some key components of the NHEJ repair pathway. For example, the fraction of HDR products increased from ~20% to around 65% or more when the DNA ligase IV was inactivated in *Drosophila* (Beumer et al., 2013). Reduced ligase IV activity is reported to promote HDR in *C. elegans*, and in cultured mammalian cells and young embryos as well.

However, NHEJ still operates in all these systems, suggesting the existence of alternative routes for NHEJ repair. In any case, many organisms cannot tolerate the loss of ligase IV activity.

Many factors, including arrest of cells in G2 or M phase, obstructing cohesion between sister chromatids and treatment with certain small molecules can improve the frequency of HDR (Porteus, 2016). For example, certain small molecules activate Rad51 protein, which is involved in HDR; this activation can enhance HDR-based genome editing in relation to NHEJ-mediated DSB repair. In addition, one can use nickases to induce nicks in the place of DSBs produced by regular SDNs. Nicks are generally not preferred substrate for NHEJ repair pathway, but a low level of NHEJ still takes place. However, the HDR level in response to nicks is also significantly depressed as compared to that with DSBs, but the reduction in the level of NHEJ is dramatic. The reporter 'Traffic Light' permits rapid discrimination between the products from HDR and NHEJ repair pathways. Some researchers consider nickases to be safer than programmed nucleases. Some other strategies, such as cell synchronization, have also been used, with some success, to enhance HDR (Carroll, 2014, 2016).

Finally, the products generated by HDR-mediated genome editing frequently serve as a substrate for subsequent CRISPR-Cas9 cleavage. This problem can be mitigated by introducing into the donor template silent mutations, *i.e.*, mutations that change the DNA base sequence without altering the amino acids encoded by the mutant codons. These mutations are the most effective when they are located in the PAM sequence (this destroys the PAM) and in the spacer sequences closest to the PAM. This strategy would be useful in such cases where the desired mutations included in the donor sequence are insufficient to block further Cas9 action (Komor et al., 2017b).

The CRISPR-Cas9 system has been repurposed to develop a 'prime editor' for the 'prime editing' procedure that exhibits similar or higher efficiency in sequence replacement and fewer by-products than HDR. It also shows much lower off-target activity than Cas9 nuclease and there is no risk of NHEJ-mediated uncontrolled sequence changes. The 'prime editor' has been dubbed as CRISPR v.2.0 or simply CRISPR 2.0, while the earlier CRISPR-Cas9 tool is referred to as either CRISPR v.1.0 or CRISPR-Cas9. The 'prime editor' comprises a long guide RNA, called prime editing guide RNA (pegRNA), that has the sequence to be copied into the genome and a Cas9 nickase fused to an engineered reverse transcriptase. The additional bases in the pegRNA are complementary to the strand not paired with the sgRNA segment as well as the sequence to be reverse transcribed into the targeted sequence. The nCas9 nicks the genome in the targeted region and the reverse transcriptase reverse transcribes the sequence from pegRNA at the nick site of the genome using the strand with the 3'-OH at the nick as primer for this reaction. Inclusion of a standard sgRNA with sequence matching the newly-edited sequence (generated by the pegRNA) improved the editing efficiency. Prime editing was used to perform over 175 edits that included all the 12 types of base substitutions, insertions and deletions. Theoretically, it would be able to correct nearly 89% of the known gene mutations associated with human diseases (Anzalone et al., 2019). Prime editing has immediately attracted immense media attention, but further optimization of the procedure may be needed for its therapeutic application. Prime editors were modified for use in plants by optimizing the codons, promoter, and the conditions of editing. These modifications enabled the prime editors to induce 'point mutations, insertions and deletions in rice and wheat protoplasts. Regenerated prime-edited rice plants were obtained at frequencies of up to 21.8%' (Lin et al., 2020).

3.11. ZFN-INDUCED MUTATIONS

ZFNs and other SDNs induce several different types of genomic alterations, namely, indels of one to few nucleotides, base substitutions, sequence replacements, gene replacements or additions, inversions, duplications, and translocations (Fig. 3.11). The type of genomic change generated by SDNs depends on the number and location of induced DSBs, and the type of cellular DNA repair pathway used for DSB repair. The DSBs are rapidly repaired either by NHEJ or HDR since unrepaired DSBs are lethal to the affected cells. NHEJ is the default DSB repair pathway that is functional in all cells and is preferred to HDR. When DSB is induced at a single genomic site, the DSB is repaired by NHEJ and is typically associated with indel of up to some tens of base pairs at the DSB site (Fig. 3.11A). On an average, one-third of the indels would be in-frame, while two-thirds are expected to lead to frameshift mutations. Frameshift is expected to lead to substitution of all the amino acids encoded by the codons located after the NHEJ site, as a result of which the concerned proteins will be non-functional. Sometimes, even addition/deletion of 3 bp or multiples thereof would lead to inactivation of the protein product since this may either create a termination codon or the affected region of the protein is critical for its function. NHEJ can generate a low frequency of substitutions of one to few bases, which may lead to amino acid substitutions and, as a result, missense mutations (Carroll, 2014, 2016).

TALENs tend to produce more and somewhat longer deletions than ZFNs probably due to the longer spacer region between the binding sites for the two TALEN monomers used for gene editing. CRISPR-Cas9 seems to produce much longer (occasionally up to several hundred base pairs) deletions and insertions, which often involve even the sgRNA-DNA hybrid regions as well as the PAM regions. Expression of exonuclease along with the SDN might promote long deletions. The NHEJ-mediated repair products from all SDNs contain occasional base substitutions at the DSB site. This indicates that an error-prone DNA polymerase participates in the DSB repair process. The insertions generated by NHEJ might result from a variety of divergent processes. Short insertions are most likely produced by nucleotide addition, without the use of a template, to the ends of the DNA strands at the DSBs. In some cases, the insertions seem to be derived from other genomic regions. At least in one case, the insertion seemed to be derived by reverse transcription of mRNA. Thus, NHEJ comprises a range of diverse processes designed for speedy repair of a DSB before it leads to more serious consequences for the affected cells (Carroll, 2014).

DSB repair by HDR requires an appropriate donor DNA with homology regions at its two ends. This donor DNA is used as template for HDR-mediated repair of the DSB (Fig. 3.11B). Generally, DSB is induced at a single genomic site, and HDR leads to replacement of a genomic sequence by the sequence included in the donor DNA. This approach can be used to correct the sequence changes responsible for mutant alleles to restore the wild type gene functions. It can also be used to introduce planned sequence changes that would lead to the desired mutations in targeted genes. HDR-mediated DSB repair can be used to insert a missing sequence of a gene/genomic region that had been lost by spontaneous deletion. It can also be used to insert an additional sequence encoding a novel epitope or a fluorescent tag to enable tracking of the endogenous protein. This process also allows replacement of the entire gene by the gene provided by the donor DNA. Finally, a novel gene can be integrated at the target genomic site either by exploiting HDR-mediated DSB repair or by microhomology-mediated end-joining (Section 3.10.4).

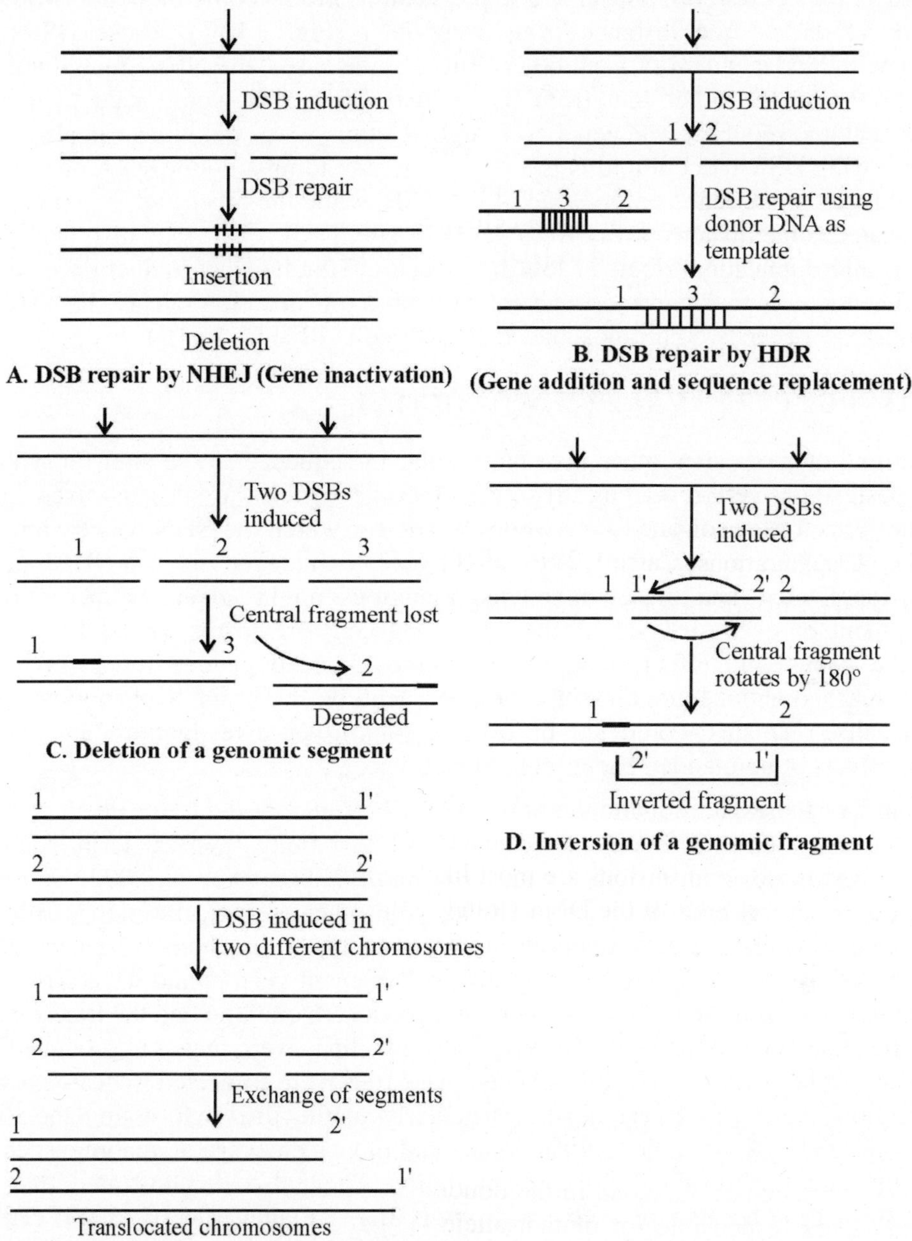

Fig. 3.11. Genomic alterations induced by DSB induction by SDNs followed by their repair by cellular DNA repair processes. **A.** DSB repair by NHEJ leading to short nucleotide insertions or deletions. **B.** HDR using double- or single-stranded donor DNA as template leads to gene/ sequence addition or replacement. **C** and **D.** Two DSBs induced by two different SDNs generate large deletions and inversions, respectively. **E.** One DSB induced in each of two different chromosomes may lead to translocation (based on Lee et al., 2016).

Large deletions of several thousand base pairs can be produced by inducing two DSBs in the chromosome at the desired distance from each other (Fig. 3.11C). These DSBs would be generally repaired independent of each other. But in some cases, the NHEJ may involve one end from one DSB and the other end from the second DSB; this would lead to a loss of the intervening genomic sequence and generate a large deletion. Large deletions can also be produced by HDR following DSB induction; this is achieved by providing a donor DNA one part of which is homologous to the sequence on one side of the DSB, while the other part is homologous to the sequence located some distance away from the DSB site. DSB repair may also lead to inversion (Fig. 3.11D) and duplication albeit in low frequencies. Simultaneous induction of one DSB in each of two nonhomologous chromosomes might lead to translocation in case the NHEJ involved one end from each of the two chromosomes that suffered DSB (Fig. 3.11E).

3.12. OFF-TARGET CLEAVAGE (ZFN TOXICITY)

Cleavage at one or more sites other than but similar in sequence to the site, for which a given SDN was designed is referred to as off-target cleavage. Often the sites involved in off-target cleavage are termed as secondary sites, while the site for which the SDN was designed is called primary site. Cleavage at secondary sites might lead to mutagenesis via NHEJ in the genes harbouring these sites, and these unintended mutations might adversely affect the function and/development or even survival of the concerned cells/organisms. In addition, they might confound the phenotypic effects resulting from the intended gene editing. As a result, the conclusions reached about function of the targeted gene on the basis of phenotypes obtained in the treated cells/organisms would not be reliable till further investigations are undertaken to exclude the effects of unintended background mutations.

In the initial studies, introduction/expression of ZFNs in cells led to the death of a proportion of the treated cells; this has been referred to as ZFN toxicity (general term, SDN toxicity). Toxicity was first discovered in studies with *Drosophila* (Bibikova et al., 2002); subsequently, it was observed by several other workers in studies with other biological systems. Bibikova et al. (2002) used a pair of ZFNs, *viz*., yA and yB, to inactivate the *yellow body* (*y*) gene of *Drosophila*. The genes encoding these ZFNs were integrated in *P* element vectors and were driven by a heat-shock promoter. Two transgenic lines of flies were produced; one line had the transgene encoding yA, while the other had the transgene for yB. These two lines were mated to produce flies having the transgenes for both yA and yB ZFNs. When these flies were subjected to heat-shock to induce ZFN expression, there was larval death, particularly in the lines expressing the ZFN yA. It seemed that the ZFN yA was able to form dimers and induce cleavage at genomic sites other than its target site and this resulted in larval mortality. They also demonstrated that the ZFNs inactivated the *y* gene because of DSB repair via NHEJ (Pattanayak et al., 2011; Carroll, 2014, Chandrasegaran and Carroll, 2016).

Initially, the off-target sites for ZFNs were either predicted by bioinformatic analysis of the genome sequence for identification of genomic sites similar in sequence to the on-target site. Alternatively, biochemical assays for DNA specificity were combined with the bioinformatic analyses. But these strategies were not able to identify all the off-target sites for ZFNs. Monitoring of off-target cleavage by ZFNs was initially based on a library of potential off-target cleavage sites for the ZFNs in question, which was subjected to cleavage in an *in vitro* system. The members of library cleaved by the ZFNs were identified by high throughput DNA sequencing. Subsequently, integrase deficient lentivirus (IDLV) was used to identify the genomic sites cleaved by ZFNs. The IDLV genome becomes integrated at those genomic sites that are cleaved by the ZFN treatment. It may be pointed out that this integration of IDLV genome does not rely on sequence homology between the cell and IDLV genomes. In simple terms, the ZFNs are transfected into cells along with IDLV, and the genomic sites with IDLV integration are identified by high-throughput sequencing of the products from linear amplification mediated PCR (LAM-PCR). The IDLV integration assay may not be able to detect off-target sites showing relatively low frequency of cleavage. Finally, whole genome sequencing for detection of indels resulting from NHEJ-mediated repair of DSBs induced by ZFNs has been used for detection of off-target genomic sites. Findings from many studies indicate that selected and validated ZFNs tend to show little or no off-target activity (Bogdanove et al., 2018).

It is known that ZFN toxicity is the result of induction of multiple DSBs at off-target sites. Obviously, this can happen only when the genome has multiple binding sites for the ZFNs or the ZFNs are able to bind to sequences that are similar but not identical to the specified binding site. The toxicity increases with ZFN concentration in the cells due to continued expression of the ZFN genes present in plasmids. The TALEN and CRISPR-Cas9 SDNs are also known to induce off-target cleavages (Carroll, 2014, Chandrasegaran and Carroll, 2016).

3.12.1. Assessment of off-target cleavage

The genomic sites likely to suffer off-target cleavage have been predicted by genome sequence analysis to identify sequences similar to the target site. Predictions have also been made based on *in vitro* cleavage of partially randomized DNA substrates and through selection for preferred binding sites. The off-target cleavage activity of ZFNs have also been predicted on the basis of DNA-binding specificities of the ZFN monomers. But the off-target activity of ZFN dimers cannot be reliably predicted by the DNA-binding specificities of the ZFN monomers. Further, target recognition by ZFNs is also influenced by the dimerized FokI nuclease domain. It may be pointed out that identification of potential off-target sites and determination of their significance to the function and development of the organism is generally problematic. Initially, the genome was scanned for sequences similar to the target sequence, followed by PCR amplification of these genomic sites and analysis of the amplification products by sequencing. The extent of change in the sequence can also be assessed by a gel electrophoresis procedure using the mismatch-specific Surveyor or T7EI endonuclease, and by high resolution melt analysis. The frequency of off-target activity can be determined by whole genome sequencing of the treated cells/organisms. It should be kept in mind that this would be quite reliable for scoring indels since errors during DNA replication generally do not lead to indels. Since base substitutions are far more frequent than indels during DNA replication, scoring them would not be very reliable. At the same time, the sequencing will have to be very deep to detect infrequent sequence changes (Carroll, 2014; Chandrasegaran and Carroll, 2016).

The recent methods of assessing off-site cleavage have been reviewed by Koo et al. (2015) and Hendel et al. (2015). These methods identify cleavage sites more broadly, and they are quite sensitive and informative. However, each method suffers from some limitations, some of which are mentioned below. In case of lentiviral capture, ends generated in the substrate DNAs by ZFNs are captured by ligation to the DNA of an integrase-deficient lentivirus. In this procedure, the off-target cleavage will go on increasing with time till cleavage of the target site is near saturation;

the same is true for GUIDE-seq and LAM-PCR HTGTS methods. Further, the lentiviral method is not applicable to all types of cells and organisms. The GUIDE-seq strategy captures individual breaks, but there is virtually no indel in the target sequence. Further, the off-target breaks increase with time as is the case with lentiviral capture. The LAM-PCR HTGTS method uses designed target sequences to capture off-target cleavage based on translocations induced by the SDNs. The captured junctions result from typical NHEJ. But it may overestimate off-target cleavage since at least two almost simultaneous breaks are needed for their detection. Digenome-seq method requires large amounts of very deep sequencing. BLESS captures breaks that are present at the point of time the cells were sampled. Thus, the detection of breaks will be affected by the harvesting time in relation to the delivery of SDNs. It is highly desirable that the efficacy of these detection methods is compared in the same set of samples (Chandrasegaran and Carroll, 2016). A more detailed discussion of off-target detection is provided in Section 5.9.1.

3.12.2. Controlling ZFN toxicity

ZFN toxicity can be minimized by designing and selecting for ZFNs that are highly specific to the targeted site. One way to achieve this is to increase the number of ZFs in the ZFNs so that they are required to bind to longer DNA sequences. This may be expected to reduce the chances of off-target binding and cleavage, but this expectation is not always fulfilled. Another strategy is to remodel the dimerization interface of FokI nuclease domain to produce two such mutant nucleases that can form functional heterodimers, but they are unable to form homodimers. Generally, the heterodimeric mutants of FokI have reduced nuclease activity (Table 3.1). These FokI mutants can be used to design a pair of distinct ZFNs that bind to sites located close to each other so that the FokI cleavage domain is able to form a heterodimer that can cleave the DNA between the two binding sites. This strategy would double the length of recognition sequence and minimize the chances of off-target cleavage. The first generation of these FokI mutants suffered from reduced cleavage efficiency. Subsequently, the amino acid substitutions in the dimerization interfaces of the mutant FokI cleavage domains were further refined. These efforts have yielded the second generation of mutants that have greatly improved cleavage efficiency. Experimental evidence shows the heterodimerization strategy based on the second generation of mutants to be highly successful in minimizing ZFN toxicity. As a result, both ZFN and TALEN SDNs in current use are based on the second generation FokI heterodimeric mutants. Both the above approaches can increase the specificity of ZFNs and minimize the frequency of off-target cuts (Ramalingam et al., 2011; Carroll, 2014, 2016).

The off-target cleavage activities of ZFNs appear to be positively associated with their DNA-binding energies. Therefore, off-target activities of ZFNs might be lowered by designing ZFNs with reduced DNA-binding affinities. The off-target cleavage can also be reduced by lowering the concentration and the period of action of ZFNs in the cells. The target sites should be carefully selected so that other related sequences in the genome differ from the target site by more than three base pairs. The eukaryotic transcription factors having ZFs bind to consensus sequences, *i.e.*, sequences that show some variation, at the promoters of the concerned genes. Thus, ZFs would have been selected for their ability to accommodate some variation in their DNA-binding sequences. Therefore, the lack of perfect specificity in target sequence binding may be inherent in the recognition capabilities of the ZFPs (Carroll, 2014).

Table 3.1. Relevant features of Fok1 nuclease mutants that function only as heterodimers (based on Wright et al., 2014)

Mutant FokI pair	Amino acid substitutions*	Salient features
RV, DA	D483R and I538V; R487D and I499A	Reduced nuclease activity
KK, EL	E490K and I538K; Q486E and I499L	-
ELD, KKR	Q486E, I499L and N496D; E90K, H537R and I538K	Nuclease activity higher than other heterodimer mutants; enhanced temperature tolerance
Sharkey (P and E)	S418P and K441E	Not a heterodimer mutant but can be combined with heterodimer mutations; increased nuclease activity
RE	S418R and K441E	Not a heterodimer mutant but can be combined with heterodimer mutations; enhanced DNA binding; reduced nuclease activity

* Amino acids substituted at critical positions of FokI, *e.g.*, D483R, aspartic acid at position 483 replaced by arginine (Asp483Arg), etc.

The level of acceptable off-target cleavage will depend on the objective of gene editing. For many investigations, *e.g.*, determination of the frequency of NHEJ and/or HDR, the desired alteration in the targeted gene/sequence is of the greatest relevance, while the off-target cleavage can be ignored so long as it does not lead to toxic effects. But when the objective is to relate a phenotypic alteration to the genetic change induced by editing of the targeted gene, the off-target cleavage should ideally be absent. Alternatively, one will have to compare the phenotypic changes in several independently derived mutants preferably by targeting different sites in the same gene using different ZFNs. The mutants may be backcrossed to the normal parental strains to transfer the mutant alleles in the parental genetic background and analyse the derived lines for their phenotypes. Finally, the normal alleles of the inactivated genes may be introduced into the mutant lines to determine if they are able to produce the normal phenotypes in them; this constitutes the complementation test. Obviously, these strategies are not applicable to all the biological systems. In case of crop plants, one may eliminate the effects of off-target cleavage by repeated backcrosses to the parental lines. For human therapeutic applications, the frequency of off-target cleavage should be zero or near zero, and a safe level for the same is yet to be agreed upon. In any case, the mutations at secondary sites and their phenotypic effects must be meticulously characterized and evaluated for safety when gene editing is used particularly for human therapy (Carroll, 2014; Chandrasegaran and Carroll, 2016).

3.13. ZFN APPLICATIONS

The ZFN SDNs function as heterodimers so that for each targeted genomic site one pair of ZFNs are used. The members of each ZFN pair are designed to bind to sequences located close to each other. Investigations with ZFNs generated critical information on several different aspects of SDN technology, including SDN delivery in several biological systems, and the idea of inducing nicks in the place of DSBs to promote HDR, which led to the development of nickases. Data on the relative frequencies of DSBs repaired by NHEJ and HDR were initially accumulated from experiments with ZFNs. The relevant features of donor DNA, *e.g.*, the lengths and double-/single-stranded donor DNAs, the length, distribution and the extent of homology of the donor with the targeted genomic region, etc., were also determined. The lengths of conversion tracts generated by HDR were defined in some biological systems. It was also shown that sequences could be inserted at the target sites by providing donors having only limited homology at their termini. In addition, the procedures that may be used to generate more complex genomic rearrangements like inversions, translocations, etc. were developed. Finally, methods for detection and quantification of SDN-induced mutations were also developed (Carroll, 2014).

The SDNs, *viz*., ZFNs, TALENs and CRISPR-Cas9, have been effectively used for gene editing in both prokaryotes and eukaryotes, including model organisms, pathogens, disease vectors, crop plants, domestic animals and mammals, including humans. Some excellent reviews on the subject are those by Segal and Meckler (2013), Carroll (2014), and Peng et al. (2014). ZFNs have been used to inactivate genes in a variety of organisms: they successfully inactivated even those genes, for which null mutations were not known before, *e.g*., in *Drosophila*, and in such organisms that were not otherwise amenable to gene manipulation, *e.g*., zebrafish. But in recalcitrant biological systems, the rate of gene inactivation was low with ZFNs, while TALENs and CRISPR-Cas9 are much more effective. SDNs have been used for gene targeting in a variety of plants, including crop plants like tobacco, maize, soybean, etc. In case of plants, one of the major issues is the delivery of SDNs. Several plants are readily transformed by coculture with *Agrobacterium tumefaciens*, but the T-DNA has limited capacity for carrying exogenous DNA sequences. SDNs have also been delivered into plants using viral vectors, but they also have limited capacity for carrying exogenous sequences (Carroll, 2014, 2016).

SDNs are particularly useful is such species that are not amenable to transgenic technology. For example, embryonic stem cells are not known in case of rats; this precludes the development of transgenic strains, including gene knock-out lines. Microinjection of ZFNs (and later TALENs) directly into the embryos was able to generate targeted gene knock-outs as well as gene knock-ins. SDNs have been deployed for gene editing in many animals including cows, pigs, rabbit, catfish, trout, etc. for increasing skeletal muscle in meat animals, making pig suitable as organ donors for transplant in humans, curing of genetic diseases in cattle, etc. A generalized procedure for silencing protein coding as well as noncoding genes was developed by Gutschner et al. (2011). They silenced the target gene by inserting into it RNA destabilizing elements using HDR promoted by ZFN treatment. This procedure enables silencing of a gene without the prior knowledge of the sequences critical for its function.

ZFNs have been used to induce HDR to enable sequence replacement in genes in several different organisms, but not in zebrafish. However, other SDNs like TALENs and CRISPR-Cas9 were able to achieve low frequencies of HDR in this organism. This might be due to the very rapid cell division in zebrafish young embryos, which could make the less demanding NHEJ the preferred repair pathway. ZFNs have been used for HDR-mediated correction of gene mutations causing haemophilia, X-linked SCID, sickle-cell disease, Parkinson's disease, etc. It has also been used for NHEJ-mediated gene inactivation of C-C chemokine receptor type 5 (CCR5) encoding gene in primary T cells to generate resistance to HIV-1 infection. Similarly, the ZFN-induced inactivation of T-cell receptor gene has yielded 'universal' T cells that have greatly reduced risk of graft-versus-host disease and are extremely useful in cancer immunotherapy (Section 9.12).

ZFNs have enabled targeted insertion of genes encoding anti-HIV restriction factors into the *CCR5* locus to generate resistance to infections by both R5- and X4-tropic strains of HIV-1. For therapeutic uses, the ZFNs should be highly specific to the targeted sequences, and should lead to low, preferably negligible, cytotoxicity. It will also be necessary to generate ZFNs by a high throughput procedure so that high affinity ZFNs for a variety of target sequences are readily generated. ZFNs have been used for HDR-based mutant sequence replacements in genes involved in human diseases, and some ZFN-based therapies are already in clinical trials (Gaj et al., 2013; Carroll, 2014, 2016).

3.14. MERITS AND LIMITATIONS OF ZFNs

The ZFN gene constructs are much smaller in size (two sequences of ~1 kb each) than those of TALENs (two sequences of ~3 kb each) and CRISPR-Cas9 (~4.2 kb for Cas9 + 0.1 kb for sgRNA). Therefore, the delivery of ZFNs into cells and organisms is relatively much more convenient than that of TALENs and CRISPR-Cas9 SDNs especially when adeno-associated virus vectors (they have limited cargo capacity) are used. Since ZFPs are derivatives of mammalian transcription factors, they may be expected to pose little if any risk of immunogenicity in humans. In comparison, TALENs and CRISPR-Cas9 SDNs are derived from bacterial systems and might present a greater risk of immunogenicity than ZFNs. However, this aspect requires a careful comparative evaluation. In additions, the information accumulated from experiments with ZFNs has facilitated the rapid progress with the other two SDNs, *viz*., TALENs and CRISPR-Cas9 (Lee et al., 2016).

The ZFNs have not been widely used mainly because developing a pair of ZFNs for each new genomic target is a challenging task. This requires considerable technical skill in protein engineering and genetic engineering to design ZFNs. Since all the engineered ZFNs are not efficient in DSB induction, their development involves lengthy, tedious and expensive evaluation and selection procedures. Therefore, a worker desiring to use ZFNs for gene editing would need to collaborate with someone capable of developing the ZFNs for the desired target sequences. Alternatively, validated ZFNs could be obtained commercially, but the price in the beginning was affordable only by laboratories having ample funds. However, the prices of ZFNs have declined with time, but they are still rather expensive for most laboratories. In addition, an opensource library of components of zinc fingers and a selection protocol for identification of highly efficient ZFNs having high affinity have been developed by an academic consortium, but these resources have not been widely used by the research community (Carroll, 2014, 2016; Lee et al., 2016).

In addition, there is some restriction on the availability of suitable ZFN targets in the genome since successful target sites generally occur in guanine-rich genomic regions comprising 5'–GNN-3' (N, any of the four nucleotides) repeat sequences. It has been estimated that a good target site may be expected to occur, on an average, every 100 bp. This limitation is not relevant when the researcher desires to inactivate a gene since he can select any one of several potential sites in the upstream regions of the genes. But when a sequence is either to be added at a specific site or a given sequence is to be replaced, this may seriously reduce the usefulness of ZFNs if the concerned site is not suitable for efficient ZFN action. At the same time, many ZFNs may show low efficiency and/or relatively high level of off-target editing (Carroll, 2014, 2016; Lee et al., 2016).

In case of the nematode *Caenorhabditis elegans,* when the SDN, including ZFN, encoding genes were delivered via plasmid transfection, gene editing was quite successful in both genomic and extrachromosomal targets in somatic cells. However, the germ line remained unaffected most likely due to RNA interference (RNAi). Delivery of ZFN encoding mRNA by microinjection into the developing gonads induced low frequency of mutations in the germ line cells. In contrast, TALEN and, more particularly, CRISPR-Cas9 SDNs delivered as mRNA, protein or ribonucleoprotein (Cas9 protein + sgRNA in the case of CRISPR-Cas9) were much more effective in germline gene targeting. In case of zebrafish, ZFNs were able to induce mutagenesis through NHEJ, but gene correction by HDR was only occasionally obtained. In this case also, TALENs were more effective in inducing HDR when the donor DNA was provided either as single-stranded oligonucleotide or as long, duplex DNA. It appears that in the embryos of this organism, there is a strong preference for NHEJ over HDR for DSB repair most likely due to the fast pace of cell division (Carroll, 2014, 2016).

3.15. HYBRID ZINC FINGER NUCLEASES

The ZFN architecture based on FokI nuclease suffers from the limitation that one pair of ZFN monomers must be engineered for each target site and delivered into the targeted cells. In addition, the FokI nuclease does not require specific sequence for cleavage, as a result of which the ZFN specificity is reduced to some extent. These limitations can be removed by using a nuclease domain that does not require dimerization for being active so that only a single ZFN monomer can be used, and it does need a specific sequence at the cleavage site so that the ZFN specificity is improved. An example of the use of a nuclease domain with specific cleavage sequence requirement is provided by the ZF-PvuII nucleases comprising ZFs linked to the restriction enzyme PvuII, which requires the sequence 5' CAGCTG 3' at the cleavage site. ZF-PvuII SDN would have higher specificity than ZF-FokI nuclease. But the potential target sites in the genome for this nuclease will be reduced due to the requirement of cleavage site sequence for PvuII, and two monomers of ZF-PvuII will have to be used (Guha and Edgell, 2017).

It will be desirable to link ZFs with a monomeric nuclease that has specific cleavage sequence requirement. A monomeric nuclease is composed of a single polypeptide that cuts both the strands of a DNA duplex. A monomeric FokI nuclease was created by linking two FokI nuclease domains by a suitable linker and this monomeric FokI architecture was linked to ZFs. This monomeric ZFN may be suitable for targeting genes in mitochondrial genomes since it has lower specificity as its target sequence is of only 12 bp. The ZFs were fused with the homing endonuclease I-TevI to construct ZFEs (zinc finger endonucleases). The ZFE recognition sites comprise two parts, namely, the target sequence for the ZFs and the sequence 5' CNNNG 3' needed for cleavage by I-TevI cleavage domain located at appropriate distance from the ZF-binding sequence; there seems to be some preference for the spacer sequence as well. These requirements impose restrictions on the potential target sites of the ZFEs (Guha and Edgell, 2017).

3.16. CONCLUSIONS

ZFNs were the first SDNs to be developed and used for gene editing in a variety of biological systems. They also enabled the collection of basic information about SDNs and gene editing that has helped rapid development of TALENs and CRISPR-Cas9, the other two SDNs developed subsequently. However, the ZFNs have not been widely used mainly because of the technical expertise and the costs involved in development of new ZFNs. In comparison, TALENs and especially CRISPR-Cas9 are relatively much easier, faster and cheaper to design and develop; designing of CRISPR-Cas9 might take only several minutes. As a result, the CRISPR-Cas9 system has rapidly overtaken both TALENs and ZFNs and become the most widely used SDN. It would be fair to conclude that ZFNs are no longer a choice SDN for routine laboratory studies mainly because of difficulties and cost associated with construction of ZFNs for new genomic targets (Carroll, 2016; Chugunova et al., 2016).

It should be noted that in many applications, a single SDN would be used repeatedly in different individuals/lines to target the same genes. This would be true for human gene therapy, and livestock and crop improvement applications. In such situations, the designing and development of the SDN is only one component of the application, which may be regarded as of smaller significance. In contrast, issues like ease of delivery and specificity of the SDN are of far greater importance. Since the ZFNs are relatively much smaller in size, their delivery will be easier at least in some circumstances. At the same time, some of the existing ZFNs are one of the most specific and effective SDNs in use (Carroll, 2016). ZFNs designed to edit the human *CCR5* gene have been in clinical trials, have proved safe and effective to the level required for Phase I trials (Tebas et al., 2014).

Some questions that remain to be answered include, why the cleavage efficiency varies greatly among different targets within the same class of SDNs, why cleavage efficiency does not appear to be related with the affinity, and how both active and inactive eukaryotic genes are targeted with comparable efficiency even though they would have drastically different chromatin states. It may be expected that in due course of time these questions will be investigated as the need for further refinements in gene editing would press for their resolution. Further, the efficiency and, especially, the specificity of cleavage, and safe and effective modes of delivery must be improved/devised to enable their therapeutic applications. These refinements would also enhance their usefulness in basic studies. It is hoped that there would be ever increasing deployment of the SDN technology in a wide range of research areas. Further, gene editing technology is entering the application phase, including the exciting field of human gene therapy. This would accelerate the research efforts in the basic aspects of SDN technology and resolve many of the outstanding issues (Carroll, 2016, Chugunova et al., 2016).

■■■■

4

TRANSCRIPTION-ACTIVATOR LIKE EFFECTOR NUCLEASES (TALENs)

4.1. INTRODUCTION

The interest in development of site-directed nucleases (SDNs) was triggered by the observation that double-strand breaks (DSBs) in genomes radically promote gene integration at the DSB sites by homologous recombination or homology-directed repair (HDR) of the DSBs; this increase in integration frequency was of at least two orders of magnitude. In addition, the repair of DSBs by nonhomologous end joining (NHEJ) pathway leads to induction of short insertions or deletions (indels) at the break sites leading to inactivation of the targeted genes. The meganucleases were the first to be used during the early 1990s to demonstrate the feasibility of induction of DSBs at genomic sites having specific sequences. This was soon followed by the development of chimeric nucleases called zinc finger nucleases (ZFNs) that are produced by fusion of DNA-binding domain from eukaryotic transcription factors called zinc finger proteins (ZFPs) and the nuclease domain from FokI endonuclease (Chapter 2). These programmable nucleases were capable of inducting DSBs at the targeted genomic sites with high specificity and very high efficiency.

One of the main limitations of both ZFNs and meganucleases is that the designing, evaluation and validation of these SDNs with new DNA binding specificities are very challenging: they require considerable expertise in highly sophisticated and technically demanding field of protein engineering, which is resource intensive and time consuming. The development of transcription-activator like effector nucleases (TALENs) in 2010 provided a SDN system that was amenable to rapid and relatively easy high throughput designing and construction of SDNs specific for novel DNA sequences. As a result, TALENs became the preferred SDN system and even small laboratories could utilize them. But before the TALENs could establish their unquestioned supremacy, the user-friendly CRISPR-Cas9 (clustered regularly interspaced short palindromic repeats-CRISPR-associated 9) SDN system was added to the genome editing toolbox in 2013. SDNs have emerged as extremely powerful tools for a variety of basic studies in numerous biological systems as well as for many very valuable applications, including generation of several gene therapy tools, some of which have already reached the clinics (Chapters 7-11). The significance of genome editing tools is evident from the fact that at the end of year 2011, the process of genome editing by programmable nucleases like ZFNs and TALENs was chosen as 'Method of the Year' by Nature Methods.

TALENs have been used by many investigators world-wide in a variety of organisms ranging from the model species like *Caenorhabditis elegans*, *Drosophila melanogaster* and

Arabidopsis thaliana, to mammals like mice, rats, cattle and human cell cultures, and several crop species. There are many excellent reviews dealing with TALENs either exclusively or along with other SDN systems. The reviews devoted exclusively to TALENs include those by Joung and Sander (2013), Sun and Zhao (2013), Wright et al. (2014), and Chaudhary et al. (2016). In addition, a book describing the TALEN methods and protocols has recently been published (Kühn et al., 2016).

4.2. TRANSCRIPTION-ACTIVATOR LIKE EFFECTOR (TALE) PROTEINS

The plant pathogenic *Xanthomonas* spp. are Gram-negative bacteria that cause devastating diseases in several important crops. At least twelve species of *Xanthomonas* are known to produce TALE proteins during the infection process. In addition, TALEs are also produced by some species of the genus *Ralstonia*. The TALEs are indeed transcription factors that are delivered by the pathogenic bacteria into their host plant cells through the type III secretion system. Within the plant cells, the TALEs dimerize in the cytoplasm, are guided by their C-terminal nuclear localization signals (NLSs) into cell nuclei where they function as transcription factors. The TALEs bind to specific sequences (denoted as EBEs) in the promoters of certain genes of their host plants and activate the expression of these genes. Generally, the expression of targeted host genes (called susceptibility genes) increases susceptibility of the host plant to and promotes colonization by the bacterial pathogen. The first TALE to be identified in 1989 was AvrBs3 produced by *X. campestris* pv. *Vesicatoria*. The first cues to the TALE function came from discoveries of a NLS in 1995, and of an acidic activation domain in 1998 in the C-terminal regions of TALEs. Subsequent studies demonstrated DNA binding and transcription activation by TALEs, and this provided the basis for their present name. A truncated version of TALEs, called iTALEs, produced by some pathovars of *X. oryzae* interferes with the resistance response mounted by some *R* genes in rice (Christian et al., 2010; Zheng-yin, 2017).

Natural TALE proteins consist of three domains, *viz.*, N-terminal translocation domain, C-terminal activation domain and central DNA binding domain. The N-terminal domain contains 287 amino acid residues, has type III secretion and translocation signal sequences, and is involved in initiation of DNA binding. The central repeat region consists of an array of TALE repeat units; this DNA-binding domain determines the DNA-binding specificity of the TALE proteins. Finally, the C-terminal domain comprises 177 amino acid residues, and has NLS sequence, followed by an acidic transcription activation domain (Fig. 4.1A). The TALENs are typically constructed by replacing the NLS and activation domain by a suitable NLS for the organism concerned and typically FokI nuclease domain. The central DNA binding domain of TALE proteins comprises a variable number (up to 30) of nearly identical repeat units arranged in tandem. Each repeat unit has 33 to 35 amino-acid residues, 34 being the most common number. But the last repeat unit is typically made up of only 20 amino acids; therefore, this repeat unit is termed as 'half-repeat' (Christian et al., 2010; Richter et al., 2016; Lin et al., 2016).

The amino acid sequence of TALE repeat units is mostly constant, but the residues at positions 12 and 13 are highly variable among the repeat units; therefore, these residues are called repeat variable di-residues (RVDs). Each TALE repeat unit recognizes one base pair of DNA, which is the function of the two amino acids comprising the RVD of each repeat unit. Thus, the DNA binding specificity of a TALE protein or TALEN is determined by the RVDs of its repeat units, while the rest of the amino acids of repeat units provide the structural framework for their function. There is some evidence that amino acid variations in some non-RVD positions might affect TALE activity, but this effect tends to vary, and the rules involved in these effects are not clear. TALE repeats with 30, 39, 40 and 42 amino acids occur in very low frequencies in nature. A TALE or TALEN with a single long or short TALE repeat in an array of 34 amino acid repeats is able to efficiently recognize two alternative sequences of the target site: one sequence is the same as the one recognised by a TALE composed of the canonical 34 amino acid repeats, while the alternative sequence has a deletion of one base pair in the vicinity of the binding site for the unusual short/long repeat. This flexibility would allow construction of TALEs with dual specificity for single base indel mutants of the target site (Christian et al., 2010; Richter et al., 2016; Lin et al., 2016).

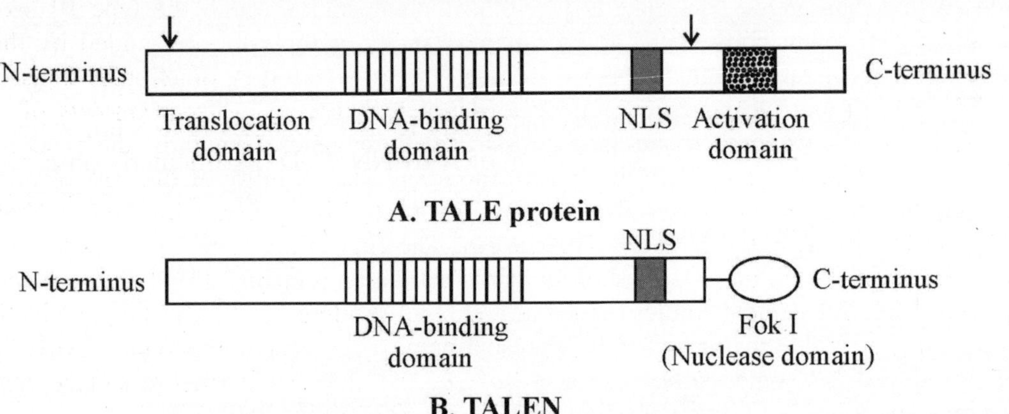

Fig. 4.1. A simplified schematic organization of a TALE protein and a TALE nuclease. **A.** TALE protein has a translocation domain at N-terminus, a DNA-binding domain in the middle, and a NLS followed by an acidic transcription activator domain at the C-terminus. The arrows indicate the points of deletion for creation of TALENs. **B.** TALEN is produced by deleting a part of the N-terminus of a TALE protein and replacing the C-terminal sequence (generally beyond the NLS) by the nuclease domain of FokI nuclease. NLS, nuclear localization signal; TALE, transcription-activator like effector; TALEN, TALE nuclease.

The DNA binding domains of TALENs and TALEs are, therefore, reprogrammed by suitably altering the RVDs of successive repeat units to enable the TALEN/TALE to bind to the target DNA sequence. The TALE proteins can also be used to construct chimeric TAL effectors for transcription activation (TALE-activator, TALEa)/repression (TALE-repressor, TALEr) by replacing the NLS and activation domain of TALEs by a suitable NLS for the organism concerned and the desired transcription activation/repression domain (Section 4.6). In case of TALEa and TALEr, the target DNA sequences are typically the promoter regions of the desired genes. In addition, the TALE proteins can be linked to enzymes involved in histone modification, DNA methylation or DNA/RNA base deamination to create epigenetic editors or base editors. Linking the TALEs to fluorescent proteins yields programmable fluorescent tags that enable genome visualization (Section 4.6; Christian et al., 2010; Lin et al., 2016).

4.3. DNA RECOGNITION CODE OF TALE PROTEINS

The TALE DNA-binding domain comprises nearly identical repeats of 33–35 amino acids, each of which binds to a specific nucleotide depending on its RVD. The relationship between the amino acid residues of different RVDs and the DNA base-pairs recognized by the TALEs have been deciphered by bioinformatics computational analyses as well as molecular biology experimental findings. Although, a given base pair may be recognized by several different RVDs in the naturally occurring TALEs, the most commonly used RVDs for the four base pairs constitute a seemingly robust and relatively simple TALE recognition code (Table 4.1): the RVD Asn Ile (NI) recognizes A, His Asp (HD) identify C, Asn Asn (NN) bind to G or A (in some contexts), and Asn Gly (NG) are specific to T in the target DNA (Curtin et al., 2012). The RVD modules HD and NN are considered to have strong DNA binding affinity, while NI is regarded to have weak binding affinity. The RVD NG is considered to have weak binding affinity by some, while others view it to have strong affinity.

According to one view, 3 or more RVDs with strong DNA binding affinity should be included for creating efficient TALENs. The RVD module HD cannot recognize and bind to 5-methylcytosine residues. In some contexts, the NN RVD recognizes A as well, which would tend to promote off-target activity. Recognition of A by NN RVD is particularly observed when the TALEs/TALENs have more than 14 TALE repeats, while in case of TALEs with only 13 repeats, the NN specifically recognizes G. The NH and NK RVDs are more specific for G than NN, but their DNA-binding affinities are considered as weak and intermediate, respectively, while RVD NN has strong DNA-binding affinity. Therefore, there could be significant reduction in overall DNA binding affinity if more than one or two copies of weak RVD modules are used in a TALEN monomer. There are some unusual RVDs, each of which recognizes two or more nucleotides; for example, the RVD NS (Asn Ser) can recognize A, G, T and C (Carroll, 2014; Guilinger et al., 2014; Chandrasegaran and Carroll, 2016).

Unlike zinc finger proteins, recognition of DNA bases by RVDs of TALE repeats seems to be largely context-independent, *i.e.*, base recognition by RVD of one TALE repeat is not affected by RVDs of the neighboring repeats. Thus, each TALE repeat binds to its specific base pair relatively independently of the other TALE repeats of a given TALE protein or TALEN. As a result, there is largely one-to-one correspondence between the RVDs and the DNA bases recognized by them. This has greatly facilitated the creation of TALENs with novel DNA-binding specificities for targeting the desired genomic sites (Curtin et al., 2012; Yin et al., 2012; Guilinger et al., 2014; Chandrasegaran and Carroll, 2016).

Analyses of crystal structures of TALE proteins bound to double-stranded DNA reveal that side chain of the second amino acid residue of a RVD, *e.g.*, I in the case of RVD NI, projects into the major groove of target DNA duplex and makes direct contact with the DNA base at that position. Therefore, the amino acid at position 13 of a TALE repeat is often termed as base-specifying residue (BSR). The side chain of the first amino acid residue of the RVD, *e.g.*, N in the RVD NI, folds back and interacts with the TALE repeat backbone to stabilize the RVD loop conformation. The recognition and binding of RVDs to only two bases, *viz.*, C and G, involves hydrogen bond formation; the RVDs for the remaining two bases, namely, A and T, recognize and bind to the respective bases by steric compatibility and van der Waals interactions. In addition, the RVDs of TALE proteins recognize and bind to the bases of only one of the two strands of their double-stranded DNA targets: they recognize and bind to the bases and phosphate residues in the backbone of the sense strand or forward strand of DNA duplex and do not interact with the antisense strand or backward strand. Sense strand of target DNA duplex is the strand that has its 5' end on the same side as the N-terminus of the concerned TALE protein (Curtin et al., 2012; Yin et al., 2012; Bogdanove et al., 2018).

Table 4.1. TALE DNA recognition code. The DNA-binding module of a TALE or TALEN comprises the amino acids in positions 12 and 13 (called RVD, repeat variable diresidues) of each TALE repeat unit (Carroll, 2014; Joung and Sander, 2013; Chandrasegaran and Carroll 2016; Kühn et al., 2016; Bogdanove et al., 2018 and other sources)

Base	Canonical RVDs**	Alternatives	Remarks
A	NI (Asn Ile)	-	Specific but low affinity
C	HD (His Asp)$	N* (Asn -)	High specificity and high affinity
G or A#	NN (Asn Asn)	NK, NH (Asn Lys, Asn His)@	High affinity, but dual specificity
T	NG (Asn Gly)	HG (His Gly)	High specificity, lower affinity than HD

* In RVD nomenclature, an asterisk (*) indicates the absence of an amino acid. This module comprises a single amino acid, *i.e.*, Asn, at position 12 but no residue at position 13; it can bind 5-mehtylcytosine, but the binding is weaker than that by HD. RVD N- is roughly similar to the RVD NG, except that it has a shorter RVD loop.

** A total of 25 natural RVDs were known till 2016. Canonical RVDs are the most commonly used. RVDs HD and NN are considered to have strong DNA binding affinity. RVDs NI, NG, NK are regarded to have weak binding. A TALE must contain at least three RVDs with strong binding affinity, *viz.*, HD and NN, for full activity of TALENs.

$ It cannot bind C residues methylated at position 5.

In some contexts, it also recognizes A; this is observed particularly when TALEs/TALENs have more than 14 TALE repeats. But in case of TALEs with only 13 repeats, the NN specifically recognizes G. In TALEN designing, NN is typically used for recognition of G mainly for want of an RVD with optimum G recognition.

@ These alternative modules are guanine-specific. The binding affinities of NN, NH and NK are considered as strong, intermediate and weak, respectively.

4.4. TRANSCRIPTION-ACTIVATOR LIKE EFFECTOR NUCLEASES (TALENs)

TALENs are chimeric nucleases that function as dimers. Each TALEN monomer is created by combining a TALE protein (Section 4.2) with the nuclease domain of FokI, a Type IIS restriction enzyme. The first TALEN was constructed by Christian et al. (2010) based on the concept and design of ZFNs. They deleted a part of the N-terminal sequences of two different TALE proteins and replaced their C-terminal sequences beyond the NLS by the nuclease domain of FokI endonuclease (Fig. 4.1B). They constructed a pair of TALENs from each of the two TALE proteins to bind two sequences in a plasmid located at variable distances from each other. The activities of the TALEN pairs were evaluated based on LacZ assay in yeast cells and their efficiencies were compared with that of ZFNs. The TALENs were as efficient as ZFNs in DSB induction, but the optimum distance, *i.e.*, spacer length, between the binding sites for the two TALEN monomers of a pair was either 15 or 21/24 bp as against the 4-7 bp for ZFNs. Further, the optimum spacer length was affected to some extent by the TALE protein used for TALEN construction.

Thus, TALENs comprise N-terminal and C-terminal domains of TALE proteins separated by the TALE repeat unit domain. The N-terminal and C-terminal domains constitute the TALEN scaffold. Generally, the TALEN scaffolds are suitably truncated versions of the N-terminal and C-terminal domains present in natural TALE proteins, but some TALEN scaffolds have been derived from the natural domains by amino acid substitutions (Table 4.2). The TALEN scaffold contributes to non-specific DNA binding, and affects TALEN specificity, optimum spacer length and the efficiency of gene editing. The N-terminal region immediately upstream of the repeat domain contains two cryptic repeat modules, referred to as repeats 0 and −1, that are like TALE repeat units in structure but differ in sequence. The repeats 0 and -1 make contacts with the base T that is generally present at position 0 in the target sequences of natural TALE proteins from *Xanthomonas* spp. In contrast, the TALEs from *Ralstonia* require a G at this position (Sun and Zhao, 2013; Carroll, 2014; Bogdanove et al., 2018).

It was once thought that each new target site for a TALEN should start with a T, but experimental evidence shows that this is not necessary. The length of C-terminal domain seems to determine the necessity of a 5' T at the target site. TALENs with a short (31 amino acid residues) C-terminal domain show higher efficiency when their target sites have a 5' T rather than some other nucleotide. In contrast, TALENs having longer (63-117 amino acid residues) C-terminal domains work equally well at target sites beginning with any nucleotide. The binding sites for the two monomers of a TALEN pair should be located 12-20 bp apart, the range being 10 to 32 bp; the spacer length mainly depends on TALEN scaffold. Further, the two binding sites of the two TALEN monomers should be oriented opposite to each other so that their FokI nuclease domains are close enough to dimerize and cleave the target DNA (Fig. 4.2A, B). A binding site having T in position 0 and 18 more base pairs is known as 18-bp site, and the TALEN specific for this site has 17.5 TALE repeats, including the final 'half-repeat' (Sun and Zhao, 2013; Carroll, 2014).

Table 4.2. A summary of some relevant features of selected TALEN scaffolds (based on Sun and Zhao, 2013; Ma et al., 2013)

TALEN scaffold	N-terminal domain	C-terminal domain*	Optimum spacer	Remarks
TALEN-I	207 aa**	31 aa	10–16 bp	High DNA cleavage efficiency in yeast and human cells
TALEN-II	207 aa	63 aa	14–32 bp	Canonical C-terminal domain; high DNA cleavage efficiency in yeast and human cells
GoldyTALEN$	136 aa	63 aa	13-20 bp	In zebrafish, editing efficiency 100% at some loci

* TALENs with short (31 amino acid residues) C-terminal domain show higher efficiency when their target sites have a 5' T than when the targets have other nucleotides at this position. In contrast, TALENs having longer (63-117 amino acid residues) C-terminal domains work equally well at target sites beginning with any nucleotide.

** aa, amino acid residues.

$ The N-terminal domain contains nine amino acid substitutions, while the C-terminal domain has five different substitutions. The central repeat domain has 15 RVDs or 14.5 TALE repeats.

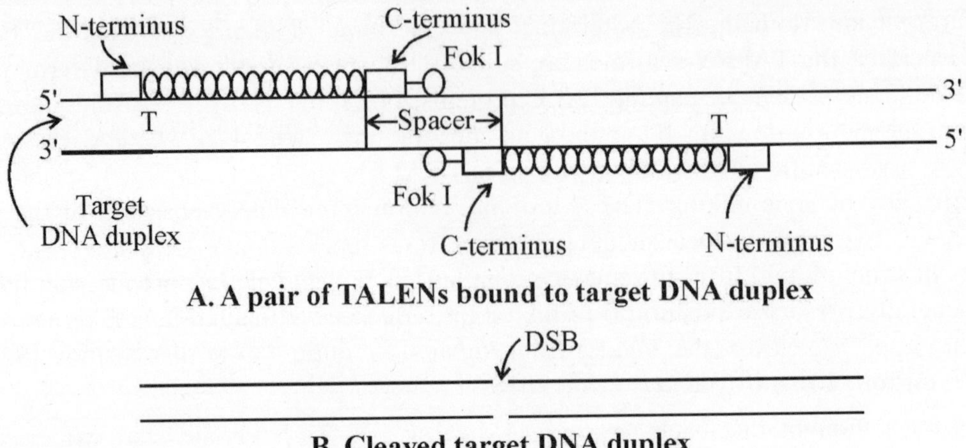

Fig. 4.2. Schematic representation of DNA cleavage by TALENs. **A.** A pair of TALEN monomers bound to target DNA. The target site comprises the base pairs contacted by the TALE repeat units. The DNA between the repeat units of the two TALEN monomers constitutes spacer DNA, which may vary from about 14-20 bp. **B.** Cleavage occurs within the spacer region. DSB, double-strand break (Modified from Feng et al., 2014).

In the central repeat domain, the number and the type of repeat units can be varied according to the target sequence and other requirements of the system. Generally, TALENs are designed for 15-20 bp target sequences, but 16-18 bp targets are the most common. In general, TALENs specific for longer than 20 bp targets tend to have reduced specificity. The number of TALE repeats included in a TALE protein depends on several factors, such as the size of targeted genome, the binding affinities of the RVDs used and the function to be performed by the TALEs. In general, shorter TALEs are more likely to have off-target sites than longer TALEs. In addition, a larger number of strong RVDs would have to be used in shorter TALEs than in longer ones. The TALENs induce cuts within the spacer region between the binding sites for the two TALEN monomers. TALENs containing wild type FokI nuclease domain can function as homodimers, *i.e.*, a single TALEN monomer can bind to two neighbouring genomic sites and induce DSBs. This can be avoided by using a pair of FokI mutants that can function only as heterodimers (Section 3.12.2). In this system, each monomer of a TALEN pair has a different member of a pair of FokI mutants, as a result of which this TALEN pair can function only as heterodimer (Sun and Zhao, 2013; Carroll, 2014; Guilinger et al., 2014).

TALEN efficiency is influenced by several factors, including the treated species, the tissue or the cell type used for treatment, the method of TALEN delivery and the molecular form (DNA, RNA or protein), in which it is delivered, architecture of the TALENs, the genomic target, and the method of detection of TALEN action. In case of rat single cell embryos, TALENs delivered as mRNA are more effective in gene editing than those delivered as DNA. In addition, the editing efficiency appears to be dose dependent. The efficiency of gene inactivation mediated by NHEJ can be increased by interfering with the NHEJ-mediated precise repair of DSBs. This has been done by expressing TALENs along with an exonuclease, like Exo1 or Trex2, which removes nucleotides from the two DNA ends at the DSBs and thereby prevents their precise repair by

NHEJ. The frequency of HDR is also affected by several factors. It appears that HDR is more efficient in some species than others, and is affected by many factors, including the donor DNA length, the position of sequence replacement in relation to the DSB site and the size of sequence to be replaced. Gene editing efficiency is affected to some extent by the composition of DNA at the targeted genomic site, the length of target sequence and the length of spacer sequence (Wright et al., 2014).

4.5. TALE NICKASES

An endonuclease that induces cuts in a single strand of a DNA duplex, *i.e.*, produces a nick in the target DNA, is called nickase. Nicks are repaired by NHEJ as well as HDR, and the frequencies of both NHEJ and HDR events are reduced as compared to those associated with DSB induction. But it is a matter of solace that reduction in the level of NHEJ is dramatic and the nicks are preferentially repaired by HDR provided a donor DNA is available. The FokI nuclease domain of TALENs must dimerise to be able to effect cleavage. Therefore, if the nuclease domain of one of the members of a TALEN pair is inactivated by mutation, the dimerised FokI will be able to induce only nicks at the target site. It is important that the two members of the TALEN pair contain separate FokI mutants that function only as heterodimers and they are unable to form homodimers so that there is no risk of DSB induction due to homodimer formation. TALE nickases show greatly improved HDR to NHEJ ratio and significantly reduced off-target editing (Ousterout and Gersbach, 2016). Nickases have been produced and used in ZFN (Section 3.7) and CRISPR-Cas9 SDN systems as well (Section 5.6). The TALEN nickases may be called TALENickases on the pattern of ZFNickases.

4.6. TALE ACTIVATORS (TALEa) AND TALE REPRESSORS (TALEr)

The TALE proteins can be fused with transcription activator (*e.g.*, VP16) and transcription repressor (*e.g.*, KRAB) domains to generate TALE-activators (TALEa) and TALE-repressors (TALEr), respectively; together, they are called TALE-transcription factors (TALE-TFs). They can also be fused with chromatin remodelling proteins like LSD1 to give rise to TALE-chromatin editors (TALE-CEs). The TALE-TFs and TALE-CEs offer the following advantages. They permit simultaneous delivery of several different TALE-effector fusions specific to many different loci. It is possible that TALE-TFs may be more effective in transcriptional activation than dCas9-TFs in view of the dynamic binding of TALE-TFs to the target loci. In addition, binding of TALE-TFs to the target DNA does not lead to unwinding of the target enhancers, which occurs in the case of dCas9-TFs. In the cases of TALEa and TALEr, the first 48 amino acids from the N-terminus of Hax3 TALE can be safely deleted without adversely affecting the level of transcription activation/repression. But the first 68 amino acids at the C-terminus of Hax3 TALE must be retained to ensure the highest level of activity since this region contains an element important for DNA binding by the TALE protein. The level of transcription activation by TALEa seems to be affected by some factors other than the TALE design; these factors include the effects of context on the DNA binding strengths of individual repeat units and the components of the complex transcription processes in mammals and other eukaryotes (Zhang et al., 2011).

TALE-TFs have been used to regulate transcription of targeted genes in a variety of organisms, such as mammalian cells in culture, *Drosophila*, plants, etc. A commonly used TALE-

TF for transcription activation is TALE-VP64 fusion protein that is used in at least four copies placed in tandem in the promoter region of the target gene for an effective activation. Another activation domain used for creation of TALEa is that from p65 subunit of the human protein complex NFκB. Programmable repressors have been created by fusing Kruppel-associated box (KRAB) and Sid4 (mSin interaction domain) repressors with TALE proteins. In addition, TALEs have been linked to enzymes involved in histone modification and DNA methylation to generate programmable epigenetic editors. In general, multiple TALE-activators need to be placed in the promoter region (within 700 bp upstream of the transcription start site) of the target gene for achieving robust transcription activation. In addition, TALE-TFs targeted to enhancers are also able to modulate transcription of the concerned genes. The TALE-TFs work well when they are targeted to genomic sites in genomic regions with either active or closed chromatin states (Thakore and Gersbach, 2016; Thakore et al., 2016).

In addition to TALE-TFs, the TALE proteins can be fused with many other types of effectors or functional domains to generate a variety of programmable chimeric effectors capable of performing diverse useful biological functions. For example, TALEs have been fused with fluorescent proteins to create programmable fluorescence tags for labelling of targeted genomic sites. Further, TALEs have been fused with the restriction enzyme PvuII, the meganuclease TevI catalytic domain as well as the Cas9 nuclease of CRISPR-Cas9 SDN to create chimeric TALENs that can function as monomers in gene editing. The TALE DNA binding domain and meganuclease fusion products. called megaTALEs, are highly specific and very promising for therapeutic applications (Bogdanove et al., 2018).

4.7. SELECTION OF TARGET SITE

Selection of the target genomic site is the first step in designing of a SDN. The chief criterion for this selection is the suitability of the genomic site for induction of the desired modification by SDN action. For example, for correction of a mutation in a gene by sequence replacement mediated by homology-directed repair, the target site should be close to the site of mutation. But when a frame-shift mutation is to be corrected by compensating for the deleted/inserted base pairs causing this mutation, the target site should be somewhere in the coding region upstream of the original mutation site. Further, in case gene inactivation by NHEJ is the objective of gene editing, the target site should be close to the transcription start site, *i.e.*, near the 5' end of mRNA, located within an exon or at an exon-intron junction. The target sequence should begin with a T unless a suitably modified version of TALE scaffold capable of working efficiently with other bases at this position is used. Further, the target sequence should have a good mix of bases. In general, the optimum spacer length depends mainly on the length and structure of the C-terminus domain of the TALE protein.

The second consideration is the selection of a unique genomic site that is the least likely to have off-target sites with high sequence homology. The selection of target sites and designing of the TALENs specific for these sites is viewed as the rate-limiting step in effective utilization of TALENs for gene editing. The selection of suitable genomic target sites is greatly facilitated by bioinformatics tools like ZiFiT Targeter, TALE-NT, E-TALEN, Mojo Hand and CHOPCHOP (Table 4.3). The ZiFiT Targeter (Zinc Finger and TALE Targeter; web server, http://ZiFiT.partners.org) facilitates high throughput design and synthesis of TALE arrays by the FLASH method. The CHOPCHOP online tool (https://chopchop.rc.fas.harvard.edu) is easy to use and it enables rapid selection of the optimal genomic sites for TALENs in a variety of organisms. It provides some additional features over the other existing tools. For example, it can accept pasted sequences, gene identifiers, and genomic regions as input data and takes comparatively much less time for target selection and rigorous prediction of off-target sites. It generates a graphical output that displays all the candidate target sites in the given gene/genomic region, and each site is color-coded according to the specificity-based quality score of the site. This visualization greatly facilitates the selection of optimal target site. This tool is also designed for selection of target sites for the CRISPR-Cas9 system. The results from CHOPCHOP can be downloaded along with a GenBank file containing the annotations of exons, intron and the target sites of the concerned gene (Montague et al., 2014).

Table 4.3. A list of some online tools for designing of TALENs (based on Bogdanove and Booher, 2016)

Tool	Available at	Remarks
CHOPCHOP	https://chopchop.rc.fas.harvard.edu	Predicts off-targets based on mismatches from the perfect target and provides their sequences, number of mismatches and locations. Output includes PCR primer sites to amplify the targeted region.
E-TALEN	http://www.e-talen.org	Predicts off-targets based on mismatches from the perfect target. Does not provide location of off-target sites
Mojo Hand	http://talendesign.org	Does not predict off-target sites
PROGNOS	http://baolab.bme.gatech.edu/Research/BioinformaticTools/prognos.html	Predicts off-targets based on modular position weight matrix; provides their sequence, score, location, etc. Allows primer design
SAPTA	http://baolab.bme.gatech.edu/Research/BioinformaticTools/TAL_targeter.html	User can specify a particular base for cleavage. Does not predict off-target sites
TALE-NT	http://tale-nt.cac.cornell.edu	Predicts off-targets based on modular position weight matrix; provides their sequence, score, location, etc.
TALEN-designer	http://www.talen-design.de/index_talen.html	Predicts off-targets based on modular position weight matrix; provides their sequence, score, location, etc. Also provides human, mouse, and rat genome browsers populated with pre-designed TALEN sites
TALENoffer	http://www.jstacs.de/ index.php/TALENoffer	It predicts off-target sites for user-provided TALEN designs based on local mixture model; provides their sequence, score, location, etc.
ZiFiT Targeter	http://ZiFiT.partners.org	Facilitates high throughput design and synthesis of TALE arrays by the FLASH method
CHOPCHOP	https://chopchop.rc.fas. harvard.edu	Easy to use; enables rapid selection of optimal genomic sites for TALENs and rigorous prediction of off-target sites in a variety of organisms; also allows selection of target sites for CRISPR-Cas9

4.8. SYNTHESIS OF NOVEL TALEs

The designing of TALENs with novel DNA binding specificities comprises assembling the TALE repeats with appropriate RVDs in the order specified by the target DNA base sequence. The repeat units with different RVDs are maintained in plasmids and used for assembly of novel TALEs/TALENs. One of the major limitations in wide use of TALENs and other TALE-effector fusions is the challenging synthesis of novel TALE proteins. This is mainly due to the repetitive nature of their DNA binding domain, which makes the creation of novel TALEs by PCR-based gene assembly or serial ligation of DNA segments quite a difficult task. Several different methods have been developed for the assembly of novel TALE repeat arrays, which are then cloned in TALE cloning framework vectors. The various methods of TALE repeat assembly can be classified into three broad groups, namely, (i) standard restriction enzyme and ligation-based cloning, (ii) "Golden Gate" cloning, and (iii) solid-phase assembly. Although commercial services are available for the synthesis of novel TALEs, they are costly for creation of large numbers of new TALEs (Joung and Sander, 2013).

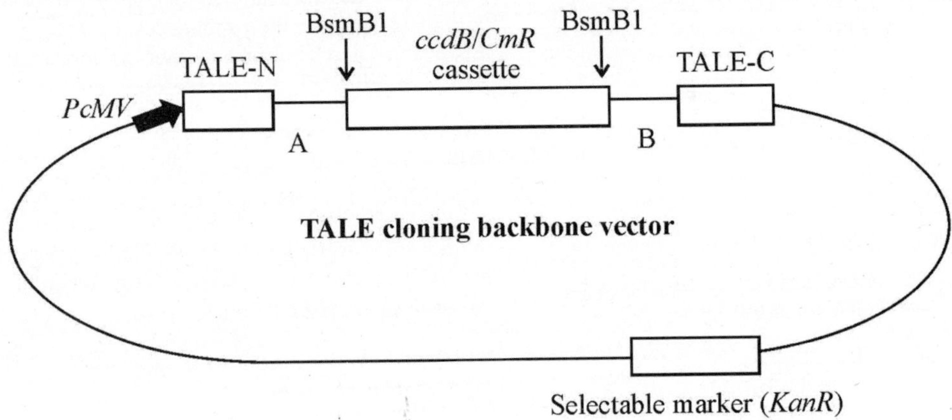

Fig. 4.3. A simplified representation of TALE cloning backbone vector. BsmB1, restriction enzyme (restriction site indicated by arrow); *ccdB*, a topoisomerase toxin from *E. coli; CmR*, chloramphenicol resistance; *KanR*, kanamycin resistance; *PcMV*, Cytomegalovirus promoter; TALE-C, TALE C-terminal domain, TALE-N, TALE-N terminal domain; A, segment with sequence GAAC; B, segment with sequence AGCA. Following digestion with BsmB1, the A and B sequences generate sticky ends appropriate for ligation of the TALE repeat array sequence into the vector.

A TALE cloning framework vector has sequences encoding the N-terminal domain including the two cryptic repeats that recognize the T at position 0 of the target site, the selectable marker (*ccdB*/*CmR*) expression cassette, and the C-terminal domain including the NLS and a selectable marker (Fig. 4.3). The vector is opened with the restriction enzyme BsmB1, and the sequence encoding TALE repeat array is integrated between the N-terminal and C-terminal domains. In addition, appropriate promoters and 3' regulatory sequences will also be included in the vector. A TALEN expression vector will have additional sequences for FokI, and a fluorescence marker like eGFP (enhanced green fluorescent protein; Fig. 4.4). In case a TALEr or

TALEa is to be synthesized, the sequence encoding FokI nuclease domain will be replaced with the sequence encoding the appropriate effector, *e.g*, activation or repression, domain.

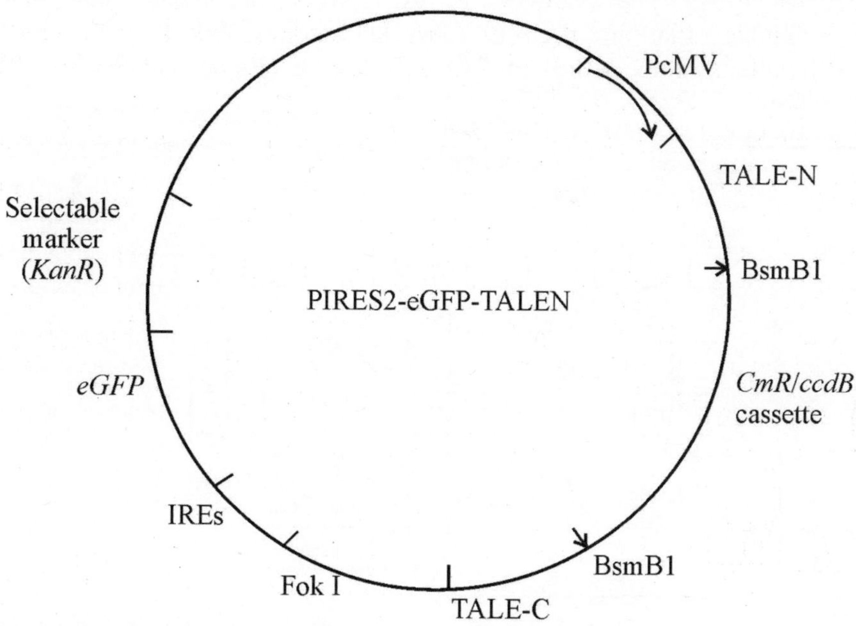

Fig. 4.4. A simplified schematic representation of a TALEN expression vector or destination vector for TALE assembly (PIRES2-eGFP-TALEN). *PcMV*, Cytomegalovirus promoter; BsmB1, restriction enzyme BsmB1 (restriction site indicated by arrow); *ccdB*, a toxin encoding gene; *CmR*, chloramphenicol resistance; FokI, FokI nuclease domain; *eGFP*, enhanced green fluorescence protein gene; *KanR*, kanamycin resistance; TALE-C, TALE C-terminal domain; TALE-N, TALE N-terminal domain. Note : a TALE-TF can be produced by replacing FokI sequence with transcription activation/repressor domain. TALE-TF, TALE-transcription factor.

4.8.1. Standard cloning-based methods

This group of methods uses standard cloning procedures for the construction of TALE repeat arrays from a plasmid archive. Each plasmid in the archive encodes either a single or multiple (2, 3 or 4) TALE repeats. The coding sequences for the repeats are separated from the plasmids by restriction digestion and the appropriate repeats are ligated in the desired order. The restriction digestion and ligation reactions are performed in a parallel hierarchical fashion. An example of such methods is the REAL (restriction enzyme and ligation) method. The first step in this method was to construct plasmids encoding different single TALE repeats; the DNA sequences encoding these repeats were produced by chemical DNA synthesis. In the next step, two DNA sequences encoding one TALE repeat each are obtained by restriction digestion of the relevant plasmids; these fragments are ligated to produce a sequence encoding one pair of TALE repeats. Then these pairs of repeats are ligated two at a time to produce sequences encoding four repeats each. In the next step, pairs of four repeat encoding sequences are ligated together to generate DNA fragments encoding eight TALE repeats each. Ligation of two eight repeat encoding sequences would give a

DNA fragment that would code for an array of 16 TALE repeats (Fig. 4.5). In the REAL-Fast version of the REAL method, the use of a large archive of plasmids containing sequences for all the possible combinations of pre-assembled multiple TALE repeats allows rapid assembly of novel TALE repeat arrays. The assembly of TALE arrays takes one week or longer. These methods use techniques that are used in most laboratories, but they are not suitable for construction of hundreds or thousands of TALEs/TALENs (Joung and Sander, 2013; Sun and Zhao, 2013).

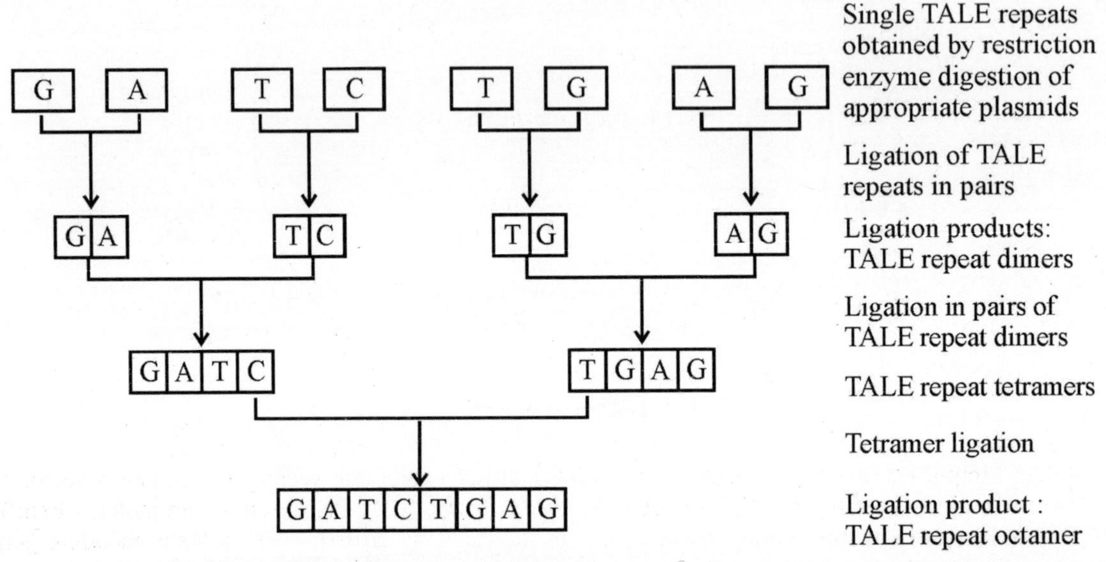

Fig. 4.5. A simplified representation of the assembly procedure REAL based on classical digestion with restriction enzymes followed by ligation. Assembly by TALE sequence with eight repeats is shown here. For longer TALEs, assembly will begin with the required number of individual TALE repeats and will proceed in a heirachical manner. The final ligation product is inserted into a framework plasmid for cloning and verification/validation. A, C, G and T, TALE repeats with RVDs (repeat variable diresidues) specific for the bases A, C, G and T, respectively (based on Sun and Zhao, 2013).

Zhang et al. (2011) developed a reliable strategy for high throughput synthesis of novel TALEs based on hierarchical DNA ligation. In this approach, base sequences of the DNA segments encoding TALE repeat units are optimized to reduce similarity among them without altering the amino acid sequences of repeat units encoded by them; this is achieved by exploiting the degeneracy of genetic code. To begin with, the sequences encoding the four RVDs (NG, HD, NI, NN) were optimized. Then the sequences encoding the amino acids Gly-Leu representing the junctions linking the repeat units were modified in a manner similar to that used for ligation of multiple pieces of DNA in the 'Golden Gate' cloning strategy (Section 4.8.2). The purpose of this sequence alteration was to permit assembly of the repeat monomers in the specified order. This was achieved by using different codons for the two amino acids to design four different 4-bp sticky-end ligation adapters; these adapters allow simultaneous ligation of four repeat monomers in the specified order. Three of these 4-mer tandem repeats can now be ligated in the given order to create a DNA sequence encoding a TALE protein with 12 TALE repeats. This sequence is then ligated into a vector with the sequences for TALE scaffold (the N- and C-terminal domains), a 'half repeat' specific for the 13th nucleotide of the target site, and the FokI nuclease. This 12.5-mer repeat TALE protein binds to a 13 bp target site. This strategy allows construction of TALE encoding DNA sequences at a fraction of the cost and in a considerably shorter period than that required for their creation via commercial DNA synthesis (Zhang et al., 2011).

4.8.2. 'Golden Gate' cloning-based TALEN assembly

Several TALEN assembly methods use the "Golden Gate" cloning technology that was first described in 2005. This cloning strategy permits simultaneous cleavage and ligation of 3 to 10 different DNA fragments in a single reaction mixture. As a result, a TALE array of 20 repeats can be assembled in merely two steps. The Golden Gate strategy uses Type IIS restriction enzymes to cut the ends of DNA fragments to be ligated together. These restriction enzymes cut the DNA duplex outside their recognition sites and generate 5' overhangs of 4 bp that are unrelated to the binding site sequences. Therefore, the end sequences of fragments to be ligated together are designed to contain recognition sites for the restriction enzyme and to generate the distinct specified overhang sequences that would ensure their ligation in the desired order and proper orientation (Fig. 4.6). Since cleavage by the restriction enzyme removes the recognition site from the two ends of the fragments, the ligation products will lack the recognition site; therefore, the cleavage and ligation reactions can be performed in a single step in the same tube (Joung and Sander, 2013; Sun and Zhao, 2013).

The `Golden Gate' cloning-based TALEN assembly was first described by Cermak et al. in 2011; all the plasmids used in this assembly system are available from Addgene (Cambridge, MA, USA) as "Golden Gate TALEN and TAL Effector Kit". The non-RVD amino acids in the TALE repeats have been modified to develop highly active TALENs called 'Platinum TALENs'; the materials for `Golden Gate' assembly of Platinum TALENs are also available with Addgene as "Platinum Gate TALEN Kit" (Sakuma and Yamamoto, 2016). In simple terms, the `Golden Gate' assembly is based on two parallel reactions for assembling a TALE comprising 16 repeats. In one reaction, 10 repeats are assembled in proper sequence into an intermediate plasmid, while in the second reaction 6 repeats are assembled. In the second-round reaction, the 10 and 6 TALE-repeat arrays are assembled into the final TALE backbone plasmid to yield a TALE array of 16 repeats plus the 'half repeat' fused with FokI or some other effector protein encoding sequence. The efficiency of TALE assembly can be increased by including the *lacZ* or the *ccdB* gene to enable blue/white screening or selection, respectively. The `Golden Gate' scheme permits creation of TALEN encoding plasmids within two weeks' time. But when preassembled TALE repeats as 3-mers and 4-mers are used, TALENs specific for 15 bp target sites can be synthesized in a single step within two days. The `Golden Gate' protocol is easy, available to public and well-suited for production of TALENs in academic laboratories. The procedure typically requires three to five days, and depends on production of intermediate vector products, which is time-consuming and cumbersome (Joung and Sander, 2013; Sun and Zhao, 2013).

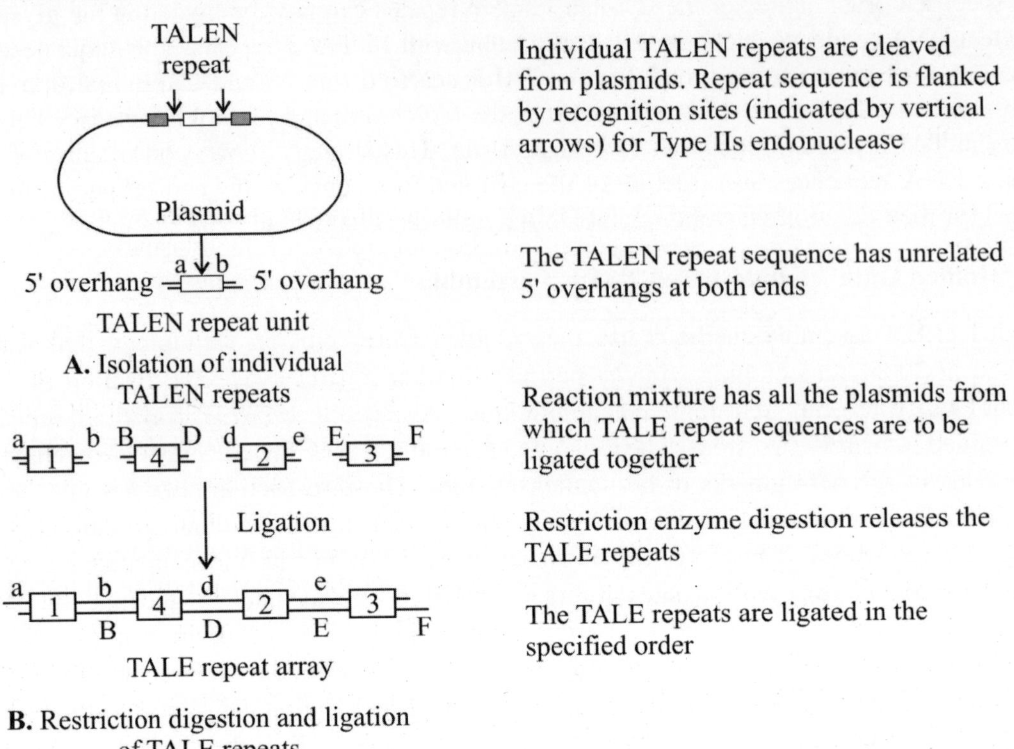

Fig. 4.6. A schematic representation of assembly of TALE repeats by `Golden Gate' cloning technology. **A.** The organization of a plasmid with a TALE repeat. The arrows indicate the site of cleavage, which generates 5' overhangs. The TALE repeats to be ligated together have complementary overhangs (indicated by A/a, B/b, etc.) at one end, while their other end has a different overhang that is complementary with the overhang of the next repeat in the array. **B.** Restriction digestion and ligation of TALE repeats to generate the desired array of TALE repeats.

A ligation-independent cloning (LIC) technique enables high-throughput assembly of TALE repeat arrays. The LIC is based on relatively longer nonpalindromic overhangs of 10–30 bp created by controlled use of the 3'-exonuclease activity of T4 DNA polymerase. The overhang sequences are so designed that they enable a highly specific annealing of the different TALE repeats in the proper orientation and the desired sequence to yield the specified TALE repeat array. Since the long overlaps of the neighbouring fragments remain annealed during transformation, there is no need for ligation; the annealing products are introduced into *E. coli* where ligation is carried out by the bacterial ligase. LIC has high fidelity and allows rapid assembly of TALE repeats; TALE arrays of 18 repeats can be assembled in two days by using TALE repeat 2-mers as the assembly blocks. But the use of 5-mers as building blocks allows automated assembly of more than 600 TALE genes encoding TALEs with 15.5 repeats in a single day (Sun and Zhao, 2013).

4.8.3. Solid-phase assembly methods

In these methods, assembly of the TALE repeat arrays is carried out on a solid phase, *e.g.*, a magnetic bead, which greatly facilitates the washing step after each ligation cycle, eliminates the need for gel electrophoresis, isolation, purification, and/or analysis of intermediate plasmids, and enables industrial scale TALEN synthesis. Reyon et al. (2012) described a speedy, cost-effective and automatable large-scale assembly of TALE repeat arrays with novel DNA binding specificities and named it as fast ligation-based automatable solid-phase high-throughput (FLASH) assembly method. This strategy is based on an archive of 376 plasmids, each of which contains a DNA insert encoding a different TALE repeat array comprising one, two, three or four repeat units designated as 1-mers, 2-mers, 3-mers and 4-mers, respectively (Fig. 4.7). There are four 1-mers representing TALE repeats with the NI, NN, HD, or NG RVDs. Similarly, the 2-mers, 3-mers and the 4-mers contain two, three and four TALE repeats representing all possible combinations of the four RVDs. In simple words, the first step in FLASH method is to ligate the desired monomer, which is biotinylated at its 5' end, to a 4-mer with the desired combination of RVDs and then attach the ligation product to a streptavidin-coated magnetic bead. This step has been modified by using streptavidin-coated magnetic beads, onto which a biotinylated DNA duplex is immobilized. This DNA fragment is an adaptor having a restriction endonuclease site at one end to permit separation of the sequence encoding the desired TALE repeat arrays after they are assembled onto the beads. The first preassembled TALE repeat array is added to the free end of the DNA adapter (Sun and Zhao, 2013).

The first ligation step is followed by sequential ligation of as many 4-mers as desired; in the end, the desired 1-mer, 2-mer, 3-mer or 4-mer is ligated to complete the desired number of TALE repeats in the TALE array. Thus, this procedure permits the assembly of TALE repeat arrays for the specified target site with any number of base-pairs. The assembled full-length TALE array is then released form the magnetic bead by restriction digestion, purified and inserted into a suitable vector for propagation/expression as TALENs. The use of magnetic beads allows the FLASH procedure comprising restriction digestion, ligation and purification steps to be automated with the help of a liquid-handling platform. The FLASH procedure has been optimized for a 96-well format for a robotic liquid-handling workstation; this enables assembly of 96 TALE arrays with different specificities in less than one day. But medium throughput manual assembly using multi-channel pipets would require one or two days for complete TALE assembly. In either case, the TALEs are cloned in expression vectors and sequence-verified TALENs are available within one week at an overall cost of ~100 US dollars. TALEN pairs were designed by FLASH method to target 96 endogenous genes in human cells, and 84 of them efficiently edited the targeted genes (Reyon et al., 2012).

The iterative capped assembly (ICA) procedure does not use an archive of plasmids with pre-assembled TALE repeats, and constructs the full-length TALE repeat arrays from the four repeat monomers that are suitably capped. The use of capping oligonucleotides prevents self-ligation of monomers and eliminates incomplete ligation, which ensure generation of correct full-length arrays of TALE repeats. The ICA can be automated to efficiently assemble TALE arrays comprising up to 21 repeats and ligate them into an expression plasmid in less than 3 h (Sun and Zhao, 2013).

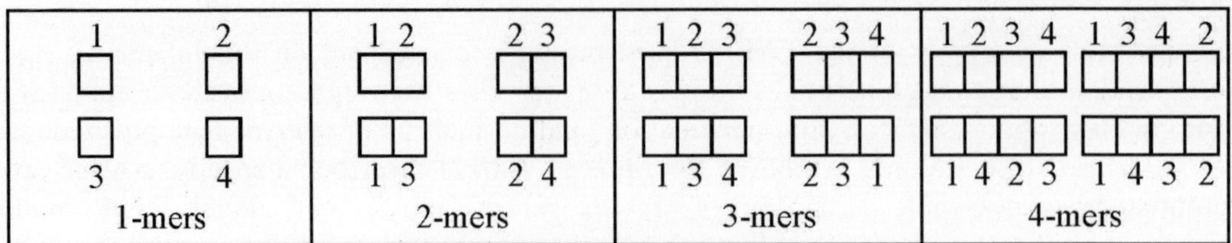

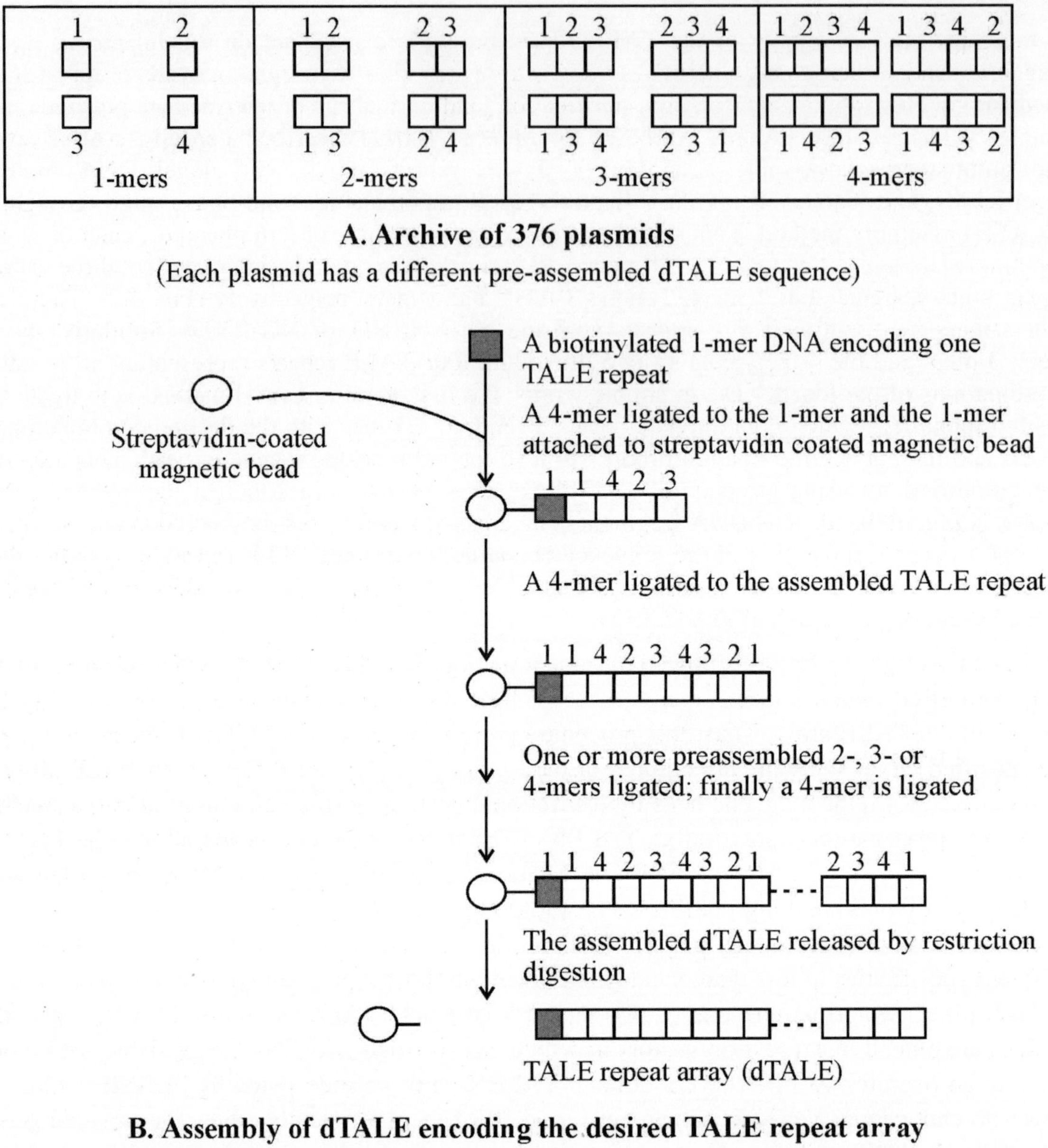

A. Archive of 376 plasmids

(Each plasmid has a different pre-assembled dTALE sequence)

A biotinylated 1-mer DNA encoding one TALE repeat

Streptavidin-coated magnetic bead

A 4-mer ligated to the 1-mer and the 1-mer attached to streptavidin coated magnetic bead

1 1 4 2 3

A 4-mer ligated to the assembled TALE repeat

1 1 4 2 3 4 3 2 1

One or more preassembled 2-, 3- or 4-mers ligated; finally a 4-mer is ligated

1 1 4 2 3 4 3 2 1 2 3 4 1

The assembled dTALE released by restriction digestion

TALE repeat array (dTALE)

B. Assembly of dTALE encoding the desired TALE repeat array

Fig. 4.7. A schematic representation of FLASH (fast-ligation-based automatable solid-phase high throughput) method of assembly of TALE repeat arrays. **A.** A library of plasmids with 1-mer, 2-mer, 3-mer and 4-mer TALE repeats. **B.** Assembly of a TALE repeat arrays using the 1-mer, 2-mer, 3-mer and 4-mer TALE repeat arrays. The assembly proceeds on streptavidin-coated magnetic beads. The assembled TALE arrays are separated from magnetic beads by restriction digestion (based on Reyon et al., 2012).

4.8.4. Simple TALE assembly reaction (STAR)

The simple TALE assembly reaction (STAR) enables individual laboratories to generate multiple TALEs in an effortless manner. STAR uses the labour- and cost-effective isothermal 'Gibson assembly' for assembling TALE arrays. The 'Gibson assembly' is a handy, fast and readily scalable procedure for assembling in the defined order multiple fragments having at least 16 bp overlapping ends. It employs an enzyme mix having T5 exonuclease, Taq DNA polymerase and Taq DNA ligase. T5 exonuclease digests DNA strands from 5' to 3' direction and generates 3' overhangs. The complementary overhangs anneal, the gaps are filled by DNA polymerase and the remaining nicks are sealed by Taq DNA ligase (Fig. 4.8). The 'Gibson assembly' reactions are performed in a single tube at 50°C for one hour (Gogolok et al., 2016).

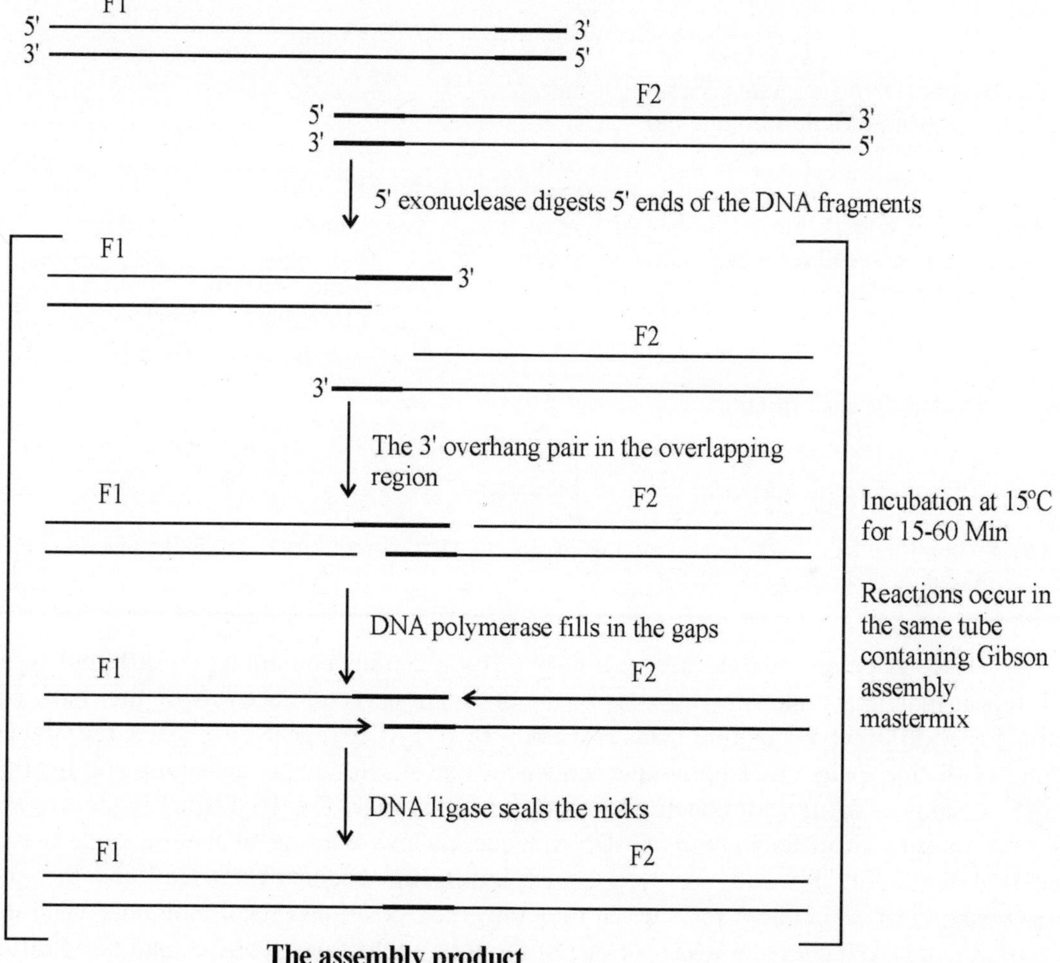

Fig. 4.8. A simplified representation of Gibson assembly reaction. The DNA fragments F1 and F2 have overlaps of atleast 16 bp; the overlapping region is indicated by dark lines. The Gibson assembly mastermix contains, in addition to the other reagents, the enzymes T5 exonuclease, Taq DNA polymerase and Taq ligase.

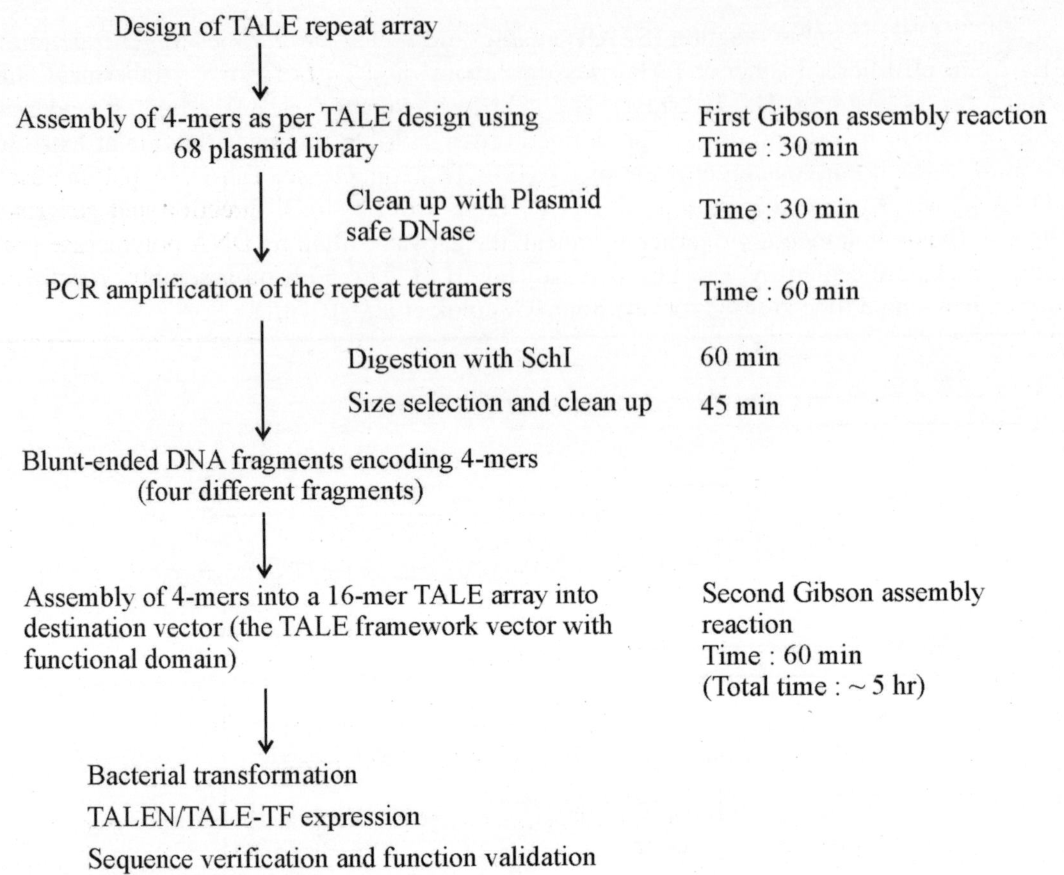

Fig. 4.9. A simplified schematic representation of the STAR (simple TALE assembly reaction) method (based on Gogolok et al., 2016).

The STAR assembly method uses a library of 68 plasmids containing 64 different types of TALE repeat monomers and four versions of the destination vector each having the 'half repeat' specific for a different nucleotide. The two ends of a TALE repeat have been redesigned to generate 16 distinct pairs of 21 bp unique sequences that would enable annealing and ligation of these 16 versions of the repeat monomers in the specified order. The 16 distinct sequences at one end of the repeat unit differ only at the DNA sequence level and all of them encode the same amino acid sequence; this was achieved by exploiting the codon degeneracy and the known natural variation in amino acid residues at the concerned positions. Each monomer version has four types, each type having the RVD for one of the four nucleotides. Thus, a total of 64 different types of repeat monomers were created and are maintained in the archive. The STAR method carries out the assembly of TALE repeat arrays in two steps as follows: (i) the assembly of TALE repeat 4-mers from the monomers, and (ii) the assembly of four 4-mers to obtain an array of 16 TALE repeats (Fig. 4.9). The four 4-mers are assembled into the TALE backbone vector having the 'half repeat' to yield a 16.5 TALE repeat array that would recognize 17 bp specified target. This method enables the assembly of TALE arrays in about 8 hr and sequence-verified TALEs are obtained within three days. The 16 sets of the two end sequences and the details of STAR procedure are available in Gogolok et al. (2016).

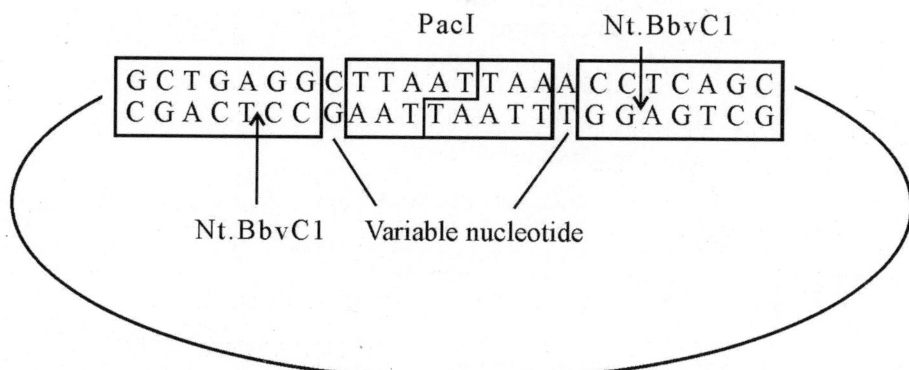

Fig. 4.10. A simplified schematic represention of USER (uracil-specific excision reagent) compatible vector, only the USER casette is shown.

The 'Golden Gate' cloning-based TALE assembly typically requires 3-5 days; it generates intermediate products, which necessitates the use of time-consuming and cumbersome procedures. Methods like FLASH and ICA use a solid-phase for the assembly of TALEs. These methods are fast and suitable for automated high throughput assembly of TALEs. However, considerable technical expertise, substantial investment in equipment and large archives of plasmids are required, particularly for large scale TALE assembly. The TALE assembly methods like ligation independent cloning, FairyTALE (an improved 'Golden Gate' method that assembles TALE arrays in a single reaction), and reTALE methods are free from some of these limitations, but they require considerable initial investment and ideally an automated liquid handling system that is technically and financially demanding. The STAR procedure allows individual laboratories to effortlessly assemble multiple sequence-verified TALENs or TALE-TFs within 3 days (Gogolok et al., 2016).

Reagent kits for some of the TALE assembly methods are available to researchers in academic institutions from Addgene (http://www.addgene.org/ TALEN/), which is a non-profit plasmid distribution service. The website http://www.TALengineering.org provides links to all kinds of information about TALE technology, including protocols, reagents, software, etc. Since different TALE architectures differ in their properties, the researchers should carefully consider the available information on efficiencies and specificities of TALENs based on various scaffolds before selection of an assembly protocol (Joung and Sander, 2013).

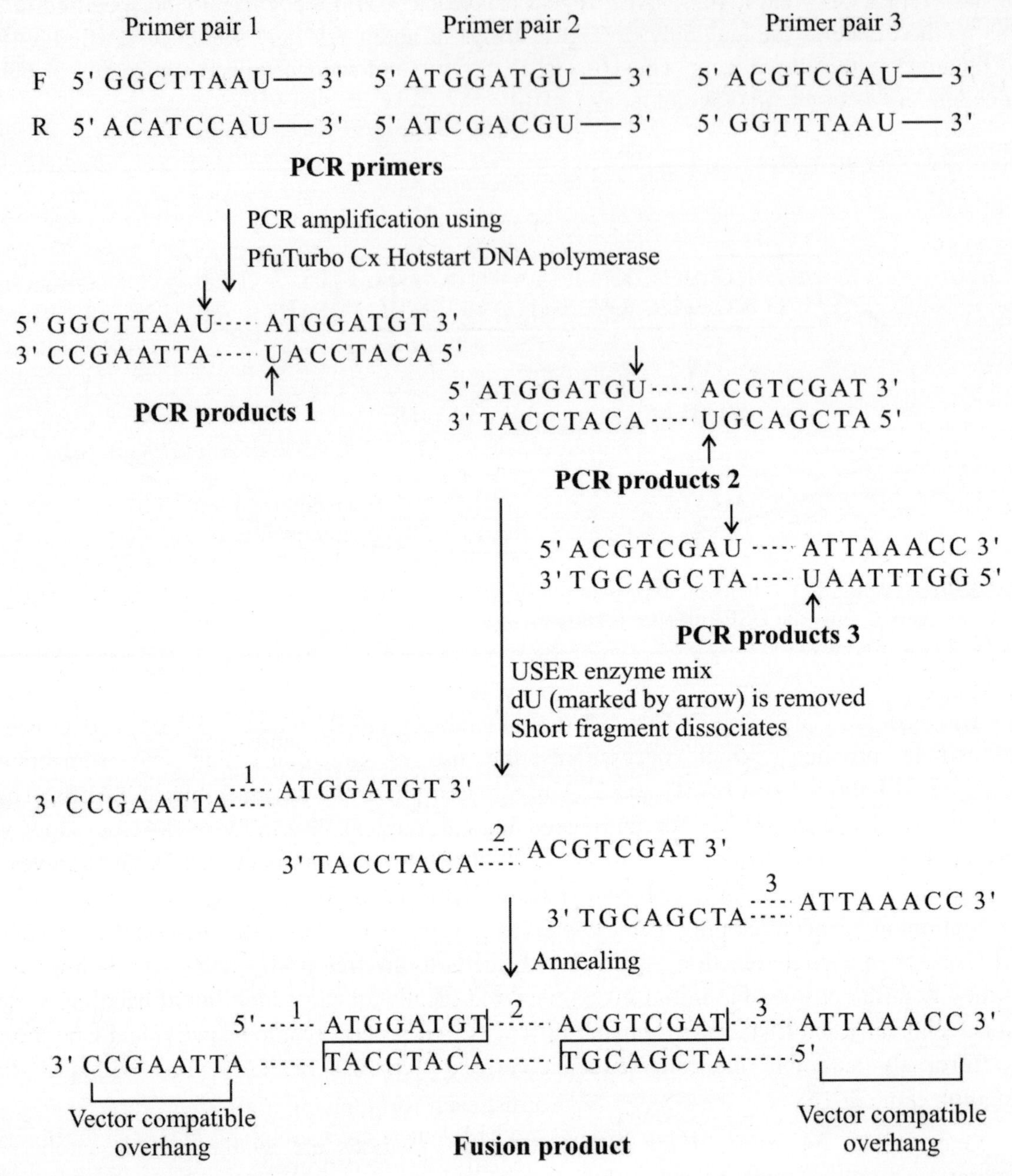

Fig. 4.11. USER fusion of PCR amplification products.

4.8.5. ULtiMATE system for TALE construction

The ULtiMATE (USER-based ligation-mediated assembly of TAL effector) system for TALE construction is a method designed for rapid assembly of TALEs with the desired DNA binding specificities. It employs the fusion technique used for USER (uracil-specific excision reagent) cloning of PCR amplification products. In simple terms, the USER cloning employs a USER-compatible vector that has a USER cassette (Fig. 4.10). The USER cassette has PacI recognition site (5'TTAATTAA3') in the middle, which is flanked by recognition sites for Nt.BbCI (5'CCTCAGC3'). The binding sites for the two enzymes are separated by a variable nucleotide (nt) that distinguishes the sequences flanking the PacI cleavage site. The variable nucleotide ensures that the PCR product is ligated in the vector in the proper orientation. For integration of the PCR amplicon, the vector is digested with PacI and Nt.BbCI; this generates 8 nt 3' overhangs at the two ends that differ for the variable nucleotide. The PCR product is amplified using a pair of specially designed primers that have, in addition to the end sequences of the template DNA, the same 8 nt sequences at their 5' termini as the two 3' overhangs of the vector. However, a deoxyuridine (dU) is included at the 3' ends of the additional sequences of the primers. Therefore, a special DNA polymerase, such as PfuTurbo Cx Hotstart DNA polymerase that can incorporate an A opposite dU present in the primers has to be used for the PCR amplification (Fig. 4.11) (Geu-Flores et al., 2007; Nour-Eldin, 2010).

The opened vector (Fig. 4.12A) and the PCR amplicon are added to the USER enzyme mix in a reaction tube (Fig. 4.12B). The USER enzyme mix contains the enzymes uracil DNA glycosidase and DNA glycosylase-lyase Endo VIII. These enzymes together act on and remove the dU residues present near the two ends of the PCR products causing a gap in the concerned strands. The short fragments beyond these gaps dissociate from the amplicon creating 8 nt 3' overhangs that are complementary to the 3' overhangs of the vector. The overhangs of PCR amplicon anneal with those of the opened vector generating a hybridization product, the recombinant DNA, that is stable enough to be used directly (without ligation) for transforming *E. coli*. The USER technique can be used for fusing two or more PCR products together; this is called USER fusion. The products of USER fusion can be cloned in the USER compatible vector. The USER strategy allows the creation of custom - made 3' overhangs at the ends of the PCR amplicons. The 3' overhangs of the PCR products are so designed that ends to be annealed together have complementary sequences that differ from the other overhangs (Fig. 4.12B). It may be pointed out that the two ends of the fusion product must have overhangs that are complementary to the overhangs of the USER-compatible vector so that it can be integrated into the vector for cloning (Geu-Flores et al., 2007; Nour - Eldin, 2010).

For the ULtiMATE system, DNA sequences for four distinct types of TALE repeat units, denoted as W-, X-, Y-, and Z-type, were designed. The DNA sequences encoding the non-RVD amino acid residues of these repeat units differ from each other. When these four TALE repeat types are combined with four different RVDs, one RVD specific for each nucleotide, a total of 16 TALE repeat units were obtained, which were produced by DNA synthesis. The TALE repeats were preassembled into 4-mers in either **WXYZ** or **XYZW** format. Since each of the W, X, Y and Z type of TALE repeats will have four forms, each form having a different RVD, a total of 512 (4x4x4x4x2) 4-mers will be obtained. These 512 4-mers are maintained in a plasmid library, and the desired 4-mers are amplified using specially designed primers. The PCR products are purified and used for USER fusion for the assembly of TALE arrays in a single reaction tube; the fusion product is used for transformation. The use of 4-mers greatly simplifies and speeds up the assembly of novel TALE arrays. It takes merely 6 hrs to assemble ~24 different TALE arrays in a single 96-well plate. In fact, the number of different TALE arrays that can be assembled depends on the capacity for PCR reactions. The ULtiMATE system is amenable to high throughput synthesis of customized TALE repeat arrays (Yang, J. et al., 2016).

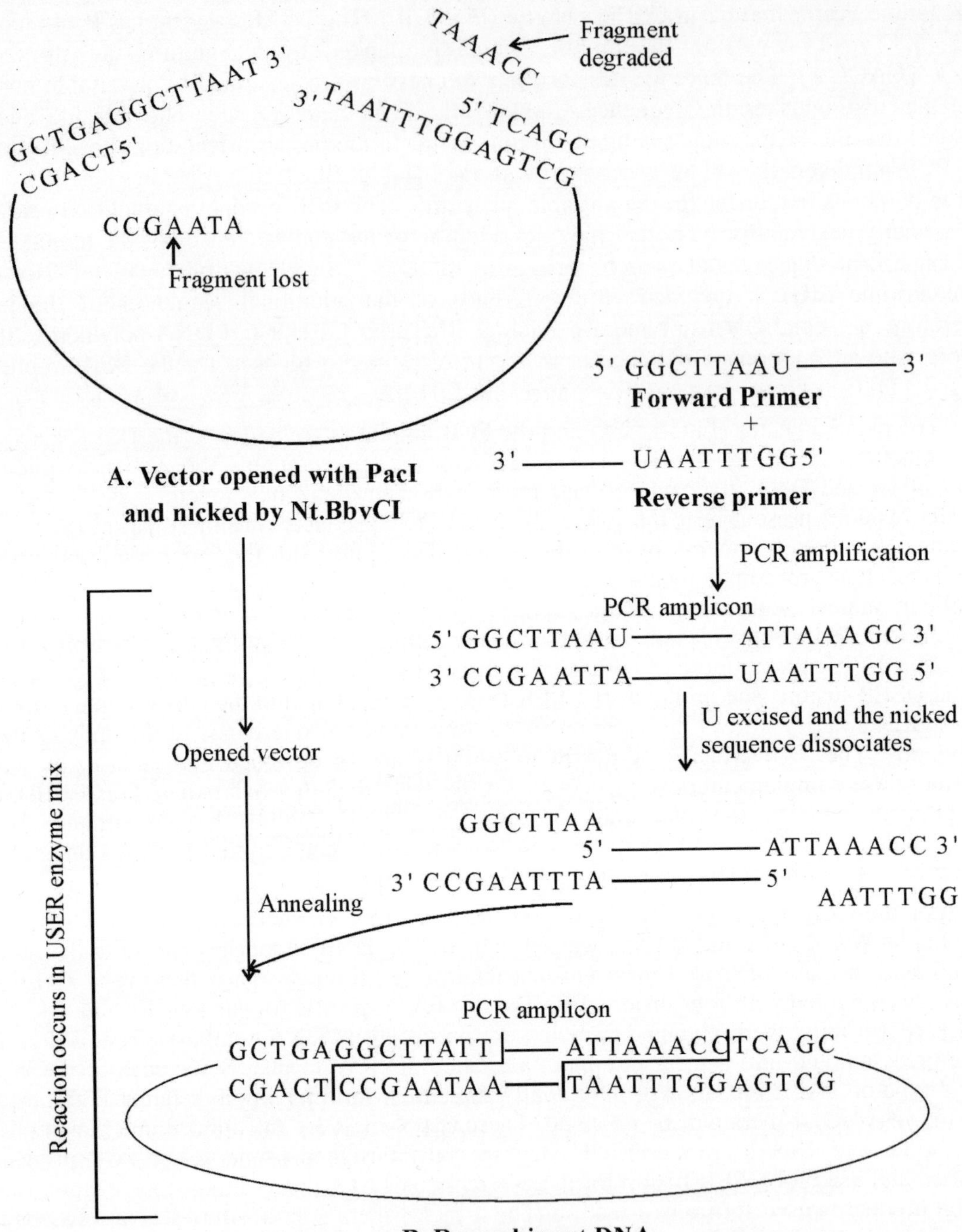

Fig. 4.12. A simplified representation of the USER cloning of PCR products. **A.** USER-compatible vector opened with PacI and Nt.BbvCI-. **B.** Integration of the PCR product into the opened vector catalyzed by the USER enzyme mix.

The 4-mers are assembled by the ULtiMATE system, but the assembly of the tetramers into TALE arrays can be done by the 'Golden Gate' method as well. The sequences of the four types of TALE repeats, and the primers used for USER-based assembly of the 4-mers and those used for PCR amplification of the 4-mers for assembly of TALE arrays (using either 'Golden Gate' or the ULtiMATE method) are available in Yang, J. et al. (2016).

4.9. DELIVERY OF TALENs

As a rule, a pair of strategically designed TALEN monomers are used together since they have the FokI nuclease domain, which must dimerize to be able to cleave DNA. Therefore, one pair of TALENs must be delivered together into cells for each targeted genomic site. The TALENs can be delivered either as DNA through a plasmid or viral vector, or as mRNA or protein. The large size of TALEN monomers and the presence of several nearly identical TALE repeats makes their delivery into mammalian cells using viral vectors a challenging task. Adenovirus vectors are suitable for the delivery of both the members of a TALEN pair, but they are highly immunogenic, and inefficient in transfection of some cell types. Helper-dependent adenoviral (HD-Ad) vectors have been derived by deleting all the coding viral sequence to minimize the immune response. The HD-Ad vectors have large enough cargo capacity to accommodate both the TALEN monomers as well as donor DNA with 4 kb homology arms (Yang et al., 2013; Richter et al., 2016; Xia et al., 2019).

The lentiviral vectors seem to be unsuitable for delivery of TALENs since the highly repetitive array of TALE repeats tends to undergo rearrangements. This difficulty can be overcome by using recoded TALEN gene constructs, called re-TALEs, which can be efficiently expressed by lentiviral vectors in mammalian cells. The re-TALE encoding sequences are first optimized for human codon usage and low folding energy at the 5' ends of mRNAs. The sequences are then refined to eliminate both direct and inverted repeats of more than 11 bp. These sequence optimizations facilitate TALE assembly, enhance expression of re-TALEs and enable their safe delivery by lentiviral vectors. However, this method may require re-optimization and synthesis of every new TALE encoding gene (Yang et al., 2013; Richter et al., 2016; Xia et al., 2019).

TALEN delivery in form of DNA necessitates decision on relevant issues like promoter type (constitutive or tissue-specific) and strength, codon bias, TALEN architecture, etc. This mode of delivery would lead to continued production of TALENs over time, which is likely to increase the frequency of off-target editing. In addition, it is associated with the risk of integration of foreign DNA into the cell genome, which would enhance the regulatory requirements. In case of crop plants, the foreign DNA can be removed by breeding, but this would require additional effort, time and financial resources. In other species, TALENs may be delivered as mRNA or protein to avoid the presence of residual foreign DNA in the final product, which would reduce the regulatory burden. TALEN delivery as mRNA is feasible in at least some species and it limits the period of TALEN action, which reduces the risk of off-target editing. But mRNA is far more vulnerable to degradation than DNA and production of high purity mRNA with 5' cap and poly(A) tail in the required quantities is much more cumbersome than that of DNA. The protein form of TALENs can permeate into cells directly or a cell-penetrating peptide, *e.g.*, R9 (polyarginine) or TAT (from HIV) motif, may be used to facilitate their entry. The TALEN proteins are covalently linked to the cell-penetrating peptides so that their cell-penetrating peptide component enables their entry into cells. This method of delivery supports efficient gene editing in cells (Wright et al., 2014; Richter et al., 2016).

The method of TALEN delivery depends mainly on the species and the tissue type involved. In general, a wider range of delivery methods is available for animal systems and the efficiency of transfection is higher than that in plants. The delivery methods for animal cells include nucleofection, lipofection and microinjection of either RNA or DNA, particle-based methods, and virus-mediated transfection. In case of plants, efficient methods of TALEN delivery are by particle gun, electroporation and *Agrobacterium tumefaciens* coculture. Generally, electroporation is used for TALENs in DNA form; it can be used in only such species that are capable of plantlet regeneration from protoplasts. Particle gun bombardment and *A. tumefaciens* coculture do not require protoplasts, are applicable to a wider range of species and typically use TALENs in DNA form. In case of *Agrobacterium* coculture, the gene constructs for the two TALEN monomers are typically integrated into the T-DNA of a single Ti plasmid; the two genes may be driven by a single promoter or by two distinct promoters. When a single promoter is used, the two genes are linked by the sequence encoding FMDV 2A ribosomal skipping peptide. In addition, modified Gemini virus system can be used to deliver TALENs into plants; this delivery method may greatly enhance gene editing frequency in comparison to that by the *Agrobacterium* coculture method (Wright et al., 2014).

4.10. OFF-TARGET CLEAVAGE

An off-target site may be considered as any genomic site other than the targeted site that has enough sequence similarity with the target site so that it can be recognized and cleaved by the TALENs specific for the target site. The methods for detection of off-target activities of SDNs have been considered in some detail in Sections 3.12.1 and 5.9.1. TALENs appear to be more specific than ZFNs as even a single mismatch between the target sequence of 16 bp and the TALEN RVD modules can significantly reduce binding of the TALENs. This might be the consequence of tight side-by-side location of the consecutive RVD modules of the TALEN repeats so that mismatch at one base may disturb the binding of neighbouring modules to their targets. But there is some evidence that TALENs can generate significant off-target cleavage in human cells even at sites that differ from the target site for eight or more mutations. Further, since binding of the TALE protein to the target DNA begins at its N-terminus, the mismatches near the 5' end of the target sequence are more disturbing than those near the 3' end. In general, the longer is the repeat domain of a TALEN, the more mismatches it can tolerate. But as the number of TALE repeats increases, the number of potential off-target sites in the genome tends to decrease. A TALE monomer seems to tolerate fewer mismatches than a pair of TALENs most likely because dimerization of the FokI nuclease domains of the two TALEN monomers stabilizes the TALEN-DNA complex (Carroll, 2014; Richter et al., 2016).

It has been suggested that specificity can be improved and off-target activity reduced by minimizing the excess DNA-binding energy of TALENs. This can be achieved by using mutant C-terminal and N-terminal domains with fewer cationic amino acid residues, increasing the number of TALE repeats, and using lower-affinity RVDs, *e.g.*, NK in place of higher affinity RVDs like NN for G. The N-terminal and C-terminal domains flanking the TALE repeat domain contain cationic amino acid residues that contribute to the energy for nonspecific DNA-binding. A reduction in the number of cationic residues in these domains increases (by 24- to >120-fold) on-target specificity and reduces off-target cleavage. An engineered TALEN variant Q3 having only 3 (in place of the normal 10) cationic residues in the C-terminal domain shows, on an average, 10-fold reduction in off-target cleavage in human cells without any adverse effect on on-target cleavage frequency. In addition, higher concentration of TALENs in the cells leads to an increased off-target cleavage *in vivo* (Guilinger et al. 2014).

In addition, the use of a small proportion of nonconventional RVDs, *i.e.*, RVDs other than NI, HD, NN and NG, in construction of TALEs increases specificity and reduces off-target activity. Juillerat et al. (2015) evaluated the 400 (20 x 20) possible combinations of the amino acid residues at the RVD positions 12 and 13 and identified a panel of nonconventional RVDs with novel specificity features. The use of a small number of carefully selected nonconventional RVDs from this panel leads to significant increase in specificity. It was suggested that this strategy in combination with other approaches like reduced number of TALE repeats and suitably engineered C-terminal TALE domain may enable designing of highly specific TALEs that may not tolerate even a single mismatch with their target sequences.

Laboratory strategies for directed protein evolution can be used to evolve TALE proteins with enhanced specificity. The phage-assisted continuous evolution (PACE) is an improved protein evolution strategy that permits continuous evolution of proteins at a rate that is ~100-times faster than those of the conventional methods. PACE has been adapted for evolution of DNA-binding proteins; this modification is called DB-PACE. DB-PACE was used together with an *in vitro* specificity assay procedure to select TALENs with improved specificity. The *in vitro* specificity profiling assays the ability of test TALENs to cleave more than 10^{12} DNA fragments with sequences related to the target sequences of the TALENs being tested. This procedure was able to select evolved TALENs with enhanced specificity and reduced off-target activity (Hubbard et al., 2015).

As in the case of ZFNs and CRISPR-Cas9 systems, the off-target cleavage induced by TALENs is likely to be the result of evolutionary requirements for the TALE proteins to be able to tolerate one to few mismatches in their binding sites located in the promoters of host genes activated by them. Since the TALE target sequences in the relevant host plant genes would be continuously evolving to avoid bacterial invasion, the pathogens will be benefited if their TALEs were able to tolerate some mismatches in their target sequences. Therefore, the TALENs derived from bacterial TALEs would also inherit this feature of TALE proteins (Chandrasegaran and Carroll, 2016). This suggestion gains support from the demonstration of improvement in specificity of TALE proteins refined for specificity by directed protein evolution.

4.11. DETECTION OF MUTANTS

The detection of mutants induced by TALENs can be based on the change at the phenotypic as well as genotypic levels. At the genotypic level, the mutants may be detected by cleavage of PCR products by a mismatch sensitive endonuclease, HRMA (high-resolution melt analysis) or HMA (heteroduplex mobility assay). In addition, the detection may be based on RFLP (restriction fragment length polymorphism) analysis of the test individuals. For endonuclease cleavage assay, the targeted sites of the parental or nonmutant line and a putative mutant individual/line are amplified, mixed together, denatured and allowed to anneal. The annealing products are then treated with Cel1 (Surveyor) or T7 endonuclease I; these enzymes recognize mismatches in DNA duplexes and cleave at the mismatch site. In case the test individual was not a mutant, only homoduplexes will be formed, which will not be cleaved; as a result, only a single band will be detectable by gel electrophoresis. But if it were a mutant, heteroduplexes will be formed, which will be cleaved by the endonuclease. The cleavage products can be detected by electrophoresis: there will be three distinct bands in the gel (Wright et al., 2014).

The heteroduplexes formed by PCR products can also be detected by HRMA and HMA. The HRMA detects heteroduplexes by the reduction in their melting temperatures as compared to those of the homoduplexes. Similarly, HMA employs PAGE (polyacrylamide gel electrophoresis) or microchip electrophoresis for detection of heteroduplexes since they show reduced gel mobility as compared to the homoduplexes. The degree of reduction in gel mobility is proportional to the extent of unpaired regions of the heteroduplexes (Wright et al., 2014). In case the DNA from T_0 individuals or their clonal progeny are used for PCR amplification, the PCR products from each individual will have both mutant and nonmutant homoduplexes since the vast majority of mutants would be heterozygous for the induced mutations. Therefore, the PCR product of each test individual can be denatured and annealed to produce heteroduplexes in the mutant heterozygotes, which are then subjected to the above assays.

In case the targeted genomic site includes recognition sequence for a restriction enzyme, RFLP analysis can be used for mutation detection. In case of mutant individuals, the recognition site for the restriction enzyme will be abolished leading to the presence of a larger fragment in their RFLP profile. In contrast, the fragment size will be smaller in the nonmutant wild type individuals (Wright et al., 2014). Alternatively, the genomic DNAs extracted from the test individuals may be digested with the concerned restriction enzyme prior to amplification of the genomic region containing the TALEN target site. In case of nonmutant individuals, the target sequence will have intact recognition site for the restriction enzyme, as a result of which it will be cleaved. Therefore, no PCR product will be generated in case of nonmutant individuals. But in case of mutant individuals, the recognition sequence will be modified, and the DNA will not be cleaved by the restriction enzyme. As a result, a PCR product will be obtained in such individuals. Finally, the PCR products of test individuals and the nonmutant/parental individuals may be subjected to DNA sequence analyses; this will permit not only the detection of mutant individuals but will also enable characterisation of the mutational changes in them.

4.12. APPLICATIONS OF TALENs

TALENs have been successfully used for gene editing in many species, including *C. elegans*, *Drosophila*, zebrafish, mice, rats, pigs, *Arabidopsis thaliana*, barley, tomato, tobacco, maize, rice, etc. They can be used for induction of every genetic modification that can be achieved by ZFNs and CRISPR-Cas SDN systems. They are useful tools for gene inactivation due to NHEJ and for specified sequence changes in the targeted genomic regions by HDR. Gene inactivation has many applications, including gene function analyses and creation of useful modification in trait phenotypes. For example, gene inactivation has been used to generate resistance to the bacterial pathogen *Xanthomonas oryzae* pv. *oryzae* in rice, resistance to powdery mildew in wheat, and to improve oil quality in soybean. On the other hand, HDR-mediated sequence replacement can be used for correction of mutated sequences in the targeted genes, induction of desired mutations by introducing specified sequence alterations, replacement of complete genes, and for targeted addition of the sequences deleted from a gene. In addition, HDR permits insertion of novel genes in 'safe' harbours, and integration of sequences encoding novel epitopes and fluorescence tags into endogenous genes. TALENs can be used to produce large deletions, inversions and even translocations of the targeted genomic regions. They have also been fused with other effector domains like transcription activation/repression domains, histone modification enzymes, fluorescence proteins, etc. to generate programmable chimeric effectors (Kühn et al., 2016).

TALENs have been used for analysis of effects generated by specific mutations and sequence changes on cell function/phenotype, and for rapid and efficient production of cellular models of human diseases. Induction of large genomic deletions has been used to investigate microRNA function. A fluorescent version of an endogenous human protein can be created by targeted insertions of a gene encoding a fluorescent protein; this would greatly facilitate analyses of the expression, translocation and distribution of the endogenous protein as well as its interaction with other cellular proteins. In addition, sequence replacement by HDR can produce isogenic human cell lines differing for only a specified SNP (single nucleotide polymorphism) or some other specific genetic modification. Such cell lines would be of considerable value for both basic and applied investigations. Gene editing by TALENs can be used to generate useful animal and plant mutants of value in animal and plant breeding (Joung and Sander, 2013; Sun and Zhao, 2013; Ousterout and Gersbach, 2016).

TALEs can be used for recognizing genomic regions that are hypermethylated. This recognition is based on the inability of RVD HD (specific for C) to recognize 5-methylcytosine, and the ability of RVD HG (specific for T) to bind to methylated cytosine. The TALEs can be designed to bind to the targeted genomic region, and the HG RVD is used for binding to the methylated C residues in the targeted sequence. The TALE binding can be detected by fusing the TALEs with fluorescent proteins like eGFP (Deng et al., 2012).

Since the TALE proteins recognize and bind to only the sense strand of target DNA duplex, they can also bind to DNA-RNA hybrid molecules, in which the DNA strand is the forward strand and the RNA strand forms the backward strand. Binding of TALE protein protects the RNA strand from digestion by RNase H. This feature might provide a potential strategy for antiretroviral therapy, *e.g.*, for HIV-1, since RNase H digestion of the RNA genome of retrovirus after synthesis of its cDNA (complementary DNA) is essential for virus replication. A TALE protein specific for the viral genome might block virus replication by preventing degradation of the RNA strand of the RNA-DNA hybrid produced as an intermediate during viral propagation. This feature might also be useful in manipulation of such cellular processes as DNA replication and transcription (Yin et al., 2012).

4.13. TALENs FOR GENE THERAPY

TALENs have been developed for targeting many human genes related to various diseases, and some of these SDNs have been used to develop gene therapies. Gene editing can be used to directly correct the gene mutation responsible for genetic diseases. For example, many cases of Duchenne muscular dystrophy (DMD) are due to frameshift mutations in the gene. TALENs were used to induce small indels in the regions upstream of the sites of mutations causing the frameshifts. On an average, about one-third of these indels would restore the reading frame and, thereby, correct the gene mutations responsible for DMD. Some other examples of correction of gene mutations responsible for genetic diseases relate to sickle cell disease, beta-thalassemia, alpha-1 antitrypsin deficiency (A1AD or AATD; leads to lung or liver disease), and xeroderma pigmentosum (a rare inherited skin disorder); these gene corrections were based on HDR (Ousterout and Gersbach, 2016).

TALENs were designed to target mutant genomes in mitochondria of patient-derived cells. In case of mitochondrial diseases, the mitochondria contain the wild type genome along with the mutant genomes. Treatment with the TALENs led to a permanent reduction in the presence of mutant genomes in the mitochondrial populations of the treated cells (Bacman et al., 2013). Similarly, TALENs have been targeted to cleave and eliminate the genome of hepatitis B virus from cultured human cells. TALENs cause efficient inactivation of human *CCR5* gene. The protein CCR5 serves as coreceptor for HIV-1 infection. Therefore, inactivation of the gene *CCR5* abolishes CCR5 function and confers resistance to HIV-1 infection (Ousterout and Gersbach, 2016).

4.14. MERITS OF TALENs

TALENs are easy to use, have high efficiency, and show low off-target activity. There is some evidence that TALENs are superior to ZFNs as they tend to cause lower cytotoxicity than ZFNs and are highly efficient in gene editing. The DNA base recognition code of TALE proteins is straightforward and robust, and is little affected by the neighbouring RVDs or TALE repeats. The designing of TALEs with novel specificities is relatively simple as compared to ZFNs. A potential limitation of TALENs relates to their organization based on nearly identical multiple TALE repeat units, which makes them unstable during PCR amplification. This limitation was overcome by several ingenious assembly schemes, which have made the synthesis of TALEs with novel specificities a routine exercise; some of these schemes are automatable for high-throughput production (Section 4.8). As a result, TALENs were rapidly adopted by research laboratories and were used in diverse biological systems. An example of their rapid acceptance and wide application is the creation of a library of TALENs specific to 18,740 protein-coding human genes (Chandrasegaran and Carroll, 2016; Ousterout and Gersbach, 2016).

TALENs have high activity and most of the designed TALENs are effective provided they can be effectively delivered. In contrast to the CRISPR-Cas9 SDNs, TALENs provide considerable flexibility in the size of target site, which can range from 10 to 20 bp or even more. Further, TALEs can be designed to target any genomic site. Although the natural TALEs require a T at position 0 in the target site, this requirement can be overcome by using suitably modified TALE scaffolds. TALEs containing a single unusually long or short TALE repeat can recognize the canonical target sequences as well as site with deletion of one base in the vicinity of the unusual repeat. Thus, a single TALE/TALEN could recognize two alleles of a locus. The TALENs can be directly injected into the cytoplasm of fertilized animal eggs to obtain gene-edited animals with relatively high efficiency (Feng et al. 2014; Ousterout and Gersbach, 2016).

4.15. LIMITATIONS OF TALENs

One of the main limitations of TALEN SDN system is that the gene encoding a TALEN monomer is about three times as large as that for a ZFN monomer. This is because a TALE repeat unit is slightly larger than a single zinc finger (ZF) but it recognizes only a single base as against three bases recognized by one ZF. Therefore, the two gene constructs encoding the two monomers of a TALEN pair have to be inserted into two plasmids for delivery into cultured mammalian cells. As a result, a cell must be co-transfected by both these plasmids to produce the TALEN pair, and enable DSB induction and gene editing. Since the frequency of co-transfected cells in the treated population is bound to be low, a suitable strategy needs to be used to enrich the cell population for the co-transfected cells. In one such strategy, genes encoding different fluorescent proteins like eGFP and DsRed (*Discosoma* red) are inserted into the two plasmids carrying the TALEN encoding genes. The co-transfected cells will express both the fluorescent proteins, and such cells can be sorted out using FACS (fluorescence-activated cell sorter). This enriches the cell population with co-transfected cells and greatly enhances the frequency of gene editing (Feng et al., 2014).

The TALEN encoding genes comprise many repeat units arranged in tandem, which presents problems in their delivery into mammalian cells/genomes using viral vectors. This problem is considerably reduced in TALEs with recoded TALE repeats. TALENs cause DSBs and gene editing at off-target sites just as other SDNs. The off-target activity of TALENs can be minimized by carefully selecting a unique target site with the help of a suitable bioinformatics webtool like www.talenlibrary.net (Chugunova et al., 2016). In addition, the TALEN design can be improved to reduce the risk of off-target activity (Section 4.10). TALENs place some restriction on the choice of target sites since the base at the 5' end of a target sequence has to be a T. Protein engineering has been effectively used to design TALENs that do not require a T at position zero in their targets. In addition, where required, the spacer region between the binding sites of the two TALEN monomers can be varied to considerable extent to cater to the need for a T at the 5' ends of the binding sites. Some flexibility is also provided by varying the number of bases in each of the two binding sites (Carroll, 2015).

The conventional TALENs cannot recognize 5-methylcytosine and, as a result, are unable to cleave genomic regions like CpG islands present in many eukaryotic promoters that have methylated cytosines. This problem can be overcome by using the RVD NG that is specific for T in the place of HD (specific for C) since 5-methylcytosine in the major groove cannot be distinguished from T. This strategy would yield a TALEN that can cleave methylated DNA, but there would be some loss of specificity. It has been suggested that the methylation status of genomic DNA should be considered while designing TALENs. In case information on the methylation status of target DNA is not available, one may design two different pairs of TALENs for each target: one TALEN pair is designed considering the DNA to be unmethylated, while the other pair is designed for methylated target sequence. Alternatively, the RVD HD can be replaced with the RVD N-, which has amino acid residue N at position 12 and the amino acid residue at position 13 is deleted (represented by – in N-); this RVD can recognize methylated cytosine as well but its affinity is lower than that of HD (Deng et al., 2012; Chandrasegaran and Carroll, 2016; Carroll, 2015).

The nearly identical repeat units in the DNA-binding domain of TALENs makes the assembly of TALEN encoding genes in *E. coli* a relatively more difficult task. However, several assembly schemes have been devised for routine creation of genes encoding TALEs with novel specificities (Section 4.8). At the same time, the large size does not allow packaging of the genes for the TALEN pair into AAV (adeno-associated virus) vectors that are very attractive for use in mammalian systems. For some reason, the TALENs are poor gene editors in rodent embryos. But when TALENs were either expressed or microinjected in combination with exonuclease 1

(Exo1), there was highly efficient gene editing in rat embryos. The initial cost of commercially available TALENs was around US $ 5000 per target site, which made it unaffordable to small laboratories. But continued refinements in TALEN assembly schemes have drastically reduced this cost. The TALENs were poised to become the favored substitute for ZFNs, when the much simpler, cheaper and highly flexible CRISPR-Cas9 system arrived at the scene. Many scientists consider TALENs to be more useful than CRISPR-Cas system for applications like gene therapy in view of their superior specificity and lower off-target activity (Mashimo, 2014; Chandrasegaran and Carroll, 2016; Carroll, 2015; Lee et al., 2016).

4.16. ACHIEVEMENTS

TALENs have been used to inactivate specific genes to produce crop plants with useful traits like disease resistance. For example, TALENs were used to disrupt the binding site for the TAL effector produced by *Xanthomonas oryzae* pv. *oryzae* to obtain rice mutants resistant to leaf blight caused by this pathogen. Similarly, simultaneous inactivation of three homeologues of wheat *Mlo* gene has yielded mutant lines resistant to powdery mildew since MLO protein function is necessary for pathogenesis by this fungal pathogen. Inactivation of soybean *FAD2* gene has yielded mutants with improved oil quality. TALEN mutagenesis has been successful in modifying rice for seed traits like fragrance and storage tolerance, and in improving potato for traits like cold storage and processing quality. In addition, TALENs have been used to mutate genes encoding noncoding RNAs as well as to introduce changes in the regulatory genomic sequences like promoters and enhancers. They have also been used to produce large deletions in the targeted genomic regions (Malzahn et al., 2017).

TALENs have been used to induce translocations involving specified chromosomal segments. Translocations are produced when two pairs of TALENs are used to induce a DSB at the targeted site in each of the two chromosomes that are desired to be involved in the translocation event. Suitably designed TALENs were used to produce certain cancer-related translocations in cultured human cells. The cells with the translocated chromosomes produced the fusion protein characteristic for the cells having the concerned translocations (Thakore and Gersbach, 2016). In addition, TALENs have been repurposed as programmable transcription factors, epigenetic editors, base editors, fluorescence tags (Section 4.6), and even nucleic acid (RNA and/or DNA) detection tools (Section 9.23).

TALENs have been used to correct several gene mutations that are responsible for human genetic diseases. For example, cases of DMD caused by frameshift mutations were corrected by induction of indels in coding regions upstream of the sites of mutations. HDR has been used for correction of gene mutations responsible for sickle cell disease, beta-thalassemia, A1AD, and xeroderma pigmentosum. TALENs were used for targeted elimination of mutant mitochondrial genomes in patient derived human cells. Similarly, TALENs have been targeted to eliminate the genome of hepatitis B virus from cultured human cells. TALENs cause efficient inactivation of *CCR5* gene; this is expected to confer resistance to HIV-1 infection. Many gene-edited gene therapies have entered clinical trials and some have reached the clinic. For example, TALEN-generated universal T cells are in clinical use. For greater detail, the reader may refer to Section 9.23.

4.17. FUTURE PERSPECTIVES

The TALEN technology has evolved rather rapidly, but many important issues remain to be adequately resolved. Since NHEJ is the major pathway for DSB repair, it is important to refine the experimental conditions and procedures to enhance the frequency of HDR in relation to NHEJ. There is an urgent need for methods capable of reliable estimation of genome-wide specificities and off-target activities of TALENs to facilitate the development of techniques for reducing frequencies of off-target effects. One approach would be to carefully screen the TALEN-edited cells by deep whole genome sequencing for assessing off-target editing. Another area that needs to be investigated relates to delivery protocols for TALENs to enable their effective use in various mammalian cell types and other biological systems. The successful fusion of recombinase domain derived from the DNA invertase Gin with TALE DNA binding domain has created TALE-recombinase (TALER), which efficiently recombines DNA in bacteria and cultured mammalian cells. The TALERs have opened exciting possibilities of sponsoring genome editing by site-directed recombination events. Similarly, TALEs have been fused with transposases to generate programmable transposases. Since recombinases and transposases carry out DNA cleavage and religation independent of DSB repair pathways, they can function in a variety of cell types at high efficiency. But TALE-transposases also exhibit significant off-target activity. In addition, the TALEs can be combined with various other effector domains to generate programmable chimeric effectors that would have novel applications. Some such chimeric effectors like TALE-TFs have already been produced (Joung and Sander, 2013; Sun and Zhao 2013; Ain et al., 2015).

There are ample indications that the DNA binding specificities of TALE proteins can be improved by fine-tuning some of their so far underexploited features; this may have both qualitative as well as quantitative impact on specificity. Some of these features are, (i) use of the most specific panel of RVDs (from among the 400 possible RVDs) for TALE construction, (ii) variation in TALE repeat backbone, (iii) alterations in amino acid residues of N-terminal and C-terminal domains of TALEs that affect DNA binding and specificity, (iv) inclusion of naturally occurring longer or shorter than the canonical repeat units, and (v) variation in the length of repeat domains of TALEs. The effects of these features on specificity and DNA binding affinity must be investigated extensively to discern the relationships between variation in these features and DNA binding specificity of TALEs. It will be useful to develop bioinformatics resources to enable dependable designing of highly specific and efficient TALEs by considering all the above features (Bogdanove et al., 2018).

4.18. CONCLUSIONS

The development of TALENs in 2010 provided a SDN system that is simpler in design, more convenient to construct and has higher success rate in different species than ZFNs. As a result, the TALEN technology was rapidly adopted by many laboratories around the world. Most custom designed TALEs are assembled using the backbone or scaffold and the repeat consensus sequence of TALEs from *Xanthomonas* sp. and the repeat units contain the four most common RVDs; some TALEs also contain one or more of the alternative RVDs listed in Table 4.1 (Bogdanove et al., 2018). The length of canonical TALE repeats ranges between 33 and 35 amino acids, but some longer and shorter repeats are also known. In general, TALENs have around 16-18 repeat units, including the half repeat, but longer and shorter TALENs are often used. Each TALE repeat recognizes one nucleotide of the target DNA, the DNA binding specificity being determined by two amino acid residues at positions 12 and 13 of TALE repeats. The RVDs most commonly used for TALE construction are NI (Asn Ile), HD (His Asp), NN (Asn Asn) and NG (Asn Gly) for A, C, G and T, respectively. The natural TALEs require a T at position 0 in their binding sites for efficient function, but TALEs with improved design do not have this limitation.

TALENs that cleave DNA with improved efficiency have been developed by optimization of TALE scaffold amino acid sequence. Similarly, novel RVDs have been identified and characterized; this has minimized the ambiguity in DNA base, especially guanine, recognition by TALENs and, thereby, reduced off-target editing. Several bioinformatics tools for selection of genomic target sites for TALENs are available online, and some of them also predict the putative off-target genomic sites. Many different methods have been developed for assembly of novel TALE repeat arrays. The assembled TALE arrays are cloned in TALE cloning framework vectors. The various methods for TALE repeat assembly can be classified into the following three broad groups: (i) standard restriction enzyme and ligation-based cloning, (ii) "Golden Gate" cloning, and (iii) solid-phase assembly. Although commercial services are available for synthesis of novel TALEs, they are costly particularly for creation of large numbers of new TALEs. The TALEN assembly methods like Golden Gate cloning-based TALEN assembly, simple TALE assembly reaction (STAR), solid-phase assembly methods, and ULtiMATE system for TALE construction have greatly simplified the construction of novel TALEs, and some of them, *e.g.*, solid-phase assembly methods, are amenable to high throughput TALE assembly. These assembly procedures have greatly enhanced the accessibility of TALEN technology and made it affordable to academic and industrial laboratories (Joung and Sander, 2013).

The TALE proteins have been fused with several different functional domains like transcription activation and repressor domains, chromatin modification domains, etc. to generate programmable transcription activators/repressors, chromatin modifiers, etc. These chimeric effectors have been used for targeted activation and repression of gene expression in a variety of biological species. Similarly, a hyperactive catalytic domain of a recombinase has been fused with TALE protein to create TALE-recombinases, which are able to efficiently recombine DNA molecules in bacteria and mammalian cells cultured *in vitro*. TALERs provide an opportunity for targeted gene editing based on recombination. TALE proteins can be fused with cytidine and adenosine deaminases to create base editors. TALEs have been linked with florescent proteins to create fluorescent tags that enable easy genome visualization and analysis of genome organization and function (Joung and Sander, 2013; Sun and Zhao 2013).

TALENs can be delivered into cells as DNA, mRNA or protein. Their delivery as mRNA and protein tends to reduce off-target effects and eliminate the risk of integration of foreign DNA into the cell genome; the latter significantly reduces regulatory requirements for the gene-edited products. Many scientists consider TALENs to be more specific and better suited for applications like gene therapy than ZFNs and even CRISPR-Cas9.

■■■■

5

CRISPR-Cas9 SYSTEM

"....the CRISPR/Cas9 system of genome editing was named the 2015 Breakthrough of the Year by Science." ----**NASEM (2017)**

5.1. INTRODUCTION

Meganucleases were the first to be used for site-specific induction of double-strand breaks (DSBs) during early 1990s. Soon thereafter, zinc finger nucleases (ZFNs) were created by linking together natural zinc finger proteins with the cleavage domain of FokI Type IIS nuclease. About a decade later, transcription activator-like effector nucleases (TALENs) were developed by joining the DNA binding part of TALE (transcription activator-like) proteins produced by certain plant pathogenic bacteria with the FokI cleavage domain. These site-directed nuclease (SDN) systems were systematically engineered to design SDNs specific for practically any user defined DNA sequence. They were used for targeted DSB-induction in many organisms for gene inactivation due to NHEJ (nonhomologous end joining), and sequence replacement or transgene insertion by homology directed repair (HDR). These SDNs are protein-based systems and each one of them offers certain advantages and challenges.

In this backdrop, the RNA-guided nuclease system based on the bacterial CRISPR-Cas (clustered regularly interspaced short palindromic repeats-CRISPR-associated) immunity system emerged as a relatively much simpler, inexpensive and highly flexible SDN system that is readily accessible to anyone interested in using it in his studies. These and other advantageous features of CRISPR-Cas9 system are rooted in its use of a relatively short RNA (and not a protein) sequence for target sequence recognition. As a result, designing of a CRISPR-Cas9 SDN for a new genomic target involves the simple designing of single guide RNA (sgRNA), which can be done with the help of any one of several open access web-based bioinformatic tools (Table 5.1; Sander and Joung, 2014). An excellent account of the discovery and deciphering of CRISPR-Cas9 system of acquired immunity of prokaryotes and its development into a programmable RNA-guided nuclease for targeted gene editing is available in Hsu et al. (2014) and Doudna and Sternberg (2017). An excellent discussion of the development and uses of the four different gene editing tools, *viz.,* nucleases, base editors, transposases and prime editors, based on the CRISPR-Cas9 system is provided by Anzalone et al. (2020).

Table 5.1. Some bioinformatics tools for designing guide RNAs (based on Tycko et al., 2016 and other sources)

Guide RNA designing tool	Website	Provides specificity score	Single/ Paired gRNAs	Supports Cas9 orthologs	Predicts off-targets	Predicts off-targets with indels
Benchling	http://benchling.com	yes	both	5	Yes	No
BLAST (on predesigned guide RNA)	http://blast.ncbi.nlm.nih.gov/ Blast.cgi	Yes (as sequence similarity)	NA	NA	Yes	Yes
Cas-OFFinder	http://www.rgenome.net/ cas-offinder	No	Single	4	Yes	Yes
Deskgen	http://deskgen.com	Yes	Both	4	Yes	No
E-CRISP	http://www.e-crisp.org/ E-CRISP/designcrispr.html	Yes	Both	1	Yes	No
EuPaGDT	http://grna.ctegd.uga.edu	Yes	Single	Custom PAM	Yes	No
GenScript gRNA Design Tool	http://www.genscript.com/ gRNA-design-tool.htm	No	Single	1	No	No
ZiFiT	http://zifit.partners.org/ZiFiT/ Disclaimer.aspx	Yes	Both	1	Yes	No
CROP-IT	http://cheetah.bioch.virginia. edu/AdliLab/CROP-IT/	Yes	Both	1	Yes	No
sgRNA Scorer 2.0	http:// crispr.med.harvard. edu/sgRNAScorerV2	Yes	-	All*	-	-
CRISPRpred	https://github.com/khaled-buet/CRISPRpred.	Yes	-	-	-	-
CRISPR-DO**	http://cistrome.org/crispr/	Yes		1	Yes	-

* All CRISPR systems with known PAM sequence and defined spacer length

** Annotates each target sequence with the PhastCons conservation score for their functional characterization where such data are available

5.2. CRISPR-Cas SYSTEM OF PROKARYOTIC IMMUNITY

CRISPR-Cas system is the basis of adaptive immunity common in bacteria and archaea. The CRISPR region in the genomes of these prokaryotes consists of a short palindromic sequence of 20–50 bp that is repeated several times. The repeat units of this sequence are separated by 20–50 bp hypervariable spacer sequences (Fig. 5.1). The spacer sequences differ from each other and are derived from genomes of viruses and plasmids that invaded these prokaryotes in the past. The short spacer segments are captured from the viral and plasmid genomes and are integrated into the CRISPR regions of the concerned prokaryotes. In this way, the CRISPR region has acquired the organisation that is aptly described by its name, *viz.*, clustered regularly interspaced short palindromic repeats. The CRISPR loci provide immunity against those viruses and plasmids whose DNA sequences were captured and integrated as spacers in these loci (Carroll, 2014; Krebs, 2014; Sander and Joung, 2014).

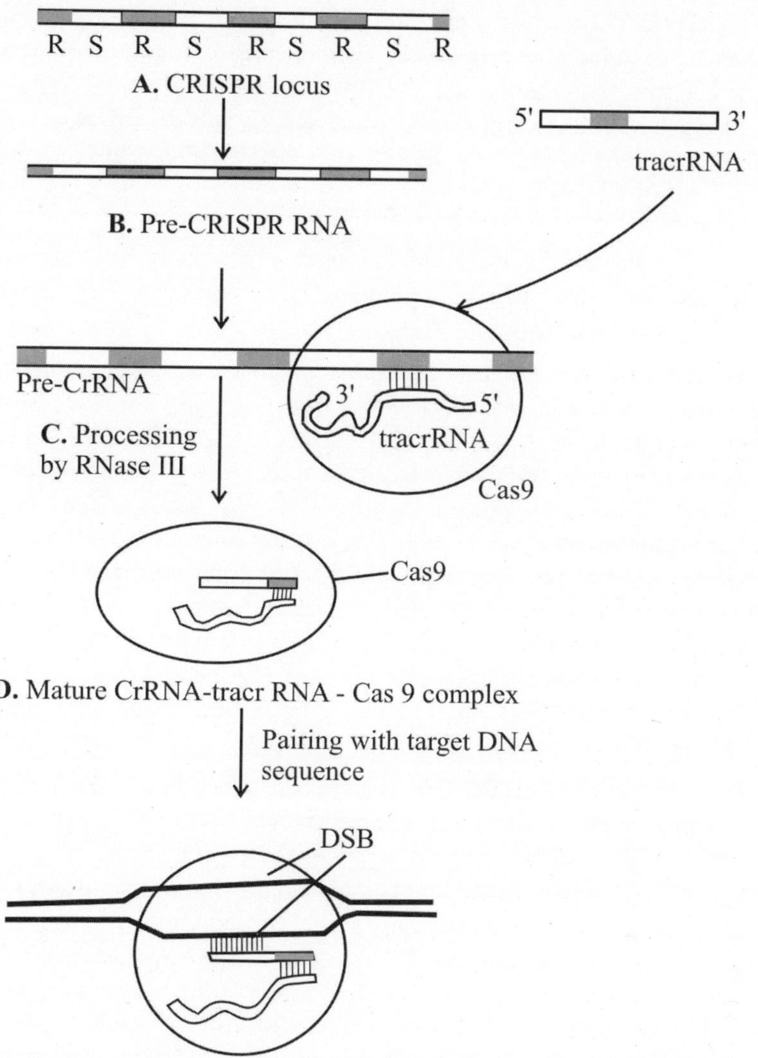

Fig. 5.1. A simplifed representation of the CRISPR-Cas9 system. Transcription of CRISPR locus produces pre-CRISPR RNA, which is processed to yield CRISPR RNA (crRNA). crRNA pairs with tracrRNA and associates with Cas9 nuclease to form crRNA-tracrRNA-Cas9 complex. The A part of crRNA base-pairs with the target DNA duplex. This places the Cas9 nuclease at the target site, where it induces a double-strand break (DSB). R, palindromic repeat : all repeats have the same sequence; S, hypervariable spacer sequence : each spacer sequence is diffrent from the rest (based on Dounda and Charpentier, 2014; Sander and Joung, 2014).

The adaptive immunity due to CRISPR involves transcription of the CRISPR loci and processing of their RNA transcripts. Several genes (eight genes in the bacterium *Escherichia coli*) located adjacent to each CRISPR locus encode proteins that are needed for processing of the CRISPR RNA transcripts. These *cas* genes are present in an independent operon close to the CRISPR locus. The Cas proteins also process the genomes of viral and plasmid invaders to enable integration of their genomic fragments into the host microbial genomes. The CRISPR locus is transcribed into a long RNA transcript called pre-crRNA (pre-crisprRNA). The pre-crRNA is processed by Cas proteins into short 57-nucleotide (nt) long crRNAs. In fact, another RNA called tracrRNA (*trans*-activating CRISPR RNA) and Cas9 protein associate with the pre-crRNA; the pre-crRNA and tracrRNA are then processed together by RNase III to generate the mature crRNA-tracrRNA-Cas9 complex. Each crRNA has one of the spacer sequences at its 5' end and a part of the conserved repeats, the CRISPR repeats, at its 3' end (Fig. 5.1; Carroll, 2014; Doudna and Charpentier, 2014; Sander and Joung, 2014).

The Cas9 nuclease first recognizes the PAM (protospacer adjacent motif) sequence that performs dual function: it recruits Cas9–guide RNA (Cas9-gRNA) complex to the possible target site and is involved in activation of the Cas9 nuclease domains. The PAM requirement enables Cas9 to identify target sequences present in viral/plasmid DNAs, and to differentiate them from the homologous spacer sequences present in the *CRISPR* locus since they lack the PAM sequence. The *Staphylococcus pyogenes* Cas9 (SpCas9) recognizes the PAM sequence 5'NGG3' located in the non-target strand, *i.e.*, the strand that does not pair with crRNA; it then unwinds the target DNA duplex upstream of the PAM and the crRNA base-pairs with the complementary sequence in the viral/plasmid DNA. The Cas9 nuclease cleaves the viral/plasmid genomes at three nucleotides upstream of the PAM sequence into the protospacer region paired with the crRNA and produces blunt ends (Fig. 5.1; Carroll, 2014; Doudna and Charpentier, 2014; Krebs, 2014; Sander and Joung, 2014).

There are three families of CRISPR-Cas system, of which the Type II CRISPR-Cas system has been used to construct SDNs. The Type I and Type III CRISPR-Cas systems involve a large complex of Cas proteins for crRNA-guided cleavage of target DNA. In contrast, the Type II systems uses a single Cas protein to achieve this end, which is very important for genome editing. Thus, the Type II CRISPR system requires only three components, *viz*., crRNA, tracrRNA, and Cas9, nuclease for cleavage of viral/plasmid DNAs. The tracrRNA is encoded by a separate gene in the *CRISPR* locus; it has a sequence toward its 5' end that is complementary to the repeat sequence of the CRISPR locus, which base-pairs with the 3' end region of crRNA. The crRNA/tracrRNA together interact with Cas9 to induce conformational changes in the Cas9 protein that enable it to bind to and ultimately cleave the target DNA. In addition, these RNAs together guide the Cas nuclease to the target sequence; for this reason, the CRISPR-Cas system is often termed as RNA-guided nuclease (Carroll, 2014; Doudna and Charpentier, 2014; Krebs, 2014; Sander and Joung, 2014).

The Cas9 has two nuclease domains, *viz*., HNH and RuvC-like domains. The HNH domain cleaves that strand of target DNA, which base-pairs with the crRNA, *i.e*, the complementary or target strand. The RuvC-like domain, on the other hand, cuts the nontarget strand, *i.e*., the strand that is not complementary to the crRNA. The sequence of the target DNA (viral/plasmid DNA) that base-pairs with the spacer sequence of crRNA is known as protospacer sequence since this is the progenitor of the spacer sequence in *crispr* locus. Another short sequence, the PAM, must be present immediately downstream of the target DNA:crRNA hybrid region for target recognition and Cas9 action. SpCas9 recognizes NGG as PAM sequence, but it also recognizes NAG with much lower efficiency. The PAM required by Cas9 nucleases from some other bacterial species (Table 5.2) and of some mutants of SpCas9 are different from NGG (Table 5.3; Doudna and Charpentier, 2014; Carroll, 2015).

Table 5.2. Some Cas9 orthologs available for gene editing (based on Tycko et al., 2016; Acharya et al., 2019)

Cas9 ortholog	Source species	Remarks
SpCas9	*Staphylococcus pyogenes*	Most commonly used enzyme
SaCas9	*Staphylococcus aureus*	Gene over 1 kb shorter than *SpCas9* gene; easily packed with sgRNA in AAV vector; target sequence 21–24 nt; uses 5'-NNGRRT* and 5'-NNGRR PAM; used in plant, zebra fish and mouse models
NmCas9	*Neisseria meningitidis*	Uses 5'-NNNNGATT PAM and 21–24 nt long target sequence; lower off-target activity and on-target efficiency than SpCas9; longer PAM limits target sites
St1Cas9	*Streptococcus thermophilus*	More specific than SpCas9 but has lower efficiency; uses 5' NNAGAAW 3' as PAM ; has been used for gene editing in mammalian cells
St3Cas9	*Streptococcus thermophilus*	Uses 5' NGGNG 3' as PAM; has been used for gene editing in mammalian cells
CjCas9	*Campylobacter jejuni*	The gene encoding CjCas9 is ~1.1 kb smaller than that for SpCas9; gene constructs for CjCas9 and sgRNA easily packed in a single AAV vector
FnCas9	*Francisella novicida*	Identified by bioinformatic analysis; high specificity; low off-target editing; staggered cuts in DNA; versatile RNA-editing system; can be useful for virus defence in eukaryotic cells; uses 5'-NGG-3' PAM

* N, any of the four nucleotides; R, A/G

Cas9 protein assumes an auto-inhibited conformation. When it associates with the crRNA and tracrRNA (or sgRNA), it undergoes large conformational changes. The conformation of Cas9 undergoes further change when the Cas9-crRNA-tracrRNA complex associates with the double-stranded DNA target. As a result, a channel is created in Cas9, and it can bind to the crRNA-target DNA hybrid and the crRNA region paired with tracrRNA. Cas9 comprises two lobes connected by a hinge that is involved in the binding of Cas9 with crRNA-target DNA hybrid. The PAM sequence is essential for recognition of target DNA; without this sequence, Cas9 cannot recognize even those DNA sequences that are perfectly complementary to crRNA (Doudna and Charpentier, 2014; Tsai and Joung. 2016).

The Cas9 C-terminal domain recognizes and interacts with PAM leading to initiation of DNA unwinding that progresses toward the upstream side from the PAM; the target strand of protospacer DNA pairs with the 5' region of crRNA or sgRNA to from an R loop structure. When there is enough complementarity between the crRNA or sgRNA and target DNA sequences, the Cas9 HNH domain undergoes a conformational change; this activates the HNH domain and triggers target DNA cleavage by both the HNH and RuvC-like nuclease domains. As a result of Cas9 binding, bases in those positions of the unwound target DNA that remain exposed may be deaminated by cellular processes; this may generate the point mutations that have been observed particularly when nCas9 (Cas9 nickase) is used for gene editing (Doudna and Charpentier, 2014; Tsai and Joung. 2016).

Table 5.3. Some genetically engineered variants of Cas9 nucleases (based on Tsai and Joung, 2016; Tycko et al., 2016; Komor et al., 2017b, and other sources)

Variant	Genetic modification	Altered features	Remarks
Nickase	Amino acid substitution in the HNH or RuvC-like domain	Cleavage of a single strand of the target DNA	Either the non-complementary (HNH mutation) or the complementary (RuvC-like mutation) strand cleaved (Section 5.6)
dCas9	Amino acid substitutions in the HNH as well as RuvC-like domains	The Cas9 is unable to cleave DNA but its binding to target DNA remains unimpaired	CRISPR-dCas9 represses targeted genes; can be fused with mutant FokI cleavage domain to minimize off-target effects; fusion with other effector domains generates programmable hybrid effector proteins
Split Cas9	Cas9 split into nuclease and alpha-helical lobes	The two parts expressed as separate peptides; reduces the size of gene construct; Cas9 activity can be regulated	The two parts assemble on sgRNA to form a functional Cas9; nuclease activity can be regulated by controlled expression of sgRNA; efficiency lower than natural Cas9
	Cas9 peptide divided into two parts and each part fused with the appropriate part of a split intein	The two parts expressed as separate peptides; greatly reduces the size of gene construct and facilitates delivery into the cells	The two parts of Cas9 joined together by trans-splicing catalysed by the split intein; efficiency comparable to natural Cas9
DD-Cas9	Cas9 fused with the destabilizing domain (DD) of a mutant human protein	DD-Cas9 destroyed rapidly by proteasomal degradation	DD-Cas9 stabilized by a small ligand molecule; period of nuclease action can be controlled by regulating the ligand availability
eSpCas9 1.1 (enhanced SpCas9 version 1.1)	Ala substitution at 3 (eSpCas9) and 4 (eSaCas9) positions	Increased genome-wide specificity; dramatically reduced off-target activity	The interaction of Cas9 with the non-target strand is likely to be weakened by the Ala substitutions; on-target cleavage efficiency comparable to wild-type SpCas9; high activity at certain off-sites
SpCas9-HF1	Four amino acids substituted by Ala	''High-fidelity'' SpCas9; increased specificity (reduced off-target activity)	Nonspecific interactions of amino acids with sugar-phosphate backbone of the target or complementary strand eliminated; high activity at certain off-sites; reduced on-target activity in association with tru-gRNAs (17-18 nt spacer)
HypaCas9	Amino acid substitutions in the noncatalytic REC3 domain	Hyper-accurate Cas9 with greatly reduced off-target activity	The REC3 domain interacts with DNA:RNA hybrid; HypaCas9 exhibits enhanced genome-wide specificity without compromising on-target activity
''VRER'' SpCas9	Four amino acid substitutions in SpCas9	Altered PAM requirement; high specificity for NCGC PAM	NCGC PAM is nearly 23-times less frequent than NGG in human genome

Variant	Genetic modification	Altered features	Remarks
Altered PAM specificity	One amino acid substitution in SpCas9	Increased specificity for PAM sequence NGG over NAG	—
Altered PAM specificity	Three to four mutations in each variant SpCas9	Each variant recognizes either NGA, NGAG, or NGCG	Bacterial selection system used to identify three new Cas9 variants; PAM recognition with high efficiency and specificity
Altered PAM specificity	Three amino acids replaced in FnCas9	5'-YG-3' PAM used by the variant (the wild type FnCas9 uses 5'-NGG-3' PAM)	Crystal structure was used to engineer this variant of FnCas9
Altered PAM specificity (KKH variant)	Three mutations in SaCas9	PAM sequence requirement changed from NNGRRT to NNA(C/T)RRT	An unbiased selection system used to engineer the variant with altered PAM requirement; off-target activity comparable to that of wild type enzyme

5.3. CRISPR-Cas SDN SYSTEM

The CRISPR sequences were discovered in 1987 in the bacterium *E. coli*, and they were shown to be involved in adaptive immunity in 2007. The CRISPR-Cas9 system was first used for genome editing in 2012 by Jinek et al. (2012). Since then, this SDN system has been used for targeted genetic modifications in a variety of organisms, including mammals, fish, nematode, plants, and bacteria. Remarkably, within a short span of less than three years, the CRISPR-Cas system emerged as the most widely used genome editing technology; some researchers consider this as 'spectacular' (Ledford, 2015a). The main reasons for this amazing popularity are, the ease with which CRISPR-Cas9 SDNs for new genomic targets are designed and the very low cost involved. Any researcher with some computer knowledge can design *in silico* a CRISPR-Cas9 SDN specific for a given target in a matter of minutes (Doudna and Sternberg, 2017).

5.3.1. Single guide RNA (sgRNA)

Jinek et al. (2012) joined the critical parts of crRNA and tracrRNA from *S. pyogenes* to create the sgRNA (Fig. 5.2). In sgRNA the constant base-pairing sequences of the crRNA and tracrRNA are replaced by an intramolecular hairpin or stem-loop structure. The sgRNA successfully guides the Cas9 nuclease to the target DNA sequence and enables DSB induction at the targeted site (Fig. 5.2). The sgRNA must have the following two features to guide Cas9 to the target DNA: (1) a 20 nt long sequence at the 5' end that base-pairs with the target DNA, and (2) a double-stranded structure at its 3' end for its association with the Cas9 nuclease. At least four different versions of guide RNAs have been used by different research workers (Fig. 5.3). The most widely used sgRNA is ~100 nt long with the following consensus organization: 22 nt long 5' guide or spacer sequence that is complementary to the target DNA, followed by 80 nt long 3' sequence derived from fusion of crRNA with tracrRNA (Fig. 5.2). This sgRNA is often termed as +85 sgRNA. The 5' 22 nt in the sgRNA comprise 20 nt that are target-specific, *i.e.*, represent the target sequence and differ among the various sgRNAs, while the remaining 2 nt represent the PAM sequence and are common to all sgRNAs. The CRISPR-Cas SDN thus comprises two components, *viz.*, sgRNA and Cas9 nuclease. Of these, *only the 5' terminal 20 nt are variable and correspond to the specified target sequences*, while the remaining 3' terminal sequence of all sgRNAs and the Cas9 nuclease are invariable (Carroll, 2014. 2015; Ramakrishna et al., 2014; Sander and Joung, 2014).

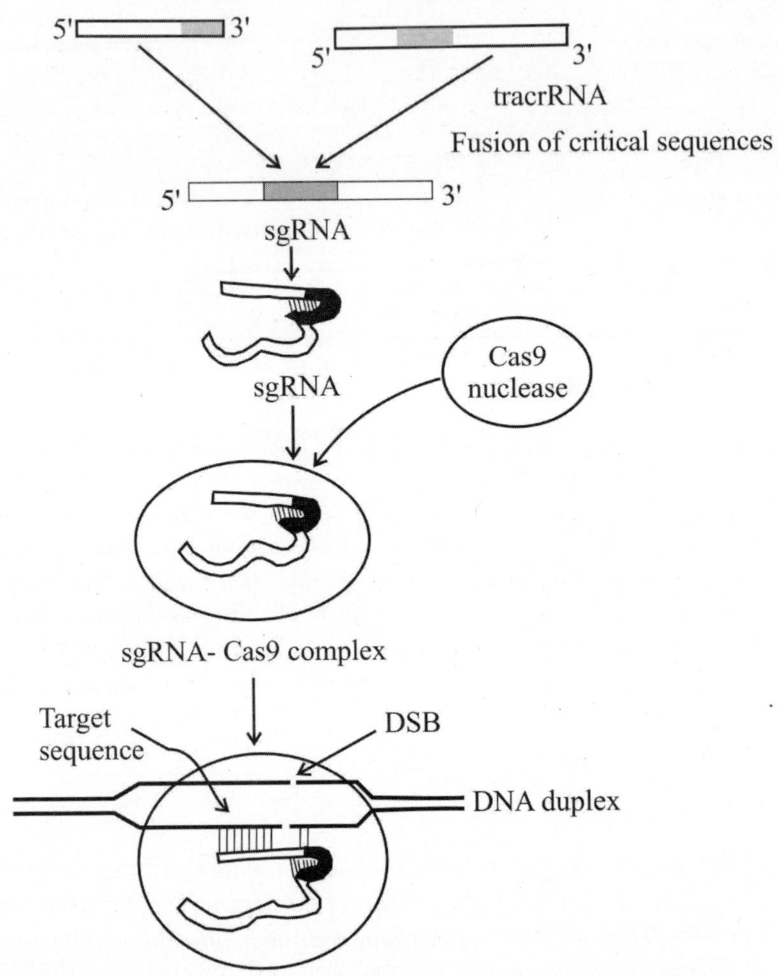

Fig. 5.2. Single guide RNA (sgRNA) produced by fusion of critical sequences from crRNA and tracrRNA associates with Cas9 nuclease and guides it to the target sequence where Cas9 nuclease induces DSB. sgRNA seems to perform better than the combination of crRNA and tracRNA (based on Doudna and Charpentier, 2014; Sander and Joung, 2014). DSB, double-strand break.

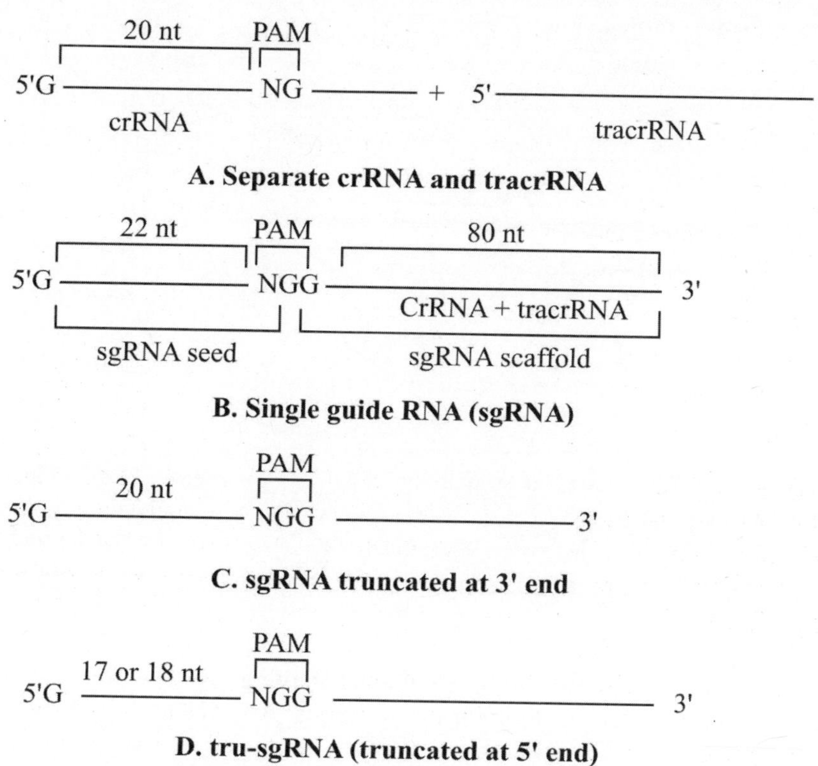

Fig. 5.3. Different types of guide RNAs used for creation of CRISPR-Cas SDNs. sgRNAs perform better than a combination of crRNA and tracrRNA. tru-sgRNA performance is comparable to that of sgRNA, but the specificy tends to be superior to the latter. The single G at the 5' end is relevant for the RNA polymerase III promoters. In case of T7, SP1, etc. promoters, two Gs (GG) will be present at the 5' end.

The sgRNA molecule has two functionally distinct regions, *viz.*, sgRNA seed and sgRNA scaffold (Fig. 5.3B). Originally, the *seed region* was defined as the 8-12 nucleotides in the 5' region next to the PAM sequence. But *it is customary to consider the sgRNA seed region as the sequence that base-pairs with the target DNA*. The *sgRNA scaffold* comprises the remaining 3' region of sgRNA; it generates the structural skeleton needed for its association with Cas9 nuclease. The natural gRNA seed region is typically 20 nt; there should be no mismatch between the sequence toward the 3'-end of sgRNA seed that is closest to the PAM and the target DNA sequence. The PAM motif NGG must be present in the non-complementary strand of target DNA immediately downstream of the base-pairing region (Fig. 5.4). It is now well known that gRNA has six functional modules that together determine the properties of the gRNAs (Xie and Yang, 2013).

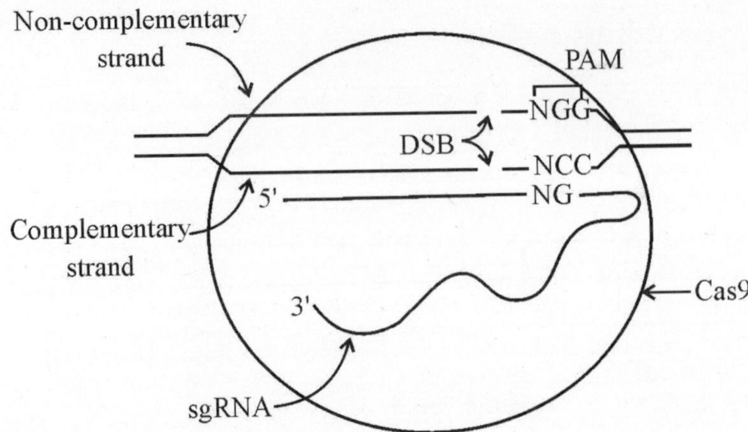

Fig. 5.4. A CRISPR-Cas SDN in association with the target DNA sequence. The 3'end of the pairing region pairs with two nucleotides of the PAM sequence in the complementary strand. The cleavage occurs three bases into the pairing region from the PAM (based on Sander and Joung, 2014).

5.3.2. Cas9 nuclease

The Cas9 protein is a multifunctional molecule having sgRNA binding, DNA sequence recognition and binding, helicase and nuclease activities. As a result, very little of its polypeptide sequence is dispensable. Cas9 crystal structures (both alone and in association with sgRNA:target DNA hybrid) and their reconstructions with the help of cryo-EM reveal that Cas9 makes extensive contacts with the sgRNA:target DNA hybrid as well as the invariant segment of sgRNA. In addition, it has two separate nuclease domains that cleave the two strands of target DNA. A knowledge of the structural and functional features of Cas9 nuclease has enabled engineering of Cas9 to modify its properties like improved specificity, changed PAM recognition, etc. (Table 5.3). The Cas9 nuclease from some bacterial species other than *S. pyogenes* are known to recognize PAM sequences different from 5'NGG3' and 5'NAG3' (Table 5.2), and use different crRNA and tracrRNA sequences; these Cas9 nucleases have been successfully used for targeted gene editing (Carroll, 2014, 2015).

Genetic engineering successfully enhanced the specificity of Cas9 and reduced its off-target effects (Table 5.3). The various approaches used include creation of CRISPR-Cas9 nickase (Section 5.6), 'dead' Cas9 (dCas9; devoid of nuclease activity) and split-Cas9 (Section 5.7), fusion of Cas9 with a destabilizing domain, and amino acid substitutions to alter Cas9 specificity or Cas9 interaction with PAM (Table 5.3; Section 5.9.2.2). The PAM sequence recognition is possibly the strictest basis of Cas9 specificity. Efforts to modify the Cas9 PAM recognition have been successful to some extent. A mutant of SpCas9, ''VRER'' SpCas9, has high specificity for the PAM sequence NCGC (Table 5.3); this PAM sequence is roughly 23-times less frequent than the regular NGG PAM. Similarly, a single mutation (D1135E) in SpCas9 leads to increased preference for the regular NGG PAM over the alternative PAM NAG. A mutant SaCas9 recognizes the PAM sequence 5'NNNRRT3' in the place of 5'NNGRRT3' recognized by the wild type enzyme (Tycko et al., 2016).

5.4. DESIGNING Of CRISPR SDNs

The CRISPR-Cas9 SDNs have two components, *viz*., sgRNA and Cas9, of which only the 5' 20 nt of the sgRNA vary with the target sequence. Therefore, designing of CRISPR-Cas9 SDN for a given genomic site involves the designing of 5' 20 nt of sgRNA. This involves selection of the target site in each gene of interest by the researcher keeping in view the following points. The target sequence should be of 20 bp that would base-pair with the 5' end region of sgRNA. In addition, PAM sequence should be located immediately downstream of this 20 bp pairing region. It is a general practice to use sgRNA, which gives considerably higher gene editing than the use of crRNA with tracrRNA presumably because two components (sgRNA and Cas9) can assemble more efficiently than three components (crRNA, tracrRNA and Cas9) to form a functional CRISPR-Cas9 SDN.

Generally, gene constructs encoding sgRNAs are generated and sgRNAs are obtained by either *in vitro* or *in vivo* transcription of these genes. Typically, RNA polymerase III-dependent U6, U3, SP6, T3, or T7 promoter is used to drive the sgRNA genes. The specific promoter used in the gene construct determines the particular nucleotide that will be present at the 5' end of the the sgRNA: it will be one G with U6 promoter, two G residues with SP6, T3 and T7 promoters, and a single A with U3 promoter. Initially, the target sequences were carefully selected to match the 5' G or GG of sgRNA with C or CC in the target sequence. Therefore, the general form of sgRNA spacer region became 5'G...NGG3' (NGG is the PAM), or 5'GG...NGG3'. These sequences may be expected to occur, on an average, every 32 and 128 bp, respectively, when both the DNA strands are taken into consideration; this would limit the choice of target genomic sites to some or even considerable extent. One way to relieve this restriction is to simply ignore the requirement for T, C or CC in the target sequence for pairing with the 5' terminal A, G or GG of sgRNA since their pairing is not important for sgRNA function (Carroll, 2014; Sander and Joung, 2014, Bortesi and Fischer, 2015).

Several web-based software, *e.g*., ZiFiT, Benchling, Cas-OFFinder, etc. are available for identification of potential CRISPR-Cas9 target sites in the given DNA sequence (Table 5.1). The web tool sgRNA Scorer 2.0 is designed to identify target genomic sites and support designing of sgRNAs for all CRISPR systems for which the PAM sequence and the spacer length can be specified. The CRISPR-DO web tool is presumably the best of the sgRNA designing tools currently available. This tool annotates each target sequence with the PhastCons conservation score for their functional characterization where such data are available. Generally, these tools rank the sgRNAs based on their predicted specificity and indicate the likely off-target genomic sites, but the accuracy of predictions can vary.

The efficiency of gene editing is affected by certain features of sgRNAs, such as base composition of sgRNA, the particular strand of DNA duplex targeted by sgRNA, length of paired region, etc. The sgRNAs having average GC content seem to be more efficient in gene editing than those with either unusually high or low GC contents. sgRNAs that target the non-transcribed strand are more effective than those targeting the transcribed strand. Further, an increase of 5 nt in the double-stranded region of sgRNAs for SpCas9 and SaCas9 enhances efficiency. Similarly, base substitutions in the stretch of four T residues located downstream of the seed sequence is reported to increase efficiency by changing transcription rate of the sgRNA (Sander and Joung, 2014; Wang, T. et al., 2014; Tycko et al., 2016). The efficiency of editing is markedly affected by the affinity of Cas9 for different sgRNAs, which is influenced by the last four bases in the spacer sequence: Cas9 nuclease associates preferentially with those sgRNAs that have purine residues in these positions.

5.5. GUIDE RNA/sgRNA LIBRARY

A gRNA library or sgRNA library is a collection of oligonucleotide-derived sgRNA vectors specific for many genes; ideally, it should cover all the protein coding genes involved if biological pathways, including developmental pathways, are to be dissected and investigated. Further, the library may have multiple sgRNAs targeting different sites within each gene. The oligos in the library are created by chip-based oligonucleotide synthesis, which is a costly, laborious and time-taking exercise. Therefore, the current approach involves synthesis of pools of sgRNAs using synthetic oligonucleotides by employing PCR-based and Gibson-assembly-based amplifications (Koike-Yusa et al., 2014; Shalem et al., 2014; Wang, T. et al., 2014). Alternatively, a ligation-independent cloning (LIC) procedure can be used to create large pools of sgRNA vectors from chip-synthesized oligonucleotides; this procedure does not require amplification of the sgRNA vectors

A sgRNA library is a powerful and valuable resource for forward as well as reverse genetic analyses and a variety of other investigations designed to probe and unravel biological functions. For example, the sgRNA library-based CRISPR-Cas9 SDNs were stably transfected into human and mouse cells. The transfected cell populations were subjected to phenotypic selection, and genomes of the selected populations were analyzed by deep sequencing. This permitted identification of the specific sgRNAs enriched in the selected cell populations; presumably, the genes targeted by the respective sgRNAs could be related with the observed phenotypic changes. Previously, short hairpin RNA libraries were used for achieving these objectives. However, the hairpin RNA libraries can facilitate gene knockdown, *i.e.*, suppression of gene expression for a period. But the sgRNA libraries, in conjunction with Cas9 nuclease, can produce both gene knockouts and gene knock-ins. sgRNA libraries with about 64,000 and 87,000 distinct sgRNAs have been created for mouse and human genomes; these libraries were useful for positive as well as negative forward genetic phenotype screens (Doudna and Charpentier, 2014; Koike-Yusa et al., 2013; Shalem et al., 2014; Schmidt et al., 2017).

A sgRNAs library was used to determine the effects of different bases at specific positions in the pairing region of sgRNAs on their gene editing activity. It was observed that C/G at base positions 17 and 18 tend to favour higher gene editing activity, while A/T is preferable at positions 14 and 15. Further, it is desirable that sgRNAs have more than 35% overall GC content. sgRNA libraries have been used for high throughput identification of genes responsible for drug resistance in cancer cells, and those crucial for cellular functions. Some sgRNA libraries are available free of cost from the concerned researchers, *e.g.*, the sgRNA **K**nock**O**ut **LIBR**ar**Y** (sgRNAKOLIBRY; Katrekar et al., 2017; Schmidt et al., 2017). Genome-wide and pathway-specific gRNA libraries for efficient screening are also available commercially.

5.6. CRISPR-Cas9 NICKASES

A CRISPR-Cas9 nickase induces cut in only one of the two strands of the target DNA duplex, *i.e.*, it induces nicks or single-strand breaks (SSBs). The nickases are based on mutant Cas9 nucleases, in which one of the two nuclease domains has been inactivated by amino acid substitution(s) so that only one cleavage site is functional. Induction of a D10A mutation in the RuvC-like nuclease domain of Cas9 produces a nickase that cuts the target strand, *i.e.,* the strand complementary to sgRNA (Fig. 5.5). but a H849A mutation in the HNH domain of Cas9 yields a nickase that nicks the non-complementary strand of target DNA. The nickase with active HNH domain seems to have higher efficiency than that with active RuvC-like domain, and it leads to highly precise HDR with little or no indels due to NHEJ. The logic for the development of nickases is that nicks are not preferred substrates for NHEJ, and they show much higher proportion of HDR than NHEJ (Cong et al., 2013; Mali et al., 2013).

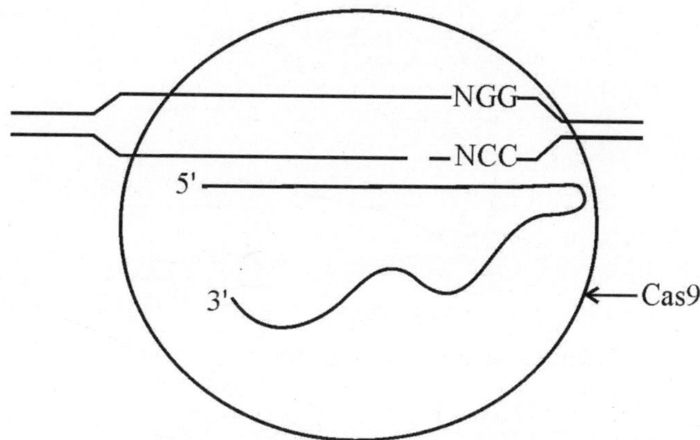

Fig. 5.5. Cleavage of single strand by CRISPR-Cas9 nickase mutated in the RuvC-like cleavage domain. The nick is produced in the strand paired with the sgRNA. A Cas9 mutated in the HNH cleavage domain will nick the other strand, the strand having the PAM sequence (based on Sander and Joung, 2014).

Experimental evidence shows that the use of single CRISPR-Cas9 nickases can induce mutations due to indels at some genomic loci. The frequency of these mutations varies with the genomic site from low to high. The indels might result from conversion of a nick into a DSB as a result of passing of the replication fork through the nicked site. The frequency of HDR, however, may be comparable to that obtained with the corresponding Cas9 nuclease at some sites, but it might be greatly reduced at other sites (Sander and Joung, 2014).

5.7. SPLIT-Cas9 NUCLEASE

Cas9 nuclease is a large protein with 1,368 amino acids and molecular mass of 160 kDa. Cas9 must be fused with nuclear localization signal (NLS) for achieving gene editing in eukaryotes; this increases the number of amino acids to 1,414 and the mass to 163 kDa (recombinant SpCas9 produced in *E. coli*). The large size of Cas9 interferes with the delivery of its gene construct via viral vectors into human cells. Therefore, attempts are being made to devise suitable strategies to facilitate Cas9 delivery and its regulated application for genome editing. These efforts include search for smaller natural Cas9 nucleases from other bacterial species, engineering of SpCas9 to reduce its size without affecting its function and creating a split-Cas9 that can be reassembled into a complete and functional enzyme inside the cells. There are two distinct approaches to create split-Cas9 based on sgRNA-Cas9 interaction and trans-splicing by a split intein.

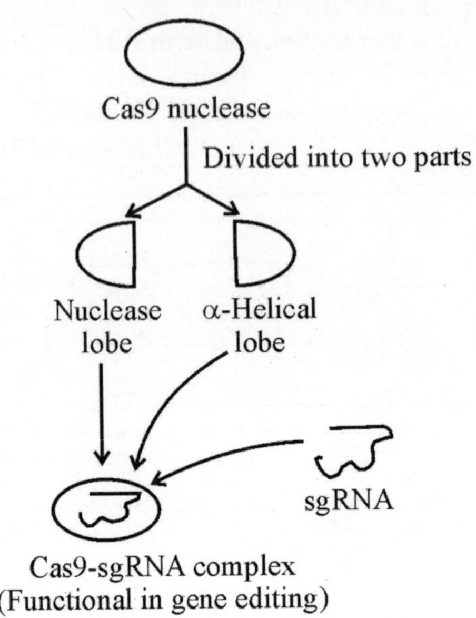

Fig. 5.6. Split Cas9 is assembled into a complete Cas9 nuclease on interaction with sgRNA. The Cas9 nuclease activity can be regulated by a controlled expression of sgRNA. A mine rebust regulation can be achieved by fusion of the two lobes with inducible dimerization domains and using then in conjunction with, sgRNAs in capable of supporting diverization of the two Cas9 lobes.

Cas9 nuclease molecule is made up of two lobes, *viz*., a nuclease lobe and an α-helical lobe. When sgRNA binds Cas9 nuclease, the nuclease lobe of the enzyme rotates ~100^0 in relation to its α-helical lobe; this generates a cleft in the enzyme, into which the target DNA can bind. The conformational change in Cas9 lobes seems to be mediated by their interaction with sgRNA, and that of sgRNA-Cas9 complex with the target DNA. Wright et al. (2015) created a split-Cas9 enzyme by separating the nuclease and the α-helical lobes of the enzyme and expressing them as separate polypeptides. When the two polypeptides were purified and mixed with full-length sgRNA, they assembled to form a stable sgRNA-Cas9 protein complex that functions in gene editing in the same way as does the natural Cas9 nuclease (Fig. 5.6). But the effectiveness of split-Cas9 enzyme is lower than that of the native enzyme, and it cannot be used with shortened versions of sgRNA. The function of split-Cas9 enzyme can be regulated in terms of time and period of activity by controlling the expression of sgRNA. For example, the two lobes may be fused with exogenous inducible dimerization domains and used with sgRNA variants lacking the ability to support dimerization of the two Cas9 lobes. This method would allow a robust regulation of CRISPR-Cas9 activity and eliminate the risk of leaky expression.

Another strategy uses split intein-mediated reassembly of the two parts of Cas9 polypeptide to recover functional Cas9 nuclease (Fig. 5.7; Truong et al., 2015). In this approach, the Cas9 polypeptide is carefully divided into two parts. The DNA sequences encoding these two parts are fused with DNA sequences encoding the appropriate parts of a split intein system; this yields two gene constructs one encoding the N-terminal part of Cas9 fused with the N-Intein, and the other producing the C-Intein fused with the C- terminal part of Cas9. When the two gene constructs are expressed together, protein splicing mediated by the split inteins will link the two parts of Cas9 by *trans*-splicing yielding a Cas9 that is fully functional in gene editing like the wild type Cas9. The split-Cas9 strategy reduces the size of individual gene constructs and greatly facilitates the delivery of CRISPR-Cas9 SDNs into the cells.

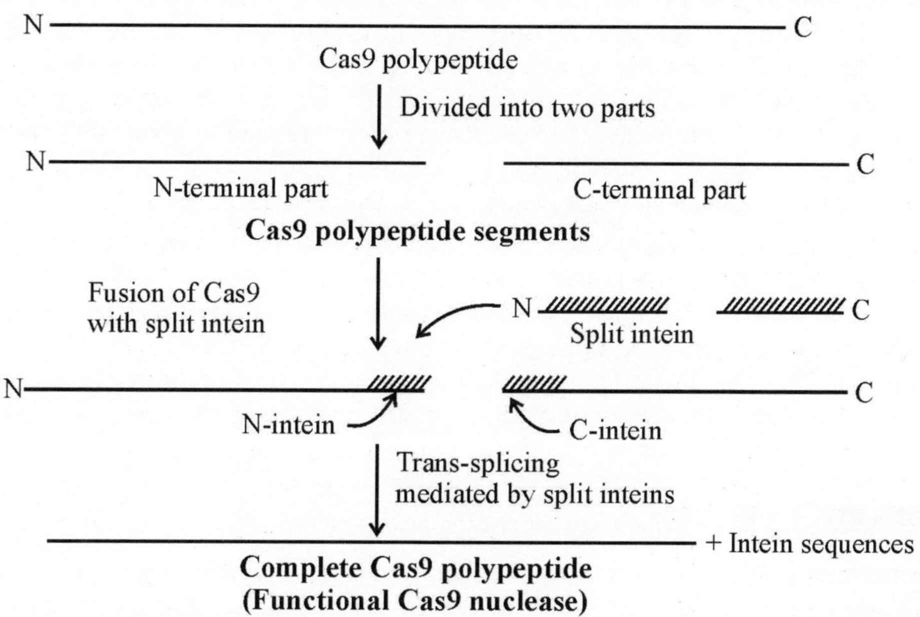

Fig. 5.7. The split Cas9 nuclease is reassembled into complete Cas9 nuclease by trans-splicing mediated by split inteins.

5.8. DELIVERY OF CRISPR-CAS9

The CRISPR-Cas9 SDNs can be delivered as DNA in form of gene constructs encoding the sgRNA and Cas9 nuclease, as mRNA encoding Cas9 and sgRNA, or as Cas9 protein and sgRNA either separately or complexed as ribonucleo-protein (RNP). The RNP has been microinjected into zygotes of several animals, electroporated into cultured human cells, introduced into cells via lipofection mediated by cationic lipids, by nucleofection or in association with cell-penetrating peptides (CPPs). The Cas9 RNP can be encapsulated in gold nanoparticles and microinjected along with a synthetic polymer that enhances endosomal escape (Brooks and Gaj, 2018). The CRISPR-Cas9 SDN has been microinjected into the gonads of roundworm in form of plasmid DNA as well as nucleoprotein. They have been delivered into animal cells in nonreplicating plasmid vectors by Lipofectamine-mediated transfection, nucleofection and electroporation for transient expression. They have also been delivered via lentiviral vectors into human and mouse cells (Doudna and Charpentier, 2014; Sander and Joung, 2014). The delivery as DNA via plasmids requires transfection protocols and can lead to plasmid integration in the cell genome, immune response to the bacterial sequences and safety concerns due to the presence of antibiotic resistance markers in the plasmids (Ramakrishna et al., 2014).

In CPP-mediated delivery, the Cas9 is conjugated with the CPP via a thioether bond, while CPP-sgRNA complex forms positively charged nanoparticles; these can be delivered either sequentially or simultaneously into the animal cells. This method was able to cause efficient gene editing in several types of human cell lines. Further, the off-target effects seem to be lower than those generated by plasmid-mediated delivery most likely due to the short period of SDN action before it is degraded. This method does not use plasmid and there is no need for transfection/microinjection/electroporation procedures. The CPP-Cas9 conjugate can be produced on commercial scale to save researchers' time and effort. This method has the potential for use in human gene therapy (Ramakrishna et al., 2014). CRISPR-Cas9 SDN has been introduced into plants like wheat, rice, sorghum, etc. in form of DNA by several methods, including PEG-mediated delivery into protoplasts, *Agrobacterium*-coculture and microparticle bombardment using particle gun. In general, transient expression of the SDN would be adequate to achieve efficient gene editing (Doudna and Charpentier, 2014; Sander and Joung, 2014).

The delivery of ZFNs is comparatively considerably easier than that of TALENs, while that of CRISPR-Cas9 SDNs is the most problematic. This is because the size of the SDN constructs is the largest in the case to CRISPR-Cas9. Some bacteria are known to produce smaller Cas9 nucleases than the commonly used SpCas9. The smaller Cas9 nucleases are effective in targeted gene editing and their smaller size might facilitate the delivery of SDNs based on them. In addition, the split-Cas9 strategy reduces individual gene construct size and facilitates their delivery into cells (Section 5.7).

5.9. OFF-TARGET EFFECTS

It is expected that a 20 bp sequence would specify a unique site in a genome as large as the human genome. But a perfect match for all the 20 bases of this sequence is not required for CRISPR-Cas9 action. The CRISPR-Cas9 SDNs can induce cleavage and mutations at off-target sequences showing one or more mismatches with the sgRNA sequences. There is no simple rule for an accurate prediction of the specificity and off-target activity of a given sgRNA-Cas9 complex. Sometimes, the observed mutation rates at many off-target sites are comparable to those at the target sites. In some studies, off-target sites with more mismatches with the targeted sequence are modified more frequently than those with fewer mismatches. It has been observed that about 98% of SpCas9 sgRNAs targeting exons and promoters of human genes have one or more off-target sites exhibiting up to three mismatches. Off-target sites with as many as six mismatches have been edited by SpCas9, but off-sites with up to three mismatches are more common (Carroll, 2014; Sander and Joung, 2014; Tsai and Joung, 2016; Komor et al., 2017b).

In general, the stringency of base-pairing requirement decreases as the distance of a base from the PAM sequence increases. Thus, mismatches are the least tolerated for bases located closest to the PAM and are the most tolerated for the bases close to the 5' ends of sgRNAs. It seems that the 8-12 bases nearest to the PAM sequence are crucial for specificity, but some bases at the 5' end of sgRNAs seem to have dramatic effect. In addition, different substitutions at a single base position may differ in their effects on specificity. It appears that the Cas9 nuclease stabilizes the sgRNA-target DNA hybrid to allow some mismatches between them, and Cas9 may also govern the base-pairing requirements (Carroll, 2014; Sander and Joung, 2014; Komor et al., 2017b).

Typically, seed sequences of sgRNAs are based on the standard reference genome of the concerned species. It is likely that there might be mutations, especially SNPs (single nucleotide polymorphisms), at few to several locations in the genome to be edited as compared to the reference genome used for sgRNA designing. This might increase the similarity of some off-target sites to the on-target site and may even create some new off-target sites. In one study, a single SNP in one off-target site reduced the mismatches from three to two, which remarkably increased the editing at this off-target site (Tycko et al., 2016). This source of off-target activity may be guarded against by whole genome sequencing of the concerned individual/cell line/tissue/inbred line and using this sequence for designing of sgRNAs.

Another source of off-target activity might be the method of production of sgRNAs. The sgRNAs can be obtained by synthesis of oligonucleotides or by transcribing gene constructs integrated in plasmid DNAs. The DNA oligonucleotides are prone to small deletions since the coupling reaction used to add individual bases during oligonucleotide synthesis is only 99% efficient. Therefore, sgRNAs should be obtained by transcription from a plasmid carrying the sequence-verified gene encoding the desired sgRNA. It may be pointed out that as compared to the error rate of 1 or 2 per 100 nucleotides during DNA oligonucleotide synthesis, the error rate of T7 polymerase is only 1 per 20,000 nt, while that of RNA Pol III is merely about one in 10^7 nt. Therefore, sgRNAs introduced in form of DNA or RNA should be carefully assessed for their specificity (Tycko et al., 2016).

Finally, the CRISPR-Cas9 system itself is a likely source of off-target activity. This system is involved in the defense of bacteria and archaea against viruses and plasmids. Since the viral genomes are rapidly evolving, it would be advantageous to these prokaryotes if the CRISPR-Cas9 system could accommodate few mismatches in the target sequences represented in the relevant spacers of the *crispr* locus (Carroll, 2014).

5.9.1. Monitoring of off-target cleavage

The various methods developed for assessing CRISPR-Cas9 off-target effects can be divided into the following two groups: (i) biased and (ii) unbiased off-target detections (Table 5.4). The biased detection generally involves prediction of off-target sites, PCR amplification of these sites after gene editing and determination of editing by a suitable procedure. The earliest and the most widely used biased detection method involves computational prediction of off-target sites based on homology with the sgRNA seed sequence, but the predictions are not very reliable. Therefore, editing at the predicted off-target sites is detected by an assay based on a mismatch-sensitive endonuclease like T7 Endonuclease I (T7E1) or targeted deep DNA sequencing. The T7E1 mutation mismatch is a relatively insensitive assay for indels induced by SDNs. The Surveyor assay analyses the band intensity of gel electrophoresis images and has a detection limit of ~5% indels; it should be used for quick initial assessment of off-target activity. Thus, T7E1 and Surveyor assays can be used only when the frequency of off-target editing is high. Targeted deep sequencing should be used when the frequency of off-target editing is low; it is the most useful for verifying editing at off-target sites identified by unbiased genome-wide methods (Tycko et al., 2016).

Table 5.4. The novel methods for detection of off-target editing (based mainly on Tycko et al., 2016 and Tasan and Zhao, 2017)

Method	Description	Remarks
Targeted deep sequencing	Sequencing of predicted off-target site amplicons; computational quantification of the percentage of reads having indels near PAM sequence	*Biased approach*; more sensitive and scalable than 'Surveyor' assay; best used as complement to unbiased detection methods
IDLV[@] capture	IDLV dsDNAs integrated into DSBs induced by CRISPR-Cas9 within cells; sites of IDLV integration determined by high throughput sequencing of products of LAM-PCR	*Unbiased approach*; low IDLV integration efficiency; false off-target sites may be detected due to random IDLV integration; can capture spontaneous DSBs, but suitable controls can filter out the background DSBs
GUIDE-seq (genome-wide unbiased identification of DSBs enabled by sequencing)	Captures DSBs with a dsODN barcode; integration sites determined by high-throughput sequencing of tag-specific PCR products	*Unbiased approach*; straightforward protocol; computational tools available online; efficient delivery of dsODN required; dsODN may be toxic to some types of cells; difficult to use for *in vivo* studies; can capture spontaneous DSBs, but suitable controls can filter the background DSBs out
BLESS (breaks labeling, enrichment on streptavidin and next-generation sequencing)	Ligation (biochemically) of NGS sequencing adapters to the genomic DNA ends exposed at DSBs; Cas9-mediated DSBs (located near PAM) separated from spontaneous DSBs (randomly distributed in the genome) by computational filtering	*Unbiased approach*; applicable to *in vivo* studies; captures only the DSBs present at the time of cell/tissue sampling, and does not detect the already repaired DSBs; current procedure requires relatively large samples
Digenome-seq (*in vitro* Cas9-digested whole-genome sequencing)	Purified genomic DNA cleaved by Cas9 associated with sgRNAs; WGS after adaptor ligation; read alignment enables DSB site identification	*Unbiased approach*; applicable to any cell type; more sensitive than GUIDE-seq; WGS may be costly; must be used with another method to confirm that the *in vitro* discovered sites are actually edited *in vivo*; the results obtained may not represent the situation *in vivo*
HTGTS (high-throughput genome-wide translocation sequencing)	An additional DSB at a "bait" site needed for translocation between the genomic sequence and the "bait"; detection of translocation by LAM-PCR using a biotinylated primer, which anneals to the bait, followed by enrichment based on streptavidin, and high-throughput sequencing	*Unbiased approach*; applicable to *in vivo* studies; translocations are rare events, and could be difficult to detect
Circle-Seq (circularization for *in vitro* reporting of cleavage effects by sequencing)	Genomic DNA sheared and circularized; remaining linear DNA, if any, digested; circles cleaved by CRISPR-Cas9 treatment *in vitro*; cleaved ends of the linearized DNA ligated with an adaptor and subjected to paired-end high throughput sequencing	*Unbiased approach*; must be used with another method to confirm that the *in vitro* discovered sites are actually edited *in vivo*; the results obtained may not represent the situation *in vivo*
WGS (whole genome sequencing)	Whole genome sequenced and compared with the control	*Unbiased approach*; preferred approach for *in vivo* studies; captures naturally occurring variations, which complicates data analysis and interpretation

[@] DSB, double-strand break; dsDNA, double-stranded DNA; dsODN, double-stranded oligodeoxynucleotide; IDLV, integrase defective lentiviral vector; LAM-PCR, linear amplification-mediated PCR; NGS, next generation sequencing; WGS, whole genome sequencing.

The unbiased genome-wide detection methods are designed for genome-wide detection of off-target editing. It has been proposed to use whole-genome sequencing (WGS) as an unbiased off-target detection method, but it is not sensitive enough to be used in genetically heterogeneous populations. In addition, the cost of deep sequencing (30-50x read coverage) of millions of cell genomes is prohibitive. Some of the other unbiased methods are as follows: breaks labelling, enrichment on streptavidin, and sequencing (BLESS); high-throughput, genome-wide translocation sequencing (HTGTS); genome-wide, unbiased identification of DSBs evaluated by sequencing (GUIDE-seq); capture of integration-deficient lentiviral vectors (IDLV capture) into DSBs; and *in vitro* nuclease-digested whole-genome sequencing (Digenome-seq). The unbiased detection methods tend to be more demanding, less sensitive and have lower throughput than the biased assays based on targeted deep sequencing. The sensitivity reported for GUIDE-seq and Digenome-seq is 0.1% and that for IDLV capture is 1%. The sensitivity is affected by the number of cells assayed, the depth of sequencing, and the method used to differentiate between the DSBs produced spontaneously from those generated by CRISPR-Cas9 (Tycko et al., 2016).

Each of the above unbiased methods requires a specific DSB enrichment protocol, might introduce some subtle artefacts, and may not be applicable to all the experiments. In addition, each method has some or the other limitation. For example, IDLV capture method has high background integration, and IDLV integration occurs near but not at the DSB site. The advantages of GUIDE-seq are experimental simplicity, very high sensitivity, and high correlation of detected off-target action with the off-target editing obtained in living cells. But it suffers from the limitation that a dsODN (double-stranded oligodeoxynucleotide) tag must be introduced into cells, which may be a challenging task in some cell types. The advantage of HTGTS is that it does not require any other component except the CRISPR-Cas9, but translocations are rare events, and the detection method is biased against DSBs located at large distances. The BLESS method does not depend on endogenous DSB repair mechanisms, but it can capture only those DSBs that exist at the moment when the cells are harvested, ~10 million cells are required, and the multistep protocol is technically demanding (Tsai and Joung, 2016).

The unbiased methods have revealed that sequence homology alone is not enough for a reliable prediction of off-target sites, and CRISPR-Cas9 can cause mutations at sites lacking the canonical PAM sequences. However, all the off-target sites identified by unbiased methods have sequence homology with the target site, but the two may differ for up to 6 nt in their protospacer and non-canonical PAM sequences. It has been suggested that the biased and unbiased methods should be used to complement each other: the unbiased methods may be used to discover the off-target sites *in vitro* and then targeted sequencing can be used to validate these sites *in vivo* (Tsai and Joung, 2016; Tycko et al., 2016).

In addition, some *in vitro* methods for assaying off-target activity have been developed and used. But the *in vitro* assays may not capture the actual *in vivo* picture since the accessibility of the sites might be affected by local chromatin environment. Similarly, the cell-based assays could fail to capture the off-target activities that affect traits like cell survival. Therefore, a combination of *in vitro* and cell-based assays for off-target effects of SDNs might be more useful (Tsai and Joung, 2016; Tasan and Zhao, 2017).

5.9.2. Strategies for minimizing off-target editing

The factors affecting efficiency and specificity of CRISPR-Cas9 SDNs include characteristics of Cas9 (including its variants and orthologs), base sequence and secondary structure of sgRNA, epigenetic state of targeted site, tissue and cell type, etc. Among these, sgRNA and Cas9 seem to be the logical choice for manipulations to minimize off-target effects since they would have general applicability (Table 5.6). For example, the inherent specificity of the smaller sized Cpf1 nucleases seems to be greater than that of SpCas9. However, the factors related to the biological systems subjected to editing, including the organism, tissue/cell type, the local epigenetic states of the genome, etc. also deserve attention, but these issues will be relevant for specific cases. The various methods used for reducing off-target activity of SpCas9-based SDNs are as follows: (i) truncation of sgRNA spacer length to 20 nt or less, (ii) reduction in the period of Cas9 activity, (iii) use of a pair of nCas9 to target two neighbouring sites, (iv) using pairs of dCas9-FokI nuclease fusions, and (v) use of Cas9 mutants with enhanced specificity and reduced off-target activity. These strategies either aim to increase the specificity of CRISPR-Cas9 (improved sgRNA design or enhanced Cas9 specificity) or to limit the period of nuclease action in cells (Tsai and Joung, 2016; Komor et al., 2017b).

5.9.2.1. *Strategies based on sgRNA*

The Cas9 nuclease can tolerate one to few mismatches between sgRNA and target DNA sequences, which leads to off-site cleavage and editing. Consecutive mismatches seem to have greater effect on sgRNA specificity than the same number of scattered mismatches. Several features of sgRNAs can be modified to enhace the specificity of CRISPR-Cas9 SDN system (Tycko et al., 2016).

1. sgRNA design and evaluation

The first and quite valuable strategy to reduce off-target effects is to select the most unique target site from all the possible sites within the gene of interest using a suitable bioinformatics tool (Table 5.1). The guide RNA has six functional modules, each of which can be altered to modify the on-target efficiency of sgRNAs. It is advisable that the overall GC content of sgRNAs is over 35%. Certain chemical modifications like nucleoside phosphoramidite modification of the three nucleotides at the 5' and 3' termini enhance sgRNA stability, increase target binding/editing efficiency and often decrease off-target effects (Sander and Joung, 2014; Wolt et al., 2016).

A reduction in the length of sgRNA spacer region from 20 to 17 or 18 nt is reported to enhance the specificity of CRISPR-Cas9; such sgRNAs are called truncated sgRNAs (tru-sgRNA; Fig. 5.3D, Table 5.5). The tru-sgRNAs seem to function as efficiently as full length sgRNAs for on-target editing, but they show reduced off-target effects since the shorter sgRNA-target DNA hybrid is more sensitive to one or two mismatches. The tru-sgRNAs also seem to increase the specificity of nickase pairs and that of the Cas9-effector fusion proteins (Fu et al., 2014; Sander and Joung, 2014). But the 5' and 3' truncations of sgRNAs might lead to some to severe reduction in efficiency of on-target cleavage; a reduction in the 5'region to 13 or 15 nt generates 'dead guides', which can guide Cas9 to the target site but fail to activate its nuclease action. Further, the tru-sgRNAs are not compatible with the enhanced specificity Cas9 mutants, *viz.*, eSpCas9 and SpCas9-HF (Tycko et al., 2016).

The use of 3' truncated versions of sgRNAs (truncation of the sequence derived from tracrRNA) also enhances specificity. Addition of two G residues at the 5' terminus of the sgRNA, *i.e.*, the GGx20 strategy, may increase specificity by 10-100-fold. The additional 5'G residues need not base-pair with the target sequence. The GGx20 strategy is likely to enhance destabilization of the sgRNA-DNA hybrid at the off-target sites with few mismatches since the extra G nucleotides are not easily accommodated into the binding pocket of Cas9 cleavage domain (Sander and Joung, 2014; Tycko et al., 2016).

Table 5.5. Features of guide RNAs (gRNAs) relevant for gene editing (based on Tycko et al., 2016)

Feature of gRNA	Modification	Effect on gene editing	Remarks
Length of gRNA seed	Long, *e.g.*, 30 nt	Processed to the natural 20 nt	-
	Truncated: 17 or 18 nt	Increased precision; reduced off-target activity	May create new off-target sites due to shorter length; not compatible with eSpCas9 and SpCas9-HF
	Length 14 or 15 nt	'Dead guide'; associated Cas9 lacks nuclease activity	Binding to target DNA not affected; suitable for transcriptional activation/ repression
Length of double-stranded region	Addition of 5 nt	Increased efficiency	Effect due to altered RNA structure; also, in case of SaCas9
Length of the 3' region	Truncation at the 3' end	Enhances specificity	-
Overall GC content	High GC contest in the seed sequence	Variable effects on specificity	The overall GC content should be over 35%
Sequence shortly downstream of the spacer	The string of four T residues mutated	Increased efficiency	Effect due to altered rate of transcription; also, in case of SaCas9
5' end sequence	(GGx20)	Increased specificity (10- to 100-fold)	Little to severe reduction in on-target efficiency
Chemical modification	Addition of certain groups at the 5' and 3' ends*	Significant improvement in editing efficiency	Increase in off-target effects at some genomic sites

*2'-O-methyl (M), 2'-O-methyl-3'phosphorothioate (MS), or 2'-O-methyl-3'-thioPACE (MSP) groups

The specificity of a sgRNA for a target sequence can be assessed by screening a library of partially degenerate variants of the sgRNA specific for the target. These sgRNA variants can be evaluated for their ability to bind to off-target sites by associating them with a dCas9 fused to a transcription activator domain (Section 10.16). Another approach uses a library of partial variants of the proposed target sequence to evaluate their cleavage *in vitro* by the Cas9-sgRNA complex. However, it is not possible to predict the number of mismatches the sgRNA for a given target site can tolerate since there seems to be some variation from one target site to the other (Sander and Joung, 2014).

A library-on-library *in vivo* assay strategy allows simultaneous evaluation of a single sgRNA across more than 1,000 loci. This approach can be used for facile high throughput assay of the various factors affecting HDR and/or NHEJ pathways involved in gene editing. Findings from studies based on library-on-library method were used to develop the software sgRNAScorer 1.0

(http://crispr.med.harvard.edu/sgRNAScorer) for identifying and/or scoring sgRNA target sites in any given sequence. The version 2.0 of sgRNAScorer has been developed that allows identification of sgRNA target sites for any CRISPR system with known PAM sequence requirement and defined spacer length (Chari et al., 2015, 2017).

2. Regulation of sgRNA function

It has been reasoned that the SDN will initially act on the targeted site, for which it has the highest affinity. But after the target site has been edited, it will continue to act on off-target sites so that the off-target effects will continue to accumulate with the duration of SDN action. Therefore, off-target activity could be minimized by limiting the period of SDN activity. This can be done by transcriptional and post-transcriptional regulation. The transcriptional control involves the use of inducible or tissue-specific promoters, but this approach does not allow a precise temporal control on gene editing and is difficult to implement in many cell types/tissues due to the lack of suitable promoters. Post-transcriptional control offers a more precise and flexible control (Tang et al., 2017).

The activity of CRISPR-Cas9 SDN can be regulated by manipulating either Cas9 or sgRNA. The regulation through the sgRNA is preferable to that via Cas9 since this would allow simultaneous but independent regulation of editing of different genes in the same cell. sgRNA regulation can be achieved by using such sgRNAs, in which the spacer is sequestered in the absence of specific ligands that bind RNA, use of aptazymes and blocking sequences that are removed by self-cleavage (catalysed by the ribozyme component of the aptazymes when activated by an appropriate ligand), employing antisense RNAs to inactivate sgRNAs by base-pairing, and recruiting ligand-responsive regulator proteins to sgRNAs. The methods of regulation of sgRNAs by means of aptazymes and blocking sequences, and by integration of aptamers in the sgRNAs are briefly described below.

The blocking sequence is complementary to and base-pairs with the sgRNA spacer sequence, as a result of which it blocks the sgRNA function. The aptazyme comprises a self-cleaving ribozyme with an integrated aptamer domain; the ribozyme is inactive till the aptamer interacts with the appropriate ligand molecule. Aptamers are single-stranded DNA (ssDNA) or single-stranded RNA (ssRNA) that bind specific ligands with high affinity and specificity. Ligand molecules that bind to Aptamers include proteins, peptides, carbohydrates, small molecules, etc. The aptamers are typically developed by selection from large pools of random sequences, but natural aptamers do occur in riboswitches. The aptazyme is added to the 5' end of sgRNA and the blocking sequence is linked to the 5' terminus of the aptazyme sequence. This aptazyme-embedded sgRNA (agRNA) has the organization 5' blocking sequence-aptazyme-sgRNA 3' (Fig. 5.8A); it is inactive because the blocking sequence will be paired with the sgRNA spacer sequence (Fig. 5.8B). When the appropriate ligand is added, it interacts with the aptamer leading to changed aptazyme conformation, which activates the ribozyme; self-cleavage by the ribozyme removes the aptazyme as well as the blocking sequence from sgRNA. As a result, the sgRNA spacer sequence is freed, and the sgRNA becomes fully functional (Fig. 5.8C). This strategy was devised by Tang et al. (2017) who fused the theophylline-dependent aptazyme and a 17-nucleotide long blocking sequence to create agRNAs that enabled theophylline-regulated genome editing, base editing, and guanine-dependent transcriptional activation.

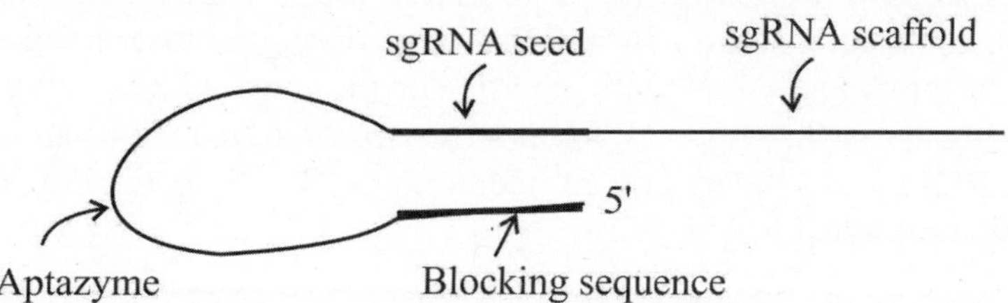

A. Aptazyme enabedded sgRNA (linear organization)

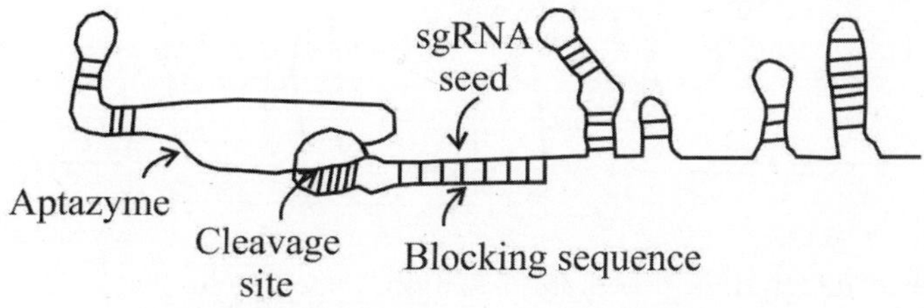

B. Aptazyme embedded sgRNA (with secondary structures)

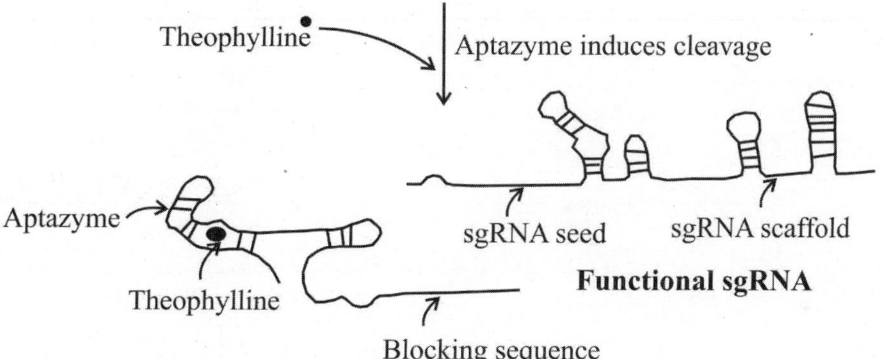

C. Self-cleavage by activated aptazyme releases functional sgRNA

Fig. 5.8. Schematic representation of the optazyme embedded sgRNA. **A.** Linear organization of the aptazyme embedded sgRNA. **B.** sgRNA with secondary structures. **C.** Theophylline ineracts with aptazyme, which becomes activated and induces self-cleavage to release functional sgRNA.

Kundert et al. (2018) integrated the theophylline binding RNA aptamer in different secondary structures of sgRNA. This aptamer interacts with high affinity and specificity with the small molecule theophylline that is not produced endogenously, is readily permeable into cells, and is used as drug for respiratory diseases. Aptamer integration within the upper stem and the hairpin regions inactivated the sgRNA so that it could not activate Cas9 (Fig. 5.9). These sgRNAs became active following interaction with theophylline and they could activate Cas9 to near normal level. In contrast, aptamer integration in the nexus region of sgRNA did not significantly reduce its activity (Fig. 5.10). But these sgRNAs were deactivated on interaction with theophylline. Thus, aptamer insertion in sgRNA secondary structure enables regulation of its activity by means of an exogenous ligand specific for the integrated aptamer. The aptamer can function both as off-to-on switch (the sgRNA is activated on interaction with the ligand) and on-to-off switch (the sgRNA is inactivated by an interaction with the ligand) depending on the site of aptamer integration in the sgRNA (Fig. 5.10).

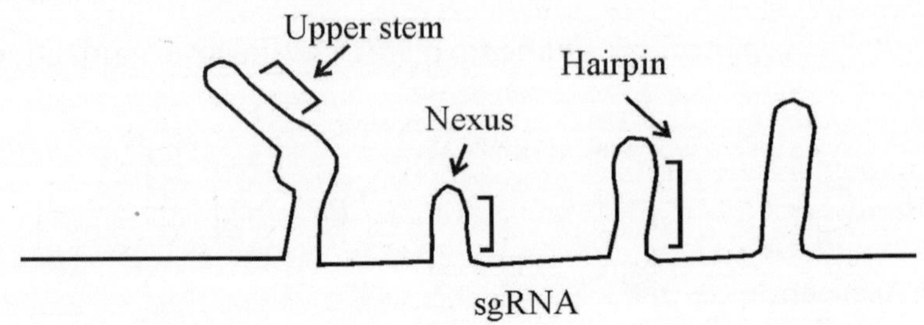

Fig. 5.9. Sites of aptamer integration. Integration within upper stem and hairpin leads to sgRNA inactivation; the sgRNA becomes active when the ligand binds aptazyme. But aptazyme insertion within nexus produces active sgRNA that is inactivated by ligand binding.

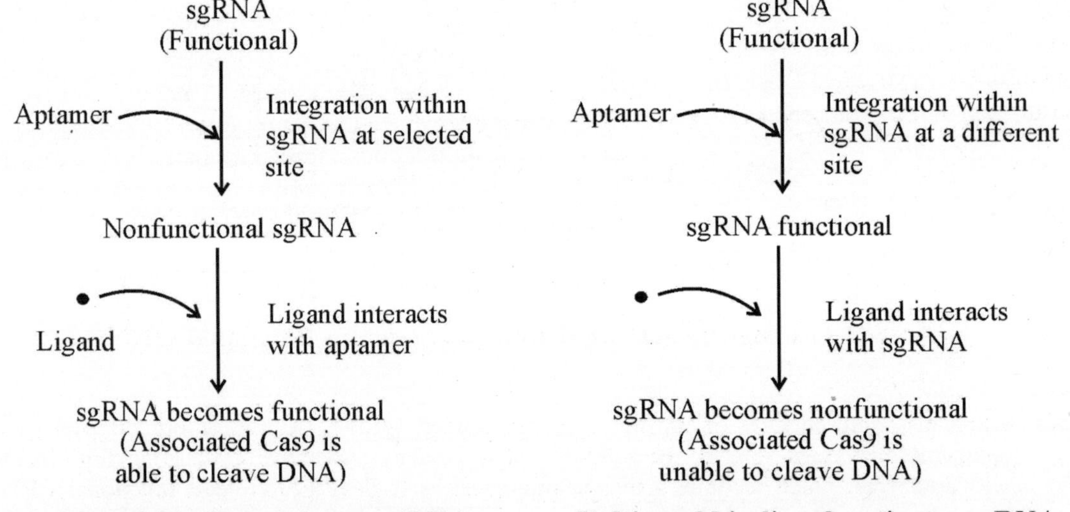

Fig. 5.10. Apatmer integration site determines the effect of ligand on sgRNA activity. **A.** Aptamer integration within upper stem and hairpin yields inactive sgRNA that is activated by ligand binding. **B.** Intgration of aptamer within nexus does not reduce sgRNA activity, while binding of the ligand deactivates the sgRNA.

Another alternative method for regulation of Cas9 action is based on the toehold mediated strand displacement technique. In this strategy, a sequence is inserted into sgRNA molecule, as a result of which the secondary structure of the scaffold region changes and the sgRNA in unable to interact with Cas9 nuclease. When a ssRNA or ssDNA molecule complementary to the inserted sequence is provided, it pairs with this sequence leading to a change in the secondary structure of the sgRNA scaffold region. This change restores the ability of sgRNA to interact with and activate the Cas9 nuclease. This method has been demonstrated *in vitro* and needs to be evaluated in *in vivo* systems (Jakimo et al., 2018).

5.9.2.2. *Strategies based on Cas9 nuclease*

It may be pointed out that Cas9 can induce DSBs at off-target sites having small indels, and it can use alternative PAM sequences, albeit with much lower efficiency than the regular PAM. Several variables related to Cas9 nuclease can be modified to increase on-target specificity and reduce off-target effects. Some of these variables are the concentration of Cas9 nuclease, the duration of Cas9 action, and the specificity of Cas9 nuclease, *i.e.*, the stringency imposed by Cas9 on the base-pairing between sgRNA and target DNA. In addition, nCas9 would necessitate the use of a pair of sgRNAs for each target site and, thereby, double the length of target sequence, which may enhance the specificity.

1. Use of CRISPR-Cas9 nickases

One of the approaches for minimizing off-target cleavage is to convert Cas9 into a nickase (nCas9; Section 5.6), which will induce only a nick at the target site (Fig. 5.5). A pair of sgRNAs can be designed for target sequences located close to each other in the targeted genomic region; each sgRNA of this pair will be complementary to a different strand of the target DNA. This sgRNA pair in conjunction with nCas9 will induce a pair of nicks close to each other (Fig. 5.11), which would behave as a DSB and support both NHEJ and HDR. It is important that nicks located up to 100 bp from each other in the complementary strands are repaired in the cells as if they were DSBs. The above strategy would require simultaneous target recognition by both the sgRNAs. As a result, it will double the length of the recognition sequence necessary for DSB induction and is expected to enhance the specificity of cleavage.

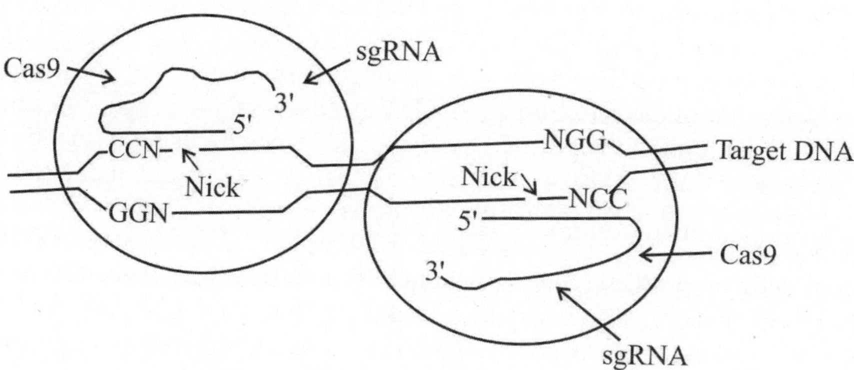

Fig. 5.11. Use of a pair of CRISPR-Cas9 nickases to induce a pair of nicks (one in each strand) close to each other. Cas9 is mutated in the RuvC nuclease site (based on Sander and Joung, 2014).

The above modification makes the CRISPR-Cas9 system analogous to ZFNs and TALENs, which must be used as SDN pairs. But this conversion has one important qualification: *the CRISPR-Cas9 nickases are not obliged to function as dimers*. In fact, nickases are so called because they function individually and induce nicks as well as mutations at on-target and off-target sites. At some off-target sites, the indel frequency is quite high although some studies do report greatly (up to 500-fold) reduced off-target editing. Therefore, the Cas9 must be converted into a nuclease that can function only as a heterodimer. One way to achieve this is to fuse dCas9 with FokI cleavage domain that has been mutated in its dimerization interface in such a way that it can function only as an obligate heterodimer.

The FokI-based strategy will increase the specificity of CRISPR-Cas9 SDN since two sgRNA binding sites will be mandatory for DSB induction. However, both the above strategies will greatly reduce the number of potential target sites since two PAM sequences must be present at strategic locations in the complementary strands of the target DNA. It may be pointed out that the binding site specificity of dCas9 seems to differ from that of Cas9 nuclease cleavage specificity. Therefore, the off-target activities of FokI-based CRISPR-Cas SDNs should be subjected to exhaustive evaluation (Sander and Joung, 2014; Bortesi and Fischer, 2015; Lee et al., 2016; Tsai and Joung, 2016).

2. Optimizing Cas9 concentration

It has been observed that off-target effects increase with the SDN concentration and this effect becomes the most pronounced at concentrations close to saturation. In general, a reduced expression of the SDN encoding genes results in increased specificity, but there is some decrease in the on-target gene editing efficiency. Therefore, the appropriate level of expression for each biological system and experimental condition needs to be optimized. The use of relatively weaker promoters, such as *H1* promoter, to drive the guide RNA encoding genes results in a greater reduction in off-target effects as compared to the on-target editing (Carroll, 2014; Tycko et al., 2016).

3. Optimizing the duration of Cas9 action

A constitutive SDN expression in transgenic individuals exposes the genome to SDN action for a long period of time. This might increase the efficiency of on-target gene editing, but it also tends to enhance off-target mutations. Therefore, use of inducible promoters, *e.g.*, heat-shock promoter, tetracycline inducible promoter, etc., to drive SDN genes would be highly desirable. Further, SDN expression can be limited to specific tissue and cell types by using tissue-specific and cell type-specific promoters (Carroll, 2014; Tycko et al., 2016). The use of inducible promoters to regulate the expression of CRISPR-Cas9 SDNs is sometimes called inducible CRISPR (iCRISPR). A tetracycline-inducible promoter has been used to drive Cas9 expression in human pluripotent stem cells to produce gene knockouts. The tetracycline inducible promoter has a tetracycline response element, which responds to doxycycline treatment and induces expression of Cas9 and Cas9 nickase. The iCRISPR systems inducible by other small molecules like rapamycin and 4-hydroxytamoxifen (4-HT) have also been created. In addition, split-Cas9-based light responsive Cas9 systems have been developed; one strategy uses the light inducible dimerization system Magnets, while another approach is based on the protein domains CRY2 and CIB1 that heterodimerize in response to light (Section 10.3.2.4; Hsu et al., 2014; Katrekar et al., 2017).

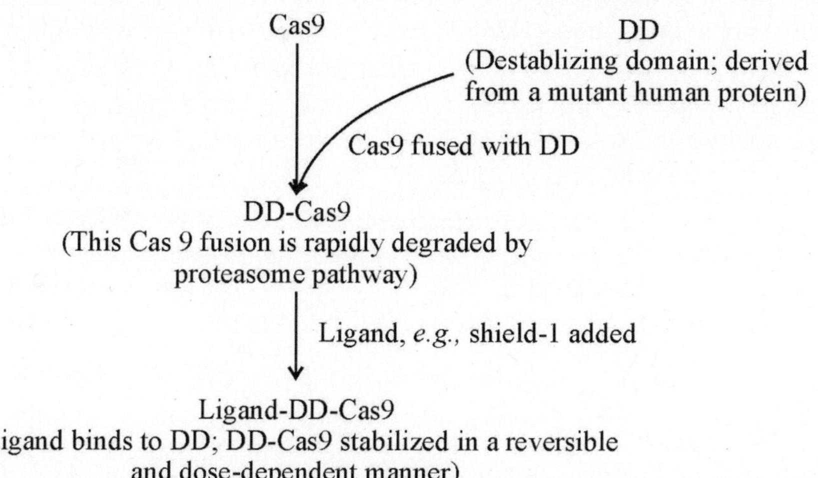

Fig. 5.12. A ligand-mediated regulation of Cas9 nuclease activity. The nuclease activity will be available in response to the ligand treatment only since DD-Ca9 will be rapidly destroyed by proteasomal degradation in the absence of the ligand.

An alternative method to limit the period of CRISPR/Cas9 action is based on conditional destabilization of Cas9. In this strategy, a destabilizing domain (DD) is fused with Cas9 to yield DD-Cas9, which is rapidly destroyed by proteasomal degradation due to the DD domain (Fig. 5.12). But when a high affinity ligand, *e.g.*, Shield-1, specifically binds to DD, the Cas9-DD is stabilized in a reversible and dose-dependent manner. Thus, the use of DD-Cas9 enables a highly efficient and easily scalable temporal regulation of nuclease activity by a controlled treatment with the ligand Shield-1. The effectiveness of this strategy was demonstrated *in vitro* as well as *in vivo*. This method for temporal control of Cas9 action is novel, very robust and is readily scalable (Senturk et al., 2017). In addition, anti-CRISPR proteins (Section 5.14) can be used to regulate the period of Cas9 action. These proteins prevent DSB induction by CRISPR-Cas9 and can be applied either as purified proteins or expressed in the cells using plasmid vectors (Shin, J. et al. 2017).

The method of SDN delivery can also help limit Cas9 activity to a relatively short period. For example, purified Cas9 and sgRNA may be delivered as RNP into animal cells or plant protoplasts by cationic lipid-mediated transfection or electroporation. The delivery as RNP leads to a burst-like Cas9 activity resulting in efficient on-target editing; the Cas9 is degraded within 24 hr and off-target editing is minimized. The RNP delivery via cationic lipid can lead to 20-fold improvement in the ratio of on-target to off-target editing. However, this mode of SDN delivery is best used *ex vivo* (Tsai and Joung, 2016; Tycko et al., 2016).

4. Enhancing the specificity of Cas9 nuclease

The specificity of CRISPR-Cas9 system depends on the ability of Cas9 to scrutinize the accuracy of and stabilize the base-pairing between gRNA and target DNA. Therefore, the Cas9 nuclease may be modified in such a way that it imposes a greater stringency in base-pairing between sgRNA and target DNA. Amino acid substitutions in those domains of Cas9 that are involved in interaction with target DNA have yielded some mutant enzymes with enhanced specificity. Substitution of 3 and 4 amino acids in the non-complementary DNA strand groove of SpCas9 and SaCas9 (*Staphylococcus aureus* Cas9), respectively, yielded mutant nucleases eSpCas9 ("enhanced specificity" SpCas9) and eSaCas9 ("enhanced specificity" SaCas9), which are more specific than the respective wild type enzymes (Table 5.3). The mutant enzymes also show a dramatic reduction in off-target cleavage, while their on-target cleavage efficiency seems unaltered. These high specificity Cas9 variants remain in an inactive state when they are bound to mismatched targets. The ''high-fidelity'' SpCas9 (SpCas9-HF) has increased target specificity, and its on-target cleavage efficiency is unchanged. This mutant has alanine in the places of those amino acids that interact with the phosphate residues in the complementary DNA strand (Tycko et al., 2016).

The REC3, a non-catalytic domain within the Cas9 recognition lobe, is responsible for the scrutiny of base-pairing accuracy. REC3 interacts with the DNA:RNA hybrid and undergoes conformational change, which determines the conformational change in the HNH domain and the overall catalytic state of Cas9. Substitutions of those amino acid residues in the REC3 domain that are involved in scrutiny of base-pairing has yielded HypaCas9, a "hyper-accurate" variant of Cas9. In human cells, HypaCas9 exhibited dramatically enhanced genome-wide specificity in comparison to the wild type SpCas9, and had on-target activity comparable or superior to those of SpCas9-HF1 and eSpCas9 (Chen et al., 2017).

The specificity of CRISPR-Cas9 system can be increased by fusing the Cas9 nuclease with a programmable DNA-binding domain (pDBD), *viz*., zinc finger proteins (ZFPs) and transcription activator-like effector (TALE) proteins. The intrinsic DNA binding affinity of Cas9 used in these fusions is reduced by appropriate amino acid substitutions. The C-terminal fusion of ZFP or TALE to SpCas9 gave the highest activity provided the DNA binding sites of the ZFP or TALE are located 3′ to the PAM sequence. The chimeric Cas9-pDBD proteins exhibit dramatically improved specificity and increased range of target sites. The increased target site range results from a decrease in the affinity of mutant Cas9 for the canonical PAM sequence, which allows it to recognize other PAM sequences. It has been suggested that the Cas9-pDBD fusions might afford a flexible SDN system that can be refined to enable very precise editing of any genomic locus. Further, this system could be optimized to achieve cleavage at a single human genome target site (Bolukbasi et al., 2015).

5. `Prime editing'

The 'prime editing' strategy uses a 'prime editor' that comprises nCas9-reverse transcriptase fusion protein and an extended sgRNA that has at its 5' end the sequence to be used as template for HDR of the nick induced by nCas9. It is reported to minimize indels at the on-target sites and greatly reduce off-target editing (Section 3.10.5; Anzalone et al., 2019).

The main strategies evaluated for minimizing the off-target editing by CRISPR-Cas SDN system are summarized in Table 5.6.

Table 5.6. A summary of some of the strategies for reduction in off-target activity of CRISPR-Cas9 system (based on Tsai and Joung, 2016; Wolt et al., 2016; Komor et al., 2017b and other sources)

Strategy	Reduction in off-target effects	Remarks
Truncation of sgRNA	5,000-fold or more reduction in editing at known off-target sites for the full-length sgRNA counterpart; 2 to 5-fold reduction in the number of off-target sites	The spacer sequence truncated by 2-3 nt from 5' end; spacer length 18 nt or even less; generates some new off-target sites
Addition of two G residues at the 5' end of sgRNA	Reduced off-target activity in case of some sgRNAs; 10- to 100-fold increase in specificity	The editing efficiency of some sgRNAs is adversely affected
Use of nCas9	Reduction by 50–1,500-fold in human cells	One pair of nCas9 targeted to two neighbouring sites to induce DSBs; certain target sites show increased frequencies of point mutations
Use of dCas9 fused with FokI nuclease domain	Reduction by two orders of magnitude	One pair of dCas9-FokI fusion targeted to two neighbouring sites to induce DSBs; mutant FokI that must form heterodimer to be able to induce DSB should be used
''High-fidelity'' SpCas9 variant	Dramatic increase in specificity and reduction in off-target activity	Genetically engineered variant of SpCas9; four amino acid substitutions
Cas9-ZFN or TALEN fusion	Dramatic improvement in specificity and increase in the range of target sites	DNA binding affinity of Cas9 attenuated; ZFN or TALEN fused to C-terminus of Cas9
Chemical modification of sgRNA	Increased sgRNA stability and specificity and decreased off-target effects	Examples, nucleoside phosphoramidite modification of nucleotides at 5' and 3' termini; *O*-Me or fluoro- substitution at 2' position, or sulphur substitution in the sugar-phosphate backbone
Reduced period of Cas9 action	Reduced off-target effects (up to 13-fold reduction in case of delivery as RNP; up to 20-fold reduction when delivered via cationic lipids)	Several different strategies, including inducible systems and delivery into cells as RNP (Cas9 is destroyed in cells within 24 hr)
`Prime editing'	Reduced off-target effects	Based on prime editor

5.9.3. Filtering out the phenotypic effects of off-target mutations

The acceptable level of off-target cleavage will vary with the objective of gene editing. Off-target cleavage may not be relevant in many investigations as long as toxicity is kept to the minimum; the chief interest of investigators in such studies is to score the desired alterations in the targeted genes/sequences. But when the objective is to relate genetic changes induced by gene editing to phenotypic alterations in the mutants, the off-target effects of SDNs should ideally be absent. Since zero off-target effect is not very likely, a suitable strategy must be used to filter out the effects of off-target mutations on phenotypes of the mutants. One approach is to compare the phenotypic changes in several independently derived mutants to determine the phenotypic expression common to all the mutants. These mutants may be preferably isolated by targeting different sites in the same gene. The second strategy involves transfer of the mutant alleles into the genetic backgrounds of respective parents by a backcross programme. The backcross-derived lines are then analysed to determine the phenotypic effects of induced gene mutations since the off-target mutations will be eliminated during the backcross programme (Carroll, 2014; Sander and Joung, 2014; Chandrasegaran and Carroll, 2016).

Finally, the normal alleles of inactivated genes may be introduced into the mutant lines by genetic transformation. In case, the normal allele is able to restore the normal phenotype in a mutant, *i.e.*, it complements the mutant allele, it may be concluded that the mutant phenotype is the result of mutation in the targeted gene. But these strategies are not applicable to all the biological systems. For example, repeated backcross to the parental lines is easily applied to annual crop species but will be problematic with perennials. In any case, for human therapeutic applications, the frequency of off-target cleavage should be zero or near zero, and a safe level for the same is yet to be agreed upon. It may be emphasized that the effects of off-target mutations must be meticulously characterized and evaluated for safety for application of SDNs in human gene therapy and for genetic improvement of such species that are used as sources of human food (Carroll, 2014; Chandrasegaran and Carroll, 2016).

5.10. MINIMIZING UNDESIRED GENETIC CHANGES AT THE TARGET SITE

It was reported that APOBEC3 catalyzes cytidine deamination in the DNA single strands generated during repair of the DSBs induced by Cas9 and the nicks produced by nCas9. This ultimately results in base substitution mutations by HDR during the repair of DSBs (Lei et al., 2018). The repair process of nicks induced by nCas9 may ultimately lead to production of indels. Thus, both APOBEC3 and the proteins involved in DNA repair are important factors contributing to indel formation during repair of DSBs and nicks. It was suggested that the level of these unwanted mutations could be minimized by using double-stranded DNA as donors for HDR since single-stranded donor DNA is vulnerable to APOBEC3 action. Further, a temporary suppression of the activity of APOBEC3 would contribute to the reduction in the levels of unwanted mutational changes surrounding the genomic sites where DBSs or nicks are induced. The CRISPR-Cas9-based 'prime editing' is reported to minimize indels at the on-target sites and greatly reduce off-target editing (Section 3.10.5; Anzalone et al., 2019).

5.11. RNA CLEAVAGE BY CRISPR-CAS9

In general, CRISPR-Cas9 system is used as a tool for DSB induction in double-stranded DNA (dsDNA) provided the substrate has the PAM sequence. But Cas9 can cleave single-stranded DNA when an appropriate PAMmer is provided along with CRISPR-Cas9 SDN. A PAMmer is an oligodeoxynucleotide with PAM sequence and a short 3' as well 18 nt long 5' extensions that base-pair with the targeted sequence of ssDNA or single-stranded RNA (ssRNA). O'Connell et al. (2014) applied the strategy of ssDNA cleavage for Cas9-meddiated cleavage of ssRNA; this version of CRISPR-Cas9 system has been referred to as RNA-targeted Cas9 (RCas9). The PAMmer pairs with the targeted region of ssRNA and generates a double-stranded region around the PAM (Fig. 5.13). The Cas9-sgRNA complex recognizes the PAM sequence of the PAMmer and associates with it. It then unwinds the double-stranded target region and the sgRNA pairs with the targeted RNA sequence. The Cas9 now cleaves the ssRNA between bases 4 and 5 from PAM on the 3' side of ssRNA. In contrast, Cas9 cleaves double-stranded DNA between bases 3 and 4 from the PAM on the 3' side of the target strand. The target RNA molecule need not have a sequence complementary to the PAM of PAMmer. The RCas9 approach is simple and far more convenient than other methods of RNA targeting since only appropriate sgRNA and PAMmer need to be designed for each new RNA target.

The RCas9 method can be used for selective cleavage of targeted RNA transcripts without cleavage of DNA sequences encoding them. It can also be used to isolate any specific native mRNA from total cellular RNA extract with biotinylated Cas9-sgRNA complex and a chemically modified PAMmer specific for the mRNA to be isolated. The Cas9-sgRNA-PAMmer-mRNA complex is pulled out using streptavidin-coated magnetic beads and the mRNA is separated by an appropriate treatment. The modified PAMmers do not activate Cas9 nuclease domain and mRNA cleavage (O'Connell et al., 2014). The need for chemical modification of the PAMmers can be avoided by using dCas9 in place of active Cas9.

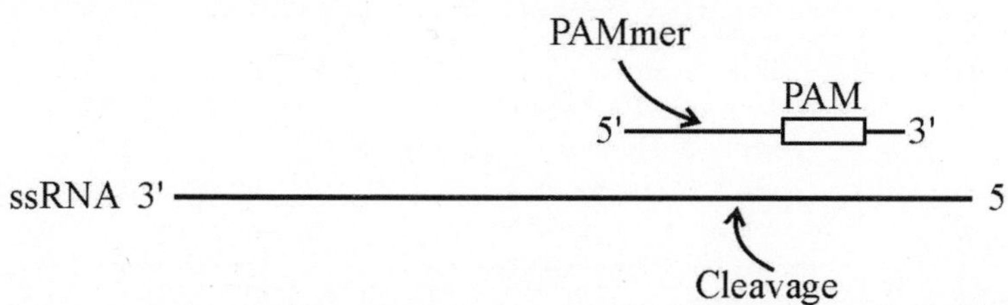

Fig. 5.13. Cleavage of single-stranded RNA (ssRNA) by CRISPR-Cas9 activated by a PAMmer oligodeoxynucleotide. The PAMmer has the PAM sequence and is complementary to ssRNA. The ssRNA need not have sequence complementary to PAM; it is cleaved between bases 4 and 5 located on its 3' side from the PAM.

5.12. ALTERNATIVE CRISPR-Cas SYSTEMS

The CRISPR-Cas9 is the most extensively characterised and the most widely used CRISPR-Cas system for gene editing, base editing and various other applications. But this system has certain notable limitations, including the requirement of a PAM sequence with a G residue. Efforts were made to identify and use other CRISPR-Cas systems that might expand the genomic regions that can be edited and might enable some novel applications. This search has resulted in the identification of certain useful variants of the Cas9 nuclease itself; one of these orthologs (CjCas9) is significantly smaller than SpCas9 so that it can be packaged together with the sgRNA into a single AAV (Adeno-associated virus) vector. In addition, two other CRISPR-Cas systems, *viz.*, CRISPR-Cas12a and CRISPR-Cas13b, have been discovered and used for gene editing (Table 5.7; Fig. 5.14).

The Cas12a (or Cpf1) was discovered and used for editing of human genes (Zetsche et al., 2015a). Cas12a has endoribonuclease activity that enables it to process its own guide RNA. In association with a single guide RNA (crRNA), it also exhibits DNase activity and cleaves the target DNA to produce staggered DSBs with 5 nt overhangs. Cas12a from *Lachnospiraceae bacterium* (LbCas12a) recognizes 5'TTTV(A/C/G) PAM located upstream of the target site, and the 23-25 nt guide sequence of crRNA pairs with the target sequence downstream of PAM. Its Nuc nuclease domain cleaves the target strand 18 bp downstream of the PAM sequence, while the RuvC nuclease domain cleaves the nontarget strand 23 bp downstream of the PAM. Two Cas12a orthologs were efficient in DSB induction in human cells. Specificity of Cas12a appears to be comparable or superior to that of SpCas9. Cas12a has been used for gene editing via NHEJ as well HDR in plants. It has also been used to modulate gene expression (Tycko et al. 2016; Swarts et al., 2017, Brooks and Gaj, 2018; Schindele et al., 2018).

Table 5.7. The various CRISPR-Cas systems used for gene editing (based on Brooks and Gaj 2018)

CRISPR-Cas system	Example bacterial species	Remarks
CRISPR-Cas9	*Staphylococcus pyogenes* and several other species	Type II CRISPR-Cas system; first to be used for gene editing; several mutants isolated; the most extensively characterised and the most widely used for gene editing and many other applications
CRISPR-Cas12a	*Francisella novicida*	Type V-A; earlier Cpf1; recognises a T-rich PAM [5'(T)TTN3'and 5'-CTA-3']; 23–25 nt spacer; can process its pre-CRISPR array to generate multiple crRNAs from a single transcript; this greatly facilitates multiplexing; used for multiple gene editing and gene regulation in mammalian cells; AsCpf1 and LbCpf1 suitable for gene editing
CRISPR-Cas13a	*Leptotrichia wadei*	Earlier, C2c2 (Class 2 Candidate 2); Type VI-B CRISPR-Cas system; discovered by bioinformatics analysis; acts on RNA; used for nucleic acid detection and tracking in live cells
CRISPR-Cas13b	*Prevotella* sp.	Used for knockdown of RNA transcripts in mammalian cells, and can be used for RNA base editing (currently limited to A to I editing; RNA base editor REPAIR uses Cas13b)
CRISPR-Cas14	Extremophile archaea	Less than half the size of Cas9; cleaves ssDNA in a nonspecific manner; used for DETECTR-Cas14 assay
CRISPR-CasF	Huge bacterio-phages	Half the mass of Cas9; uses a signle crRNA active in human and plant cells

(A) Cas9

(i) Cas9 + sgRNA ⟶ Target DNA ⟶ DSB in target DNA ⟶ Gene editing

(ii) CD - nCas9 - UGI + sgRNA ⟶ Target DNA ⟶ C ⟶ T base editing

(B) Cas12a

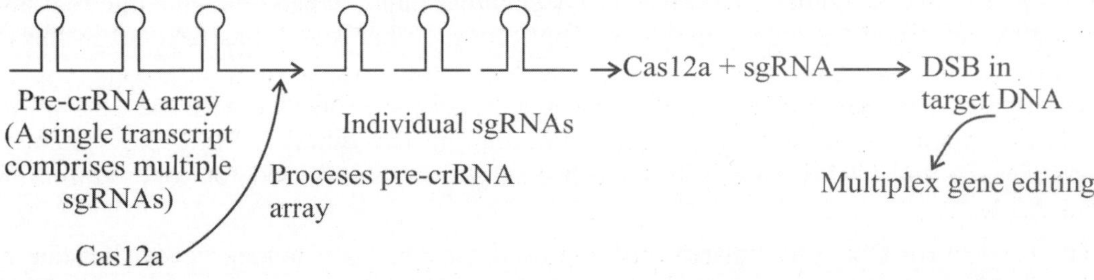

(C) Cas13

(i) Cas13a + sgRNA ⟶ Target RNA (including mRNA) ⟶ RNA cleavage ⟶ RNA transcript knockdown

(ii) Adenine deaminase - dCas13b + sgRNA ⟶ Target mRNA ⟶ Base editing (A-to-I)

Fig. 5.14. A comparative depiction of the relevant salient features of CRISPR-Cas9, CRISPR-Cas12a and CRISPR-Cas13 systems. CD, cytidine deaminase; nCas9, nickase Cas9; UGI, uridine glycosylase inhibitor; dCas13b, `dead' Cas13b (lacks ability for RNA cleavage).

After associating with double-stranded or single-stranded target DNA, the CRISPER-Cas12a nuclease induces indiscriminate cleavage of ssDNA present in its neighbourhood. This property was used to develop a DNA detection method, called DNA endonuclease targeted CRISPR *trans*-reporter (DETECTR). This method can detect rare DNA molecules present in a sample. A ssDNA-fluorophore quencher reporter is cleaved by the Cas12a when it is activated by binding to the target DNA that may be present in the sample; this causes fluorescence (Fig. 5.15). Isothermal amplification enables detection of target DNA present in attomolar concentrations (Schindele et al., 2018). This DNA detection method has been refined to enable easy and rapid GMO detection and disease diagnosis; the test uses filter paper-based DNA extraction and lateral flow assay with recombinase polymerase amplification

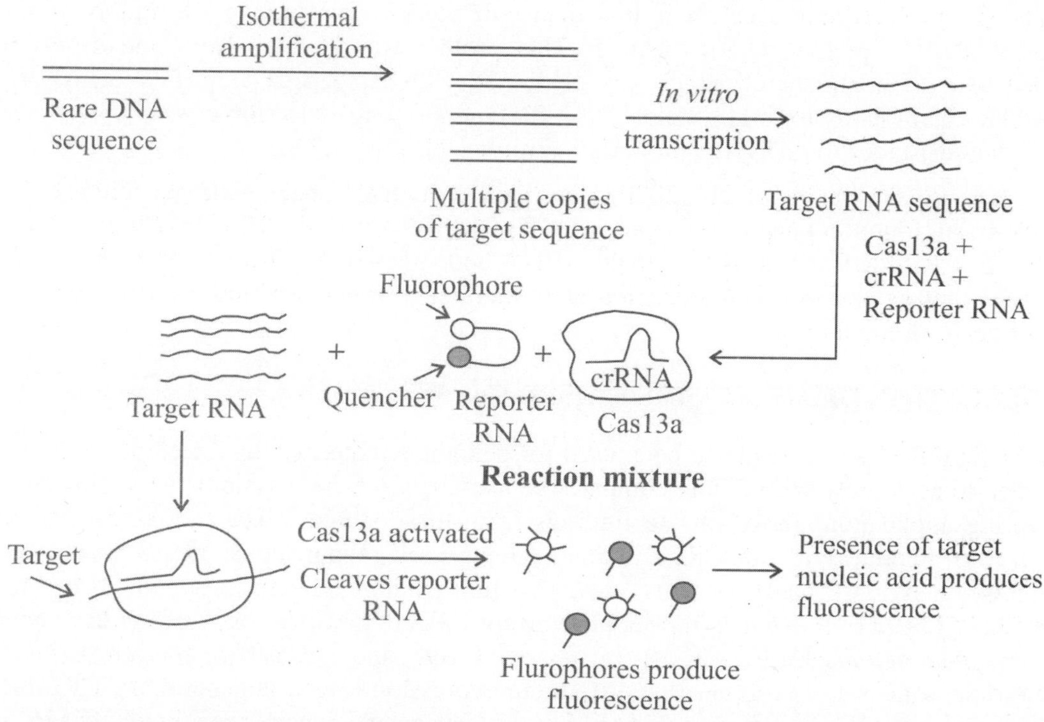

Fig. 5.15. Schematic representation of detection of rare nucleic acid seqeunces in samples using Cas13a (based on Schindele et al., 2018).

The Cas13a (earlier, C2c2) is an RNA-guided RNase that uses a single crRNA to guide it to ssRNA targets. Target recognition is based on complementarity to a 28-nt sequence of the crRNA. In *in vitro* systems, the C2c2 RNase can cleave RNA molecules indiscriminately once it has been primed by an interaction with its RNA target (Abudayyeh et al., 2016). This property of Cas13a permits detection of attomolar quantities the target RNA present in a sample or cell. The presence of target RNA is manifested as cell death *in vivo* and as fluorescence *in vitro* as activated Cas13a degrades all cellular RNA *in vivo* and nontarget labelled RNA molecules *in vitro*. This method of RNA detection has been termed as Specific High-Sensitivity Enzymatic Reporter UnLOCKing (SHERLOCK) that is used for *in vitro* RNA/DNA detection. SHERLOCK has attomolar sensitivity and can be used in combination with nucleic acid amplification for a variety of purposes, including detection of specific strains of viruses, distinguishing pathogenic bacteria, genotyping human DNA, etc. SHERLOCK can be stored for long periods in lyophilized state and rapidly reconstituted on a piece of paper for applications in the field (Gootenberg et al., 2017).

Cas13b has also been used for knocking down specific RNA transcripts for post-transcriptional gene silencing similar to RNAi, interfering with infection by RNA viruses and imaging of specific RNA transcripts. More recently, dCas13b (a 'dead' version of Cas13b) has been used for RNA base editing by coupling it with the deaminase domain of ADAR2 (ADAR2DD) to create the base editor REPAIR (Section 6.5; Brooks and Gaj, 2018; Schindele et al., 2018).

The Cas14 (530 amino acids) is less than half the size of Cas9 (1368 amino acids) and cleaves ssDNA in a non-specific manner. This protein was isolated from and seems to be confined to extremophile archaea. It was combined with isothermal amplification to develop DETECTR-Cas14 assay for high-fidelity DNA SNP genotyping, detection of ssDNA viruses, and engineering resistance to ssDNA plant viruses (Aquino-Jarquin, 2019).

In July 2020, Pausch et al. (2020) described a hypercompact genome editor based on CRISPR-CasΦ found in huge bacteriophages. The CRISPR-CasΦ is ~70 kDa, which is only half the mass of CRISPR-Cas9. It uses a single crRNA and is active *in vitro* and in human and plant cells. The compact size of CasΦ or CasF greatly facilitates its delivery and it further expands the range of gene editing tool-box.

5.13. SELECTION OF A SUITABLE CRISPR-Cas SYSTEM

Several CRISPR-Cas systems have been used for genome engineering by different researchers. It is difficult to accurately and reliably compare the usefulness of these systems for various genome engineering applications based on the findings from these studies since they differ for one or more factors unrelated to the CRISPR-Cas systems used. Wang et al. (2018) compared the usefulness of five of these systems, *viz.*, AsCpf1 (*Acidaminococcus* sp. BV3L6 Cpf1 or AsCas12a), LbCpf1 (*Lachnospiraceae* bacterium ND2006 Cpf1 or LbCas12a), NmCas9 (*Neisseria meningitidis* Cas9), SaCas9 (*S. aureus* Cas9), and SpCas9 (*S. pyogenes* Cas9), for editing of the same set of 90 genes that differed in expression levels. In general, the CRISPR-Cas system showing higher specificity exhibited lower cleavage efficiency. The editing efficiencies of the smaller Cas9 nucleases, *viz.*, NmCas9 and SaCas9, were much less influenced by the level of gene expression than those of the other CIRSPR-Cas systems. The AsCpf1 and LbCpf1 enzymes prefer ssODNs (single-stranded oligonucleotides) with sequence of the non-target strand as donor templates; the homology arms of these templates can be only 17 nt long. In contrast, SpCas9 has a preference of the ssODNs with sequence of the target strand and is the most suitable enzyme when linearized plasmid donor templates are used. The editing efficiencies of the Cpf1 and SpCas9 nucleases can be increased by using asymmetric donor templates having longer arms on the 5' side of the desired change in the sequence.

The authors (Wang et al., 2018) concluded that SpCas9 and LbCpf1 are equally useful for NHEJ-mediated induction of indels to generate routine gene knockouts when distinct genomic sites are targeted. But when a repetitive genomic region or genes with two or more paralogs are to be targeted, AsCpf1 is preferable since it has lower tolerance for mismatches between sgRNAs and the targeted sequences than the other four nucleases. In case of HDR-mediated genome editing, both AsCpf1 and LbCpf1 are the most suitable for efficient and precise genome editing when ssODNs are employed as donor templates. In addition, SpCas9 and Cpf1 work equally well with linearized plasmids serving as donor templates (Table 5.8).

Table 5.8. The most suitable CRISPR-Cas system for different applications (based on Wang et al., 2018)

Type of genome editing	Type of target site or donor template*	The most appropriate CRISPR-Cas system**
NHEJ-mediated	Distinct genomic locus	SpCas9 of LbCpf1
NHEJ-mediated	Many similar loci	AsCpf1 or SaCas9
HDR-mediated	Asymmetric ssODN	SpCas9 or LbCpf1
HDR-mediated	Linearized plasmid	SpCas9 or LbCpf1

* In case of NHEJ-mediated genome editing, the type of target site is relevant, while the donor template is relevant for HDR-mediated genome editing.

** AsCpf1, *Acidaminococcus* sp. BV3L6 Cpf1 or AsCas12a; LbCpf1, *Lachnospiraceae* bacterium ND2006 Cpf1 or LbCas12a; SaCas9, *Staphylococcus aureus* Cas9; SpCas9, *Staphylococcus pyogenes* Cas9.

5.14. ANTI-CRISPR PROTEINS

Anti-CRISPR proteins are a group of phage-encoded proteins that inhibit the prokaryotic CRISPR-Cas immune system. First described in 2013, these proteins enable the concerned phages to escape CRISPR-Cas-mediated destruction and successfully infect their hosts. The CRISPR-Cas immune systems and the anti-CRISPR proteins seem to have coevolved in relation to each other. Different anti-CRISPR proteins use diverse mechanisms to inhibit CRISPR-Cas action. For example, the anti-CRISPR protein AcrIIA4 interferes with DNA binding, target DNA unwinding and Cas9 nuclease action. But binding of anti-CRISPR protein AcrIIC1 to Cas9 disables its nuclease action. In contrast, the anti-CRISPR protein AcrIIC3 prevents binding of Cas9 to the target DNA by promoting dimerization of Cas9. The anti-CRISPR proteins may be useful in controlling CRISPR-Cas9-mediated gene editing *in vivo* (Harrington et al., 2017, Zhu et al., 2018).

The 87–amino acid long AcrIIA4 is the most potent inhibitor of most widely used SpCas9. AcrIIA4 associates with sgRNA-Cas9 complex and interferes with recognition of and binding to PAM sequence by Cas9. AcrIIA4 has highly acidic surface and it binds in 1:1 ratio to that region of Cas9, which interacts with the PAM sequence. AcrIIA4 also inhibits the unwinding of target dsDNA, blocks the entry of non-complementary strand into the RuvC-like domain nuclease site, and stops action by the HNH nuclease domain by interfering with the change in its conformation. This protein does not associate with either sgRNA alone or with free Cas9 and has no effect on sgRNA-Cas9 complexes already bound to target DNA. The AcrIIA4 also inhibits several orthologs of SpCas9 (Shin, J. et al. 2017; Zhu et al., 2018).

The AcrIIA4 protein can be delivered into cells either as protein, or in an expression plasmid to modulate CRISPR-Cas9 activity by preventing new sgRNA-Cas9 complexes from binding to the target DNA. Thus, AcrIIA4 allows a control on the period for which the CRISPR-Cas9 carries out gene editing. AcrIIA4 was shown to effectively stop Cas9-mediated gene editing at several loci in human cells. The reduced period of Cas9 action would permit on-target editing and reduce the frequency of off-target mutations. It is hoped that precisely timed addition of anti-CRISPR-Cas9 proteins would help reduce off-target effects and may foster therapeutic applications of CRISPR-Cas9 SDN (Shin, J. et al. 2017; Zhu et al., 2018).

5.15. HYBRID SITE-DIRECTED EFFECTORS

Several researchers have combined components from different SDN platforms to generate hybrid site-directed effectors. For example, meganucleases have been fused with DNA binding domains from zinc finger or TALE proteins to enhance the flexibility in target recognition by the hybrid SDNs. Similarly, SDNs have been fused with diverse effector domains to create programmable hybrid transcription regulators, DNA/histone methylases, histone acetyltransferases, recombinases, base modification enzymes, etc.; this has enormously expanded the range of dCas9 applications (Chapters 6, 10). In cases of ZFNs and TALENs, their DNA binding domains are fused with the relevant effector domains. But in case of CRISPR-Cas9 system, either dCas9 is fused with the concerned effector domain or the effector domain is attached to sgRNA via RNA hairpin and viral protein interaction (Carroll, 2014, 2015; Chandrasegaran and Carroll, 2016). Targeting of dCas9 alone to promoters represses gene expression in *E. coli* as well as human cells by interfering with initiation and/or elongation phase of transcription; this is termed as CRISPR interference (CRISPRi). CRISPRi can simultaneously repress several genes, the repression effects are reversible, there seems to be no off-target effect (Doudna and Charpentier, 2014).

The programmable transcription activators and repressors have been shown to activate and repress, respectively, gene expression in bacteria, and human and mouse cells. In human cells, activators based on CRISPR-dCas9 appear to be less effective than those based on TALE proteins. Targeting the same promoter with dCas9 based activators using 2-10 different sgRNAs leads to substantial increase in gene expression levels; TALE-based activators also show a similar effect. The effector domains may be fused with dCas9 or attached to suitably modified sgRNAs. For example, an RNA hairpin may be linked to the 3' end of the sgRNA to which the desired effector fused with the MS2 phage coat protein binds. The modified sgRNA is used in conjunction with dCas9: the MS2 coat protein-effector domain fusion becomes attached to the RNA hairpin in the sgRNA and is carried to the genomic site specified by the sgRNA. CRISPR-dCas9 has been used to guide fluorescent proteins like GFP (green fluorescent protein) to visualize specific genomic loci. This is greatly facilitated by the use of multiple sgRNAs for a single locus or systems like dCas-SAM and SunTag that can load multiple effector domains at a single site (Section 10.3.2.2; Sander and Joung, 2014; Katrekar et al., 2017).

There are conflicting reports on the specificity of dCas9 as compared to that of the wild type Cas9 depending mainly on the method employed for assessing specificity. Further, the dCas9-activator fusions are efficient in gene regulation only when their binding site is not more than 200 bp upstream of the transcription start site (TSS) of the concerned gene. In contrast, the dCas9-repressor fusions are effective over a longer region, *i.e.*, 200 bp upstream to 800 bp downstream of the TSS. Some dCas9-effector fusions may function in gene regulation even without the sgRNA (Tycko et al., 2016; Pulecio et al., 2017).

5.16. APPLICATIONS OF CRISPR-Cas9 SDN SYSTEM

The CRISPR-Cas9 system has been used for DSB induction and DNA repair-associated mutagenesis in diverse organisms, such as bacteria, yeast, human cells, zebrafish, rice, wheat sorghum, etc. These studies showed that the efficiency of CRISPR-Cas9 SDNs in gene editing is as good as TALENs and ZFNs. Further, simultaneous delivery of several different sgRNAs enables simultaneous targeting of multiple genes. The DSBs induced by CRISPR-Cas9 SDN support both NHEJ and HDR. HDR can use double-stranded donors supplied in a plasmid as well as single-stranded oligonucleotides. CRISPR-Cas9 system can induce precise chromosomal rearrangements associated with different malignancies. For example, some researchers have induced in human cells chromosomal rearrangements associated with the development of lung cancer, acute myeloid leukaemia, etc.

CRISPR-Cas9 has been used for systematic genome-wide unbiased analysis of gene functions using a genome-wide library of pooled lentivral sgRNAs to develop a genetic screen for loss-of-function mutations in human cells (Wang, T. et al., 2014). It was also employed to identify such genes that are necessary for cell survival in cancer cells and stem cells. Earlier studies on loss-of-function screens used RNAi, which temporarily and often partially reduces the expression of targeted genes; further, RNAi produces considerable off-target effects and is limited to loci that are transcribed. In contrast, CRISPR-Cas9 system can be used for functional analysis of both coding and noncoding regions of genomes. Further, the use of sgRNA library stably inactivates the targeted genes and produces gene knockouts. CRISPR-dCas9 system can also be used to suppress/activate the expression of targeted genes (Section 5.15). This system may help discover distant enhancers, provide further insights into promoter architecture, identify novel elements affecting gene expression levels, identify novel drug targets and contribute to disease management and possibly cure (Doudna and Charpentier, 2014; Katrekar et al., 2017).

In case of plants, Xie and Yang (2013) designed two RNA-guided Genome Editing vectors (pRGE3 and pRGE6) for gene targeting in rice. These vectors have sequences encoding the sgRNA scaffold region and the Cas9 nuclease driven by appropriate promoters. The Cas9 is engineered to have a nuclear localizing signal (NLS) sequence. There is a multiple cloning site (MCS) adjacent to the sgRNA scaffold encoding sequence to allow integration of the sgRNA seed region encoding sequence in such a way that transcription generates the complete sgRNA with both seed and scaffold regions. This vector system provides the general scheme for vectors with sgRNA gene constructs. PAM motif occurs, on an average, after every 10 bp, and about 90% of the rice transcription units can be targeted by this system.

The use of single-stranded DNA donors complementary to the non-target strand, *i.e.*, the strand not involved in base-pairing with sgRNA, in conjunction with Cas9 markedly enhanced the rate of HDR. Surprisingly, the use of ssDNA donors with dCas9 produced recombination frequencies of <1% without any DNA cleavage. In this process, the target DNA duplex is locally unwound by the CRISPR-dCas9 allowing the ssDNA donor to base-pair with the nontarget strand. This leads to integration of the donor DNA by a recombination process. CRISPR-Cas9 has also been used to achieve NHEJ-mediated homology independent targeted integration (HITI) of donor DNA to generate gene knock-ins. In HITI, the donor DNA has the PAM sequence; as a result, it is also cleaved along with the target DNA by CRISPR-Cas9, and the donor DNA integrates in the genomic DSB at the targeted site. Notably, HITI is effective in both mitotic and post-mitotic cells, while HDR works only in mitotic cells (Suzuki et al., 2016; Katrekar et al., 2017).

The CRISPR-Cas9 SDN system has been modified to develop the EvolvR mutagenesis platform that induces high frequency semi-random mutations in the targeted genomic sites. The EvolvR tool comprises nCas9 fused with an error-prone DNA polymerase (PolI3M) and the appropriate sgRNA that guides it to the targeted site. The nCas9 was mutated to reduce its affinity for DNA so that it dissociates from the target after nicking, while the PolI3M was mutated to increase its error rate. The action of EvolvR, in simple terms, involves nicking of one strand of the target DNA by nCas9, which dissociates from the target DNA. The PolI3M associates with the genomic DNA and extends the 3' end at the nick; the single strand displaced by this process is cleaved by the flap endonuclease domain of PolI3M. EvolvR generated a 212,000-fold increase in mutation rate over the wild type *E. coli*. This platform can be optimized for applications in eukaryotes similar to the application of CRISPR-Cas9 SDN system (Halperin et al., 2018)

5.17. ADVANTAGES OF CRISPR-Cas9 SDN SYSTEM

The CRISPR-Cas9 SDN system is highly versatile, far simpler to design, and far easier, faster and cheaper to implement than the other genome editing tools. The effectiveness and efficiency of this SDN system is not affected by the methylation state of the target sequence. This is particularly relevant in plants where ~70% of CpG/CpNpG sequences present in promoters and proximal exons of genes are methylated. Since Cas9, the protein component of this SDN, does not vary with change in the target sequence, there is no need for protein engineering unlike in the case of other SDN systems. Further, only the sgRNA, in fact the first 20 nucleotides from the 5' end of sgRNA, need to be designed for each new target sequence since complementary base-pairing of this sgRNA region is the basis for target sequence recognition. RNA designing is simple and requires knowledge of only the simple rules for complementary base-pairing between RNA and DNA, *viz.*, A:U and G:C. The production of sgRNA for any specified target is very easy; it does not require cloning and can is achieved by transcription either *in vitro* or *in vivo*. As a result, large sgRNA libraries can be easily constructed and the CRISPR-Cas9 can be used for high throughput large scale genomics investigations at affordable costs (Carroll, 2014, Bortesi and Fischer 2015).

The CRISPR-Cas9 research community has adopted an open access policy, which has promoted the widespread adoption of this technology. It allows relatively easy access to plasmids through Addgene (a non-profit repository), web tools for selection and specificity prediction for the designed sgRNAs and sponsors active discussion groups (*e.g.*: https://groups.google.com/forum/#!forum/crispr). The CRISPR-Cas9 is far cheaper than ZFN and TALEN systems: the CRISPR-Cas9 system is estimated to cost less than 1% of the ZFN cost (US$ ~30 as against about US$ 5,000 for ZFN); this makes the CRISPR-Cas9 system affordable to even small laboratories with limited funds. Further, it does not require expensive equipment, prolonged training and very high technical expertise. It has been suggested by many that 'this platform democratized genome engineering' (Carroll, 2014; Bortesi and Fischer 2015; Ledford, 2015a).

A unique advantage of the CRISPR-Cas9 system is that it allows simultaneous targeting of multiple even thousands of genes, *i.e.*, multiplexing. Multiple gene targeting requires designing of sgRNAs specific for each selected target and their delivery into the cells along with Cas9. The feasibility of this approach has been amply demonstrated in zebrafish, rats and human cells. Multiplexing is also feasible with ZFNs and TALENs, but this is considerably more demanding. In addition, the frequency of off-target cleavage tends to increase due to the formation of wrong ZFN/TALEN heterodimers. This risk is absent in the case of CRISPR-Cas9 since it works as a monomer (Carroll, 2014; Chugunova et al., 2016).

5.18. LIMITATIONS OF CRISPR-Cas9 SDN SYSTEM

A common limitation of CRISPR-Cas9 system is that the target site must include a specific PAM sequence at its 5' end (at 3' end of sgRNA seed/spacer sequence). This restricts, albeit modestly, the choice of target sites. Several mutants of Cas9 nuclease exhibit altered PAM sequence requirements (Table 5.3), and Cas9 proteins from some other prokaryotes require other PAM sequences (Table 5.2); identification of such Cas9 nucleases will relax this constraint. In addition, fusion of Cas9 nuclease with zinc fingers (ZFs) yields Cas9-ZF SDNs, which have broad target range as they do not require the wild type PAM sequence. An important limitation of the CRISPR-Cas9 system is the off-target effects that seem to be rather high at least in some cases. Several strategies, including Cas9 mutagenesis, have been implemented to minimize these effects (Section 5.9.2; Carroll, 2015; Chugunova, 2016; Tsai and Joung, 2016; Guha and Edgell, 2017).

The large size of CRISPR-Cas9 SDN encoding sequences (~4.2 kb for Cas9 + ~500 bp for sgRNA) presents problems in delivery especially into animal cells through some commonly used viral vectors, *e.g.*, AAV vectors (Section 5.8). The AAV vectors are considered as highly promising vectors for gene therapy applications. In fact, a gene therapy based on an AAV vector has recently been approved for Leber's congenital amaurosis (a form of inherited blindness). Strategies to resolve this problem include the development of split-Cas9 (Section 5.7) and the search for smaller Cas9 orthologs. The Cas9 encoding gene from *Campylobacter jejuni* (*CjCas9*) is ~1.1 kb smaller than the *SpCas9* gene. As a result, the *CjCas9* as well as the sgRNA gene constructs can be accommodated in a single AAV vector. However, delivery of this SDN system into plant cells (via *Agrobacterium* and microparticle bombardment using a particle gun) and protoplasts (by polyethylene glycol-mediated transformation) presents little or no difficulty (Lee et al., 2016; Brooks and Gaj, 2018).

5.19. APPLICATIONS OF TARGETED GENE MODIFICATIONS

The gene editing tools enable introduction of specified modifications in the targeted genomic loci of living cells as well as whole organisms. This makes them a powerful tool for both basic biological investigations as well as practical applications, including therapy of human genetic, epigenetic and even pathological diseases. The SDN technology enables inactivation of specified genes as well as restoration of mutated genes to the wild type by targeted sequence replacement. Targeted gene inactivation allows determination of gene functions, and the phenotypes produced due to their inactivation. In addition, the introduction of predetermined sequence changes at the specified sites of targeted genes permits investigation of the effects of specific sequence changes on the functions of targeted genes. Both the above approaches allow reverse genetic analyses and assignment of functions to various genes and genomic loci.

HDR also allows integration of transgenes, intragenes and cisgenes at specified sites in the genome. This procedure eliminates the risk of unintended activation of undesirable genes and inactivation of some essential genes, which may result from random transgene integrations. It would also allow replacement of the endogenous allele of a gene with an intragene or cisgene. These features of targeted gene integration would reduce the biosafety concerns and evaluation requirements for transgenic lines and might enhance their acceptance by consumers. HDR can also be used to insert novel and useful sequences into the genes of interest to generate novel fusion protein products with additional specified properties. For example, a protein could be fused with a specific epitope tag or a domain with novel function.

HDR can be used to introduce specified amino acid substitutions in the target proteins. These alterations might be useful for treatment of genetic diseases and may even enable creation of commercially useful proteins. Targeted gene editing can be used to produce animal models for various investigations on human diseases including their treatment and possibly cure. Efforts are being made to develop gene-edited strains of animals like pigs that could serve as sources of organs for xenotransplant in humans. Gene editing is being used to improve the performance and/or resistance to abiotic/biotic stresses of our agricultural crops as well as livestock (Doudna and Sternberg, 2017) and some successful examples are already available (Section 7.18). Gene editing could be used to modify biological pathways to generate novel useful materials and to enhance the efficiency of biofuel crops (Hsu et al., 2014).

5.20. SELECTION OF A SUITABLE SDN SYSTEM

A comparative summary of various relevant features of the three SDN systems are presented in Table 5.9. The majority of newly designed ZFNs and TALENs have low specificity; commercially available ZFNs perform better since they are subjected to stringent procedures of selection and validation. Some bioinformatics tools have been developed for predicting the specificity and efficiency of newly designed sgRNAs and TALENs. But the predictions of editing specificity and efficiency of newly designed nucleases are difficult, and the predicted values might differ from those observed *in vivo*. The *in-silico* predictions may be used to select the most promising sgRNAs and TALENs for *in vivo* evaluation; this approach will considerably reduce the work involved in *in vivo* evaluation. The efficiency of mutagenesis shows a wide variation ranging from 1% to over 80% for the various SDNs. The specificity and efficiency of SDNs are influenced by a variety of factors, including organism, cell type, and the method of SDN delivery (Chugunova et al. 2016; Lee et al., 2016).

All the three SDNs generate off-target effects in variable frequencies. TALENs appear to possess the highest intrinsic specificity, but some rigorously selected ZFNs are highly specific. The CRISPR-Cas9 SDNs can tolerate mismatches of up to 5 bases, and they are extremely insensitive to mismatches at the 5' termini of the sgRNAs. Cas9 has been engineered to develop some high specificity/fidelity variants that show reduced off-target action. Further, an increase in the binding affinity of an SDN for its specified target sequence might enhance the chances of off-target cleavage. Several different strategies for reducing off-target activities, especially of CRISPR-Cas9 SDNs have been developed. ZFNs tend to introduce comparable frequencies of off-target deletions and insertions, while TALENs cause more deletions than insertions. In contrast, CRISPR-Cas9 SDNs mainly lead to insertions of one or two nucleotides (Chugunova et al. 2016; Lee et al., 2016; Yang and Wu, 2018).

Table 5.9. A comparison of programmable nuclease platforms [based on Bortesi and Fischer, 2015; Chandrasegaran and Carroll, 2016; Chugunova et. al., 2016; and other sources]. All programmable nucleases are capable of high efficiency editing of genes irrespective of their epigenetic and active/inactive states.

Programmable nucleases	**ZFNs**	**TALENs**[@]	**CRISPR-Cas9**
Year of introduction	1996	2010	2012
Derived from	Eukaryotic transcription factor	Eukaryotic transcription factor[@]	Adaptive immune system of bacteria and archaea
Functional molecule	Protein	Protein	Ribonucleoprotein
DNA binding domain	Zinc fingers	TALE protein	Guide RNA
Nuclease domain	FokI cleavage domain	FokI cleavage domain	Cas9 nuclease
SDN function provided by	One pair of ZFNs	One pair of TALENs	A single CRISPR-Cas9 SDN
Sequence recognition involves	Protein–DNA interaction	Protein–DNA interaction	RNA–DNA interaction
Recognition code	One ZF recognizes three bases; recognition context-dependent	One set of two amino acids recognizes one base; recognition rules simple and independent of context	One base of sgRNA recognizes one base of the complementary strand of target DNA
Design and selection	Tedious and time consuming (protein engineering)	Easier than ZFNs (protein engineering)	Simple (RNA sequence design)
Commercial pricing	Very expensive	Expensive	Cheap
Targeting efficiency	Variable[a]	Moderate	High[b]
Off-target cleavage	Variable[a]	Low	Moderate[b]
Viral delivery	Easy	Moderate	Moderate
Length of binding site	18-24 bases	30-40 bases	22 (20 target + 2 PAM) bases[$]
Target site requirement	G-rich; estimated to occur, on an average, every ~100 bp of genome sequence	Begins with T and ends with A; modified versions of TALENs do not have this requirement	Ends with NGG or NAG (to a lesser extent); mutant Cas9 have altered PAM requirements; other Cas9 orthologs require other PAM sequences
Length (in kb) of DNA encoding the SDN	1	3	4.2
Suitability for targeting multiple loci (multiplexing)	Low	Low	Very high
Ease in targeting multiple targets	Difficult	Difficult	Easy
Success rate	Low	High	High
Cut ends at DSBs	Staggered with cohesive overhangs of variable lengths; useful for precise integration of donor DNA	Staggered with cohesive overhangs of variable lengths	Blunt; staggered ends can be generated by using a pair of Cas9 nickases for DSB induction

[@] Produced by pathogenic bacteria.

[a] Depending on ZFN constructs, targeting efficiency and off-target cleavage can fluctuate from high to low.

[b] As with ZFNs, targeting efficiency and off-target cleavage vary with the CRISPR-Cas9 construct and the target site.

[$] For the widely used *S. pyogenes* Cas9.

ZFNs work better when their target is guanine-rich and comprises 5'-GNN-3' triplets. This restricts, to some extent, the potential target sequences in the genome for ZFNs, which can be detected with the help of computer programs. The initially developed TALENs required a T as the first base in their binding sequences, but TALE variants that can recognize other bases at the first position are now available. Regular TALENs are unable to recognize sequences having methylated C, but the TALE domain with RVD Asn-Gly (specific for T) can recognize methylated C. In case of CRISPR-Cas9, the PAM sequence must be either 5'NGG3' or 5'NAG3'. Further, A, C or CC residues may be required for pairing with the A, G or GG at 5' terminii of sgRNAs, but the current approach is to just ignore it. These requirements would limit the availability of target sites for this SDN. CRISPR-Cas9 SDNs, however, can recognise methylated DNA, but the cleavage efficiency varies with target site (Chugunova et al. 2016).

The choice of SDN system for use in clinical or agricultural applications will be influenced by several considerations. One of the first considerations will be the ease and the speed of development of SDNs specific for new target sites. The designing of ZFNs and TALENs is technically demanding, time-consuming and quite expensive. In contrast, CRISPR-Cas9 requires only redesigning of sgRNA for each new target, which is extremely simple, rapid and inexpensive. However, when SDN designed for targeting a gene is to be repeatedly used in many individuals/lines, the cost and effort involved in designing the SDN may not be that important. In comparison, efficiency and specificity of the SDNs are of far greater significance, and they will have to be very high at least for clinical applications (Carroll, 2014. 2016).

In addition, SDN delivery into cells should be relatively easy and applicable to a variety of organisms. The delivery of ZFNs may be much easier at least in some systems as they are considerably smaller than TALEN and CRISPR-Cas9 SDNs. But the Cas9 from some bacteria are smaller than SpCas9, and their use may facilitate the delivery of CRISPR-Cas9 SDNs. The CRISPR-Cas9 system is amenable to multiplexing, which is much more difficult with the other two SDN systems (Carroll, 2014. 2016).

5.21. ACHIEVEMENTS

The CRISPR-Cas9 system has proved to be an efficient tool for targeted gene editing to achieve NHEJ-mediated gene inactivation as well as HDR-mediated sequence replacement. The technology is being used for a variety of basic investigations, including identification of genes involved in diseases and novel targets for therapeutic agents. This technology is being deployed for the improvement of crop species (Chapter 7) and domestic animals (Chapter 8), and for tackling various human health related issues including treatment of genetic diseases (Chapter 9). The CRISPR-Cas9 system has been modified to generate programmable hybrid effector proteins that enable targeted gene activation/suppression, chromatin modification, base editing and imaging of specific chromosomal regions. Strategies for application of this technology to modify the various features of natural populations are being developed.

The CRISPR-Cas9 system has been used to produce a browning resistant strain of mushroom by targeted inactivation of the gene encoding polyphenol oxidase. Gene inactivation by targeted editing has been used to develop some disease resistant lines of crop plants. These and other achievements of the CRISPR-Cas9 technology in the improvement of different crop species are considered in some detail in Chapter 7 (Section 7.18). In addition, therapeutic *ex vivo* gene editing for cancer treatment based on immunotherapy is undergoing clinical trials.

5.22. FUTURE DIRECTIONS

The pace of developments in the field of CRISPR-Cas9 has been stunning, and they have paved the way for its numerous basic and applied researches. However, some notable issues need to be resolved to expand the range, reliability and the safety of this powerful technology. The first issue is the need for a specific PAM sequence. Cas9 nucleases from some other bacteria need different PAM motifs and search for more such nucleases should continue to identify smaller enzymes with different PAM requirements. Efforts to engineer Cas9 with enhanced specificity and capability of using alternative PAM sequences have been successful; these efforts need to be intensified to develop Cas9 variants with increased specificity and negligible off-target effects. In addition, robust, efficient and easy methods for the delivery of CRISPR-Cas9 SDNs need to be developed to facilitate its therapeutic use (Doudna and Charpentier, 2014; Sander and Joung, 2014).

Off-target cleavage remains a major concern. Simple and easy to implement strategies for an unbiased global assessment of off-target effects would greatly facilitate such refinements in the system that would minimize (hopefully eliminate) off-target effects. Some of the possible approaches could be to fuse dCas9 with FokI mutant that can form only heterodimers. A pair of CRISPR-dCas9-FokI SDNs would be needed for each target site, which is expected to suppress off-target effects. Some other approaches would be to combine tru-sgRNA CRISPR-Cas9 nickases, to use protein engineering to modify Cas9 properties, and to suitably modify the sgRNA nucleotides critical for target specificity. Off-target effects can also be reduced by decreasing SDN concentration in cells by using inducible or tissue-specific promoters to drive SDN gene constructs, or by using DD-Cas9 or applying anti-CRISPR proteins (Section 5.14; Sander and Joung, 2014). A simple, robust and easy to use technology for regulating the level and the period of Cas9 expression would greatly facilitate the usefulness of this SDN system in various applied fields.

Strategies for efficient and simultaneous expression of several CRISPR-Cas9 SDNs, preferably from one vector, will permit simultaneous editing of many genes. For many applications, it is critical to minimize or eliminate the risk of NHEJ and allow only HDR. This is particularly relevant when the objective is to induce in the target gene a specified base sequence substitution via HDR. In such a case, the HDR products will be almost as susceptible to SDN action and NHEJ as the native gene; this will, therefore, reduce the yield of HDR products. Near complete suppression of the NHEJ pathway in the cells being edited might adversely affect cell survival. The 'prime editing' method eliminates the risk of NHEJ without suppressing this pathway and ensures high HDR rates (Sander and Joung, 2014; Anzalone et al., 2019). There is an urgent need for standardization of the genome editing practices hopefully leading to the development of best practices to be followed for clinical applications of the technology. The establishment of a genome-editing consortium might hasten the standardization process. One area that needs improvement is the accuracy of prediction of off-target sites using bioinformatics tools since the whole genome unbiased detection of off-target effects is technically demanding and rather costly (Tasan and Zhao, 2017).

The effects of chromatin structure and its epigenetic states on Cas9 binding to target sites and cleavage induction need to be elucidated. But dCas9-transcriptional activator fusions can enhance transcription of genes lacking DNase I hypersensitive sites; this indicates that CRISPR-Cas can successfully bind chromatin that is not in active state. CRISPR-Cas cleaves target DNA *in vitro* irrespective of the state of CpG methylation; it would be quite useful to establish if the same is true about *in vivo* cleavage. The proteins involved in homologous recombination are mainly expressed during the G_2 phase of cell cycle. As a result, HDR-based editing of genes in quiescent G_0 cells like neurons, cardiac myocytes, etc. is difficult to achieve. The 'prime editing' overcomes this need and it can be used in both mitotic and post-mitotic cells. Finally, the suitability of other CRISPR immune systems for gene editing needs to be carefully examined. Cmr-CRISPR system has been used for RNA targeting but its suitability for RNA targeting *in vivo* needs to be evaluated (Hsu et al., 2014).

Researchers are eagerly searching for alternative tools for gene editing to expand the range of targets, enable easier delivery and increase editing specificity as compared to those for the currently available SDNs. Recently, a protein from the Argonaute (Ago) family, NgAgo (*Natronobacterium gregoryi* Argonaute), was reported to be able to cleave DNA at predetermined sites. NgAgo is guided to the target DNA by a ~24 nt single-stranded guide DNA (gDNA). It was used to edit genes in human cells and to insert genes at predetermined genomic sites. This protein does not require a PAM sequence and is reported to cleave only the target site. However, several researchers have failed to confirm these findings. Another system called lambda Red is programmable and does not require a guide RNA, but it works only in bacteria. Some scientists are exploring the possibility of using integrase and recombinase enzymes for gene editing (Cyranoski, 2016).

A structure-guided endonuclease (SGN) has been created by fusing flap endonuclease-1 (Fen-1) with the cleavage domain of FokI endonuclease. The SGN is guided to the targeted sequence by a guide DNA (gDNA), which is complementary to the target sequence, except for its 3' end region which forms a 3' flap when the gDNA pairs with the target. The SGN recognizes 3' flap formed by the gDNA and cleaves the target DNA following FokI cleavage domain dimerization. This structure-guided DNA cleavage system can edit any genomic target (Xu et al., 2016).

There are numerous ethical questions related to gene editing that need to be resolved to the satisfaction of all the stake holders. A rational discussion on this aspect will require the public at large to be aware of the basics of gene editing technology and to appreciate its powers and limitations. In addition, the regulatory agencies should develop procedures that foster further research and development in the technology and ensure its responsible application to promote human welfare (Doudna and Charpentier, 2014).

5.23. CONCLUSIONS

The SDN technologies have enabled precise and defined modifications of selected genes in a wide range of organisms. Among SDNs, CRISPR-Cas9 system is simple, inexpensive, has high efficiency and is widely applicable; as a result, it has rapidly acquired wide acceptance for both basic and applied biological and biomedical researches. One measure of the impact of CRISPR-Cas9 SDN is the number of publications, which increased from merely ~50 in 2010 to about 1400 in 2015 (Songstad et al., 2017). This SDN has greatly facilitated both forward and reverse genetic analyses by enabling specified base sequence modifications in the targeted genes. Large sgRNA libraries can be used to simultaneously alter several targeted genes to enhance the scope of these studies. But off-target effects are a matter of concern, and they need to be at least minimized to acceptable levels to pave it's way for therapeutic applications. Several strategies based on sgRNA and Cas9 have been devised to minimize off-target action of the CRISPR-Cas9 system. There have been marked improvements in the sgRNA designing, engineering of Cas9 nuclease for enhanced specificity and reduced off-target activity, split-Cas9 for facilitating delivery, nickases for reducing off-target effects, DD-Cas9 for controlling the period of nuclease action, methods for detection of off-target effects, etc. Two high specificity mutants of SpCas9, *viz*., eSpCas9 and Cas-HF, have been developed. Some SpCas9 orthologs and non-Cas9 nuclease enzymes have been used for gene editing, but much needs to be done to refine their characteristics and applications. The methods of delivery must be refined and improved at least for therapeutic applications (Sander and Joung, 2014; Tycko et al., 2016).

The continued research efforts are expected to refine the CRISPR-Cas technology to a level that will enable expansion of its range of applications to therapeutics. It may enable the identification of mechanisms of disease development, design of precision drugs and selection of novel therapeutic targets. *Ex vivo* therapeutic gene editing for cancer treatment based on immunotherapy is undergoing clinical trials. The insights gained from these trials will stimulate research and development for other therapeutic applications of gene editing technology (Katrekar et al., 2017).

The Cascade (CRISPR-associated complex for antiviral defense)–Cas3 system has the potential for development of a tool for induction of targeted deletions of large genomic sequences, including complete genes, clusters of genes, prophages, etc. The Class 2 CRISPR-Cas systems like CRISPR-Cas9 have so far proved inefficient in performing this function. Cas3, a processive nuclease, was used in combination "with a minimal Type I-C Cascade-based system" for induction of targeted deletions of 7–424 kb in the bacterium *Pseudomonas aeruginosa*; the efficiency of this system was nearly 100%. In contrast, Cas9 generated only small deletions and point mutations. The Cascade-Cas3 system induces deletions in both the directions from the targeted site; the boundaries of the deletions were specified by the donor DNA provided as template for HDR. The system Type I-C Cascade–Cas3 (*Pae*Cas3c) system of *P. aeruginosa* "facilitates rapid strain manipulation with applications in synthetic biology, genome minimization and the removal of large genomic regions" (Csörgő et al., 2020).

■■■■

6

BASE EDITING

6.1. INTRODUCTION

Gene or genome editing comprises *in vivo* induction of specified alterations in the base sequences at targeted genomic sites. In simple terms, a site-directed nuclease (SDN) is used to induce either a double-strand break (DSB) or a nick at the targeted site. The DSBs/nicks are repaired by cellular DNA repair processes, which generally lead to sequence changes at the site of DSB/nick. Repair by the nonhomologous end-joining (NHEJ), microhomology-mediated end joining (MMEJ), and single-strand annealing (SSA) often lead to insertion or deletion of one to few bases at the break site, and sometimes may even produce translocations. Therefore, NHEJ leads to inactivation of the targeted genes, which can serve many basic as well as applied research objectives. But for sequence correction of a mutant gene or introduction of a specified mutation in the selected gene, an appropriate donor DNA must be provided along with the SDN. The donor DNA is used as template for homology-directed repair (HDR) of the break/nick, leading to replacement of the endogenous gene sequence by that of the donor DNA. Despite the various refinements and modifications of the SDN technology, the frequency of HDR is rather low under conditions relevant for its therapeutic applications, and NHEJ predominates, especially in the nondividing or quiescent cells. The 'prime editing' procedure offers a very promising technology for targeted sequence replacement with little risk of NHEJ (Komor et al., 2017a; Anzalone et al., 2019).

Efficient correction of mutated genes as well as inactivation of targeted genes by creating early termination codons can also be achieved without induction of DSBs in the genomic DNA. One strategy uses the concepts and techniques from SDN technology and chemical biology to selectively modify the targeted bases without causing a DSB. The components of a suitable SDN, *e.g.*, CRISPR-Cas9 SDN, are combined with base modifying enzymes to create base editors that catalyse targeted *in vivo* conversion of one base pair into another; this is called base editing. *Base editing directly and irreversibly converts one base into another at the targeted genomic site without induction of a DSB.* Although base editing is the most recent tool in the gene editing toolbox, this potent technique has already unfolded many capabilities and found several different applications (Komor et al., 2016, 2017a; Section 6.14).

6.2. BASE EDITING

Base editing was first reported by Komor et al. (2016) who demonstrated programmable irreversible *in vivo* conversion of a C:G base pair into a T:A base pair. Base editing reaction does neither involve DSB or nick induction nor HDR. Base editing is catalysed by specifically designed recombinant proteins called base editors, which are created by fusion of either dCas9 ('dead' Cas9) or nCas9 ('nickase' Cas9 with active HNH nuclease domain) with the appropriate single-stranded DNA (ssDNA)-specific cytidine or adenosine deaminase enzyme, and some other relevant proteins in the case of C-to-T base editors (Table 6.1). The C-to-T base editors in current use are third or fourth generation base editors (BE3 or BE4), which comprise nCas9, cytidine deaminase rAPOBEC1 (rat apolipoprotein B mRNA editing complex 1) and uridine glycosylase inhibitor (UGI, obtained from *Bacillus subtilis* bacteriophage PBS1) (Section 6.3.2). In comparison, the A:T-to-G:C base editors (designated as adenine base editors, ABEs) comprise dCas9 and a tRNA adenine deaminase engineered to catalyse deamination of adenine residues present in DNA.

Table 6.1. Constituents of different base editors

Constituent	Function	Remarks
C-to-T base editor		
i. Cas9 (with appropriate sgRNA)	Guides the base editor to the target site	Generally, nCas9 (Cas9 nickase) used; promotes replacement of G opposite the edited U leading to C:G to T:A transition
Cas12a (Cpf1; with appropriate sgRNA)	Guides the base editor to the target site	dCas12a used; acts on DNA downstream to the PAM sequence (TTTV); lower off-target activity than Cas9
ii. Cytidine deaminase	Deaminates cytidine to uridine	Deamination may or may not be dependent on the sequence context of C within the base editing window; in some editors, two copies used
iii. Uracil glycosylase inhibitor	Inhibition of the repair enzyme uracil glycosylase	Uracil glycosylase preferentially removes U from G – U mismatches; UGI inhibits this process and, thereby, promotes C-to-T transition
iv. Gam protein	Binds to DNA ends at DSBs	Reduces frequency of indels
A-to-G base editors		
i. Cas9 (with appropriate sgRNA)	Guides the other constituents to the target site	Generally, nCas9 (Cas9 nickase) used; promotes the replacement of T opposite the edited I leading to A:T to G:C transition
ii. TadA*	Deaminates adenosine to inosine	Two copies of TadA* used; inosine is read as G by cellular polymerases
RNA base editors		
i. Cas13b (with appropriate guide RNA)	Guides the adenosine deaminase to the target site	Cas13b is an RNase; dCas13b is used in the RNA base editors REPAIRv1 and REPAIRv2
ii. ADAR2DD*	Deamination of adenosine to inosine	The DD* is a hyperactive mutant of ADAR2 deamination domain

TadA*, *E. coli* tRNA adenosine deaminase mutated to enable it to deaminate adenosines in DNA; ADAR2DD, adenosine deaminase acting on RNA 2 deamination domain.

The base editors work with suitably designed sgRNAs (single guide RNAs) that associate with the Cas9 component and guide the base editor to the targeted genomic site. The Cas9 recognizes the PAM (protospacer adjacent motif) sequence and begins to unwind the target DNA to allow pairing between the 'spacer' region of sgRNA and the target DNA. This generates the protein-RNA-DNA ternary complex and exposes a small editing window of single-stranded DNA. The cognate base must be present within this editing window to enable the deaminase component of BE to gain access to and deaminate the base and, thereby, carry out base editing. The cognate base pair is the particular base pair, in this case C:G base pair, that is the target of base editing. All the other C:G base pairs at the targeted site, especially within the editing window, are non-cognate base pairs.

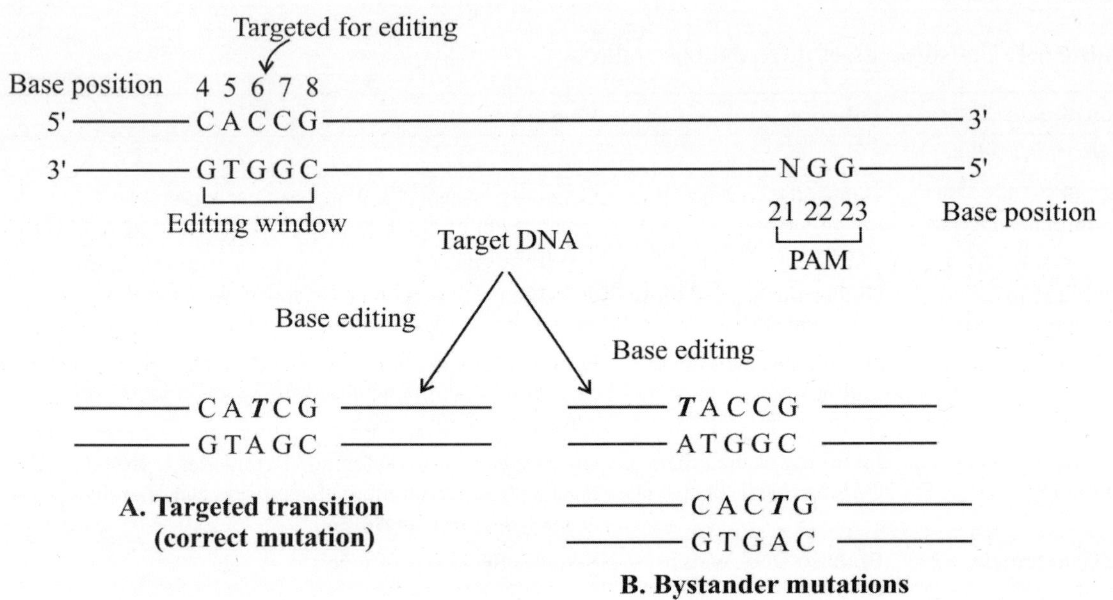

Fig. 6.1. Correct and bystauder mutateosis induced by base editors. **A.** Corect mutation. **B.** Bystander mutations. Edited base shown in bold italics. PAM, protospace adjacent molif.

Editing window width is "the number of nucleotide positions at a given site, for which editing efficiency exceeds the half-maximal value for that target site" (Kim, D. et al., 2017). Thus, in simple terms, editing window is the region of target DNA that is accessed by the BE for editing the cognate base. The editing window is typically of 5 bp located in the non-target DNA strand downstream of the PAM sequence. Some engineered BEs have a narrower window of 3 bp or even 1 bp. Generally, the editing window spans the base positions 4 to 8 of the non-target strand when the PAM sequence, NGG, is counted as positions 21 (N) to 23 (the second G; Fig. 6.1). But the boundaries of the editing window can vary depending on the targeted loci, and the width of editing window may vary to some extent from one organism to the other. For example, the window width was reported as 7 bp for BE3-based rBE3 in rice (Ren et al., 2017) and 13 bp (positions 1 to 13) for BE3 in silkworm (Li, Y. et al., 2018) as compared to the 5 bp for BE3 in mammalian cells. *It may be emphasized that base editors act on the non-target strand,* i.e., *the DNA strand not involved in pairing with the sgRNA or guide RNA.*

The C-to-T base editors were the first to be developed and several versions of these base editors are now available (Section 6.3). Recently, base editors capable of A-to-G transition were developed using an engineered tRNA adenosine deaminase that acts on adenosines present in DNA. These two types of base editors together can catalyse all the four types of transitions, *viz.*, C-to-T, T-to-C, A-to-G, and G-to-A. Although T-to-C and G-to-A base editors are not available, these transitions can be indirectly induced by editing their complementary bases by using sgRNAs complementary to the DNA strands having the T/G residues of interest.

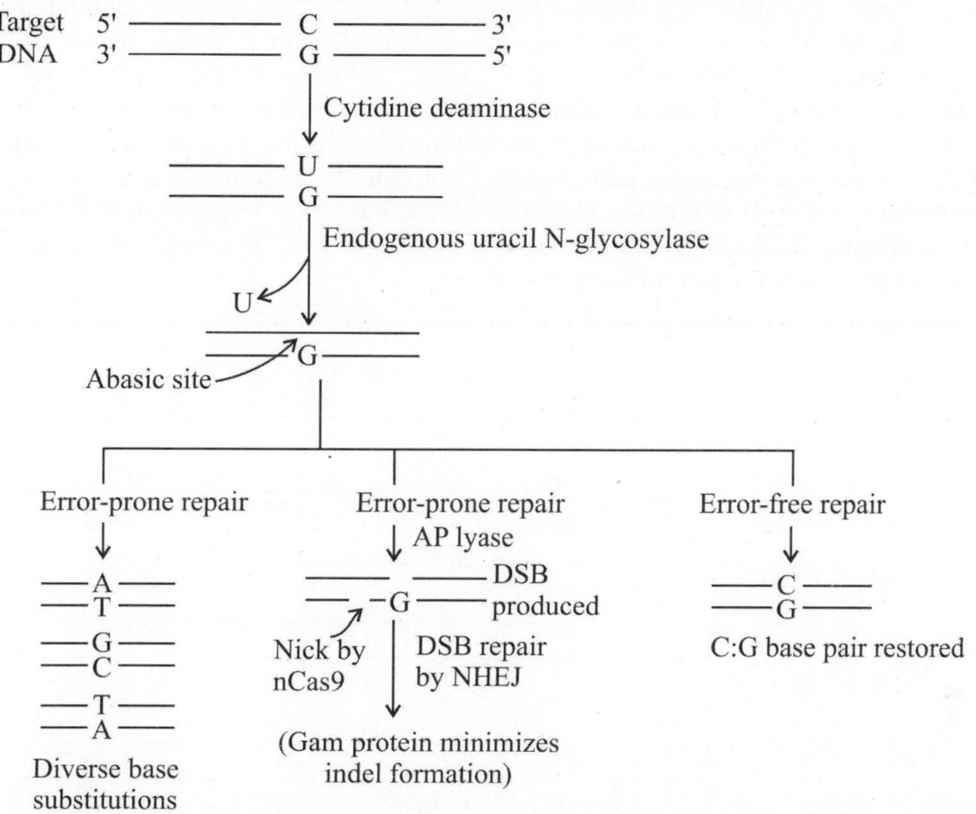

Fig. 6.2. Base editing outcomes depend on the type of DNA repair following removal of U by endogenous uracil N-glycosylase action when the base editor does not have uracil glycosylase inhibitor. Error-free repair reactors the C : G base pair, while error-prone repair leads to indel formation and diverse base edits. AP lyase, DNA-apurinic or apyrimidinic site layse (based on Hess et al., 2017b).

6.3. G:C-to-A:T BASE EDITORS

Spontaneous deamination of cytosine is the chief source of transition of G:C-to-A:T base pairs; this transition is responsible for half of the human genetic diseases produced by single gene mutations. Therefore, G:C-to-A:T base editors were the first to be created by joining naturally occurring cytidine deaminases with dCas9 or nCas9. All base editors are used in conjunction with appropriate sgRNAs, which guide the base editors to the targeted genomic sites. The dCas9 or nCas9 begins to unwind the target DNA and pairing of sgRNA with the target strand generates the editing window. The deaminase component deaminates the cytidine in the nontarget strand to generate a U:G mismatch base pair. The U:G will be converted to A:T base pair by DNA replication if it were not repaired by the DNA repair pathwys. The outcomes of C-to-U base editing depend mainly on the type of DNA repair process implemented by the cell following removal of U from the edited strand (Fig. 6.2).

The first generation of C-to-T base editor (denoted as BE1) was produced by joining the cytidine deaminase rAPOBEC1 to the N-terminus of dCas9 by the 16 amino acid linker sequence XTEN (Fig. 6.3A). APOBECs catalyse hydrolytic deamination of cytidines present in RNA and ssDNA generated during DNA replication, DNA repair and transcription, and convert C residues to uridines (Us). These deaminases contribute to defence against viruses and are involved in cancer development by mainly causing deamination of Cs in the TpC doublets. They seem to cause heritable germline mutations and, thereby, contribute to evolution (Yang et al., 2017). BE1 was able to convert C to U to produce G:U mismatched base pair (Fig. 6.4). It was found that the *in vivo* base editing efficiency of BE1 was disappointingly low (7.7% or less), while its average *in vitro* base editing efficiency was 44%.

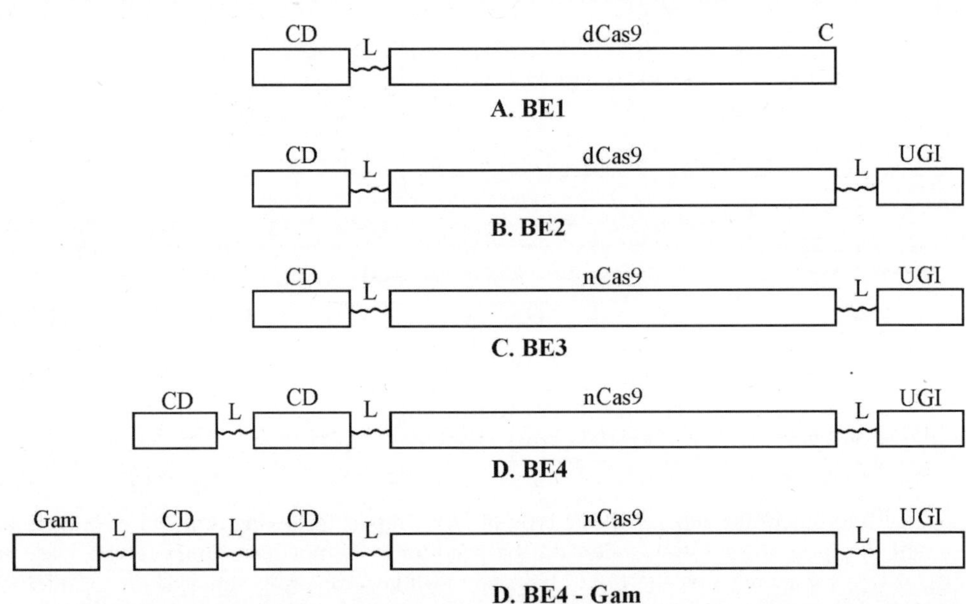

Fig. 6.3. A schematic representation of the organization of C : G to T : A base editors BE1, BE2, BE3, BE4 and BE4-Gam. C, C-terminus; CD, cytidine deaminase; dCas9, `dead' Cas9 nuclease; Gam, phage Mu encoded protein binds to double-strand breaks; N, N-terminus; nCas9, nickase Cas9 (one nuclease site active); L, linker sequence; UGI, uracil glycosylase inhibitor.

It was reasoned that the drastic reduction in base editing efficiency *in vivo* was due to the cellular DNA repair pathway involving uracil DNA glycosylase (UDG), which excises uridine residues from G:U mismatches and restores the G:C base pair at the mismatch sites. Therefore,

UGI was fused with the C-terminus of BE1 (essentially, the C-terminus of dCas9) to generate the second-generation C-to-T base editor, BE2 (Fig. 6.3B). UGI directly binds to UDG, prevents the excision of uracil and blocks the BER (base excision repair) pathway that would invariably follow this excision. BE2 (rAPOBEC1-dCas9-UGI) showed a 3-fold higher *in vivo* base editing efficiency (average 20%) than BE1. It is important that the frequency of indels (insertions/deletions) generated by both BE1 and BE2 was 0.1% or less (Table 6.2; Komor et al., 2016).

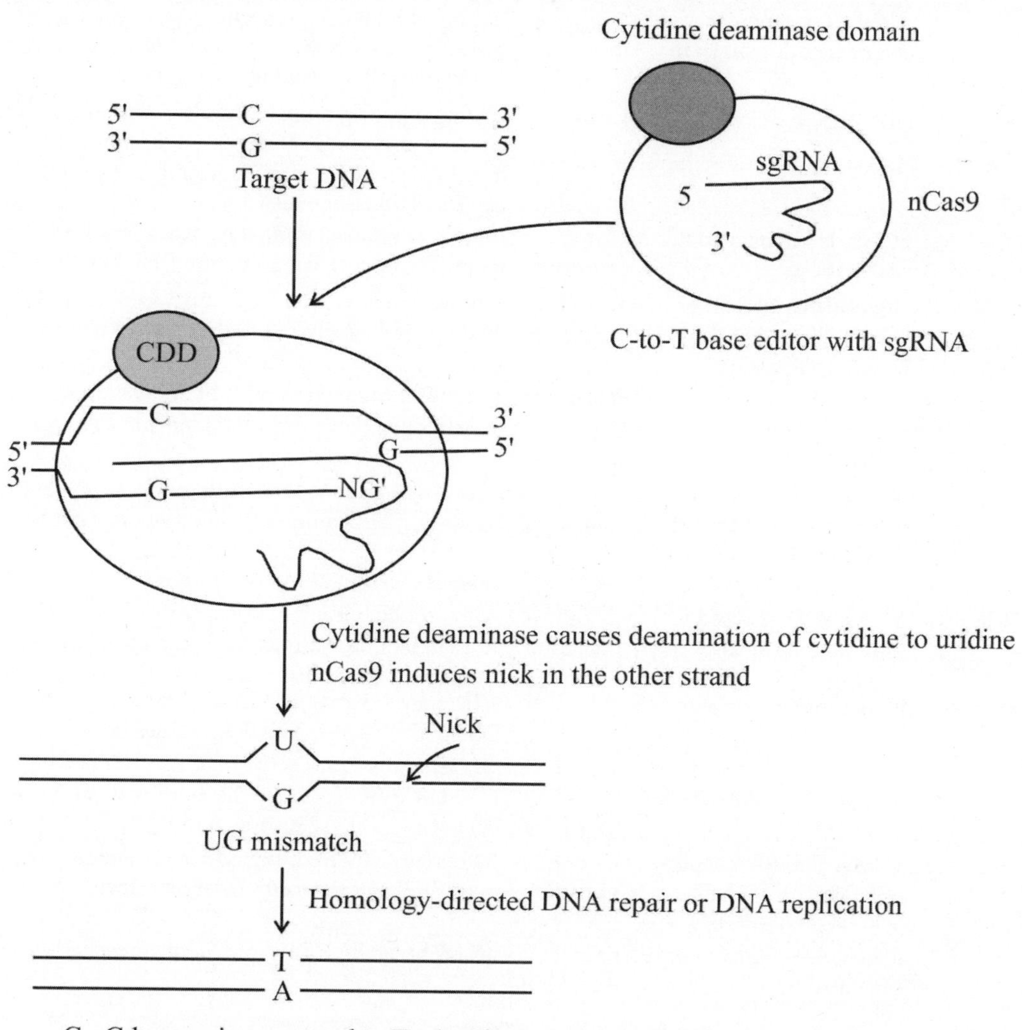

Fig. 6.4. Deamination of C-to-U and conversion of C : G base pair to T : A by nCas9-based base editors. CDD, cytidine deaminase domain; NGG, PAM sequence.

Table 6.2. Summary description of some base editors

Base editor[#]	Constituents[@]	Function	Remarks
DNA base editors			
BE1	APOBEC1 cytidine deaminase, dCas9	C-to-T transition	Editing window, 5 bp (positions 4-8 from the 5' end of sgRNA); cytosine deaminated to uracil; base editing efficiency <10%; bulk of uracil removed by uridine glycosylase
BE2	APOBEC1 cytidine deaminase, dCas9, UGI	C-to-T transition	Editing window, 5 bp; uridine glycosylase inhibited by UGI; C-to-T transition increased, on an average, 3-fold over BE1 reaching up to 20%
BE3	rAPOBEC1, nCas9, UGI	C-to-T transition	Editing window, 5 bp; edits any C within the editing window; use of nCas9 increased the yield of C-to-T transitions (editing efficiency ranges between 15–75%); in current use
YE1 BE3	nCas9, UGI, mutated rAPOBEC1	C-to-T transition	Editing window, 3 bp in most cases; edits any C within the editing window; limits target range
HF-BE3	High fidelity mutant of nCas9, UGI, rAPOBEC1	C-to-T transition	Editing window, 5 bp; edits any C within the editing window; drastically reduced off-target editing
eA3A-BE3	nCas9, UGI, eA3A	C-to-T transition	Edits C mainly in the TCR[$] context within the editing window; editing window, 5 bp; limits target range
SaBE3	nSaCas9, UGI, rAPOBEC1	C-to-T transition	nCas9 based on Cas9 from *Staphylococcus aureus*
Target-AID	dCas9 or nCas9, AID ortholog PmCDA1, with or without UGI	C-to-T transition	AID ortholog, PmCDA1 from sea lamprey; editing window, 5 bp; editing efficiency about 1.5-fold and 2.1-fold lower than that of BE3 and BE4, respectively; enables creation of mutant libraries
CRISPR-X	dCas9, AID*Δ fused to MS2 protein, sgRNA has two MS2 RNA hairpins in its scaffold region	C-to-T transition	Also produces transversions; conversion of G:C to A:T slightly higher than other changes; large editing window (-50 to +50 in relation to PAM); editing efficiency 1.5-fold and 2.1-fold lower than that of BE3 and BE4, respectively; enables local sequence diversification
TAM	dCas9, AIDx with or without UGI	C-to-T transition	Like CRISPR-X, except that editing window lies between -16 to -12 upstream of PAM
BE4	nCas9, UGI (two copies separated by a 9 amino acid linker, rAPOBEC1	C-to-T transition	Increased C-to-T editing efficiency and product purity over BE3, and reduced off-target effects
SaBE4	nSaCas9, UGI (two copies separated by a 9 amino acid linker, rAPOBEC1	C-to-T transition	nCas9 based on Cas9 from *Staphylococcus aureus*
BE3-Gam, BE4-Gam, SaBE3-Gam, SaBE4-Gam	Gam protein added to the respective base editors	C-to-T transition	Gam protein from Mu phage; it binds to DSBs and increases the base editing efficiency/indel ratio, which was BE4-Gam > BE3-Gam >BE4 > BE3
HF-BE3	nCas9-HF1 variant in place of nCas9 in BE3	C-to-T transition	Enhanced specificity (average 37-fold increase) over BE3

Base editor#	Constituents@	Function	Remarks
HF2-BE2	nCas9-HF2 variant in place of nCas9 in BE2	C-to-T transition	Higher base editing efficiency than BE2; up to 100% biallelic mutations in mouse embryos; low frequency of indels
ABE	dCas9, TadA* (mutant tRNA adenosine deaminase that acts on DNA)	A-to-G transition	The *E. coli* tRNA adenine deaminase (ecTadA) engineered to act on adenines in DNA; ecTadA evolved for seven generations by protein engineering; two copies of the evolved enzyme (TadA*) fused in tandem in the ABE
RNA base editors			
REPAIRv1	dCas13b, ADAR2DD*	A-to-I	Cas13 cleaves RNA, *i.e.*, is an RNase; ADAR2DD* is a hyperactive mutant of ADAR2 deaminase domain; the guide RNA has a C opposite the A to be edited in the RNA; I (inosine) is read as G for translation and pre-mRNA splicing
REPAIRv2	dCas13b, ADAR2DD**	A-to-I	ADAR2DD** is the hyperactive mutant DD of ADAR2; mutated and selected for improved specificity

BE3, third generation base editor; eA3A-BE3, engineered-A3A-BE3; ABE, adenine base editor; REPAIR, RNA editing for programmable A to I replacement

@ nCas9, nickase Cas9; UGI, uridine glycosylase inhibitor; rAPO1BRC1, rat APOBEC1 (single-strand-specific cytidine deaminase from rat); APOBEC1 (apolipoprotein B mRNA editing enzyme, catalytic polypeptide-like 1); A3A, human APOBEC3A; ADAR2, adenosine deaminase acting on RNA 2; ADAR2DD*, hyperactive mutant of ADAR 2 deaminase domain; ADAR2DD**, ADAR2 hyperactive DD mutated for enhanced specificity; AID, activation-induced cytidine deaminase; AIDx, AID with deleted C-terminal domain; AID*Δ, a hyperactive variant of AID with deleted NLS (nuclear localization signal); TAM, targeted AID-mediated mutagenesis

$ R, A/G

It was hypothesized that the G in nonedited strand (the complementary or target strand) located opposite the U in the edited strand would be substituted by A if the nonedited strand were nicked. It may be expected that the nick would be repaired by long-patch BER and the edited strand would be preferentially used as template for this repair. This would yield a U:A pair that would become a T:A pair following the subsequent DNA replication (Fig. 6.4, 6.5). Therefore, dCas9 in the BE2 architecture was replaced by nCas9 (D10A mutant; Asp at position 10 replaced by Ala) having the active HNH catalytic domain; this modification yielded the third-generation base editor BE3 (Fig. 6.3C). The BE3 (rAPOBEC1-nCas9-UGI) showed 2- to 6-fold increase in editing efficiency as compared to BE2, the average efficiency being 37% for the loci targeted in human cells, and the highest efficiency was ~75%. However, the frequency of indels was higher (average 1.1% in case of BE3) than in cases of BE1 and BE2.

The BE1, BE2 and BE3 all have a 5 bp window from base positions 4 to 8 of the target strand when the PAM sequence NGG is counted as positions 21 to 23 (Fig. 6.1); this would place the editing window from the 13th to 17th bases upstream of the second G of the PAM sequence NGG. The BE3 and similar base editors have been used for a range of base editing activities in plants, bacteria, yeast, zebrafish mammalian cells, mice and mouse embryos, and also to produce gene knockouts (Komor et al., 2016, 2017a).

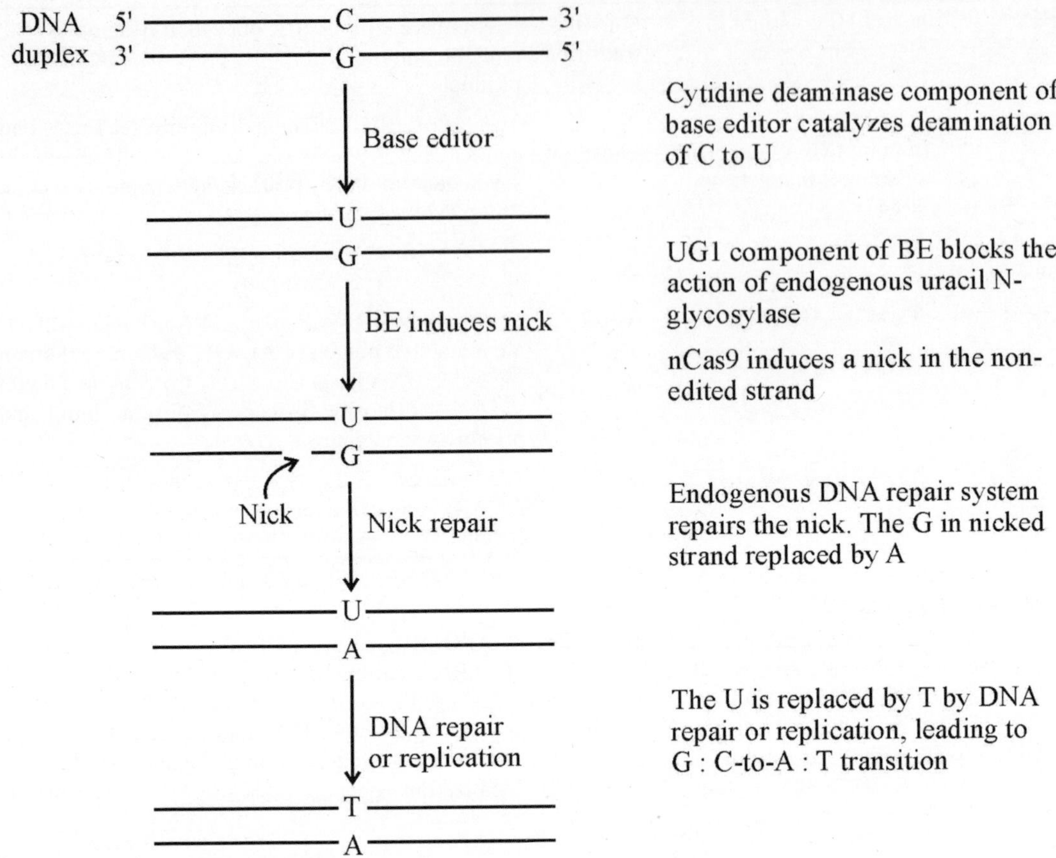

Fig. 6.5. Schematic representation of the reactions involved in conversion of C : G base pair to A : T base pair by a base editor like BE3 or BE4. BE, base editor; UG1, uracil N-glycosylase inhibitor, nCas9, Cas9 nickase (based on Hess et al., 2017b).

Nishida et al. (2016) designed another C-to-T base editor, called Target-AID, by combining dCas9 with AID (activation-induced deaminase) ortholog PmCDA1 from sea lamprey. This editor has editing window of 5 bases (positions -16 to -20 upstream of the PAM sequence), and generally over 90% of the C-to-T transitions are induced at -16 to -18 positions. The use of nCas9 in the place of dCas9 increased the editing efficiency in yeast, but this editor also induced indels in mammalian cells. Further, linking of UGI with this editor supressed indel formation and improved the efficiency of base editing.

6.3.1. G:C-to-A:T base editors with different PAM sequence requirements

The PAM sequence (NGG) requirement for BE3 imposes a restriction on the sites in the human genome that are accessible for efficient editing since most genomic sites of interest do not have this sequence 13–17 nt (nucleotides) downstream of the target C. Further, BE3 efficiently edits all the C residues located within the ~5-bp editing window; this is likely to induce bystander mutations (Section 6.8) with undesirable consequences. Therefore, BE3 variants with favourable modifications in their characteristics were produced. In one modification, SpCas9 (Cas9 from *Staphylococcus pyogenes*) variants with altered PAM sequence requirements and SaCas9 (Cas9 from *Staphylococcus aureus*) were used (Table 6.3). The SaCas9-based BE3 (SaBE3) requires NNGRRT PAM in place of NGG for the SpCas9-based BE3. SaBE3 edited the targeted C with very high efficiency (~50–75% of the total number of Cs converted to Ts) in six human genomic loci. Notably, SaBE3 was more efficient at the sites having NNGRRT PAM than BE3 at sites with NGG PAM sequence.

Table 6.3. A list of different C-to-T base editors and their PAM requirements (Based on Kim, Y.B. et al., 2017)

Base editor	PAM sequence	Remarks
BE3	NGG	The canonical C-to-T editor; editing window 5 bp; all C residues within the editing window typically converted to T; may show lower efficiency for some 5'GC3' doublets; commonly used
VQR-BE3	NGAN	Editing efficiency up to 50%; editing window similar to that of BE3; substantially lower off-target editing at off-target sites of VQR SpCas9
EQR-BE3	NGAG	Editing efficiency up to 50%; editing window similar to that of BE3; substantially lower off-target editing at off-target sites of VQR SpCas9
VRER-BE3	NGCG	Editing efficiency up to 50%; editing window similar to that of BE3
SaBE3	NNGRRT	More efficient than BE3 at the appropriate genomic sites; editing window (from 2 to 15 bp upstream of PAM) exceeds the limits for BE3; off-target base editing at known off-target sites for Cas9
SaKKH-BE3	NNNRRT	Editing efficiency up to 62%; off-target base editing at known off-target sites for Cas9
BE2-HF2	NGG/A	Editing efficiency higher than BE2, and reached 100% biallelic mutations in mouse embryos; edits Cs in both target and nontarget strands, and at sites proximal to the sgRNA binding region; ~4.5% indels induced
dCpf1-BE	TTTV	Cpf1, also known as Cas12a, cleaves DNA downstream to the PAM sequence and has lower off-target activity than Cas9
ZF-AID*	None	ZFs bind target DNA; human AID deaminates C; base editing efficiency 13% in *E. coli* and 2.5% in human cells, off-target editing up to 150 bp from the target site
TALE-C3-AID*	None	TALEs bind target DNA; human AID deaminates C; base editing efficiency 13% in *E. coli* and 2.5% in human cells, off-target editing up to 150 bp from the target site

* AID, activation induced deaminase; TALE-C3-AID, the C-terminus of TALE truncated to 63 amino acids.

The range of PAM sequence requirements by base editors was further expanded by using 'dead' Cas12a (dCas12a) instead of dCas9 in the BE2 architecture. Cas12a uses a single CRISPR RNA that pairs with the target DNA on the downstream side of PAM sequence (Fig. 6.6, Table 6.3). The base editor dCas12a-BE (APOBEC1-dCas12a-UGI) has base editing window of 6 nt from position 8 to 13 on the downstream side of the PAM, and it showed base editing efficiency comparable to that of BE3. Further, the proportion of C-to-T transitions among the total base transitions induced by this editor was nearly as high (~88–99%) as that for BE3. This base editor does induce low frequencies of indels, base substitutions other than C-to-T as well as off-target base editing, but the levels of these actions were lower than those for BE3. In addition, the use of YE mutant of APOBEC1 (Section 6.3.2) yielded the base editor dCas12a-BE-YE. This editor had an editing window of 3 nt (positions 10–12), exhibited ~30–90% editing efficiency, and a higher proportion of single C-to-T transitions than dCas12a-BE (when the editing window had two or more Cs). This base editor extends the base editing to target sites having the PAM sequence TTTV (V = A, C or G) in their upstream regions (Li, X. et al., 2018).

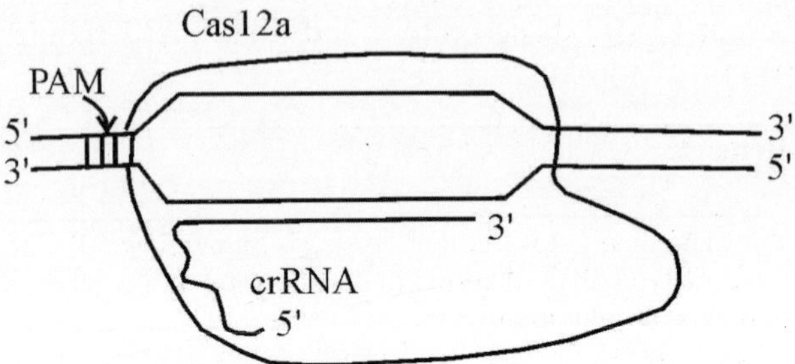

A. Cas12a in association with DNA

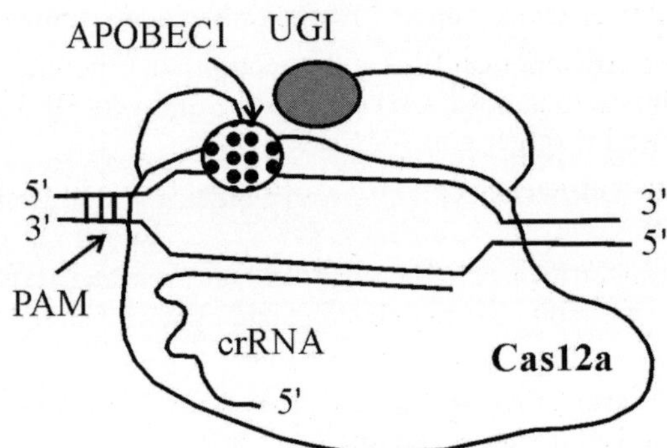

B. d Cpf1-BE associated with target DNA

Fig. 6.6. Cas12a associates with target DNA on the downstream side of PAM sequence. **A.** Cas12a nuclease associated with target DNA. **B.** dCpf1-BE (base editor based on dCas12a). APOBEC1, apoliproprotein B editing complex 1, UGI, uridine glycosylase inhibitor; Cpf1, Cas12a; CrRNA, CRISPR RNA.

Finally, two C-to-T base editors, namely, ZF-AID, and TALE-C3-AID, have been developed, which have no PAM sequence requirement. These editors comprise human AID cytidine deaminase joined to the C-terminus of ZFs or TALEs by an eight amino-acid long linker (Table 6.3). The ZF and TALE components of the editors guide them to the targeted DNA sequences where AID causes cytidine deamination. In the TALE-C3-AID, the C-terminus of TALE protein is truncated to 63 amino acids in the place of normal 178 amino acids. Under optimum conditions, these base editors were able to convert specific Cs to Ts with 13% efficiency in the bacterium *E. coli*, but the base editing efficiency was only 2.5% in human cells. They also performed off-target editing up to more than 150 bp from the targeted genomic site. Thus, they need to be further refined to improve their base editing efficiencies and, more particularly, to minimize off-target base editing (Yang, L. et al., 2016).

Base editing within mitochondria is challenging because delivery of sgRNA essential for the function of Cas9-based editors into these organelles is quite problematic. The interbacterial toxin DddA catalyses deamination of cytidines without unwinding of the target DNA duplex. The DddA was split into two parts, each of which was fused with a TALE protein and UGI, the TALE pair being designed to target a specific site in the mitochondrial DNA (mtDNA). The two TALE-DddA-UGI fusions bind to the target DNA and restore the complete functional DddA, which edits the C-to-T 'with high target specificity and product purity'. These base editors, called RNA-free DddA-derived cytosine base editors (DdCBEs), perform editing of duplex DNA without unwinding it (Mok et al., 2020).

6.3.2. G:C-to-A:T base editors with improved specificity

In order to minimize the risk of bystander mutations, attempts were made to reduce the width of editing window of BE3. Genetic engineering was used to create rAPOBEC1 variants with mutations in those amino acids of the deaminase domain that are not essential for deaminase activity. Three amino acid substitutions, *viz.*, W90Y, R126E and R132E (denoted as Y, E1 and E2, respectively), appeared promising; they were combined to generate three double mutants (YE1-BE3, YE2-BE3 and EE-BE3) and one triple mutant (YEE-BE3) (Table 6.4). The mutant base editors have base editing window width of ~2 bp as compared to the 5 bp for BE3. The editing efficiency of the double mutants was either comparable to or slightly lower than that of BE3, but it was about 2.9-fold lower than that of BE3 in the case of triple mutant. Significantly, *off-target editing by the mutant base editors was about 3.6-fold lower than that by BE3.* The precision of base editing by BE3 was improved by replacing rAPOBEC1 with engineered human cytidine deaminase APOBEC3A (A3A) modified by protein engineering. The eA3A-BE3 editor based on the engineered A3A (eA3A) edits C residues located in the sequence context TCR (R = A or G) within the editing window, and shows greatly reduced bystander effects (Komor et al. 2017a,c; Kim, Y.B. et al., 2017; Gehrke et al., 2018).

BE3 induced, on average, 7.4-fold higher frequency of two C-to-T transitions than single C-to-T transition. But the mutant base editors YE1-BE3, YE2-BE3, EE-BE3, and YEE-BE3 sponsored substantially higher frequencies of single C-to-T transitions than transitions of two C residue most likely due to reduced processivity of the mutant cytidine deaminases. For example, the triple mutant YEE-BE3 showed, on an average, 14-fold higher frequency of one C-to-T transition than transitions of two C residues. In addition, the mutant base editors showed, on an average, 3.6-fold lower off-target base editing than BE3 over 21 off-target sites. However, most of the narrow-window BE3 mutants could not be productively combined with the Cas9 variants having altered PAM requirements (Table 6.3; Kim, Y.B. et al., 2017).

Cas-9-based base editors induce off-target edits due to binding of their Cas-9 component to off-target sequences; such edits can be minimized by using Cas-9 variants with improved specificity to create BEs. In addition, they can induce off-target edits at random genome wide sites due to Cas-9-binding-independent action of the deaminase domain; the frequency of these edits may be as high as 10^{-7} per base pair. The YE1-BE3 was further engineered to generate variants of the cytidine deaminase component with robust on-target editing activities combined with, on an average, ~10–100-fold lower off-target editing. These variants are particularly promising for such applications where off-target editing must be minimal (Doman et al., 2020).

The specificity of BE3 was sought to be improved by using Cas9 variants like Cas9-HF generated by genetic engineering. The use of nCas9-HF1 in the place of nCas9 in BE architecture generated HF-BE3 (high fidelity BE3). HF-BE3 showed enhanced specificity (average 37-fold increase), but there was detectable off-target base editing (average, 0.03±0.005%) at both repetitive and nonrepetitive loci when delivered into the mammalian cells as plasmid DNA. But when HF-BE3 was delivered as ribonucleoprotein (RNP), off-target editing was negligible at nonrepetitive loci and greatly reduced at repetitive loci. The delivery of base editors as RNP seems to uncouple on-target editing from off-target editing, and this mode of delivery would be particularly beneficial for *in vivo* base editing (Rees et al., 2017).

Table 6.4. Narrow window mutant base editors based on BE3 C-to-T base editor (Based on Kim, Y.B. et al., 2017)

Mutant base editor*	Window width (bp)	Base editing efficiency	Off-target editing	Remarks**
YE1-BE3	~2	Comparable to BE3	About 3.6-fold lower than BE3	Two amino acid (W90Y + R126E) substitutions; further engineering of YE1-BE3 has yielded mutant BEs with 10-100-fold lower off-target editing
YE2-BE3	~2	Moderately lower than BE3	About 3.6-fold lower than BE3	Two amino acid (W90Y + R132E) substitutions
EE-BE3	~2	Comparable to YE2-BE3	About 3.6-fold lower than BE3	Two amino acid (R126E + R132E) substitutions
YEE-BE3	~2 (1 at some loci)	About 2.9-fold lower than BE3	About 3.6-fold lower than BE3	Three amino acid (W90Y + R126E + R132E) substitutions
eA3A-BE3	5	Comparable to that of BE3	Greatly reduced bystander effects	Has engineered human cytidine deaminase APOBEC3A (eA3A) in the place of rAPOBEC1; edits C in the sequence context TCR

* Mutations in the deaminase domain involving amino acids not essential for deaminase activity.

** BE3 shows little preference for the different Cs present in the editing window. In contrast, the mutant base editors showed marked preference for one C in the editing window.

Another high-fidelity base editor, BE2-HF2, was produced by using the 'dead' version of Cas9-HF2 variant in the place of nCas9 in the architecture of BE2 (Liang et al., 2017). The Cas9-HF2 has one more amino acid substitution in addition to the four substitutions contained in Cas9-HF1. The PAM sequence required by the base editor BE2-HF2 was NGG/A in the place of NGG for BE2. It can edit C residues present in both target and nontarget strands in the targeted genomic region; this feature is also shared by BE2. Both BE2-HF2 and BE2 could deaminate the C residues proximal to the binding of sgRNA; further refinements are needed to eliminate this feature. The base editor BE2-HF2 had higher base editing efficiency than BE2, and the efficiency could reach 100% biallelic mutations in mouse embryos. Surprisingly, it induced low frequency of indels in 4.6% of the embryos although dCas9 is unlikely to induce either a nick or DSB. Possibly, the indels were generated during base excision repair following deamination of C-to-U. In case C deamination occurred in both the strands at sites close to each other, base excision would create abasic sites in both the strands, which may ultimately lead to DSB formation, followed by NHEJ repair.

The above modifications either relieved the restriction imposed by a specific PAM sequence requirement or improved the product purity or precision of base editing. Product purity refers to the percentage of the targeted C-to-T transition among all the different types of genetic changes, including indels generated by NHEJ. The term precision of base editing may be defined as the ratio of the frequency of desired C-to-T transition to that of all the other types of base transitions, *e.g.*, C-to-G or C-to-A. The efficiency of base editing may be defined as the proportion of all the targeted genomic sites exposed to the base editor that have undergone the desired base editing. Another term, the base editing product distribution describes the frequencies of various types of genetic changes generated by the base editing procedure.

6.3.3. G:C-to-A:T base editors with increased efficiency

The BE3 architecture was further modified to enhance its base editing efficiency. It was reasoned that since the natural cytidine deaminase functions as a dimer, an additional copy of this enzyme in the base editor might enhance base editing efficiency. Therefore, a second copy of cytidine deaminase was fused with the existing copy in BE3 as well as SaBE3 (Table 6.2), and the two copies were separated by a 9 amino acid linker. This yielded the fourth-generation base editors, BE4 (Fig. 6.3D) and SaBE4, which showed about 50% increase in C-to-T base editing efficiency over those of BE3 and SaBE3, respectively. In addition, the frequencies of unintended mutations were reduced to half as compared to BE3.

The Gam protein from bacteriophage Mu competes with the DSB repair proteins and binds to the DNA ends at DSBs. This essentially prevents DSB repair, which leads to death of the affected cells; as a result, the cells with potential for indel formation are eliminated from the population of edited cells (Hess et al., 2017). The Gam protein was fused at the N-termini of BE3, BE4, SaBE3 and SaBE4 to yield the base editors BE3-Gam, BE4-Gam (Fig. 6.3E), SaBE3-Gam and SaBE4-Gam, respectively. The BEs with Gam showed much reduced (average ~2-fold reduction) indel frequencies, while base editing efficiency was almost unaffected. The overall product purity obtained was in the following order: BE3 < BE3-Gam < BE4 < BE4-Gam in cultured human cells (Komor et al., 2017c).

The efficiency of base editing as well as the product distribution tends to vary with the genomic target. In the base editor architecture, the type of Cas9 protein used, inclusion of uridine glycosylase inhibitor, and the presence of bacteriophage Mu protein Gam have marked influence on base editing outcomes; the Gam protein causes a drastic reduction in the indel frequency. UNG (uracil N-glycosylase) activity is necessary for the formation of by-products, and its inhibition is essential for minimizing by-product formation. In general, genomic sites having multiple C residues in the editing window show higher product purity than those with single C possibly be due to easier action of UNG on isolated single U:G mismatches. Joining the rAPOBEC1 deaminase to nCas9 by a 32 amino acid long linker in place of a 16 amino acid linker increases the base editing efficiency (by about 1.2-fold) most likely because it allows better access to the editing window. As a result, the multiple Cs present in the editing window are more uniformly deaminated leading to enhanced base editing efficiency.

The optimum length of the linker joining UGI to nCas9 C-terminus seems to be 9 amino acids as it noticeably improves product purity. Addition of a second copy of UGI at the C-terminus by another 9-amino acid linker markedly improves product purity. Further improvement in product purity can be achieved by adding the Gam protein to the base editor. When Gam protein is linked at the N-terminus of base editor by a 16 amino acid XTEN linker, there is up to nearly 2-fold reduction in indel frequency. It has been suggested that genetic engineering may enable creation of a modified Gam that would further reduce indel frequencies (Komor et al., 2017c).

6.3.4. G:C-to-A:T base editors for diversifying base editing

The above discussion (Sections 6.3.1 to 6.3.3) pertains to base editors used for precision base editing. In precision base editing, the objective is to edit a single C at the targeted site either to correct a mutation or to induce a desired point mutation. In addition to such base editors, two C-to-T base editors, *viz*., targeted AID-mediated mutagenesis (TAM) and CRISPR-X (Fig. 6.7), have been developed that generate local sequence diversity. These editors generate mutations in several different C residues located in the targeted localized region of the genome; this type of base editing is called diversifying base editing. The diversifying base editors are used to create a library of diverse mutations located in specifically targeted genomic regions (Ma, Y. et al., 2016; Hess et al., 2016).

The TAM base editor has a copy of human AID that lacks the C-terminus. The C-terminus of AID includes a nuclear export signal (NES), which usually limits its activity. The AID lacking NES is designated as AIDx, which is linked to the C-terminus of dCas9 to generate the base editor TAM. TAM has been shown to cause >20% mutations in the GFP locus when targeted by a pool of sgRNAs; these mutations were due to C-to-T and G-to-A transitions, and transversions of C and G to the other three bases. Further, the frequencies of C-to-T and G-to-A transitions were only slightly higher than those of the other changes. But when UGI was co-expressed with TAM and targeting was done by a single sgRNA, the C-to-T transitions were induced within a window of 5 nt (positions -16 to -12 upstream of the PAM; Ma, Y. et al., 2016).

The CRISPR-X editor has a design that is quite different from those of other base editors. It has a hyperactive mutant of AID, *i.e.*, AID*, generated by three amino acid substitutions (K10E, T82I, and E156G); AID* has higher deamination activity than wild type AID. The NES sequence of AID* was deleted to obtain AID*Δ, which was fused with the MS2 coat protein (denoted here as MS2). The AID*Δ-MS2 fusion protein is used with dCas9 and specially designed sgRNAs that have two MS2 RNA hairpins or MS2 aptamers integrated into their backbone or scaffold regions. Thus, the AID*Δ is not fused with dCas9; instead, it is recruited to the targeted genomic region by the specially designed sgRNAs. The CRISPR-X editing frequency is over 20% and is spread over a very large window of 101 nt (-50 to +50 positions relative to the PAM). Like TAM, CRISPR-X induces transitions of C-to-T and G-to-A as well as transversions of C and G residues to the other three bases. Further, the C-to-T and G-to-A transitions were only slightly more frequent than the other types of changes. It also induced relatively much lower frequencies of mutations of A and T residues (Ma, Y. et al., 2016).

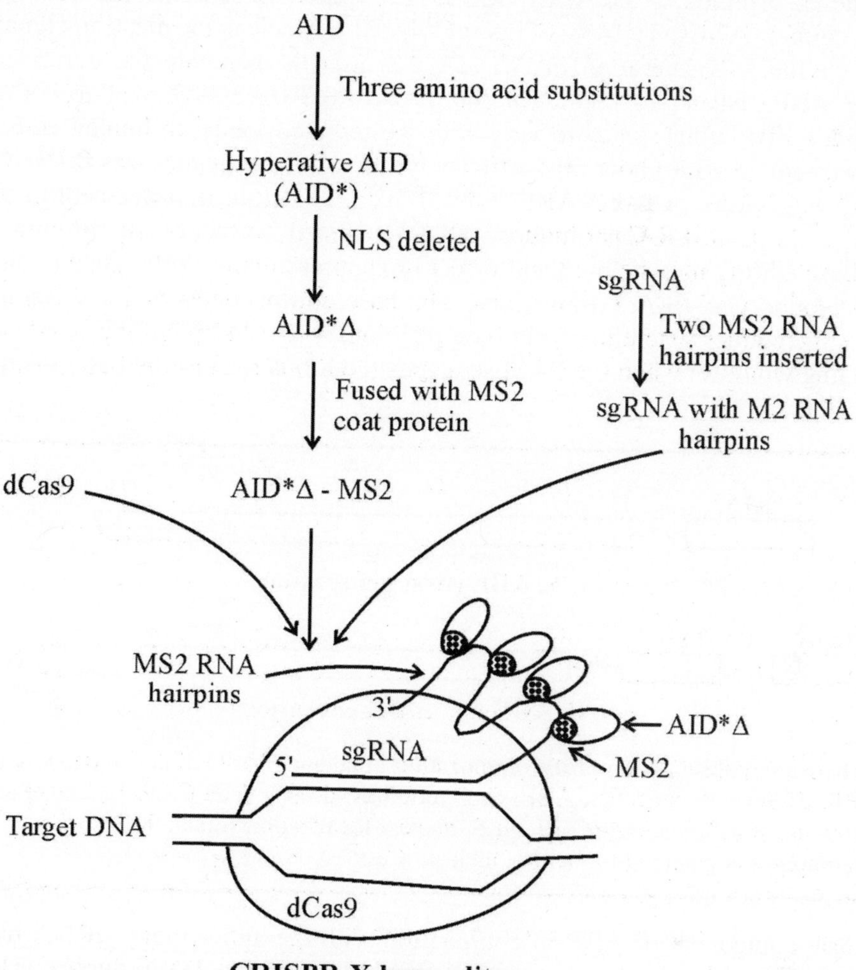

Fig. 6.7. Schematic representation of CRISPR-X base editor. The AID variant AID* has three amino acid substitutions (K10E, T82I, E156G); the NLS is deleted from AID* to generate AID*Δ, which is fused with MS2 coat protein. The appropriate sgRNA is modified by insertion of two MS2 RNA hairpins in the scaffold region. The three components (AID*Δ-MS2, sgRNA and dCas9) are encoded by separate genes. When dCas9 associates with sgRNA and the target DNA, the AID*Δ-MS2 fusion protein is recruited to the MS2 RNA hairpins of the sgRNA (based on Hess et al., 2017). AID, activation induced deaminase; MS2, MS2 coat protein.

6.4. A:T-to-G:C BASE EDITING

Deamination of adenine produces inosine that is read by both DNA and RNA polymerases as guanine. However, there is no known natural deaminase that catalyses deamination of adenines present in DNA. The tRNA adenine deaminase (TadA) of *E. coli* catalyses adenine deamination in tRNA. Gaudelli et al. (2017) genetically engineered the *E. coli* TadA (ecTadA) to evolve an enzyme capable of deaminating adenines present in DNA. Initially, a single copy of TadA* was fused with the N-terminus of nCas9 by the XTEN linker to generate the early generation of adenine base editors (ABEs; TadA*–XTEN–nCas9–NLS, nuclear localization signal; Fig. 6.8A). The seventh engineered generation of ecTadA was used to generate the current adenine base editors; these ABEs have two copies of the TadA* and have high A:T-to-G:C base editing efficiency (Fig. 6.8B), which tends to vary with the targeted locus. In human cells, the targeted A:T-to-G:C was edited with about 50% efficiency; the indel frequency was 0.1% or less and the product purity was 99.9% or more. ABE showed over 1,000-fold improvement in product purity as compared to the CRISPR-Cas9-induced HDR mediated correction of the mutant base pair. A:T-to-G:C base editing may help either correct or suppress many of the genetic diseases caused by single nucleotide G:C-to-A:T transitions. The base editing window for different variants of ABE was 4-6 nucleotides extending from base positions 4 to 7 (4 bp editing window) and from 4 to 9 (6 bp editing window) when the PAM sequence (NGG) is read as the base positions 21 (N) to 23 (Gaudelli et al., 2017).

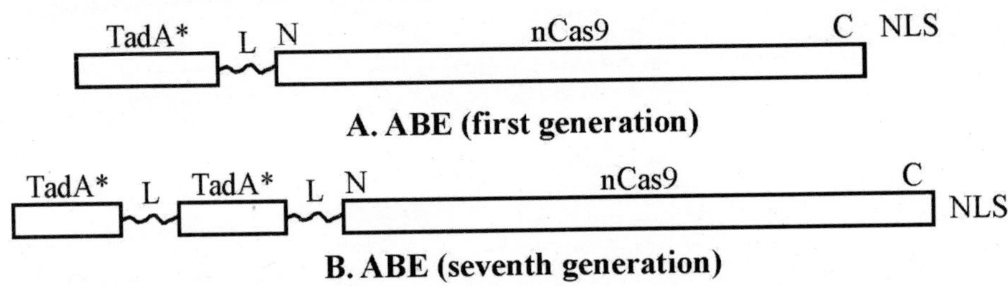

Fig. 6.8. A schematic representation of the organization of adenine base editors (ABEs). **A** First generation ABE. **B.** Seventh geneation ABE. C, C-terminus; dCas9, dead Cas9; L, Linker sequence; N- N-terminus, nCas9, `nickase' Cas9, NLS, nuclear localization signal; TadA*, *E. coli* tRNA adenine deaminase a engineered for acting `nickase' Cas9.

The nCas9 component of ABE in conjunction with the appropriate sgRNA places the ABE at the targeted site and nCas9 induces local unwinding of the DNA duplex. This allows the sgRNA to pair with the targeted strand and make the cognate A present in the nontargeted strand accessible to TadA*. TadA* deaminates this A-to-I, which is read as G by the relevant cellular machineries. The I would be converted to G during the subsequent DNA replication or by DNA repair (DNA repair tends to favour replacement of A:T by G:C base pairs). The ABEs have expanded base editing capabilities to include all the four possible transitions. The base editors BE3 and BE4 enable C-to-T and G-to-A transitions. Similarly, the ABEs can produce A-to-G and T-to-C transitions. Thus, the BE3/BE4 and ABE base editors together can correct any mutational defect associated with SNPs (single nucleotide polymorphisms; Gaudelli et al., 2017) provided it is not impeded by the PAM sequence requirement. It is encouraging that several base editors requiring PAM sequences different from NGG have been developed making more mutant SNPs accessible to precision base editing.

6.5. RNA BASE EDITORS

RNA editing is the irreversible conversion of one base present in an RNA molecule into another base after it has been produced by transcription. RNA editing occurs naturally in many RNA species, including tRNAs. RNA editing enables the introduction of discrete and specific changes in the nucleotide sequences of concerned RNAs that may be critical for their proper function. In humans, the most frequent modification is the conversion of A-to-I mediated by the deaminases ADAR1 (adenosine deaminase acting on RNA 1) and ADAR2. The I in the RNA transcript is read as G by the splicing and translation machineries. As a result, base editing can affect both splicing patterns and translation processes (Matsoukas, 2018). An ADAR has a N-terminal double-stranded RNA binding domain and a C-terminal deaminase domain (DD). The adenosines present in RNA duplexes and mis-paired with cytidines are preferentially edited by ADARs.

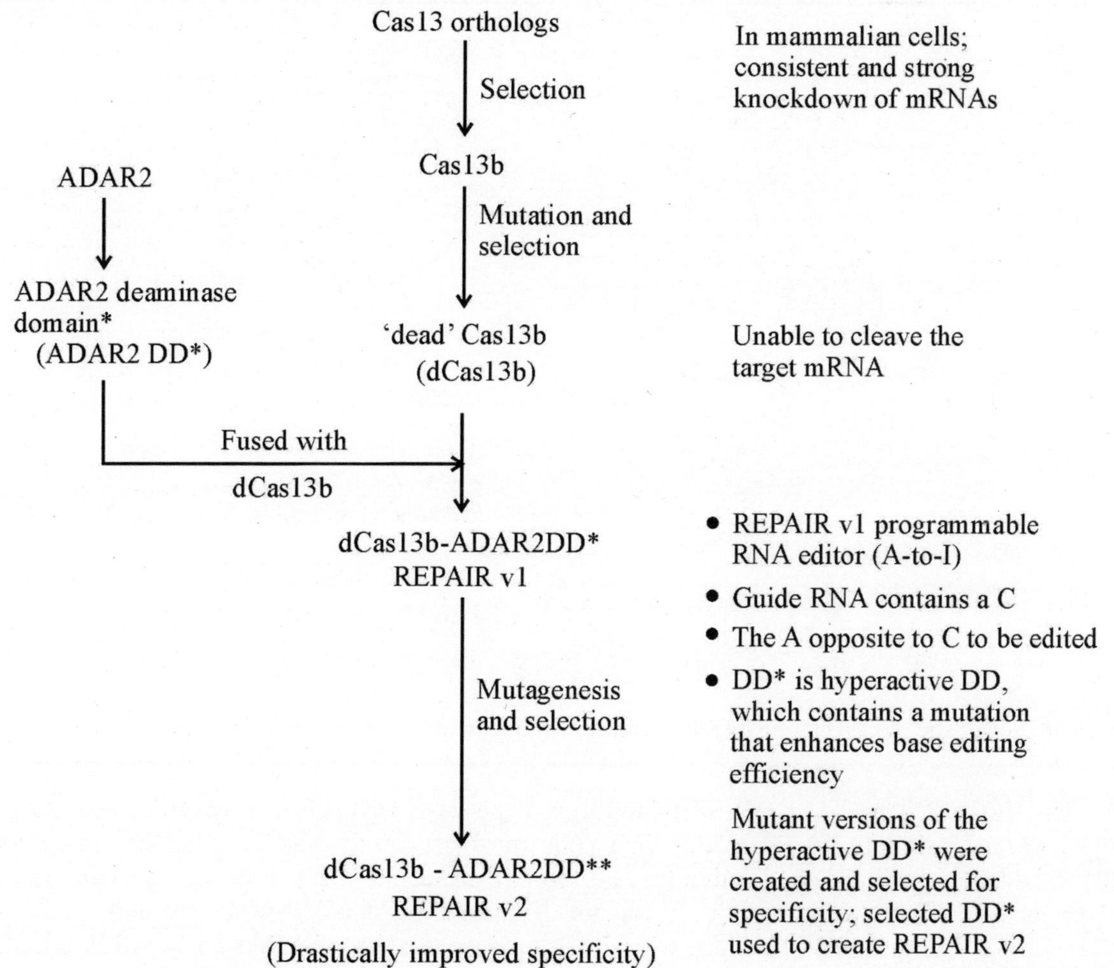

Fig. 6.9. The development of RNA base editors REPAIR v1 and REPAIR v2. REPAIR, RNA editing for progammable A-to-I replacement; ADAR2DD* hyperactive mutant of deaminase domain of ADAR2; ADAR2, adenosine deaminase acting on RNA2; ADAR2 DD** hyperactive DD mutated for increased specificity.

Stafforst and Schneider (2012) covalently linked ADAR to a guide RNA to create a hybrid deaminase that converts A-to-I in a sequence-specific and selective manner. This hybrid enzyme could repair point mutations in mRNAs *in vitro*. Subsequently, Montiel-González et al. (2016) designed an efficient site-directed RNA editing system by non-covalently linking antisense RNA (for correct positioning of DD) with the DD of an ADAR. This system was highly efficient with ~70% conversion of the targeted UAG codon to UGG. However, it did induce several off-target changes in the targeted mRNA. Recently, the natural process of ADAR-mediated RNA editing was ingeniously repurposed into a programmable RNA editing system called REPAIR (RNA editing for programmable A-to-I replacement; Cox et al., 2017).

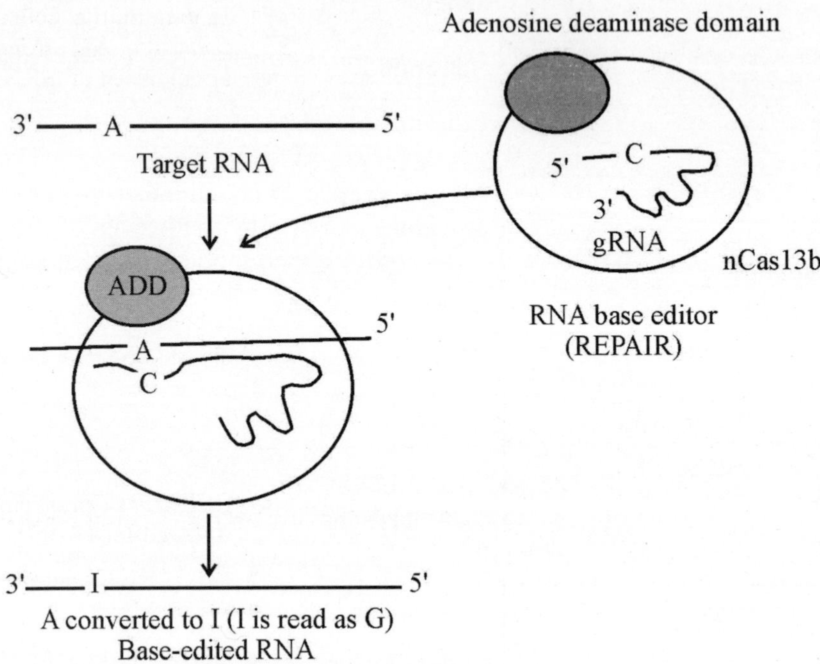

Fig. 6.10. Schematic representation of RNA editing. ADD, Adenosine deaminase domain.

The REPAIR is based on CRISPR-Cas13b, a Type VI-B CRISPR-Cas system, since Cas13b can cleave RNA molecules in mammalian cells most efficiently and specifically. Cox et al. (2017) evaluated several Cas13 orthologs and found PspCas13b (Cas13b from *Prevotella* sp.) to be the most promising. They produced mutant dPspCas13b ('dead' PspCas13b) and used it to create REPAIR (Fig. 6.9) by fusing it with the deaminase domain of ADAR2 (ADAR2DD). The dCas13b component (together with the appropriate guide RNA) of the REPAIR base editor guides the latter to the target site and places ADAR2DD at the proper location for editing the desired adenosine. The ADAR2DD then catalyses the conversion of A-to-I without cleavage of the RNA molecule (Fig. 6.10).

The efficiency of the REPAIR was improved by introducing a C in the guide RNA opposite the A to be edited since this mismatch promotes editing of the mismatched A. Further, a hyperactive mutant of DD was used in the place of wild type DD; this modified base editor was denoted as REPAIRv1 (Fig. 6.9). The REPAIRv1 was shown to correctly edit two human mRNAs with efficiency of 35% and 23%. The DD component of REPAIRv1 was mutated and selected for enhanced specificity, which resulted from a weakening of the interaction of ADAR2DD with the target RNA molecule. The mutant ADAR2DD was used to generate the improved version REPAIRv2 (Fig. 6.9). REPAIRv2 was 919-times more specific than REPAIRv1 and it consistently and correctly edited up to 52% of the mutant mRNAs. REPAIRv2 showed merely 20 off-target edits in the whole transcriptome, while this value was 18,385 for REPAIRv1 (Cox et al., 2017).

REPAIR base editor has broad applicability and can be used for both RNA knockdown and RNA editing. It could be useful in investigation of the mechanisms of alternative splicing, RNA editing and mRNA translation. RNA editing might be useful in the treatment of ailments caused by temporary conditions like localized inflammation. Further, RNA editing may be a more desirable option than gene editing for therapeutic applications since the undesirable off-target effects of RNA editing will be short-lived due to the rapid RNA turnover. Further, the ethical issues concerning gene editing would not be relevant for RNA editing since the changes induced by editing are not heritable. But RNA editors would need to be administered at regular intervals to maintain the effects produced by editing (Matsoukas, 2018). However, it may take several years for implementing the refinements in base editing-based therapeutics that are necessary for their clinical trials.

6.6. BASE EDITORS FOR PLANTS

The base editing techniques as well as the reagents were initially developed and optimized for animal, particularly human, cells. Subsequently, base editing was extended to plant systems, and there are already some examples of targeted base editing for inducing mutations in useful genes of rice and some other crop species. In 2017, four groups of workers (Li et al., 2017; Lu and Zhu, 2017; Ren et al., 2017; and Zong et al., 2017) generated gain-of-function mutations in rice by using base editors designed as per the scheme of BE3. The BE3 is based on CRISPR-Cas9 nickase, rAPOBEC1 and UGI. The base editor rBE3 (rice base editor 3) has essentially the same architecture as the BE3 editor, *viz.*, (APOBEC1-XTEN-nCas9-UGI-NLS), but the sequences encoding rBE3 were codon-optimized for expression in rice. The efficiency of base editing by rBE3 was more than 17% and reached up to 38.9%. The bulk of editing involved C in the 5'TC3' doublets, while in some genes a low frequency (2.4%) of 5'AC3' doublets were also acted upon. The editing window was of 7 bp from base pair -19 to -13 counted upstream from the PAM sequence (Ren et al., 2017). The base editor devised by Li et al. (2017) had similar architecture as rBE3 (APOBEC1-XTEN-nCas9-UGI). The editing efficiency of this base editor was up to 40%, and the editing window was presumably of 5 bp. But the C residues were substituted by T, A or G residues.

Lu and Zhu (2017) modified BE3 by replacing UGI at the C-terminus of nCas9 with a NLS (APOBEC1-XTEN-Cas9(D10A)-NLS). The UGI was excluded as its stable integration in the plant genome might have some adverse effects on the base-edited plants. The base editor was introduced into rice cells by *Agrobacterium*-mediated genetic transformation. The editing efficiency was tested on two useful genes, *viz.*, *NRT1.1B* (encodes a nitrogen transporter) and *SLR1* (encodes a DELLA protein), of rice. This editor induced C-to-T (1.4% to 11.5%) as well as C-to-G (1.6%-3.9%) base substitutions, and the editing efficiency was dependent on the targeted locus. It also induced a high frequency (~10%) of indels, while BE3 typically induces <1% indels in mammalian systems. The pnCas9-PBE (NLS-APOBEC1-XTEN-nCas9-UGI-NLS codon optimized for cereals) designed by Zong et al. (2017) is similar to rBE3, except that it has an additional NLS upstream of APOBEC1. This editor had editing window of about 7 bp (positions 3 to 9 upstream of PAM when the NGG PAM sequence is counted as 21 to 23), and base editing efficiency of up to ~ 7%. Both editing window and editing efficiency varied to some extent with the plant species and the targeted locus. Single C-to-T transitions were the most frequent, but multiple (2-5) C-to-T transitions were also recovered. Significantly, off-target editing and indels were not detected.

The rat APOBEC1 has a striking preference for C residues having a T on their 5' side (doublets 5'TC3') and the editing efficiency of 5'GC3' doublets is quite low. The rice genome, however, has high GC content. Therefore, rBE system was modified by replacing rat APOEBC1 with hAID (human AID), that has a clear preference for C residues having G or A on their 5' sides (doublets, 5'GC3'and 5'AC3'). AID is one of the 11 members of the human APOBEC family that catalyse deamination of cytosine residues in gene segments encoding the variable regions of antibodies. An engineered variant of hAID that lacks the nuclear localization signal (NLS) sequence (hAID*Δ) has improved catalytic activity, and was codon optimized for expression in rice. The base editor, rBE5 (hAID*Δ-XTEN-nCas9-NLS) acts on C residues in the GC, AC, TC as well as CC contexts. Base editing of the rice gene *OsFLS2* mainly generated C-to-T transitions, but C-to-G and C-to-A substitutions were also observed. In addition, there was a high (71.9%) frequency of indel production, most likely due to NHEJ following uridine glycosylase action. Therefore, UGI was added to the 3' end of rBE5 to generate rBE9, which induces a greatly reduced frequency of indels (Ren et al., 2018).

The Target-AID base editor originally optimized in yeast and used in mammalian cultures was extended to rice and tomato. Shimatani et al. (2017) used Target-AID (dCas9-PmCDA1 and nCas9-PmCDA1) together with the appropriate sgRNAs for induction of targeted mutations at several genomic target sites in rice and tomato. Target-AID with nCas9 seemed to be more efficient than that based on dCas9. The base editor also induced C-to-G transversions and variable frequencies of indels, while off-target editing was negligible (0.14–0.38%). The editing window was 3 nt wide spanning the base positions −19 to −17 upstream of the PAM sequence.

Li, C. et al. (2018) optimized the adenine base editor ABE7.10 for use in plants and designated it as PABE (plant adenine base editor). The most successful editor was PABE-7, which had two copies of adenosine deaminase (ecTadA-ecTadA*) at the N-terminus and three copies of nuclear localization signal at the C-terminus of nCas9. The PABE-7 base editor induced A-to-G transitions at high frequencies (up to 7.5% in protoplasts and up to 59.1% in regenerated plants) at several loci of rice and wheat. At the same time, it induced negligible (< 0.02%) frequency of undesired mutations at target loci and very low (<0.1%) frequency of indels; the action window of the PABE system is 5 bp. Base editing of rice gene *OsACC-T1* encoding acetyl-coenzyme A carboxylase resulted in plants resistant to herbicides of the aryloxyphenoxy-propionate, cyclohexanedione and phenylpyrazoline chemical groups.

Another A-to-G base editor, rBE14, was designed by Yan et al. (2018) for use in rice. The editor rBE14 has one copy each of ecTadA and ecTadA*7.10 at the N-terminus of nCas9 linked by two copies of XTEN linker. rBE14 has editing window width of 4 bp. The editor strongly preferred A residues located at the -15 position for conversion to G. It induced only A-to-G transitions in rice, no indels were detected and there was no evidence of off-target activity. Although ecTadA deaminated A in the triplet UAC, the ecTadA:ecTadA* heterodimer base edits A residues in a sequence context-independent manner.

The enzyme acetolactate synthase catalyses the first step in the synthesis of branched chain amino acids. This enzyme is targeted by herbicides like imazamox. The conversion of C at position 287 to T (mutation C287T) of the gene encoding this enzyme leads to substitution of the amino acid Ala at position 96 with Val (A96V amino acid substitution), which makes the enzyme insensitive and rice plants resistant to the concerned herbicides. Shimatani et al. (2017) used Target-AID for induction of the mutation C287T. The hygromycin-resistant calli were cultured on medium containing the herbicide imazamox. More than 3% of the calli were tolerant to the herbicide, and 50% of these callus lines had the A96V mutation. Two plants carrying this mutation were finally regenerated. There was no evidence of off-target editing.

Hua et al. (2018) designed two A-to-G base editors, *viz.*, ABE-P1 and ABE-P2, for plants; these editors are based on the PABE-7.10 base editor (Table 6.5). ABE-P1 comprised one copy of ecTadA linked to TadA*7.10 by the 32-amino-acid long XTEN linker; the TadA-TadA* heterodimer was, in turn, linked to the N-terminus of nSpCas9 by the same XTEN linker. A strong NLS sequence was fused to the C-terminus of nCas9 (Fig. 6.11). The base editor ABE-P2, on the other hand, had nSaCas9 in the place of nSpCas9 in order to access AT-rich genomic sites. The editing window width for ABE-P1 was 6 nt, the editing efficiency ranged from 4.8% to 26% depending on the targeted genomic site, and no off-target editing was detected. In comparison, ABE-P2 had editing window width of 10 nt; the editing efficiency ranged from 17% to 61.3% and there was no evidence for off-target base editing.

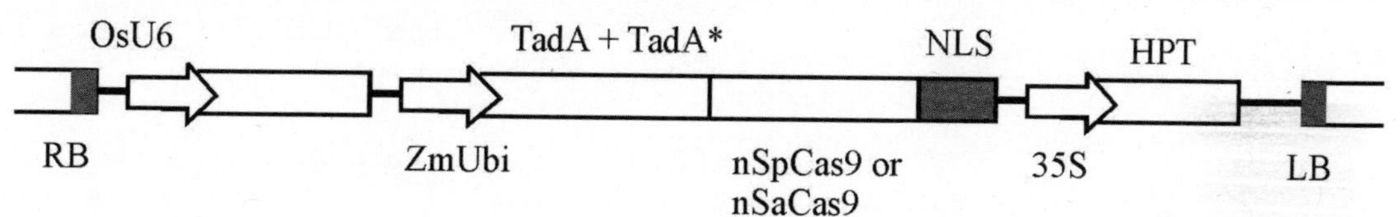

Fig. 6.11. Schematic representation of the A : T to G : C base editors ABE-P1 (has nSpCas9) and ABE-P2 (has nSaCas9). RB, right border of TDNA; LB, Left border of TDNA; OsU6, *Oyza sativa U6* promoter; ZmUbi, *Zea mays Ubiquitin* promoter, TadA, *E. coli* tRNA adenosine deaminase; TadA*, engineered TadA; NLS, nuclear localization signal sequence from *nos* gene of *Agrobacterium*; 35S, 35S promoter from CaMV; HPT, hygromycin phosphotransferase.

6.7. DELIVERY OF BASE EDITORS

The mode of delivery of base editors depends on the organism and cell type to be used for editing. The introduction of plasmid DNA containing gene constructs encoding the base editor and the guide RNA by a suitable genetic transformation or transfection method is applicable to most biological systems. But this strategy leads to integration of plasmid sequences as well as selectable markers into the cell genome. Further, base editor expression level will be subject to position effects generated by the site of integration into the genome. In addition, the base editor and guide RNA will continue to be produced in the cells, which would increase the level of off-target editing due to the prolonged exposure of genomic DNA sequences to the editor. Therefore, it would be desirable to implement an expression system that allows temporal regulation of base editor action to minimize off-target editing.

Table 6.5. Base editors for plants

Base editor	Constituents	Function	Remarks
rBE3	rAPOBEC1, nCas9, UGI, NLS	C-to-T transitions	*APOBEC1* and *UGI* sequences codon-optimized for expression in rice; editing window width 7 bp; edits Cs mainly in TC, followed by AC doublets; efficiency of base editing over 17%
-	rAPOBEC1, nCas9, UGI	C-to-T transitions	Editing window, presumably 5 bp; editing efficiency up to 40%; U replaced by T, A or G during subsequent DNA replication
-	rAPOBEC1, nCas9, NLS	C-to-T transitions	Induces C-to-T and C-to-G substitutions; editing efficiency up to nearly 15%; indel frequency about 10%
rBE5	hAID*Δ, nCas9, NLS	C-to-T transitions	Editor sequences codon-optimized for expression in rice; edits C residues in the GC, AC, TC and CC doubles; C-to-T substitutions; C-to-G and C-to-A conversions also occur; high indel frequency
pnCas9-PBE*	rAPOBEC1, nCas9, UGI	C-to-T transitions	Editing window 7 bp; editing efficiency nearly 7% (~44% in regenerated plants); only C-to-T transitions; predominantly single transitions; no indels and off-target editing
rBE14	TadA:TadA*, nCas9,	A-to-G transitions	Editing window width 4 bp; nucleotides at positions -16 to -14 (the first nucleotide upstream of PAM counted as -1) strongly preferred; only A-to-G transition; no indels induced
PABE-7	Two copies of adenosine deaminase, nCas9, NLS	A-to-G transitions	Used with esgRNA; editor codon optimized for expression in rice; base editing efficiency up to 59.1% in regenerated plants
ABE-P1	TadA:TadA*, nSpCas9, NLS	A-to-G transitions	Editing window width 6 nt; editing efficiency 4.8% to 26%; no off-target editing detected
ABE-P2	TadA:TadA*, nSaCas9, NLS	A-to-G transitions	Editing window of 10 nt; editing efficiency range, 17% to 61.3%; no off-target editing detected

* NLS, nuclear localization signal; esgRNA, engineered sgRNA; hAID, human activation-induced deaminase; hAID*Δ, hAID that lacks the nuclear localization signal sequence; pnCas9-PBE, plant codon-optimized nCas9.

Alternatively, the base editor may be delivered as RNP complex comprising sgRNA and base editor. Negatively charged proteins and RNPs can be delivered into cells with the help of cationic lipid reagents like the commercially available Lipofectamine 2000. Typically, the editor protein is complexed with the guide RNA prior to the lipid-mediated delivery. The RNPs can also be directly introduced into cultured animal cells, young animal embryos and plant protoplasts. This method allows a built-in control on the period of base editor action since the cellular machinery would rapidly degrade them; this is expected to enhance specificity and minimize off-target editing. The delivery of base editors as RNP into cultured human and mouse cells resulted in a dramatic reduction in off-target editing as compared to delivery of the same base editors as plasmid DNA (Rees et al., 2017).

6.8. BYSTANDER MUTATIONS

Base editors (BEs) work within an editing window, which comprises the bases that are liable to be edited by a BE bound to a specific target site (Fig. 6.1). A bystander mutation is change in a base located within the editing window but other than the one targeted for editing. It is noteworthy that bystander mutations occur at the targeted site, while off-target mutations are induced at genomic sites other than the one targeted for editing. The editing window for the common C-to-T BEs (BE3 and BE4) is typically five base-pairs. Thus, these BEs will efficiently edit all the Cs present in the nontargeted strand within this window. When two or more C residues are present in the editing window, the modification of C residues other than the one targeted for editing would generate bystander mutations (Fig. 6.1). It may be noted that some of these unwanted mutations may be potentially deleterious.

Efforts have been made to minimize the occurrence of bystander mutations. The first strategy was to reduce the width of editing window of BE3 by mutating rAPOBEC1. One of the mutants, YE1 (used in base editor, YE1-BE3) has editing window of 3 bp and editing efficiency comparable to that of BE3. However, YE1-BE3 does not discriminate between cognate and bystander C residues. At the same time, it reduces the editing window, which limits the range of mutations that can be targeted for base editing. Another approach was to replace the rAPOBEC1 with a different cytidine deaminase that exercises some discrimination in selection of C residues for editing. The human cytidine deaminase A3A was engineered to produce a mutant denoted by eA3A that discriminates between the C residues on the basis of their sequence context, the order of preference being TCR>TCY>VCN (R = A or G; V = G, A or C; Y = C or T). The base editor eA3A-BE3 exhibits base editing of C residues in the TCR motifs at efficiencies comparable to that of BE3, but efficiency of editing Cs in other sequence contexts is greatly reduced. The precision of editing by eA3A-BE3 is over 40-fold higher than that of BE3; in addition, it shows reduced off-target mutations (Gehrke et al., 2018).

6.9. OFF-TARGET EFFECTS

Base editors are known to carry out off-target editing, *i.e.*, editing of the concerned bases located at genomic sites other than but similar in sequence to the target site. It has been observed that Cas9 binds to a greater number of off-target sites than the number of sites at which it induces cleavage. It would, therefore, be expected that in a given genome Cas9-based editors would exhibit higher frequencies of off-target editing than that of Cas9 induced off-target cleavage. However, comparisons of frequencies off-target base editing with those of off-target Cas9 cleavage suggest that generally the two events follow the same specificity rules with some notable exceptions. For example, BE3 seems to be more tolerant to DNA-sgRNA mismatches than Cas9. However, at all such off-target sites that are devoid of a C residue within the editing window, Cas9 induces cleavage, while the BE3 is unlikely to produce base substitutions.

The frequency of off-target editing varies with the type of base editor used, the repetitive or non-repetitive nature of the targeted locus, the method of delivery of the base editor, and the duration of action by the editor. The various approaches available for minimizing off-target base editing include improved sgRNA design, use of Cas9 variants with enhanced specificity, use of a different Cas protein, refinements of the deaminase component of the base editor, optimization of delivery method, and development of dependable methods for detection of genome-wide off-target editing (Hess et al., 2017). There is considerable evidence that the levels of off-target editing are much higher when the base editor is delivered into cells as plasmid DNA than when the same editor is introduced as RNP. This seems to be due to the relatively short period available for action by the base editor before the RNP is rapidly degraded by the cellular machinery. In contrast, the base editor is continuously produced by the gene constructs in the plasmid and the base editing continues unimpeded over a prolonged period. The delivery of the base editor as mRNA (encoding the Cas9, deaminase, linker, etc. sequences) seems to be comparable to their delivery as RNP particles.

Several different methods have been developed for the detection and scoring of off-target DSBs, but they are not suitable for detection of off-target base editing. Therefore, it is important that suitable methods for unbiased and rapid detection of off-target base editing are developed. Kim, D. et al. (2017) modified the Digenome-seq method for *in vitro* detection of genome-wide DSBs for the detection of off-target base editing. They used the enzymes USER (uracil DNA glycosylase and DNA glycosylase-lyase endonuclease VIII) to remove the uridines from all such sites where the C-to-T base editor had converted the Cs to Us. At the same time, the nCas9 component of the base editor would nick the nonedited strand. This would generate DSBs at the edited sites and enrich the DNA sample for free DNA ends at all the edited sites. The sequence data from these samples are analysed to identify the base-edited off-target sites. It may be noted that DSB production at the edited sites depends on the activity of DNA glycosylase; therefore, *this approach can be applied to only such base editors that lack UGI in their architecture.* Modified BE3 (devoid of UGI) exhibited, on an average, lower off-target activity than Cas9, and the base edited off-target sites were often different from the Cas9 cleavage sites.

In addition, whole genome sequencing can be used to detect base conversions at genome-wide sites, and off-target sites showing higher frequencies of base changes than the background noise level may be regarded to represent off-target editing. This approach has been used in yeast where it revealed a slight genome-wide increase in the frequency of single nucleotide variants. But WGS did not detect any off-target editing in mouse embryos, suggesting that base editors may exhibit a very high specificity in some biological systems. It may be noted that WGS is applicable to only such biological systems that are genetically homogeneous and do not harbour within sample endogenous single nucleotide variations. But in case of some biological systems like plants, WGS seems to be the only feasible strategy for monitoring off-target editing. In any case, where both WGS and the modified Digenome-seq methods are applicable, the latter is more sensitive and can detect such sites that have base substitutions at frequencies as low as 0.1%.

There are many sgRNA design tools for Cas9 nuclease, but only one of them (https://www.benchling.com) permits the ranking of sgRNAs for base editors on the basis of predicted off-target editing. It is important that more efficient sgRNA design tools be developed that take into account off-target sites with small indels as well as those having non-canonical PAMs. In addition, the sgRNA should be designed based on the genome sequence of the populations to be subjected to base editing in place of the standard reference genome since it may differ from the target genome sequence for some genomic sites of relevance. Further, some alterations in sgRNA architecture might improve specificity of base editing. For example, addition of two extra G residues to the 5' end of sgRNAs (sgRNA, 5'GGN_{20} 3') is reported to enhance specificity without affecting efficiency of base editing (Hess et al., 2017). Finally, the Cas9 and the deaminase components of base editors can be engineered to reduce the width of editing window to reduces bystander mutations and/or improve editing specificity. There is some evidence that the use of base editors like dCpf1-BE that are based on other Cas proteins may show greater specificity than those based on Cas9 (Section 6.3).

It may be emphasized that bystander mutation and off-target editing are a serious concern for *in vivo* base editing particularly for therapeutic applications. But they are virtually of little concern in plant breeding since they can be easily eliminated by backcross breeding, which plant breeders use routinely in their breeding programmes.

6.10. ESTIMATION OF MUTATION FREQUENCY

The estimation of frequencies of mutations induced by base editing is challenging particularly in case of diversifying base editors that induce low frequencies of mutations over a relatively larger genomic region. There are mainly two approaches for quantifying the induced mutational changes, *viz*., assessments based on phenotype, and those utilizing analysis of base sequence data. Estimations from changed phenotypes generally use such loci that govern easily scorable traits, such as fluorescence (*e.g*., *GFP*), pigmentation, or resistance to specific antimetabolites like canavanine, 6TG, etc. (Table 6.6). These assays are rapid, simple and straightforward, but they usually tend to underestimate the mutation rates (Hess et al., 2017). For example, base editing using the TAM (Targeted AID-mediated mutagenesis) system rescued GFP fluorescence in 4-5% of the cells, while *GFP* locus sequence analysis revealed at least one mutation in 20% of the *GFP* loci (Ma, Y. et al., 2016).

Table 6.6. Some genes used for estimation of base editing frequency based on their changed phenotypic expression

Locus	Mutant phenotype	Type of mutation
General:		
GFP	Fluorescence	Gain-of-function
	Lack of fluorescence	Loss-of-function
Yeast (S. cerevisiae):		
CAN1	Resistance to canavanine	Loss-of-function
ADE1	Red pigmentation	Loss-of-function
LYP1	S-aminoethyl-L-cysteine tolerance	Loss-of-function
Mammalian cells:		
HPRT	Resistance to 6TG	Loss-of-function

The base sequence data of targeted loci are analysed to obtain estimates of base editing frequencies. The high throughput sequence data are typically filtered for quality of sequence reads, and the selected reads are then aligned against a reference genome. The sequence of the targeted locus of the base edited population is then compared with that from the unmutated or parental population; the altered bases in the edited population are scored to determine the frequencies of base substitutions at various positions in the locus. These estimates are regarded to be more sensitive and reliable than those from phenotype change data. But they are subject to certain errors integral to the sequencing procedure, *viz*., presence of SNP heterogeneity in the targeted genomic region of the cultured cell lines, errors introduced in the base sequence by PCR used for preparation of sequencing libraries, and erroneous scoring of the bases by the sequencing procedure used. It may be pointed out that the error rates for the various high throughput sequencing systems are generally <1%.

An estimate of the error rate inherent in the sequence data can be obtained by sequencing the concerned genomic region from unmutated cell population. The estimates of mutation rates obtained from the base edited populations may then be normalized against the estimated error rates. Often high coverage sequencing may be necessary to estimate frequencies of less frequent mutations. In any case, detection of co-occurring mutations is challenging, and it is important to analyse sufficiently large numbers of cells to obtain reliable estimates. In general, mutation frequency estimation is much easier for precision base editing than that for diversifying base editing (Hess et al., 2017).

6.11. REGULATION OF BASE EDITING

Since base editing is based on the CRISPR-Cas9 system, the various strategies devised for regulation of this SDN system can, in principle, be applied to base editing as well. The various regulation strategies (Section 5.9.2) manipulate either the Cas9 or the sgRNA component of the CRISPR-Cas9 SDN system. Tang et al. (2017) devised a sgRNA-based system that enables regulation of base editing by application of a small ligand. This system uses an aptazyme-embedded sgRNA, denoted as agRNA, which contains a blocking sequence and a self-cleaving aptazyme; this sgRNA remains nonfunctional so long as the blocking sequence is not removed. The aptazyme specifically interacts with a small ligand; this interaction activates the ribozyme activity of the aptazyme (Section 5.9.2.1). The activated aptazyme cleaves itself and, thereby, removes the blocking sequence; this allows the sgRNA to be functional. Appropriate agRNAs were used in conjunction with BE3 to edit two loci in cultured human embryonic kidney cells. The use of agRNA without the ligand resulted in base editing at a frequency of 1.0 to 2.1%. The editing frequency went up by nearly 3-fold when the ligand was applied. Thus, the use of agRNA system appears to enable a highly specific control of base editing.

6.12. GENE INACTIVATION

The CRISPR-Cas9 SDN has been used to generate genome-wide gene knockouts and gene function interrogation. This tool has unprecedented power compared to the other gene knockout technologies, but it suffers from certain important limitations. The gene knockouts are generated by induction of DSBs by Cas9 nuclease, followed by DSB repair either by NHEJ or HDR. NHEJ induces uncontrollable random indels that may or may not lead to frame-shift; they may also generate proteins with novel functions when the number of added or deleted bases is in multiples of three. Although HDR introduces controlled changes in gene sequences, it occurs at rather low frequencies. Further, DSBs may lead to excessive DNA damage, especially when multicopy genes/genomic regions are targeted, and cause variable cell toxicity. In contrast, base editors based on CRISPR-Cas9 system can be used for targeted gene inactivation by generating premature stop codons that would knockout genes by producing truncated nonfunctional proteins or leading to stop codon-mediated mRNA decay (Kuscu et al., 2017; Billon et al., 2017).

Kuscu et al. (2017) used the C-to-T base editor BE3 for efficient and specific silencing of targeted genes without DSB formation. This base editor would efficiently edit the codons for Arg (***C***GA), Gln (***C***AA, ***C***AG) and Trp (U***G***G) into STOP codons UGA, UAA, UAG and UAG, respectively, by editing a single C or G (written in bold italics in the targeted codons) in each codon. This can be achieved by designing a sgRNA that is complementary to the noncoding strand and that places one of the above codons in the editing window for BE3. The sgRNA will have the general design 5'-$N_{(3 \text{ to } 7)}$-(CGA/CAG/CAA)-$N_{(14 \text{ to } 10)}$-NGG (PAM sequence) for the codons CGA,CAG and CAA. The N residues in the sgRNA will be the same as those surrounding the relevant codon in the coding strand of the gene since the targeted C residue will be in the coding strand. However, for the UGG (Trp) codon, the sgRNA design will be 5'-$N_{(2 \text{ to } 7)}$ -(CCA)-$N_{(15 \text{ to } 10)}$-NGG (PAM) and the N residues will be the same as those surrounding the sequence representing the relevant codon in the noncoding strand since the C residue to be edited is located in the noncoding strand.

The above sgRNAs will direct BE3 to the targeted C residue; the editor will deaminate this C to produce a mutant gene with a premature stop codon at the targeted site. Thus, the C-to-T base editor can inactivate genes by creating premature or early STOP codons; therefore, this method was termed as CRISPR-STOP. This procedure inactivates genes at frequencies comparable to that by CRISPR-Cas9 SDN at single and low copy number genomic regions, while it is more efficient at genomic regions with high copy number. More importantly, the CRISPR-STOP technology causes much less DNA damage and cell toxicity than the CRISPR-Cas9 SDN procedure. Computational prediction showed that ~17,000 human genes can be inactivated by introduction of early stop codons. Genome-wide screening of gene functions by introducing loss-of-function mutations with the help of a suitably designed sgRNA library to introduce early stop codons in human genes of interest has been successfully implemented (Kuscu et al., 2017).

Subsequently, Billon et al. (2017) devised the induction of STOP codons (iSTOP) procedure for gene inactivation by creating early stop codons. They have made available a database of over 3.4 million sgRNAs (denoted as sgSTOPs) that target between 97% and 99% of the genes present in genomes of eight eukaryotic species; the database also includes sgSTOPs for induction of 32,000 nonsense mutations associated with cancer. They also developed an assay based on restriction fragment length polymorphism (RFLP) to rapidly detect the products of iSTOP-mediated editing. The iSTOP strategy is a precise and potent technique for gene inactivation and is supported by comprehensive resource database for its application in several eukaryotic species.

The CRISPR-STOP and iSTOP techniques for gene inactivation are essentially the same, and they are efficient and precise. In contrast, genome editing relies on DSB induction and error-prone NHEJ-mediated repair of DSBs. Therefore, gene editing might cause excessive DNA damage and generate deleterious effects chiefly in high copy number genomic regions. In comparison, the CRISPR-STOP and iSTOP methods cause much less DNA damage in such genomic regions and lead to much lower cytotoxicity than gene editing (Kuscu et al., 2017). The chief limitation of these technologies is the requirement for a CAA, CAG, CGA or TGG codon in the 5' region of the mRNA and the presence of PAM sequence NGG within 13 to 17 bp from the base to be edited. An estimated 40% of all available CAA, CAG, CGA, and TGG codons cannot be subjected to CRISPR-STOP/iSTOP due to the absence of PAM motif within the specified distance from the targeted bases. Another limitation is the inefficient editing by BE3 of the Cs located in 5'GC3' doublets. Base editors with different PAM sequence requirements have already been developed (Section 6.3.1), and some base editors can efficiently edit the C residues in 5'GC3' doublets.

6.13. INDUCTION OF TARGETED MUTATIONS IN POST-MITOTIC OR QUIESCENT (G_0) CELLS

In many situations, *e.g.*, in genetic correction of mutations involved in human diseases, it is critical to replace the mutated base(s) with those found in the wild type alleles at the relevant sites of the concerned genes. One approach is to use a donor DNA with the wild type sequence to replace the mutated bases by HDR following DSB or nick induced by a SDN like CRISPR-Cas9 nuclease or nickase. But the frequency of HDR is very low as compared to NHEJ and HDR is limited to mainly *S* and *G2* phases of cell cycle since post-mitotic cells (G_0 cells) do not express adequate amounts of one or more proteins involved in homologous recombination (HR) between sister chromatids, and the HR is required for HDR. The base editing procedure is considered to be independent of HR machinery. Therefore, nondividing cells can be efficiently base edited. This expectation was validated by targeted base in terminally differentiated cells of mouse inner ear. The base editor BE3 injected as RNP:lipid complex into the cochlea of mice successfully edited the gene encoding β-catenin (*CTNNB1* in humans) in these differentiated cells, the editing efficiency being around 13% (Yeh et al., 2018).

6.14. APPLICATIONS

The technique of base editing was developed in 2016 (Komor et al., 2016), and within 3 years more than 300 research papers have been published in this fast evolving field. Base editing has been used in a variety of biological systems ranging from bacteria to human cells to achieve several different objectives, which range from precision base editing for therapeutic applications to localized base sequence diversification. In their pioneering work, Komor et al. (2016) demonstrated precise correction (in mouse astrocytes) of the mutations in *apoE4* gene that are associated with Alzheimer's disease, and of mutations in the *TP53* gene in a mouse breast cancer cell line. Since then base editing has been used in many mammalian cell lines and also used for *in vivo* editing in many organs, including mouse embryos leading to the production of base-edited mice.

According to Komor et al. (2017a), around 3,000 of the genetic diseases listed in ClinVar are due to T to C transitions, and these mutations can be corrected by a precision C-to-T base editing at the mutated sites. They predicted that only 300 to 900 of these mutations would be accessible to Cas9-based C-to-T base editors due to the requirement for a specific PAM sequence, and the presence of more than one C in the editing window, while only the single mutant T must be edited for correction of mutations involved in different diseases. Therefore, several base editors with different PAM sequence requirements and two editors without the need for a PAM sequence (editors based on ZFs and TALEs) have been developed (Table 6.3). The problem posed by multiple Cs in the editing window has prompted the development of base editors with narrow width (1-2 nt) editing windows (Table 6.4).

Both C-to-T and A-to-G base editors have been developed for use in plants; these editors are mainly based on the BE3 design. Base editing has been shown to induce the desired mutation in several targeted genes. For example, multiplex base editing was used to induce targeted mutations in two rice genes (*ALS* and *FTIP1e*) to obtain double mutants that are resistant to two herbicides. The genes encoding base editing reagents have been generally delivered by *Agrobacterium*-mediated transformation, but particle gun has also been used for this purpose. Precision base editing can be used to produce gene knockouts by CRISPR-STOP and iSTOP systems for screening gene functions. These procedures use BE3 to produce early stop codons in the targeted genes (Section 6.12; Hess et al., 2017).

The sequence diversifying base-editing has been used for directed protein evolution and for saturation mutagenesis. In directed protein evolution, sequence diversification is induced in those regions of endogenous genes that encode protein domains known/expected to be involved in the functions of interest. The treated cell populations are then subjected to suitable selection procedures and the cells possessing the desired protein modifications are isolated and analysed. For example, human cells expressing the wild type GFP were subjected to diversifying base editing using CRISPR-X system in association with a pool of sgRNAs targeting those regions of GFP gene that encode domains involved in spectrum shifting. The cells were then subjected to FACS (fluorescence-activated cell sorting) for isolation of cells with enhanced fluorescence by GFP. This permitted the isolation of enhanced GFP (eGFP; up to 30x higher fluorescence) variants. Genome sequencing of the *GFP* locus in cells with the eGFP phenotype revealed that the mutant *GFP* loci contained mutations that are known to give rise to the eGFP phenotype (Hess et al., 2016).

Diversifying base editing can be used to map the protein domains involved in protein-drug as well as protein-protein interactions. This approach enables identification and characterization of mutations in the concerned proteins that generate resistance to specific drugs. For example, Imatinib is used for chemotherapy in several types of cancer. This drug binds to and inhibits the ABL protein kinase. Mutations in the ABL protein kinase are known to be involved in 50% of the cases of resistance to Imatinib. TAM system with a pool of suitably designed sgRNAs was used for diversifying base editing of the *ABL* gene and human cells resistant to Imatinib were selected. The *ABL* locus from the selected cells was subjected to deep-sequencing. It was observed that ~30% of the read counts had the C944T base substitution (this corresponds to a T315I amino acid substitution in the ABL kinase). This is known to be the most common mutation in the patients resistant to Imatinib. In addition, several novel base substitutions conferring resistance to Imanitib were also identified; these mutations may represent the unidentified ABL kinase mutations responsible for Imatinib resistance in patients (Ma, Y. et al., 2016; Yang et al., 2017).

Similarly, diversifying editing of *PSMB5* gene was carried out for identifying mutations that confer resistance to bortezomib. This drug is a proteasome inhibitor and is used for cancer treatment. Diversifying editing followed by selection of cells resistant to bortezomib and deep sequencing of their *PSMB5* locus identified 11 mutations. Two of these mutations were already known, and five novel mutations were experimentally confirmed to generate resistance to the drug. This type of information would be useful for gaining an insight into disease development, and help in disease diagnosis, treatment of diseases and designing of drugs. This approach has some useful features as compared to the alternative approaches, including CRISPR-Cas9-mediated gene editing (Hess et al., 2016; Yang et al., 2017).

6.15. ACHIEVEMENTS

Although base editing is merely three years old, it has been used for targeted gene mutations in many organisms extending from microorganisms to plants and mammals. Precision base editing has been used in mammalian cell cultures to precisely correct mutations due to single base substitutions suggesting its potential for *in vivo* correction of disease causing mutations. In addition, diversifying base editing has been used to generate sequence diversity for protein evolution as well as for identification of base substitutions responsible for resistance of cancer cells to chemotherapy agents (Section 6.3.4). The information generated by base editing on protein-protein and protein-drug interactions, genetic changes responsible for drug resistance, etc. might greatly facilitate drug designing, and disease diagnosis and treatment. Remarkably, C-to-T transitions were induced in nearly 100% of *Dmd* and *Tyr* genes when BE3 mRNA was micro-injected into mouse embryos (Komor et al., 2017a,c), and base-edited mice have been produced.

The first successful application of base editing in plants was in rice, where BE3 was used to induce C-to-T transitions in genes *OsNRT1.1B*, *OsSLR1*, *OsPDS*, and *OsSBEIIb* with editing efficiency of 12 to 20%. In addition, multiple genes were edited (C-to-T) using BE3 in maize and wheat (efficiency up to 44%). Similarly, C-to-T transitions leading to herbicide resistance were induced in rice by Target-AID (efficiency up to 32%; Li et al., 2017; Lu and Zhu, 2017). Crop plants have been subjected to base editing for improvement of economically important traits and the products are expected to reach field trial soon.

Initially, the C-to-T base editors were created and improved in various ways to enhance editing efficiency and specificity, and to relax the restriction imposed by PAM sequence requirement. Subsequently, A-to-G base editors were developed for animals and then extended plants. Together, the two types of base editors enable all the four types of transitions, *viz*., C-to-T, T-to-C, A-to-G and G-to-A. In addition, RNA base editors for conversion of A-to-I (read as G by cellular machinery) have also been developed (Sections 6.3-6.5).

6.16. ADVANTAGES

Base editing does not involve DSB induction and HDR to generate the desired alterations in base sequences at the targeted genomic sites, and there is no need for a donor DNA. The DSB induction during gene editing may result in extensive DNA damage that may lead to cell death; this is avoided in the case of base editing. Base editing can be implemented in nondividing cells as well, while HDR is confined to dividing cells. The efficiency of base editing is high reaching up to 50% or even more in at least some cases, *e.g.*, 100% in mouse embryos. At the same time, the frequencies of random indels at the targeted site and translocations are greatly reduced as compared to those generated by the other gene editing methods (Yang et al., 2017). However, there are some reports of induction of quite high frequency (~10%) of indels in certain plant systems. It has been argued that in plant breeding situations, indels and other unintended mutations are not of serious consequence since they can be easily eliminated by a suitable breeding strategy like backcross method (Lu and Zhu, 2017).

6.17. LIMITATIONS

One major limitation of base editors built on dCas9 or nCas9 is the requirement of the PAM sequence NGG located at a definite distance from and on the downstream side of the targeted base. This requirement limits the range of genomic sites of interest that can be accessed by base editors. This limitation can be relieved by using a Cas9 mutant that recognizes a different PAM sequence. For example, a Cas9 mutant was used to construct an eA3A-BE3 base editor that was able to edit sites having the NGA PAM sequence. The use of Cas9 orthologs like SaCas9 or a different Cas protein, such as Cas12a, would enable the editor to use a different PAM sequence (Table 6.3). Another limitation of most base editors is that they produce bystander mutations. Mutant base editors with increased base editing specificity have been produced but they tend to restrict the target range either by narrowing the editing window or by making the editing sequence context-dependent (Section 6.3.2).

The available base editors produce off-target mutations at variable frequencies. In addition, sometimes they do produce indels in addition to base transitions. The efficiency of cognate base editing, frequency of bystander mutations and possibly off-target effects tend to be locus-specific (Komor et al., 2017c). It seems likely that a huge library of base editors may be created, each member of which would be highly specific and efficient for editing a locus/group of loci. In any case, a truly 'single-base editor' remains an ideal yet to be realized.

6.18. COMPARISON BETWEEN BASE EDITING AND GENOME EDITING

Base editing differs from genome editing in a number of important features, some of which are summarised below (Table 6.7). Genome editing uses either nucleases or nickases that introduce DSBs or nicks in the target DNA. Repair of DSBs or nicks is the basis for induction of genetic alterations at the targeted sites. Base editing, on the other hand, utilizes either dCas9 or nCas9, which nicks the nonedited strand and improves base editing efficiency by promoting DNA repair using the edited strand as template so that the base change induced in the edited strand is imported into the nonedited strand as well. Most of the genetic changes induced by genome editing are indels that generally lead to gene inactivation. A directed genetic change requires an appropriate donor DNA that is used as template for HDR leading to replacement of the targeted sequence by the desired sequence from the donor DNA. Nick induction for genome editing aims to reduce NHEJ and promote HDR, but a low proportion of indels is also induced. Significantly, HDR is confined to mitotically active cells. In comparison, base editing involves targeted modification of the targeted bases of the specified gene directly to the desired base by either cytidine deaminase or adenosine deaminase. This modification does not need donor DNA, and the role of HDR-mediated mismatch repair is only to enhance base editing efficiency. However, there is generally a very low frequency (up to 1%) of indel formation, but in rice, ~10% indels were reported in one study.

Base editing has been extended to editing of RNAs using dead mutant of Cas13b, a natural RNA-guided RNase. Genome editing, on the other hand, can be applied to DNA alone. Both gene editing as well as base editing induce off-target changes, but their frequencies are relatively much lower in base editing. Finally, genome editing can be used for transgene integration at specified genomic sites, but this objective cannot be achieved by base editing.

Table 6.7. A comparison between salient features of base editing and gene editing.

Feature	Base editing	Gene editing
DNA binding domain	Cas9/Cas12a/Cas13b with guide RNA	Zinc fingers, TALEs*, Cas9 with guide RNA
Cas9/Cas12a/Cas13 variant used	dCas9, nCas9*, dCas12a or dCas13b*	Cas9 or nCas9*
DSB induced*	No	Yes
Mutation induced by	Specific base deamination (C-to-A or A-to-I)	Repair of the DSBs/nicks by cellular DNA repair processes
Type of mutations induced	Mainly transitions, very low frequency of indels	Mainly indels; sequence replacement by HDR based on donor DNA; also very low frequencies of translocations and inversions
Mutation correction	Single nucleotide mutations corrected; this was the chief motivation for the development of this technique	Appropriate donor DNA must be provided; few to several bases can be replaced by HDR* using donor DNA as template
Effects of the induced mutations	Amino acid substitutions; premature termination codons (yield gene knockouts) generated by single base changes (precision base editing) and localized diversification in base sequences (diversifying base editing)	Indels lead to frame-shit mutations (gene knockouts), and disruption of binding sites for transcription factors and/or splicing elements: HDR produces in-frame protein variants, precisely corrects deleterious mutations and enables tagging of endogenous genes
Induction of specific mutations	Single nucleotide mutations	Mutations included in the donor DNA sequence
Off-target action	Comparatively much lower than that for gene editing	Comparatively much higher than that for base editing
Bystander mutations	Occur in variable frequencies	Not relevant in gene editing
Undesired products like indels, translocations, etc.	Very low frequency; generally, <5% for BE3; for some improved editors, 1% or less; in rice it may be nearly 10%	Much higher than HDR-mediated sequence replacement
Transgene integration at specified genomic locations	Not possible	Yes; the donor DNA includes the transgene construct

*dCas12a, 'dead' Cas12a; dCas13b, 'dead' Cas13b; DSB, double-strand break; HDR, homology-directed repair; nCas9, Cas9 nickase; dCas9. 'dead' Cas9; BE3, base editor 3; TALEs, transcription activator like effectors.

6.19. COMPARISON BETWEEN PRECISE SEQUENCE ALTERATIONS BY HDR AND BASE EDITING

Predetermined changes in base sequences can be induced by HDR as well as base editing. The two procedures, however differ for several features, and some of the important differences are listed in Table 6.8. HDR is based on a donor DNA that must be provided with the SDN, but no donor DNA is needed for base editing. The HDR can be used to introduce sequence changes for few to several bases, while base editing induces transition of a single base at the targeted site. In general, the frequency of HDR is low and it is confined to dividing cells. In contrast, base editing efficiency is quite high, and it can be implemented in nondividing cells as well. HDR is accompanied by considerably high frequencies of indels and there is invariably some off-target activity. Base editing also results in indels, bystander mutations, unintended base substitutions, and off-target editing, but the overall product purity is much higher for base editing than that for HDR. HDR is associated with DSB formation, while base editing avoids DSB induction. DSB induction might lead to excessive DNA damage particularly in high copy number genomic regions and cause cell death (Yeh et al., 2018).

Table 6.8. A comparison of sequence alterations by base editing and HDR-mediated precise sequence replacement

Feature	Base editing	HDR-mediated precise sequence replacement
Target molecule	DNA or RNA	DNA
Reagents	Base editors based on a Cas protein and sgRNA or on ZFs/TALEs	Site-directed nuclease/nickases based on a Cas protein with sgRNA or ZFNs/TALENs and donor DNA*
Cas protein variants used	dCas9, nCas9*, dCas12a or dCas13b*	Cas9, nCas9* or Cas12a
DSB induced	No	Yes
Mutation induced by	Specific base deamination (C-to-A or A-to-I)	Use of donor DNA for homology-directed repair of DSB or nick
Type of mutations induced	Mainly single base transitions, very low frequency of indels	Mainly indels; sequence replacement by HDR based on donor DNA; also very low frequencies of translocations and inversions
Mutation correction	Single nucleotide mutations corrected	Few to several bases can be replaced with that of donor DNA sequence
Effects of the induced mutations	Amino acid substitutions; premature termination codons and localized diversification in base sequences	Indels result in gene knockouts; HDR produces in-frame protein variants, precisely corrects deleterious mutations
Induction of specific mutations	Single nucleotide mutations	Mutations included in the donor DNA sequence
Local sequence diversification	Yes; by specifically designed base editors (TAM, CRISPR-X)	No
Off-target action	Comparatively much lower than that for HDR	Comparatively much higher than that for base editing
Bystander mutations	Occur in variable frequencies	Not relevant in HDR
Undesired products like indels, translocations, etc.	Very low frequency; generally, <5% for BE3; for some improved editors, 1% or less; in rice it may be nearly 10%	Much higher than base editing
Transgene integration at specified genomic locations	Not possible	Yes; the donor DNA includes the transgene construct
Efficiency	Relatively high	Relatively low
Product purity	Much higher than HDR	Much lower than base editing
Applicable to post-mitotic cells	Yes	No
Cell death	Unlikely; some cell death occurs when Gam protein is used in the base editor	DSBs often lead to cell death

* Extended guide RNA in case of prime editing; the RNA is converted to DNA sequence by reverse transcriptase linked to Cas9 protein.

Several approaches have been used to improve the efficiency of HDR. These strategies include suppression of specific proteins (*e.g.*, DNA ligase IV, DNA-dependent protein kinases, etc.) involved in NHEJ pathway that competes with HDR, promotion of HDR by treating cells with certain molecules like RS-1 (it activates Rad51, which is involved in homologous recombination), use of donor DNAs with asymmetric design, use of donor DNAs with silent (same sense) mutations to eliminate the risk of re-cleavage by CRISPR-Cas9, use of sgRNAs having optimal design, and delivery of Cas9 along with sgRNA as ribonucleoprotein and not as plasmid DNA. However, suppression of NHEJ pathway components is likely to weaken the DNA repair in edited cells. In any case, the ratio of targeted base substitution frequency to the total frequency of other changes remains relatively low despite the various strategies used to improve the efficacy. Overall, base editing seems to be more promising for therapeutic applications than HDR-mediated precise base substitutions (Yang et al., 2017).

6.20. FUTURE DIRECTIONS

The developments in the field of base editing have been phenomenal within the short span of time since its first description by Komor et al. (2016). The available base editors induce variable frequencies of bystander mutations, other base substitutions, indels and off-target effects. Further, the accessibility of many mutations for correction by base editing is limited by specific PAM sequence requirements for the major base editors. Therefore, several improvements must be made to make base editing safes and to expand the range of its applications. Considerable progress has been made in increasing the specificity of base editing (Table 6.2), reducing the width of editing window (to minimize bystander mutations; Table 6.4) and to relax the PAM sequence restrictions (Table 6.3). Base editors using ZFs and TALEs to target the desired genomic sites have been developed; these editors do not have any PAM sequence requirements, but their efficiency and specificity need to be further improved.

Efforts are being made to engineer deaminases with different sequence specificities to enable editing of the cognate bases present in different sequence contexts. Some variants of APOBEC1 having reduced window width and enhanced single C-to-T conversions have been produced, but the ideal of single targeted C-to-T or A-to-G transitions essential for therapeutic applications is yet to be achieved. It would also be highly desirable to develop base editors that would induce controlled targeted substitutions of a given base, say C, to a base other than T, *e.g.*, A or G. Such base editors will enable correction of transversions involved in genetic diseases. Further, the method of delivery of base editors for therapeutic applications needs to be optimized. Base editors can be delivered as RNP (lipofection) or mRNA (microinjection); these two methods are comparable in effectiveness. They can also be delivered as gene constructs in suitable viral vectors, the adeno-associated virus (AAV) vectors being the preferred vehicles. However, base editors are large fusion proteins and their gene constructs are difficult to package in AAV vectors; this issue needs to be satisfactorily resolved before this route of delivery can be used for therapeutic base editing (Hess et al., 2017).

The applications of base editing can be expanded to several important novel areas for seeking answers to difficult questions. For example, most of the human genetic diseases are associated with single base substitutions in specific genes. But in many cases, it is not clearly established whether a given point mutation is the cause of a specific diseased phenotype. A base editor could be used to correct such mutations in cultured cells or animal models. In case, the correction of a mutation restores the wild type phenotype, the causal relationship between the mutation and the disease phenotype will be clearly established. Diversifying base editing can be used to generate sequence variations in the regions of interest of selected genes. This is expected to permit the identification of sequence changes that affect such features as levels of gene expression, alternative splicing, etc. Base editing can also be applied to investigate sequence-function relationships of the noncoding genomic regions, including promoters and enhancers. It seems that the potentials of this powerful technology are just beginning to unfold, and many novel applications would emerge with the passage of time (Hess et al., 2017).

6.21. CONCLUSIONS

Base editing has emerged as a powerful tool for targeted correction of single nucleotide mutations in genes. The development of this method was motivated by the desire to correct single base mutations associated with human genetic diseases. Base editing uses recombinant proteins comprising nCas9 (for DNA editing) or dCas13b (for RNA editing) fused with deamination domains of cytidine or adenosine deaminase enzymes and other proteins where required. Initially, base editors for C-to-T transitions in DNA were developed, followed by those for A-to-G transition in DNA and for conversion of A-to-I in RNA. Thus, it is now feasible to induce all the four possible types of transitions, *viz*., A-to-G, G-to-A, C-to-T and T-to-C as well as to convert A-to-I (=G) in RNA. Base editing has been applied in a variety of organisms ranging from microorganisms to plants and animals (Cox et al., 2017; Komor et al., 2017a).

The base editing methods provide a relatively safer and highly precise tool for irreversible conversion of single base pairs without the need for induction of DSBs. The gene editing methods based on nucleases, in contrast, induce DSBs that are lethal if unrepaired, and may lead to deleterious effects on the cells even after repair. The genetic effects generated by gene editing methods arise from the repair of DSBs either by NHEJ (the predominant mode of DSB repair) or HDR (when appropriate donor DNA is provided). The NHEJ primarily leads to random insertion or deletion of one to few base pairs and generally leads to inactivation of the targeted genes. Sequence replacement for correction of mutations requires HDR that is confined to dividing cells and occurs in much lower frequencies than NHEJ. The base editing technology avoids these limitations of gene editing, but it is applicable to only mutations arising due to single base changes (Komor et al., 2017a).

Base editing has now been adapted to induce stop codons in genes and thereby inactivate them. The CRISPR-STOP and iSTOP methods of base editing were developed for specific and precise gene inactivation by creating early stop codons in the targeted genes. However, there are certain limitations of base editing, some of which are consequence of the CRISPR-Cas9 system used in base editors. For example, the need for specific PAM sequence limits the sites that are amenable to editing. Some other limitations result from the properties of base editors, *e.g.*, the width of editing window, the production of bystander mutations and off-target base editing. Efforts are being made with some notable successes to minimize these undesirable attributes of base editing.

■■■■

7

GENE EDITING FOR PLANT BREEDING

7.1. INTRODUCTION

It has been estimated that by 2050, the global yields of rice, maize and wheat should increase by 70% for feeding the human population that is expected to go past nine billion by that time. The need for increased food production becomes more imperative since an increasing proportion of food grains is being diverted for meat and biofuel production. In the past, the increased crop production resulted from a combination of several factors, such as improved crop varieties, enhanced use of agricultural inputs like fertilizers, irrigation water, etc., and increased total area of crop cultivation. But the freshwater reserves are declining, petroleum resources are progressively depleting, and the available cultivable area is decreasing due to several factors, including increasing salinization caused by chemical fertilizers and irrigation water. At the same time, global climate change is predicted to lead to a steady rise in temperature and unpredictable precipitation that would generate moisture stress and alter the dynamics of insect pests and plant diseases resulting in reduced crop yields (Reynolds, 2010; Bailey-Serres et al., 2019). Thus, the only available option for increasing global food grain productions seems to be the development of superior crop varieties that yield more of better-quality produce under less favorable conditions. In fact, the development of crop varieties with high yield potential, and enhanced and stable performance under various abiotic and biotic stresses is considered as one of the biggest challenges before the plant scientists of the present century.

7.2. MUTATION BREEDING

Plant breeding, the technology that generates new crop cultivars more suited to the human needs, thrives on genetic resources, which provide the required genetic variation for the desired improvements in crop varieties. Generally, elite lines from the primary gene pool of a crop are hybridized to produce simple or complex crosses, but often unadapted germplasm is also involved in hybridization programmes. The segregating generations from these crosses are handled according to one of the conventional breeding schemes. Where required, genes/alleles from close relatives of crop species are transferred into crop species. The genetic variation available in the germplasms of crop species has arisen due to spontaneous mutations and natural selection over a very long period of time. But following the discovery of physical and chemical mutagens during 1920s and 1940s, respectively, mutation breeding has been used to create the desired genetic variation mainly for qualitative traits. This variation had been exploited to develop 3,281 mutant varieties (these include somaclonal variants as well) that have been released for commercial cultivation (listed in IAEA Mutant Variety Database, https://mvd.iaea.org/, as on October 30,

2018). It is important to note that the activities in mutation breeding have declined drastically, and merely 94 mutant varieties were registered in the IAEA database after 2010.

Mutation breeding is the easiest and the most successful when the traits are governed by one or few genes and are highly heritable. Mutation breeding for quantitative traits, as a result, is relatively more complicated and has been less successful although over one thousand mutant varieties with improved yield and yield traits have been developed. The chief limitations of mutation breeding are the random nature of mutations and the very low frequency (around 0.1%) of useful mutations. Generally, mutation breeding is based on phenotypic screening of huge populations, which is often tedious and time consuming. Novel screening methods like TILLING (Targeting Induced Local Lesion in Genomes) use molecular strategies to detect mutations at the genotype level. But these strategies require profound knowledge of genes governing the traits of interest and involve considerable amount of relatively technically sophisticated and resource demanding activities. Generally, induced mutations are associated with many unintended mutations. For example, a plant subjected to EMS treatment may carry up to 100,000 different mutations, *i.e.,* changes in base sequence. Often the elimination of these unwanted mutations may require backcross breeding that will consume time and resources. In addition, the targeted genes may be present in multiple copies in the genomes of, especially, the polyploid species and these copies would have redundant functions. In such cases, all the copies of the targeted gene must be mutated to generate modifications in the relevant phenotypes, which is highly unlikely with any of the mutagen treatments.

7.3. MODERN BREEDING TOOLS

In view of the global climate change, it is highly desirable that modifications in various characteristics of crop varieties are carried out in a controlled manner and within increasingly shorter time frames than is common with the conventional plant breeding approaches. Marker-assisted plant breeding enables effective and reliable selection for genes and quantitative trait loci (QTLs), and markedly reduces the time required for variety development, but it is not able to generate novel genetic variation (Singh and Singh, 2015). The transgenic technology generates novel genetic modifications; crop varieties developed by this technology are in commercial cultivation since 1996 and the total area under these varieties was 191.7 million hectares in 2018. Despite the proven usefulness of these varieties, the consumer acceptance remains at best lukewarm presumably due to many safety concerns. But the opposition to GM crops seems to be mainly due to the proprietary ownership of the transgenic varieties by the companies that develop them, and the clearly unfair deal given to farmers by these companies (Chandrasegaran and Carroll, 2016).

Therefore, attempts have been made to develop technologies that generate novel genetic variation by directed changes in the endogenous genes of crop plants. These novel technologies are based on site-directed nucleases (SDNs) and chimeric oligonucleotides. SDNs induce either a double-strand break (DSB) or a nick at the targeted genomic site that is either repaired by nonhomologous end joining (NHEJ) leading to gene inactivation or by homology-directed repair (HDR) enabling directed sequence replacement at the targeted site. Chimeric oligonucleotides, on the other hand, enable replacement of one to few bases at the targeted genomic sites. These activities are called gene editing or genome editing. Most of the reported gene editing in plants concerns gene inactivation due to NHEJ. SDNs have also been adapted to achieve permanent modification of bases at specific sites of the targeted genes, *i.e.*, base editing. These technologies are active areas of research and patent acquisition, especially in USA and EU. It is important that in these technologies the transgenic state is either transient and does not persist in the end-product or it can be completely avoided. It was hoped that the varieties developed by these technologies would be more readily acceptable to the consumers than the transgenic varieties (Lusser et al., 2012). But the regulation of these varieties varies among different countries even within European Union.

There are several excellent reviews on the various aspects of these technologies, including those by Puchta (2003, 2017), Durai et al. (2005), Paul and Ma (2011), Curtin et al. (2012), Lusser et al. (2012), Tzfira et al. (2012), Gaj et al (2013, 2019), Podevin et al. (2013), Osakabe and Osakabe (2014), Sainsbury and Lomonossoff (2014), Kamburova et al. (2017), Songstad et al. (2017), Jung et al. (2018), Gao (2019) and Montecillo et al. (2020).

7.4. REQUIREMENTS FOR USEFUL GENE EDITING IN PLANTS

Editing of plant genes for generating useful mutants for traits of significance requires a wealth of knowledge, including genetic and genomic information, and considerable technical expertise. First, inheritance of the trait of interest should be clearly established and the gene(s) involved in development of the trait should be known. In case the trait of interest is governed by a QTL, it is more likely that the QTL would contain more than one gene, which are likely to govern different traits. This may sometimes be the case for certain oligogenes as well, *e.g.*, the *RIN* locus of tomato (Section 7.9.6). In such cases, it will be necessary to identify with reasonable confidence the specific gene that participates in control of the desired trait. Further, a knowledge of the biosynthetic pathways and regulatory circuits involved in the development of the trait would be helpful. All this information would enable the selection of gene(s) for targeted editing to achieve the desired modification in phenotype of the trait of interest. Second, the availability of a high-quality reference genome sequence is necessary for designing effective SDNs having the minimum off-target activity. The ZFNs and TALENs require a high level of technical expertise in protein engineering and complicated selection and validation processes for designing SDNs specific for the targeted sites of the desired genes. In comparison, the CRISPR-Cas9 system involves designing of only the spacer region of sgRNAs specific for the selected genomic targets since the Cas9 component remains the same for all the targets. The designing of the sgRNAs is relatively simple and straightforward, and several robust computer programmes are available for the purpose (Table 5.1).

The easiest and the most straightforward gene editing aim is gene inactivation by NHEJ-mediated indel (insertion/deletion) induction at the DSB site. This would be useful when inactivation of the targeted gene is expected to generate the desired phenotype without a negative effect on plant performance. For example, the MLO protein is used by the powdery mildew pathogen for infection of its host plants. Therefore, inactivation of the *Mlo* gene will prevent the pathogen from infecting the host plants. However, it is critical to ensure that the function of the targeted gene is either not essential for the plant or is provided by another redundant gene (Section 7.9.8). When the gene to be targeted is a member of a multigene family, it may be necessary to edit all the genes of the family. In such a case, a conserved region of the gene should be targeted so that a single SDN is able to edit all the members of the gene family. When several different genes need to be edited, multiplex editing would be necessary. Finally, it may be desirable to replace a sequence of the endogenous gene with a mutated sequence. In such cases, strategies like intron targeting, HDR-mediated DSB repair or ODM (oligonucleotide-directed mutagenesis) must be used. The construction of appropriate donor DNA/chimeric oligonucleotide would require precise and validated information on the sequence changes that would ultimately generate the desired alteration in the trait phenotype.

The last, but not the least, requirement is the capability of the targeted plant species/ genotype to regenerate complete plantlets from tissues/cells cultured *in vitro*. This is because the SDNs and donor DNA must be introduced into plant cells and plantlets must be regenerated from the transformed cells. *Agrobacterium*-mediated transformation can also be achieved by floral dip method (Pyott et al., 2016), which obviates the need for regeneration from cultured cells. In addition, gene-edited plants can be generated without tissue culture by inducing *de novo* shoot meristems in soil-grown dicotyledonous species. This involves *A. tumefaciens*-mediated delivery of developmental regulator genes like *WUSCHEL (WUS), SHOOT MERISTEMLESS (STM)*, etc. along with the reagents for gene editing into somatic cells of the plants. The transgene expression leads to induction of adventitious shoot meristems that give rise to shoots with the desired edits in their genomes. The seeds produced on these shoots contain the targeted genomic modifications (Maher et al., 2020). But SDN delivery via biolistic method or PEG (polyethylene glycol)-mediated delivery requires dependable plantlet regeneration protocols. In the case of PEG-mediated delivery, protoplasts should be amenable to plantlet regeneration, which may be difficult in at least some plant species.

7.5. THE SDN TECHNOLOGY

The use of SDNs for induction of sequence changes at the targeted genomic sites is known as SDN technology. These technologies exploit endogenous DNA repair pathways for the induction of targeted genetic changes in genomes. The SDNs induce DSBs at the targeted genomic sites, which trigger the DNA repair pathways. Generally, the NHEJ-mediated repair predominates and it typically leads to insertion or deletion of one to few base pairs and, thereby, inactivation of the targeted genes. Sometimes, DSBs may be repaired by homologous recombination resulting in insertion of the supplied donor DNA sequences into the genomes. Thus, the logic of SDN technology is as follows: creation of site-specific DSBs in the genomic DNA activates cellular DNA repair mechanisms, which lead to mutational changes at the DSB site. The sequence changes at the DSB sites are either random or they are directed by sequences of the donor DNAs provided with the SDNs. There are basically three types of SDN technologies, *viz*., SDN-1, SDN-2, and SDN-3 technologies (Lusser et al., 2012, Podevin et al., 2013; Table 1.3).

7.5.1. Generalized procedure for SDN technology

The SDN technology for targeted genetic modification of the selected plant trait includes the following steps.

(i) The first step comprises identification of the gene that should be modified to generate the desired trait phenotype. A decision is then made about the site of this gene to be targeted and the specific genetic change that is to be induced.

(ii) The next step consists of selection of the SDN system to be used and designing of the SDN specific for the selected site of the targeted gene. The CRISPR-Cas9 system is the easiest to design and implement and the cheapest to use.

(iii) The designed SDN can be delivered into plant cells as DNA, RNA, RNP (ribonucleoprotein) or protein (Section 7.5.2). In case the SDN is to be used as DNA, gene constructs encoding the designed SDN are developed. If it is to be delivered as RNA, *in vitro* transcription is used to generate RNA copies of the SDN. The protein component of SDNs is typically produced in *E. coli*.

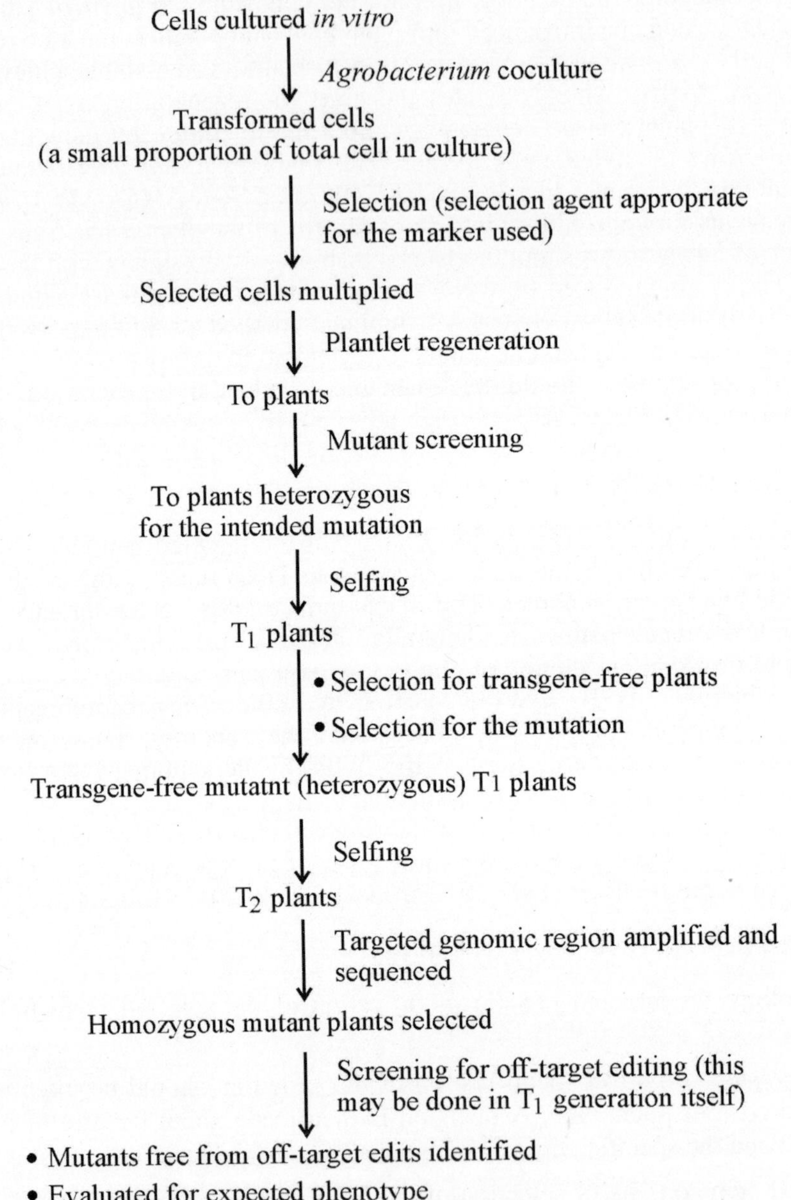

Fig. 7.1. A simplified schematic flow chart for gene editing in plants.

(iv) The SDN gene construct is delivered either with or without donor DNA (depending on the SDN technology used) into plant cells. This is typically achieved by coculture with

Agrobacterium tumefaciens or particle gun delivery. The expression vectors can be designed for either transient expression or for stable transformation. The use of expression vectors based on virus genomes, *e.g*., Tobacco Rattle Virus (TRV) genome, allow efficient delivery of SDNs and regeneration of mutant plants that are free from foreign DNA (Section 7.5.2).

(v) The SDN gene construct produces the SDN, which ultimately leads to mutational changes (SDN-1 and SDN-2) or gene insertion (SDN-3) at the targeted site.

(vi) Finally, plantlets are regenerated and subjected to screening for identification and selection of plants for the desired changes in their DNA sequences. A rigid selection for foreign DNA-free mutant plants is applied in the sexual progeny of mutants.

After introduction of the SDN gene constructs, the cultured plant cells are subjected to selection against the selection agent appropriate for the selectable marker used. The selected cells are multiplied, and plantlets are regenerated (T_0 plants) and screened for the intended mutation. The genomic region surrounding the targeted region is amplified and used for a suitable assay to detect the targeted mutation (Section 7.6). The mutant T_0 plants are self-pollinated, and the T_1 progeny are screened for the presence of transgenes as well as the intended mutations. Since it is relatively simpler to identify heterozygous mutant plants by molecular techniques, transgene-free heterozygous mutant T_1 plants are selected and self-pollinated. The targeted genomic regions of T_2 progeny plants are amplified and sequenced; this permits the selection of homozygous mutant plants as well as characterization of the mutational events. The homozygous mutants are then evaluated for off-target editing (Section 7.7), and plants free from off-target edits are selected. The assessment for off-target editing can be undertaken on T_1 plants themselves as this would save one generation. Finally, the mutants are subjected to evaluation for the intended trait modification as well as agronomic performance (Fig. 7.1).

The mutant frequency may be as high as 10% or higher, and up to 4% of these mutants may be due to homologous recombination. Further, the mutational events may occur up to 2,000 to 3,000 bp from the DSB site (Tzfira et al., 2012). The SDN technologies enable the induction of genetic changes with great precision, and the desired mutations can be readily identified by phenotypic screening and/or appropriate molecular analyses (Section 7.6). Once the desired genetic changes are identified, any associated undesirable features, if present, can be removed by backcrossing with the parent strain combined with selection. The mutant alleles/genetic changes can be readily transferred into other elite varieties by a suitable breeding scheme to develop a commercial product.

7.5.2. Delivery of SDNs and donor DNAs into plant cells

The SDNs can be delivered as DNA, RNA, RNP or protein. They can be introduced into plant cells by *Agrobacterium*-mediated transformation or particle gun bombardment, or into protoplasts by PEG-mediated delivery. The SDN gene constructs are generally delivered into plant cells by *Agrobacterium*-mediated transformation. The genes encoding the SDNs are placed within the T-DNA of, usually, a binary vector and generally a selectable marker is also included for selection of the transformed cells (Fig. 7.2). In case of CRISPR-Cas9 system, the genes encoding Cas9 nuclease as well as the appropriate sgRNA are integrated into the same vector (Fig. 7.2B). The expression vectors for CRISPR-Cas9 system contain the Cas9 gene construct, and the experimenter must incorporate only the sequence for the appropriate sgRNA (actually, only the spacer region of sgRNA) for each targeted site. But for ZFNs and TALENs, one pair of gene constructs are needed for each target site since these SDNs function as dimers.

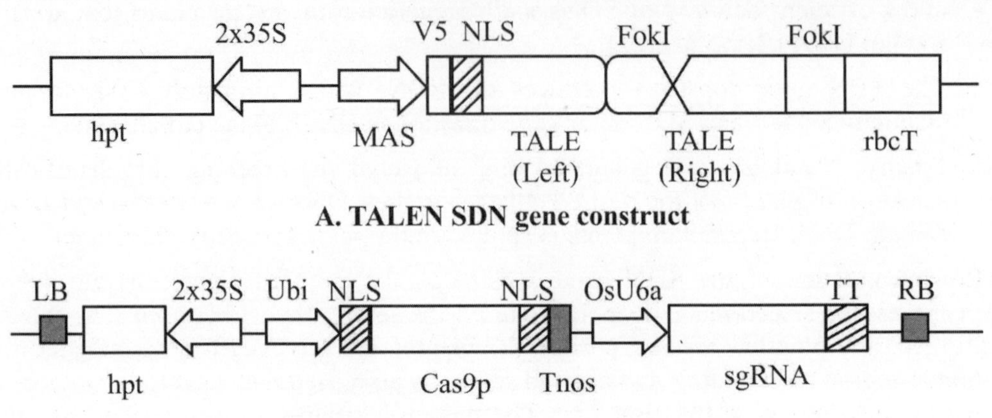

Fig. 7.2. Schematic representation of SDN gene constructs. **A.** TALEN SDN gene construct (modified from Sauer et al., 2016). **B.** CRISPR-Cas9 SDN gene construct (based on Wang et al., 2016). LB, RB, Left and right borders of T-DNA; V5, V5 epitope tag; NLS, nuclear localization signal; rbcT, *rbc* terminator; hpt, hygronycin phosphtransferase (selectable marker); MAS, mannopine synthease promoter; 2x35S, 2 copies of CaMV 35S promoter; Ubi, Ubiquitin promoter; OsU6a, rice U6a gene promoter; Cas9p, Cas9 gene codon optimized for plants; sgRNA, single guide RNA; Tnos, terminator nopalive synthase gene; TT. transcription terminator.

The two gene constructs encoding the SDN pair (in case of ZFNs and TALENs) can be introduced into the cells by co-delivery of two separate expression vectors. The SDN gene constructs can also be delivered through a single vector expressing both the monomers as either separate genes or as a bicistronic gene construct (in case of ZFNs and TALENs). In the bicistronic gene, the DNA sequences encoding the two SDN monomers are linked in tail-to-tail orientation and are separated by a DNA sequence encoding a spacer peptide sequence (Fig. 7.2A). The SDNs may be delivered either with (in cases of SDN-2 and SDN-3 technologies) or without (in the case of SDN-1 technology) appropriate donor DNA: the SDNs and the donor DNA are generally delivered by co-transformation through *Agrobacterium* T-DNA. The SDN gene constructs, including T-DNA become integrated into the plant genome. As a result, SDNs are continuously produced, which is likely to increase the frequency of off-target effects. It is important to minimize off-target editing to minimize the risk of unintended mutational changes in the mutant lines. Further, the SDN transgenes, marker genes and T-DNA sequences will have to be removed by selection following segregation to obtain foreign DNA-free mutant plants (Tzfira et al. 2012). But the use of vectors based on virus genomes allows efficient delivery of ZFNs into a wide range of organs of tobacco and petunia plants; this system can be used in other plant species as well. Vectors based on a positive-sense strand RNA virus, *e.g.,* TRV, were used to deliver ZFNs and sgRNAs into plant cells, and gene edited plants were recovered by regenerating plantlets from infected cells using tissue culture. In contrast to the limited cargo capacity of these viruses, negative-sense strand RNA viruses like Sonchus yellow net nucleorhabdovirus (SYNV) were able to deliver both Cas9 and sgRNA gene constructs into *N. benthamiana* cells. Plantlets regenerated from the SYNV-infected cells showed heritable gene editing (Dinesh-Kumar and Voytas, 2020). A negative-sense strand RNA genome cannot itself serve as mRNA; it must be used as template to synthesize mRNA that can be translated.

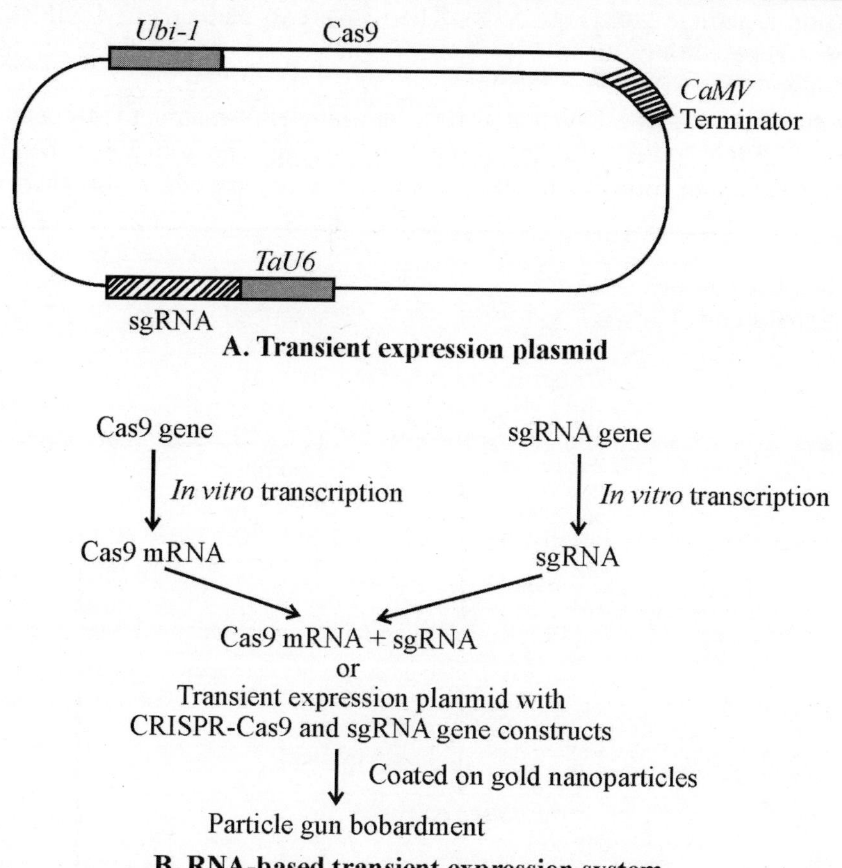

Fig. 7.3. Transient expression systems for CRISPR-Cas9 SDN system. **A.** Schematic representation of a plasmid for transients expression of CRISPR-Cas9 DNA (TECCDNA). The CRISPR-Cas9 gene construct is not likely to be integrated into the plant genome. **B.** RNA-based transient expression system. Genes encoding Cas9 and sgRNA are transcribed *in vitro* and used for transformation. *CaMV,* Cauliflower Mosaic Virus; *TaU6, Triticum aestivum U6* promoter; *Ubi-1 Ubiquintin-1* promoter (based on Zhang et al., 2016)

Zhang et al. (2016) designed a transient expression plasmid to avoid the integration of SDN gene constructs into plant genomes (Fig. 7.3A). This plasmid carries the Cas9 and sgRNA genes and can be introduced into plant cells by biolistic method or into protoplasts by PEG-mediated delivery. Once inside the cells, the genes are transcribed and translated to generate the Cas9-sgRNA complex for gene editing. But the gene constructs and the plasmid DNA would soon be cleaved by the cellular enzymes, and some of the resulting fragments might become inserted into the plant genome as a part of DSB repair process. It will be very difficult to detect such insertions and to eliminate them from the mutant plants produced by gene editing. The inclusion of foreign DNA segments into the genomes of gene-edited plants is a matter of biosafety concern and needs to be avoided. In view of this, the authors developed the procedure for RNA-based transient expression of CRISPR-Cas9 SDNs. In this strategy, the Cas9 mRNA and the sgRNA are produced by *in vitro* transcription, mixed together and coated onto gold nanoparticles that are delivered into plant cells using a particle gun (Fig. 7.3B). The above two methods of CRISPR-Cas9 delivery enable efficient gene editing: up to 5% of the T_0 plants regenerated from the treated embryos without selection for transformed cells/plants showed gene editing. There was no editing at off-target sites with 2-5 mismatches; even at a site with a single mismatch, the frequency of off-target editing was several-fold lower than that of on-target editing. The above two methods were developed for wheat immature embryos, but they can be used for gene editing in other plants as well.

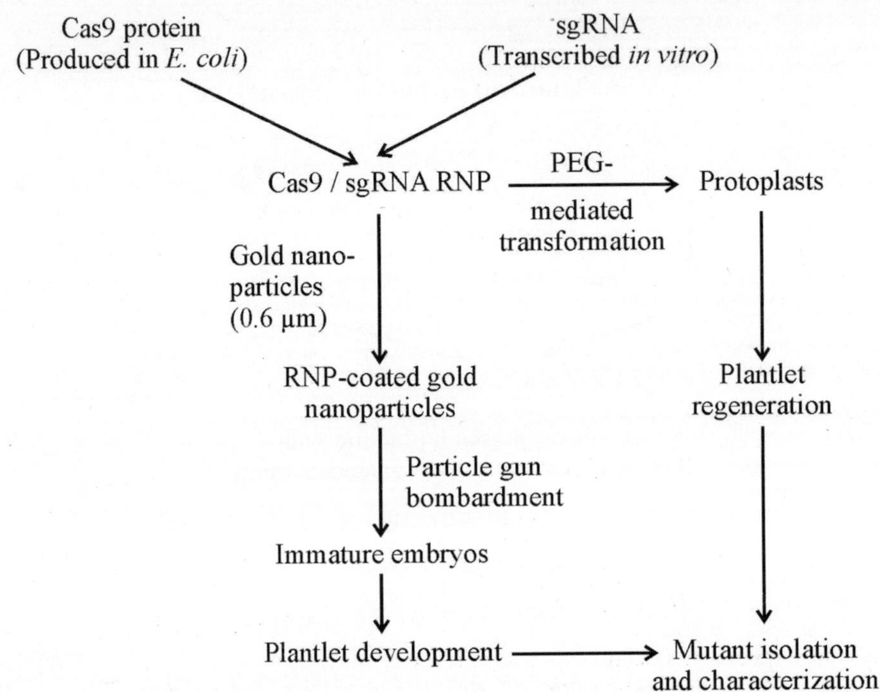

Fig. 7.4. Schematic representation of gene editing using Cas9/sgRNA RNP delivered into protoplasts (PEG-mediated delivery) and immature embryos (particle gun delivery) (based on Liang et al., 2017). Particle can be used to deliver the RNPs into intact cells and even organs like short tips. PEG, polyethylene glycol; RNP, ribonucleoprotein.

The delivery of SDNs as protein (ZFNs and TALENs) or RNP (CRISPR-Cas9) also eliminates the risk of integration of foreign DNA into the crop genome. CRISPR-Cas9 SDN has been delivered as RNP into protoplasts by PEG-mediated delivery and into immature embryos by particle gun or biolistic delivery (Fig. 7.4). Since plantlet regeneration from protoplasts is very difficult in some crop species, transformation of young embryos is highly desirable since they would readily germinate or form embryogenic calli that regenerate plantlets, *e.g.*, in wheat. The purified Cas9 protein and sgRNA were mixed together in Cas9 reaction buffer, and then coated onto gold nanoparticles (0.6 μm diameter). The RNP-coated gold nanoparticles were accelerated using a particle gun device and bombarded into immature wheat embryos (Fig. 7.4). The RNP would be delivered into the nuclei of at least some cells where it would carry out editing of the targeted genomic site. The CRISPR-Cas9 SDN was designed to edit the wheat gene *TaGW* (for grain weight). The mutation frequency was high (4.4%) and comparable to that obtained by biolistic delivery of the same SDN in form of DNA in the transient expression plasmid. However, the frequency of off-target editing by the SDN delivered as RNPs was less than one-fifth of that produced by its delivery as DNA in the transient expression plasmid (Liang et al., 2017).

Generally, the CRISPR-Cas9 and the sgRNA gene constructs are delivered together. But in the case of '*sequential transformation technique*' the plants are first transformed with the Cas9 gene construct driven by a suitable promoter and the resulting transgenic lines are transformed with gene constructs for sgRNAs, a selectable marker and donor DNA. It was reported that in *Arabidopsis*, the frequencies of gene knock-ins and HDR-mediated sequence replacements at several genomic loci were improved when the Cas9 gene was driven by an egg cell- and early embryo-specific promoter from the *DD45* gene (Miki et al., 2018). Similarly, Wolter et al. (2018) found that editing in *Arabidopsis* was substantially higher when Cas9 expression was controlled by the egg cell-specific promoter *EC1.1* fused with the *EC1.2* enhancer.

7.5.3. The SDN-1 technology

The SDN-1 technology uses SDNs without any donor DNA (Table 1.3). The SDN binds to the specified genomic sequence and produces a DSB that is generally repaired by NHEJ. NHEJ often leads to small (one to few base pairs long) deletions, insertions, or substitutions. The SDNs can be used to produce two DSBs at nearby sites. This may lead to a loss, duplication or inversion of the DNA segment located between the two cuts; it may also result in translocation of this segment to another genomic site. This strategy can be used to delete complete endogenous genes, but this would require the expression of four specifically designed ZFN or TALEN monomers. This would be greatly facilitated by using the vector system designed by Tovkach et al. (2009) that allows four separate expression cassettes for ZFN monomers to be located and expressed in a single T-DNA.

Thus, SDN-1 strategy would ordinarily inactivate the target genes. SDN-1 is preferable to the other strategies, including antisense RNA and RNAi (RNA interference) technologies, for gene inactivation since it enables predictable and stable inactivation of the targeted genes. It may be expected that this approach would replace the other gene suppression strategies for both basic and applied researches.

7.5.4. The SDN-2 technology

Plant cells show extremely low frequency of integration of introduced DNA segments into their genomes by homologous recombination, which is increased several thousand-fold by induction of DSBs. The SDN-2 and SDN-3 technologies exploit this feature of plant cellular DNA repair systems. In case of SDN-2 technology, the vector carrying the SDN gene also contains a short donor DNA segment that serves as template for HDR-mediated DSB repair. The end regions of donor DNA are complementary to the sequences flanking the DSB site to allow homologous recombination. But the middle region of the donor DNA contains the sequence changes desired to be introduced in the genome. Thus, donor DNA contains well-planned and precisely placed alteration in the DNA sequence for inducing the desired mutation. The use of Cas9-VirD2 fusion protein-based CRISPR-Cas system for gene editing enhances the rate of HDR since VirD2 binds to donor DNA and brings it to the DSB site.

In this approach also, the SDN binds to the targeted genomic sequence and induces DSB that will be repaired in some cells by HDR using the donor DNA as template so that the donor DNA sequence is copied into the genome (Durai et al. 2005). This will generate the desired mutation at the DSB site. The SDN-2 strategy can be used to correct deleterious mutations (targeted gene correction) and induce specified mutations in the target genes to create new alleles leading to new phenotypes (gene/allele replacement). ZFNs, TALENs and CRISPR-Cas9 SDNs have been used to induce such mutations in several genes (Lusser et al., 2012, Podevin et al., 2013).

7.5.5. The SDN-3 technology

The SDN-3 technology involves integration of complete gene constructs into plant genomes at specified target sites. In this technology, SDN gene along with a long donor DNA containing the gene of interest is introduced into plant cells. The SDN will induce a DSB at the target site and there will be HDR-mediated DSB repair in at least some cells. This would lead to integration of the gene construct contained in donor DNA into the plant genome at the DSB site. Thus, the SDN-3 technology allows targeted integration of additional copies of the desired endogenous genes or novel transgenes into the plant genomes. In addition, it can be used for replacing the existing alleles of endogenous genes with new alleles of the same genes, and for trait stacking, *i.e.*, integration of genes for two or more traits at the same locus.

The targeting of desired genes to specific genomic sites would eliminate the risk of 'position effects' on transgene expression associated with random transgene integration in the genome. It would also minimize the risk of unintended inactivation of an essential gene or activation of a detrimental gene due to transgene integration within or close to endogenous genes. SDNs have been used for integrating transgenes at targeted genomic sites in several other species. In addition to HDR, NHEJ pathway of DSB repair has been recruited to achieve transgene integration at predecided genomic sites. In this strategy, both the genomic site and the donor DNA are cleaved by the same nuclease. Therefore, the two ends of donor DNA have the same sequence as the targeted genomic site. When the genomic site and the donor DNA are cleaved simultaneously, repair of the DSB by NHEJ would ligate the donor DNA into the genome. This approach has permitted the integration of large (up to 14 kb) transgenes at the specified genomic sites (Lusser et al., 2012, Podevin et al., 2013; Gaj et al., 2013).

7.5.6. Applications of the SDN technology

The use of ZFNs for gene editing in plants began in few molecular biology laboratories and the advent of TALENs enabled more laboratories to use this novel tool. However, the availability of CRISPR-Cas9 system has placed the gene editing within the reach of any molecular biology lab desiring to use this technology. As a result, there has been a spurt in gene editing activities in plants and several reviews devoted to the applications of CRISPR-Cas9 SDN in plants, including those by Puchta (2017), Belhaj (2015), Bortesi and Fischer (2015), Schiml and Puchta (2016), Arora and Narula (2017), and Schindele et al. (2018), have been published. PennState and Arizona Genomics Institute support the web portal CRISPR-PLANT that has been designed to facilitate use of CRISPR-Cas9 for gene editing in plants. In addition, several web-based tools assist in selection and prediction of specificity of sgRNAs for plant genome editing; some of these tools are available at the following websites: (http//cbi.hzau.edu.cn/cgi-bin/CRISPR; http://www.genome.arizona.edu/crispr/; and http:// www.rgenome. net/cas-offinder, http://www.e-crisp.org/E-CRISP/index.html.

The SDN technologies can be used for improving many important traits of crop plants. For example, SDN-1 and SDN-2 could be used to modify endogenous genes to generate herbicide tolerance, resistance to biotic or abiotic stresses, improved nutritional composition, reduced levels of allergens and/or anti-nutritionals, enhanced industrial usefulness, etc. The SDN-3 technology facilitates targeted integration of transgenes leading to more predictable and stable transgene expression and minimizes the risk of unintended changes in the expression of endogenous genes that may be located at or close to the transgene integration sites. It can also be used for stacking of transgenes at the same locus or at different predetermined genomic sites. Trait stacking comprises integration of two or more transgenes into the genome close to each other and/or a pre-existing transgene. Significantly, trait stacking cannot be achieved either by traditional plant breeding strategies or the conventional transgenic technology. Stacking of transgenes minimises the risk of their separation by recombination and segregation and they tend to be inherited together as a single locus. As a result, they are introgressed together into new lines/varieties, and selection for any one of them would select for all of them. The SDN-3 technology can also be used to integrate transgenes for metabolic engineering or for molecular farming in such genomic regions that serve as 'safe harbour loci' and support strong and stable expression of the transgenes placed in them (Podevin et al. 2013, Bortesi and Fischer, 2015).

This technology can be used for removing specific transgenes like selectable markers by using pairs of SDNs specific for the DNA sequences flanking the targeted transgenes. This can be achieved by crossing the concerned transgenic line/plant with a plant/line having the SDN gene constructs targeting the sequences flanking the marker gene. The expression of SDNs in the F_1 plant will lead to deletion of the marker gene, and transgenic plants free from the marker can be recovered from its progeny. The removal of unwanted marker genes from transgenic plants would have significant regulatory benefits. This strategy can also be used to delete selected genomic regions with highly repetitive DNA or some undesirable loci, and this might generate new variations in relevant traits of crop species (Curtin et al. 2012).

Among the SDNs, ZFNs were the first to be used for site-directed mutation induction in plants. Lloyd et al. (2005) used ZFNs for induction of random mutational changes in a reporter gene integrated into the genome of *Arabidopsis thaliana* (SDN-1 technology). In the same year, Wright et al. (2005) applied SDN-2 strategy to introduce HDR-mediated mutation in a reporter gene inserted into the genome of *A. thaliana.* Zhang et al. (2010) applied the SDN-1 approach using ZFNs and observed mutation frequencies of 7% or 16% for the genes *ADH1* and *TT4* of *A. thaliana.* About 20% of the mutant plants were found to be homozygous for mutations, indicating that DSBs were produced in both the alleles of the concerned genes in the cells from which these mutant plants were obtained.

7.6. MUTATION DETECTION

In this book, the term mutation generally means the intended sequence change induced by the gene editing procedure. Similarly, the term mutant refers to an individual that contains the mutation in homozygous or heterozygous state. In gene editing studies, mutation detection is ordinarily undertaken in the T_0 generation. The T_0 plants are the plants regenerated from transformed tissues either without any selection pressure or after they are subjected to selection appropriate for the selectable marker used in the study. A proportion of T_0 plants are expected to be heterozygous for the intended mutation, but a variable proportion of the mutant plants may be homozygous. There are several approaches, including CEL1 or T7E1 assay, restriction enzyme PCR assay, heteroduplex mobility assay, and DNA sequencing for the detection of induced mutations. In addition, homozygous mutant plants can be identified by their altered phenotypes in case the mutations in question generate phenotypic changes.

The SDN-induced mutations can be detected by the restriction enzyme digestion suppressed PCR (RE–PCR) procedure. In RE–PCR assay, the genomic DNA of the test plants and the parental line are digested with the selected restriction enzyme before they are used for PCR amplification (Fig. 7.5). This restriction enzyme has a recognition site in the parental genome that includes the site of DSB induced by the SDN used in the study. The unique genome sequences flanking this recognition site are used to design a pair of primers, which are used to amplify the targeted genomic region from restriction enzyme-digested genomic DNAs of the parental line and the test plants. In case a DSB was produced by the SDN system and the repair of DSB had changed the base sequence at the DSB site, the recognition sequence will be abolished. As a result, there will be amplification of the genomic region and a PCR product will be detected. In contrast, there will be no detectable PCR product in the parental line and in all the test plants lacking mutation at the target site since the DNA will be cleaved by the restriction enzyme, and no amplification will be detected (Xie and Yang, 2013).

Alternatively, the RE–PCR assay may be used like CAPS (cleaved amplified polymorphic sequences) marker system. In this case, the genomic DNAs are used for PCR amplification without digestion with the concerned restriction enzyme. The amplification products from all the plants are digested with the restriction enzyme and then subjected to gel electrophoresis. The heterozygous mutant plants will show two shorter bands produced by digestion of the nonmutant amplicon in addition to a longer band representing the mutant amplicon. In contrast, the PCR products from parental line and the nonmutant plants will yield two shorter bands generated by restriction digestion of the amplicons. Homozygous mutant plants, on the other hand, will show only one longer band representing the uncleaved amplicon.

In case of CEL1 (surveyor nuclease) or T7E1 assay, the DNAs are generally extracted from leaf tissues of T_0 plants and the parental line, and separately amplified using a pair of PCR primers specific for the unique sequences flanking the targeted site of the genome. The PCR products are denatured, annealed and then treated with either CEL1 or T7E1 endonuclease, which are specific for mismatches in DNA duplexes. DNA samples from the parental line, and T_0 plants that are either nonmutant or homozygous for the same mutation will be resistant to CEL1 or T7E1 endonuclease and will yield a single band after electrophoresis (Fig. 7.6). But heterozygous mutant plants, and homozygous mutant plants having different sequence changes in the two alleles of the targeted gene will form heteroduplexes, *i.e.*, DNA duplexes with few to several mismatched base pairs. These heteroduplexes will have bulges at the mismatch sites, which will be recognized and cleaved by CEL1 or T7E1 endonuclease. This cleavage will generate two shorter bands in addition to the larger band formed by the two homoduplexes representing the mutated and the normal alleles of the gene (Fig. 7.6). It has been reported that T7E1 assay is more sensitive than CEL1 assay. In any case, this assay can detect only heterozygous mutants and fails to distinguish homozygous mutants from nonmutant plants (Pyott et al., 2016; Jung et al., 2018).

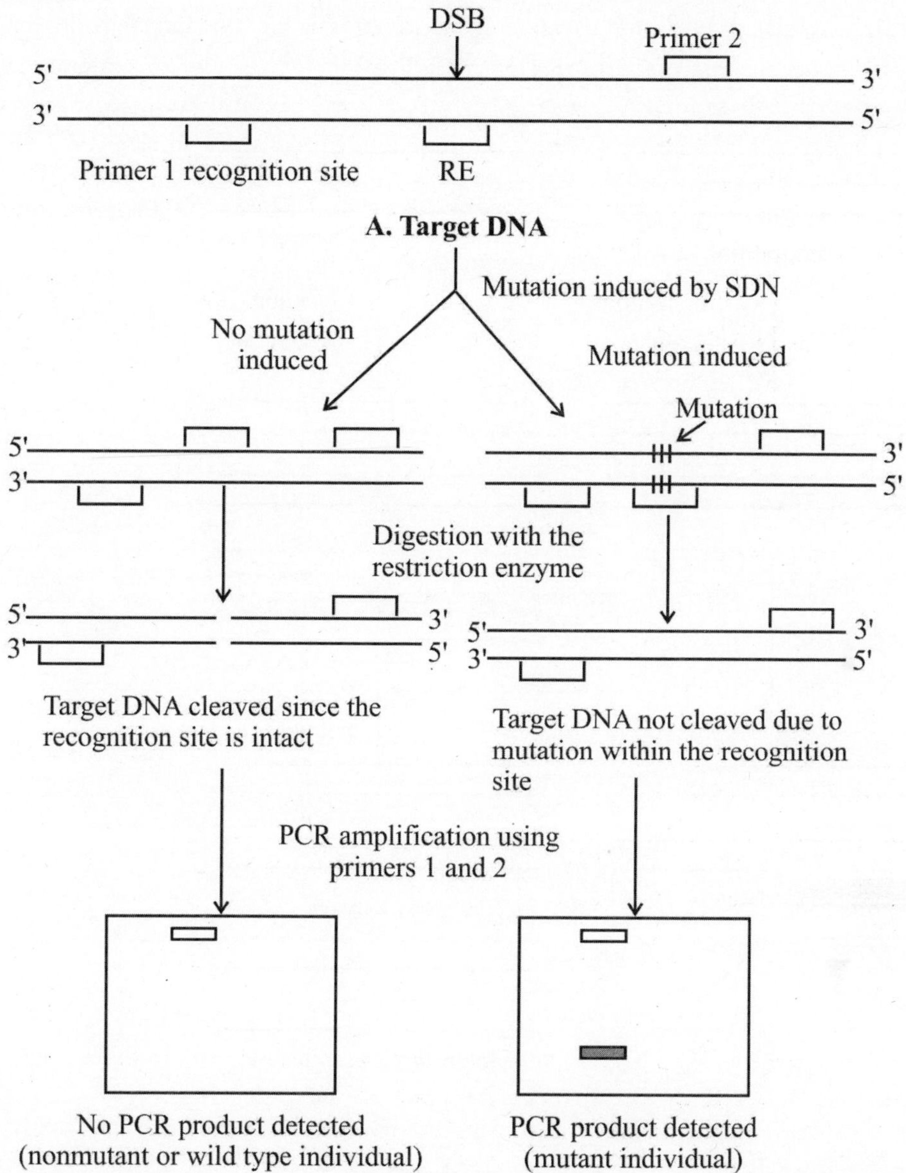

Fig. 7.5. A schematic representation of RE-PCR (restriction enzyme - PCR) used to detect mutations induced in specific genomic regions. In this case, SDN is used to induce mutations at a site that functions as the recognition site for a restriction enzyme (RE). This RE is used for RE-PCR to identify mutant individuals. DSB, double-strand break; PCR, polymerase chain reaction; SDN, site-direrted nuclease.

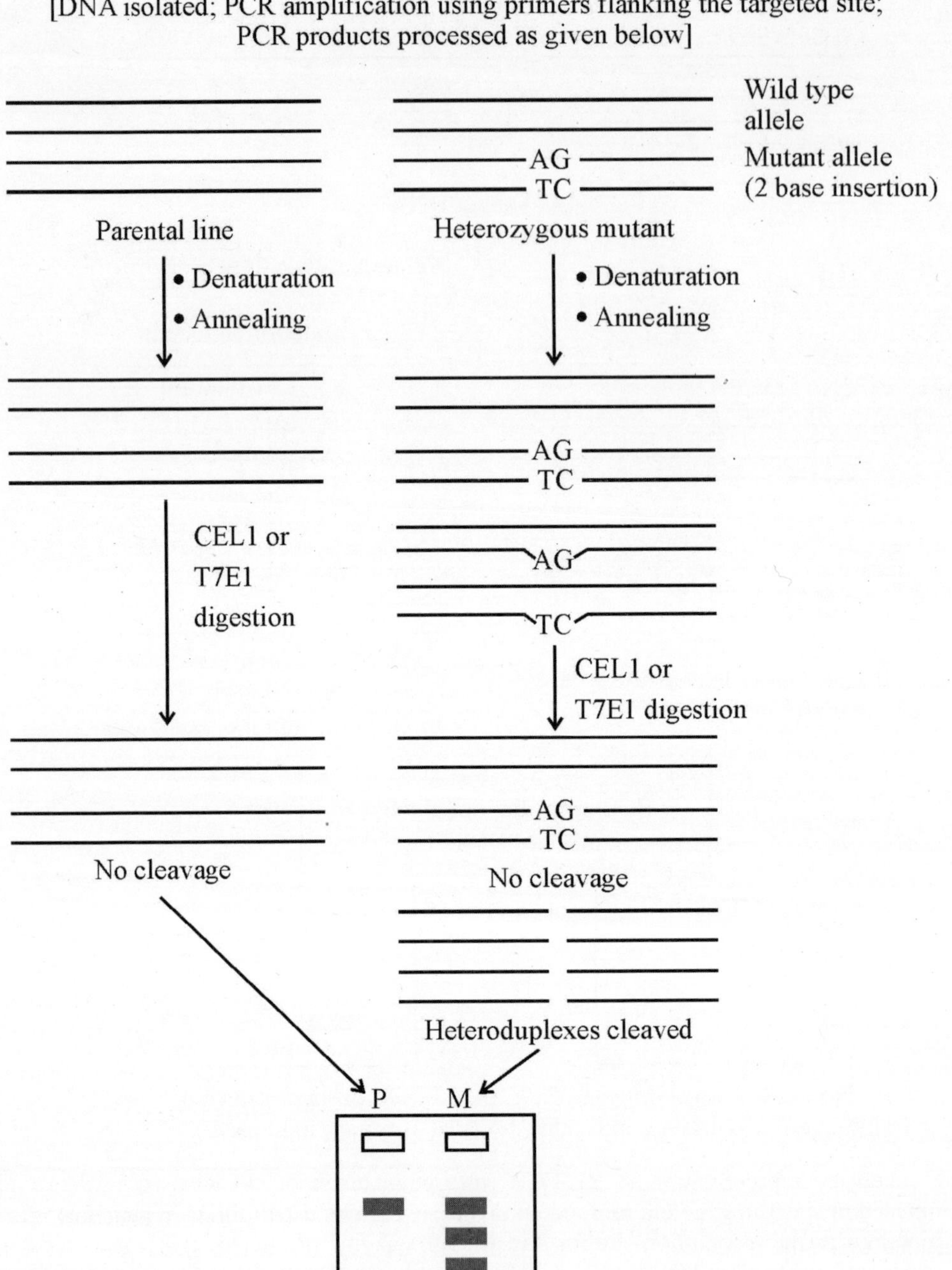

Fig. 7.6. The CEL1 or T7E1 assay. Parental line and nonmutant T_0 plants will show a single band, while heterozygous mutant will show at least three bands. P, Parental line, M, heterozyous mutant.

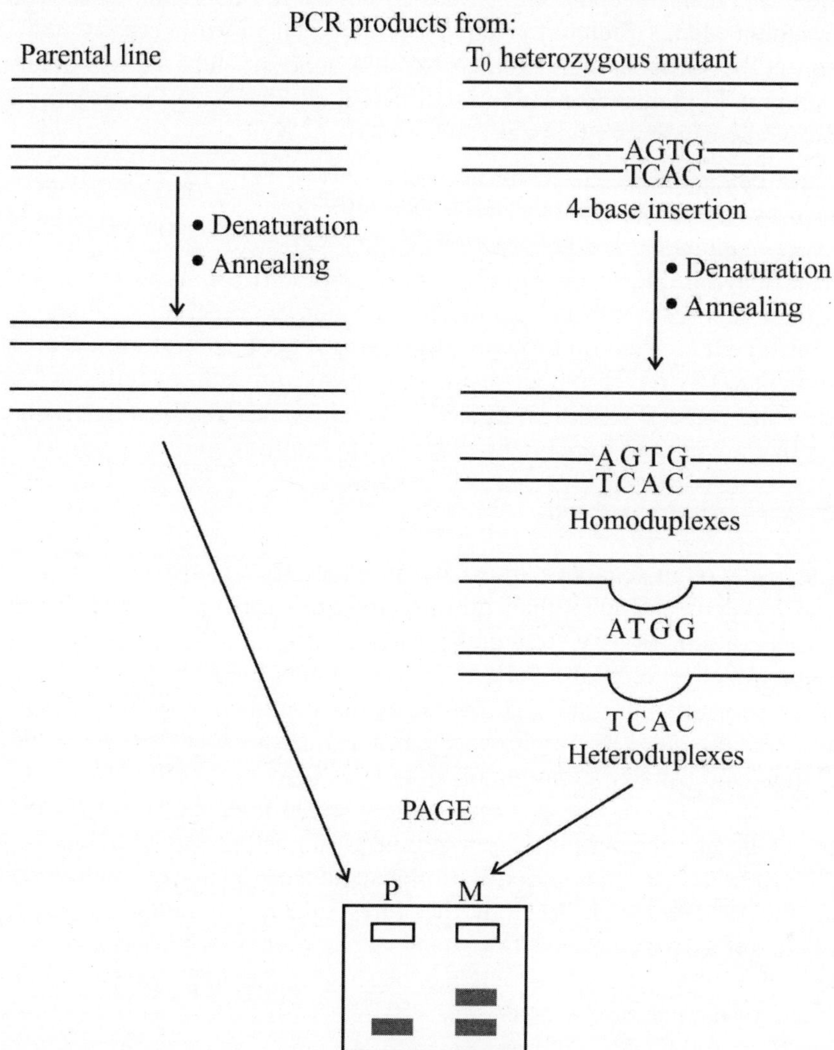

Fig. 7.7. Heteroduplex mobility assay (HMA). The parental line and nonmutant. T_0 plants will exhibit a single band, while the heterozygous mutant will show at least two bands. Indels of two or more base-pairs detected by HMA. PAGE, polyacrylamide gel electrophoresis.

The heteroduplex mobility assay (HMA) is based on differential mobilities during gel electrophoresis of homoduplexes and heteroduplexes having the same size. The DNA samples from T_0 plants and the parental lines are used for PCR, amplification products are denatured and annealed as in the case of CEL1 or T7E1 assay. The annealed products are now subjected to polyacrylamide gel electrophoresis and the resulting bands are scored. The parental line and the nonmutant T_0 plants will show a single band (all amplicons will be homoduplexes). In contrast, heterozygous mutant T_0 plants will show two bands as they will have both homoduplexes and heteroduplexes, and the latter would move slower than the former (Fig. 7.7). This assay can detect mismatches of 3 bases or more and would be unable to differentiate between homozygous mutant and nonmutant plants (Delwart et al., 1993; Ito et al., 2015). Finally, the PCR products can be subjected to DNA sequencing, and their base sequences compared with that of the parental line to identify mutant T_0 plants. Sequencing will detect all kinds of sequence changes, including base substitutions and will identify homozygous mutants as well.

Attempts are being made to develop noninvasive imaging methods for detection of mutations induced by SDNs. Feng et al. (2017) used NIR (near infrared) hyperspectral imaging of rice seeds to detect mutants for seed characteristics like perimeter, area, length, width and 1000-grain weight. The imaging data were subjected to chemometric analysis to successfully identify mutant seeds. Chemometrics comprises application of mathematical and statistical procedures to infer the relationships between quality or physical parameters and the data obtained from analytical instruments. The development of imaging-based procedures will obviate the need for lengthy extraction and wet-lab analytical procedures and would greatly simplify mutant detection using techniques that are easy to implement and have high throughput.

7.7. OFF-TARGET GENE EDITING

An off-target site has enough sequence similarity with the targeted site to allow the SDN to act at this site albeit at lower frequencies than the targeted site. Base editing at an off-target site is termed as off-target editing. Several strategies for monitoring of off-target gene editing have been developed for animal cells (Section 5.9.1). In case of plants, monitoring of off-target activity is generally based on sequencing of selected off-target sites predicted *in silico* based on homology with the targeted genomic site. In simple words, bioinformatics tools are used to predict the off-target genomic sites, and usually a sample of these sites showing the highest sequence similarity with the target site are selected. The genomic regions containing the selected off-target sites are amplified using PCR and the amplification products are subjected to DNA sequencing. The sequences of PCR products are compared with the sequences of corresponding genomic regions of the parental line to identify the regions that have undergone mutation and to estimate the frequency of off-target activity of the SDN. A more comprehensive approach to score off-target editing is based on whole genome sequencing of the regenerated plants and comparing their sequences with that of the parental line (Puchta, 2017).

Several strategies for minimizing off-target editing have been devised and evaluated; a discussion of these strategies for CRISPR-Cas9 system is provided in Section 5.9.2. These strategies are based on Cas9 mutants, sgRNA design, restricting the period to SDN activity, etc. The strategies based on Cas9 include the use of nickases, Cas9 mutants having improved specificity and reduced off-target effects, and Cas9 variants that allow regulation of the period of Cas9 action. The design and function of sgRNAs has been actively investigated and has resulted in several refinements and innovations that improve specificity and reduce off-target activity. In addition, careful selection of the target site might reduce off-target activity: it has been proposed that a genomic sequence should be selected as target site that is unique and has no homology to an off-target genomic site, especially in regions up to the first 10 bp upstream beginning from the PAM sequence location.

A major factor affecting the off-target activity is the duration of SDN activity in the cell: the longer this duration, the greater is the chance of off-target editing. Therefore, several strategies based on either Cas9 or sgRNA component have been devised to regulate the time as well as the period of expression and/or action of Cas9. Another strategy to limit the period of SDN action is based on the mode of their delivery: delivery of SDNs either as protein (ZFNs, TALENs), RNA (all SDNs) or RNP (CRISPR-Cas9) limits the period of their activity since they are degraded relatively rapidly by the cellular machinery (Jung et al., 2018; Section 5.9.2). In addition, efforts to develop C-to-T base editors with reduced off-target editing have been successful (Section 6.3.2; Doman et al., 2020). In several studies, analyses of whole genome sequence data revealed only very low or negligible off-target editing. In case of *Arabidopsis*, even off-target sites that were highly homologous to the target sites exhibited no editing (Belhaj et al., 2015).

7.8. BASE EDITING IN PLANTS

Base editing directly and irreversibly converts one base into another at the targeted genomic site without induction of a DSB. The base editing techniques as well as the reagents therefor were initially developed and optimized for animal, particularly human, cells cultured *in vitro*. These developments provided the motivation and the experimental framework for subsequent extension of this technology to plant systems. Several groups have explored base editing in plants using the editors developed for animal cells; sometimes, the editors were used without any modification but often some optimization for the plant systems was done. One of the modifications concerns codon-optimization of the various proteins used for gene and/or base editing. Further, the regulatory sequences like promoters are also optimized for expression in plants. It is encouraging that there are already some examples of targeted base editing in plants for inducing mutations in useful genes of rice and some other crop species.

7.8.1. Base editors for plants

In 2017, four groups of workers generated gain-of-function mutations in rice by using C-to-T base editors (Section 6.6) based on the animal system base editor BE3 (base editor, third generation; Section 6.3). The BE3 comprises CRISPR-Cas9 nickase, rAPOBEC1 (rat cytosine deaminase) and UGI (uridine glycosylase inhibitor). The rBE3 (rice base editor 3) has essentially the same architecture as BE3, except for codon-optimization for rice and promoters for plants. Other base editors either did not have the UGI sequence, or had an additional NLS (nuclear localization signal; Ren et al., 2017; Lu and Zhu, 2017; Zong et al., 2017). These base editors exhibited editing efficiencies up to 44% within a variable editing window width (Section 6.6). Since rAPOBEC1 edits C residues present in 5'GC3' doublets with quite low efficiencies and the rice genome has high GC content, this deaminase was replaced with hAID (human activation-induced deaminase) that prefers C residues in 5'GC3' and 5'AC3' doublets. An engineered variant of hAID (denoted as hAID*Δ) was codon optimized for rice and used to create the base editor rBE5, which edits C residues in 5'GC3', 5'AC3', 5'TC3' as well as 5'CC3' doublets. The base editor rBE5 mainly generated C-to-T substitutions, but C-to-G and C-to-A substitutions were also observed. In addition, it induced a high (71.9%) frequency of indels. The rBE9, created by adding UGI to rBE5, induced greatly reduced indel frequencies (Ren et al., 2018). The Target-AID base editor (Section 6.3.4) that was originally optimized in yeast and used in mammalian cell cultures was extended to rice and tomato. Target-AID having nCas9 seemed to be more efficient than that based on dCas9 ('dead' Cas9). This base editor also induced C-to-G conversions and indels, but off-target action was negligible (0.14–0.38%).

The adenine base editor (ABE) was optimized for use in plants and designated as PABE (plant adenine base editor). PABE-7, the most successful editor, was codon optimized for expression in cereal plants. This base editor induced A-to-G transitions at high frequencies (up to 59.1% in regenerated plants) in rice and wheat. At the same time, it induced negligible (< 0.02%) frequency of undesired mutations at target loci and very low (<0.1%) frequency of indels (Li, C. et al., 2018). Another A-to-G base editor, rBE14, induced only A-to-G transitions in rice in a sequence context-independent manner, and no indels and off-target editing were detected. Two more A-to-G base editors, *viz*., ABE-P1 and ABE-P2, are based on PABE-7 (Table 6.5). The ABE-P2 contains nSaCas-9 so that it can access AT-rich genomic sites. The editing efficiencies of ABE-P1 and ABE-P2 ranged from 4.8% to 61.3%, depending on the targeted genomic site, and no off-target editing was detected (Section 6.6).

7.8.2. Some examples of base editing in plants

Both C-to-T and A-to-G base editors have been used for editing useful genes in several plant species, including crop plants (Table 7.1). The traits modified by base editing range from resistance to herbicides plant architecture and disease resistance. A selection of examples is described in some detail in the following subsections.

Table 7.1. Selected examples of base editing in rice

Locus/Gene	Base editor	Type of mutation	Phenotypic change	Remarks
Acetolactate synthase	Target-AID	C-to-T transition (C287T)	Resistance to herbicides like imazamox	Complete plants regenerated
SLR1	ABE-P1	T-to-C substitution	Dwarf plants	SLR1 is a repressor in the gibberellic acid (GA_3) signalling pathway
OsSPL14	ABE-P1	T-to-C transition	Ideal plant architecture	OsmiR156 unable to bind to *OsSPL14* transcript; this prevents OsmiR156-mediated degradation of the transcript
An allele of *Pi-d2*	rBE5	G-to-A transition	Resistance to rice blast	Pathogen, *Magnaporthe oryzae*

7.8.2.1. *Herbicide resistance*

The enzyme acetolactate synthase catalyses the first step in synthesis of branched chain amino acids. This enzyme is targeted by herbicides like imazamox. The conversion of C at position 287 to T (mutation C287T) of the gene encoding this enzyme leads to substitution of the amino acid Ala at position 96 with Val (A96V amino acid substitution), which makes the enzyme insensitive and rice plants resistant to the concerned herbicides. Shimatani et al. (2017) used Target-AID along with the appropriate sgRNA to induce the mutation C287T in rice. The base editor and sgRNA gene constructs were transferred into rice calli by *Agrobacterium*-mediated transformation. The hygromycin-resistant calli were cultured on medium containing the herbicide imazamox. More than 3% of the calli were tolerant to the herbicide, and 50% of these callus lines had the A96V mutation. Two plants carrying this mutation were finally regenerated, and there was no evidence of off-target base editing.

7.8.2.2. *Plant architecture*

The rice gene *SLR1* encodes a DELLA protein that functions as a repressor in the gibberellic acid (GA_3) signalling pathway. The TVHYNP motif of protein SLR1 is involved in interaction with the GA receptor protein GID1; this interaction is the first step in GA signalling. Certain amino acid substitutions in TVHYNP motif of SLR-1 proteins considerably reduce its interaction with GID1 and lead to dwarf plants (Asano et al., 2009). Hua et al. (2018) used the ABE-P1 base editor to induce a T-to-C substitution in such a way that it resulted in a V92A substitution (Val at position 92 replaced by Ala) in the TVHYNP domain. Five out of the 40 transgenic lines obtained following ABE-P1 action in rice calli contained intended T-to-C transition and the V92A substitution. Significantly, the editing window had only a single editable T, and no other base substitutions were detected in the target locus.

The rice gene *OsSPL14* is important for ideal plant architecture. The OsmiR156 (*Oryza sativa* microRNA 156) binds to RNA transcript of this gene and leads to its degradation. It is known that a point mutation in the region of *OsSPL14* that is involved in binding of OsmiR156 protects its RNA transcript from cleavage mediated by this miRNA since it can no longer bind to the transcript. This allows expression of *OsSPL14*, which ultimately leads to an ideal architecture of rice plants and enhanced grain yields. Hua et al. (2018) used ABE-P1 base editor to target for editing a sequence of *OsSPL14* that is involved in binding of OsmiR156. A total of 23 independent transgenic lines of the rice variety Nipponbare were regenerated; 6 of these lines had the desired T-to-C transition in the targeted region.

7.8.2.3. *Disease resistant plants*

Base editing can be used to produce disease resistant plants provided the resistance is governed by a single gene and disease susceptibility is the result of a single base substitution in this gene. Obviously, the base sequence of the concerned gene, and the site and the type of base substitution required to generate resistance must be clearly established. Rice blast caused by *Magnaporthe oryzae* is a globally important disease, and several major genes conferring blast resistance are known. Gene *Pi-d2* is an important blast resistance gene, and a recessive allele of this gene, *pi-d2*$_{kit}$, causes susceptibility to the pathogen. The mutation in this allele is due to A-to-G transition, which leads to a single amino acid substitution in the concerned R protein. The base editor rBE5 was used to edit the mutant G to induce G-to-A transition, and 26 transgenic plants were regenerated. Sequences of the gene *pi-d2*$_{kit}$ of these plants revealed eight of them to be heterozygous for the desired G-to-A substitution (Ren et al., 2018).

7.9. APPLICATIONS OF GENE EDITING IN PLANT BREEDING

Gene editing, including base editing, has many different applications in plant breeding. These tools can be used to investigate the functions of specific genes by generating loss-of-function mutations of targeted genes by either DSB production and NHEJ-mediated DSB repair or generation of early stop codons by targeted base editing (Section 6.12). Phenotypic evaluation of the mutant plants is expected to reveal the functions of the mutated genes. Such studies have confirmed the functions of most of the genes concluded from earlier studies. But in some cases, the earlier conclusions had to be revised in the light of findings from SDN-induced targeted gene inactivation. For example, inactivation of the gene *ABP1* by CRISPR-Cas9 in *Arabidopsis* revealed this gene not to be needed for auxin signalling or development as was concluded earlier. But many genes may be essential for plant survival and it may not be possible to recover their loss-of-function mutations. In such cases, mutant alleles with altered functions can be produced usually by base editing, more particularly diversifying base editing (Section 6.3.4), or sequence replacement by HDR. Analysis of such mutants might permit identification of mutant alleles with altered phenotypic effects, which will indicate functions of the targeted genes and might even lead to recovery of some useful mutations (Malzahn et al. 2017; Jung et al., 2018).

Genomic regions encoding structural RNAs and the regulatory regions of genes can be edited to investigate their roles in regulation of gene function. In case of cultured mammalian cells, multiplexed assay systems for screening of edited regulatory regions have been developed. These assays have permitted the identification of new regulatory elements that differ from the typical regulatory elements. Once such elements are identified and their functions are elucidated, gene editing can be used to introduce base substitution at selected genomic sites to either generate, abolish or alter the functions of these sequences to modulate the expression of genes affecting yield, produce quality and other agronomic traits. In addition, the CRISPR-Cas9 SDN has been used to induce mitotic recombination that enabled gene mapping without making crosses (Sadhu et al., 2016; Puchta, 2017).

Gene editing can be used to replace selected genomic regions with the desired DNA sequences using SDN-3 technology (Section 7.5.5). A pair of DSBs may be induced at sites flanking the region to be replaced and the donor DNA having the sequence to be integrated in its place is provided for HDR-mediated repair of the DSBs. The end regions of the donor DNA must be homologous to the genomic sequences located immediately outside the genomic region being replaced. This approach can be used to replace a gene with another gene, a copy of the same gene that has suitably modified/optimized sequence, or a transgene or a regulatory sequence like promoter that has some desirable regulatory feature(s). The SDN-3 technology can also be used to integrate transgenes at selected genomic sites to minimize/eliminate the the risk of gene silencing and the chances of unintended endogenous gene activation/inactivation. SDNs can be used for simultaneous editing of multiple members of gene families by targeting a sequence conserved among all the members of family. CRISPR-Cas9 system can be used for simultaneous editing of several different genes by expressing multiple sgRNAs in the same cell, each sgRNA being specific for a single gene (Jung et al., 2018).

The CRISPR-Cas system has been repurposed for targeted transcriptional activation or repression of selected genes. For example, dCas9 fused with a histone acetyltransferase enzyme is used to acetylate H3 histones at critical positions in mammalian cells to achieve transcription activation of the targeted genes. Fusion of dCas9 with DNA methyltransferase enables targeted DNA methylation of genes, which leads to suppression of their transcription. In addition, transcription activation and suppression domains can be linked to dCas9 to achieve activation and suppression, respectively, of the targeted genes (Chapter 10; Jung et al., 2018).

7.9.1. Flowering time

Time of flowering and photoperiod sensitivity are two of the important factors that affect geographical distribution of crop plants. Tomato gene *SELFPRUNNING 5G* (*SP5G*) is located in a QTL known to affect flowering time. This gene is homologous to *Arabidopsis FT* gene and is known to suppress flowering so that plants overexpressing *SP5G* flower very late. Inactivation of *SP5G* by gene editing using CRISPR-Cas9 enabled isolation of two mutants that were compact and determinate in habit. These mutants flowered earlier than the parental line under long-day conditions (Table 7.2). Early flowering is desirable in crops like tomato since early lines produce fruits early in the season when there would be greater demand in the market. However, under short-day conditions, the mutants flowered at the same time as the parental line. These findings indicate that gene *SPG5* might be responsible for the daylength sensitivity remaining in tomato germplasm (Soyk et al., 2017).

Table 7.2. Selected examples of gene editing for agronomic traits in crop plants

Crop	Locus/Gene	SDN	Type of mutation	Phenotypic change	Remarks
Maize	*MTL*	TALEN	Loss-of-function	Haploid production	Maternal haploid induction when used as pollinator; MTL encodes a sperm-specific phospholipase
Rice	*CSA*	CRISPR-Cas9	Loss-of-function	Reverse photo-period-sensitive genic male sterility (rPGMS)	Male fertility under short-day conditions as compared to the long-day needed for the first discovered PGMS line
	Tms5	CRISPR-Cas9	Loss-of-function	Temperature-sensitive genic male sterility (TGMS)	TGMS lines comparable to the parental lines in morphology; promising for hybrid seed production
*Setaria viridis**	Homologue of maize gene *ID1*	-	Loss-of-function	Late flowering	Field trials permitted
Tomato	*SP5G*	CRISPR-Cas9	Loss-of-function	Early flowering	Located in a QTL; suppresses flowering; compact and determinate in habit
	SlS	-	-	Inflorescence branching	A range of different Inflorescence branching patterns generated
	SlSP	-	-	Shoot architecture	A range of different shoot architectures produced

*QTL, quantitative trait locus

7.9.2. Parthenocarpy

Parthenocarpy is the development of fruits without fertilization. It is an important trait in many horticultural crops, but it is available in only some crops. In tomato, pollen production is poor under unfavourable environments like moderately low or elevated temperatures. This has adverse effects on fruit development and prevents its year-round cultivation. In addition, seedless fruits are preferred by consumers as well as food industry. Auxin is known to initiate parthenocarpy, and the protein AUXIN-INDUCED 9 (Aux/IAA9 or IAA9) is important for fruit development. Ueta et al. (2017) inactivated gene *SlIAA9* (*Solanum lycopersicum IAA9*) by CRISPR-Cas9-mediated gene editing. The mutation frequency was high (30–90% of plants) in the T_0 generation, and no off-target effects were observed. The mutant plants had seedless fruits and simple leaves in the place of normal compound leaves; the changed leaf morphology did not affect plant growth (Table 7.3). The fruit development began before fertilization, and the phenotypic alterations in the mutants were inherited.

Table 7.3. Targeted editing of genes for quality traits of crop plants

Crop	Gene(s)	Function	SDN system	Mutation	Mutant phenotype	Remarks
*Camelina sativa**	*C3008a, C3008b* and *C3009*	-	CRISPR-Cas9	Loss-of-function	Improved oil yield and quality	Enhanced omega-3 fatty acid content
Maize	*IPK**	Phytic acid biosynthesis	ZFN, CRISPR-Cas9, TALEN	Gene inactivation	Reduced phytic acid content in seeds	Phytate cannot be digested by humans
	*Waxy1**	Granule-bound starch synthase; amylose synthesis	CRISPR-Cas9	Gene inactivation	Starch grains having only amylopectin	-
Potato	*VASCULAR INVERTASE*	TALEN	Converts starch into glucose and fructose during cold storage	Loss-of-function	Reduced browning and acrylamide level in chips and French fries	Lower levels of reducing sugars produced during cold storage of tubers
Soybean	*FAD2-1A* and *FAD2-1B*	Desaturation of oleic acid	TALEN	Loss-of-function	High (80%) oleic acid oil	Both *FAD2-1A* and *FAD2-1B* must be inactivated
Tomato	*SlIAA9*	Participates in fruit development	CRISPR-Cas9	Loss-of-function	Parthenocarpy	Mutation frequency 100% in T_0; produces seedless fruits and simple leaves
	SlAGL6	MADS-box family transcription factor	CRISPR-Cas9	Loss-of-function	Parthenocarpy	Mutant plants produce seedless fruits even under heat stress; fruits size, weight and pollen fertility unaffected
	LeMADS-RIN	MADS-box family transcription factor	CRISPR-Cas9	Loss-of-function	Incomplete ripening of fruits	Mutants have normal sepals and determinate inflorescence

*Declared as nonregulated article by APHIS, USDA, U.S.A.

Another tomato gene, *SlAGL6* (*Solanum lycopersicum AGAMOUS-LIKE 6*), encodes a transcription factor belonging to the MADS-box family, which is involved in fruit development. The cultivar MPK-1 has the spontaneous mutation *Pat-k*; it shows stable parthenocarpy and produces fruits that are nearly seedless (they do contain few seeds). An EMS (ethyl methane sulfonate)-induced parthenocarpic mutant produces high quality seedless fruits. Both the spontaneous and the EMS-induced parthenocarpic mutations have been mapped to the *SlAGL6* locus. Klap et al. (2016) targeted the second exon of *SlAGL6* to produce gene knockouts, which were parthenocarpic so that they can produce fruits even under heat stress environments. This confirmed the earlier finding that this gene is involved in fruit development and its inactivation leads to parthenocarpy (Table 7.3). The fruits produced by the parthenocarpic mutant are comparable to the parental line fruits in size and weight. Further, pollen fertility of the mutants is not reduced, and they reproduce sexually under normal environmental conditions. *Significantly, the parthenocarpic trait in gene-edited line is facultative; this feature is important for seed production of the parthenocarpic cultivars.*

7.9.3. *In vivo* haploid production

Inbreds are critical for heterosis breeding since they serve as perpetual parental lines of hybrid varieties. Conventionally, inbreds are produced by continuous inbreeding, selfing where feasible, combined with selection; this procedure may take over 6-7 years to obtain inbred lines. One quick way to develop inbreds is to isolate haploid plants and double their chromosome number to produce homozygous doubled haploid (DH) plants. Haploids can be produced by anther culture, ovary culture, interspecific hybridization and, at least in maize, the use of an 'inducer' strain for pollination. The haploid production based on inducer strains is much simpler and easier to use than the other routes of haploid production (Singh, 2018). The maize inducer strains are derived from 'Stock 6' that was identified in 1959, and induces 2-5% maternal haploid production. Some inducer lines derived from 'Stock 6' induce up to 13% haploids. Kelliher et al. (2017) used marker assisted fine mapping to identify the QTL for haploid induction. Further, they carried out genome sequencing and RNAi knockdown of the RNA transcripts to infer that the gene *MATRILINEAL* (*MTL*) is involved in maternal haploid induction. The *MTL* gene encodes sperm-specific phospholipase. They observed that a loss-of-function mutation due to a 4-base insertion leads to frameshift mutation and generates the mutant allele *matrilineal* (*mtl*) that induces maternal haploid production.

The above findings were confirmed by inactivating the *MTL* gene by targeting a pair of TALENs at a site close to the site of 4 base insertion in the spontaneous mutant and recovering the *mtl* mutant alleles. The gene-edited mutant lines induced maternal haploid induction when they were used as pollinators (Table 7.2). These findings suggest that inducer strains could be produced in other cereals by inactivating their *MTL* homologues since this gene is highly conserved in cereals.

7.9.4. Environment-sensitive male sterility

Hybrid seed production is greatly facilitated by a male sterility system that permits easy maintenance of the male sterile line and can be readily and effectively restored to produce male fertile hybrids. The cytoplasmic-genetic male sterility system generally exhibits both the above features so that it has been widely used for hybrid seed production. But this system requires three lines, *viz.*, A (male sterile), B (maintainer) and R (restorer) lines, for hybrid seed production. In rice, two-line genetic male sterility systems have been developed and used since they reduce the work and facilitate hybrid seed production. These male sterility systems are called photoperiod-sensitive genic male sterility (PGMS) and temperature-sensitive genic male sterility (TGMS) since the expression of male sterility and fertility is regulated by the prevailing photoperiod and temperature regimes, respectively (Singh, 2018).

The majority of PGMS lines in rice are based on the PGMS line Nongken 58S (NK58S), the first PGMS line discovered in 1973. Therefore, it is desirable to isolate new PGMS (male sterile under long-day conditions) or reverse PGMS (rPGMS; male sterile under short-day conditions) lines for hybrid rice seed production. It is known that mutant genes *pms1*, *pms2* and *pms3* are responsible for PGMS of NK58S. In addition, a mutation in the gene *carbon starved anther* (*CSA*) produced a rPGMS line. The *CSA* gene regulates sugar allocation during pollen development in rice. The transfer of mutant *csa* allele into new rice cultivars would require years of backcross breeding. Therefore, Li, Q. et al. (2016) used CRISPR-Cas9 to edit the first exon of *CSA* gene in two rice varieties. Several lines with loss-of-function mutations in the *CSA* gene and the rPGMS phenotype were isolated, some of which had promising features (Table 7.2).

In China, more than 95% of male sterile lines used for hybrid seed production using the two-line system are TGMS lines. The first TGMS rice line, Annong S-1, was discovered in 1987; this TGMS is due to the mutant gene *tms5*, which encodes a ribonuclease. TGMS is also produced by cosuppression of *Ugp-1* (*UDP-glucose pyrophosphorylase1*), which makes the splicing of *Ugp-1* transcripts temperature-sensitive. Thus, most of the genes involved in TGMS and PGMS affect RNA processing or RNA metabolism. Zhou et al. (2016) used the CRISPR-Cas9 SDN system to inactivate *Tms5* gene. The frequency of on-target editing was high and, in some cases, reached up to 100%. There was no evidence for off-target editing at the nine predicted off-target sites. They were able to isolate TGMS lines free from any foreign DNA from 11 different rice cultivars. These TGMS lines were comparable to the corresponding parental lines in morphological features and appeared to be quite promising for two-line hybrid seed production (Table 7.2). This procedure for TGMS isolation requires 1-2 years as compared to around 5-6 years needed for transferring the TGMS trait to new varieties by backcross scheme.

7.9.5. Yield and yield traits

Grain yield per plant in rice is determined by the component traits panicle number per plant, grain number per panicle, and grain weight. These yield traits are affected by several different genes. The four genes, *IPA1* (*IDEAL PLANT ARCHITECTURE 1*), *Gn1a* (*GRAIN NUMBER 1a*), *GS3* (*Grain Size 3*) and *DEP1* (*DENSE AND ERECT PANICLE 1*), affect important yield traits. The CRISPR-Cas9 SDN system was used to induce loss-of-function mutations in these genes (Table 7.4) of rice cultivar Zhonghua 11. The mutation frequency ranged from 27.5% to 67.5, depending mainly on the targeted gene. The frequency of off-target editing ranged from no off-target activity to 67.5% (Li, M. et al., 2016).

Inactivation of gene *Gn1a* due to a frame-shift mutation resulted in an increase in plant height, size of panicle and flower number per panicle. Inactivation of *DEP1* led to reduced plant height, shorter panicles, and smaller seed size, but enhanced flower number per panicle. The mutants with inactivated *GS3* showed increased grain length and weight, and had long awns. Finally, frame-shift mutation in gene *IPA1* produced dwarf plants with increased tiller number per plant. But plant with a 12 or 21 base deletion that changed the OsmiR156 target site in the *IPA1* mRNA but did not affect the function of the encoded protein showed reduced tiller number, but increased plant height, panicle length and flowers per panicle. Thus, the mutant phenotype resulting from *IPA1* inactivation depended primarily on the site and type of mutation induced by gene editing. In general, the phenotypic effects of loss-of-function mutations in the four genes were essentially comparable to those reported earlier. A notable difference, however, was the decrease in plant height and grain size in *DEP1* mutants induced by gene editing, while these traits are not affected in spontaneous mutations. Another difference relates to the dependence of phenotypic effect produced by a mutant allele of the gene *IPA1* on the site and the type of mutational change generated by gene editing (Li, M. et al., 2016).

Table 7.4. Targeted editing of yield and yield trait genes in rice by CRISPR-Cas9 SDN syatem

Gene	Function	Trait(s) affected	Mutation	Mutant phenotype
Gn1a	Cytokinin dehydrogenase 2	Grain number per panicle	Inactivation	Increased grain number per panicle
DEP1	Gamma-subunit of G protein	Plant height, erect panicle, grain size	Inactivation	Dense and erect panicles
GS3	Gamma-subunit of G protein	Seed size	Inactivation	Larger grains
*IPA1***	Squamosa promoter binding protein	Plant height, tiller number	Inactivation	Dwarf plants; increased tiller number**

* IPA1, IDEAL PLANT ARCHITECTURE 1; Gn1a, GRAIN NUMBER 1a; GS3, Grain Size 3; DEP1, DENSE AND ERECT PANICLE 1

** The mutant phenotype resulting from *IPA1* inactivation depended primarily on the site and type of mutation induced by gene editing.

7.9.6. Quality traits

There are several instances of quality trait modification by gene editing (Table 7.3). The fatty acid composition of oils determines their quality, including shelf-life and nutritional quality, and their suitability for various uses. Soybean oil has high contents of polyunsaturated fatty acids, particularly linoleic acid, which reduce its shelf life as they are prone to oxidation. Therefore, soybean oil is often partially hydrogenated to reduce polyunsaturated fatty acid content, and to increase shelf life and oxidative stability. But hydrogenation of unsaturated fatty acids produces *trans*-fatty acids, which pose a health threat since their molecules stack together to form a solid mass at room temperature. In contrast, the molecules of *cis*-fatty acids do not stack together so that solid mass is not produced. In natural oils, the *cis*-fatty acids predominate. The FAD2 (Fatty acid desaturase 2) enzyme catalyses desaturation of oleic acid, a monounsaturated fatty acid, to linoleic acid (a di-unsaturated fatty acid). Soybean has at least five members in the *FAD2* family of genes; of these, *FAD2-1A* and *FAD2-1B* are expressed mainly in developing seed tissues and are the most likely to determine polyunsaturated fatty acid content of seed oil. Plant breeders have isolated mutants of *FAD2-1A* and *FAD2-1B* genes and used them in breeding programmes to develop high oleic acid soybean varieties (Haun et al., 2014). In addition, RNAi has been used to suppress the expression of these genes, and two transgenic lines with high oleic acid seed oil have been approved for commercial cultivation (http://www.unitedsoybean.org/topics/high-oleic-soy/).

The production and deregulation of RNAi transgenic lines is time-taking and costly, while gene transfer to new varieties requires several years of backcross breeding. The use of SDN to selectively inactivate the targeted gene(s) is an effective technology that requires much less time than either of the above approaches. Therefore, Haun et al. (2014) used TALENs specific for conserved sequences in the *FAD2-1A* and *FAD2-1B* genes and successfully mutated them with high efficiency (about 21%) with no off-target activity. Plants homozygous for the mutant alleles of either *FAD2-1A* or *FAD2-1B* gene did not show significant change in their oleic acid contents, indicating that the two genes are redundant. This contrasts with the previous findings that spontaneous or induced mutations in *FAD2-1A* alone enhanced the level of oleic acid. The gene-edited double mutant (both *FAD2-1A* and *FAD2-1B* genes mutated) plants produced oil containing approximately 80% oleic acid as compared to merely 20% in the parental lines. On the other hand, the linoleic acid content in the double mutants was only 5% in comparison to about 50% in the parental lines. This represents an almost four times increase in oleic acid content and a 10-fold decline in the linoleic acid content in the oils from the double mutant lines as compared to the nonmutant parental varieties. The mutant lines were free from foreign DNA and had only small deletions in the coding regions of the two genes. In 2019, the FDA, USA approved the gene-edited high oleic acid soybean for human consumption.

Fruit ripening is an important trait in tomato as it affects shelf-life and, consequently, storage, transportation, and industrial handling of fruits. A spontaneous mutation, *rin* (*ripening-inhibitor*), inhibits fruit ripening including the respiratory burst and ethylene evolution typical of climacteric type of fruit ripening. This mutation is used to produce hybrid tomatoes that ripen slowly and remain firm over a period of time. The *rin rin* homozygotes show large sepals and indeterminate inflorescence in addition to inhibition of fruit ripening even in response to exogenous ethylene. It is known that the *rin* locus comprises two genes, *viz*., *LeMADS-RIN* and *LeMADS-MC* (MC, macrocalyx) that encode transcription factors of MADS-box family. The gene *LeMADS-RIN* controls fruit ripening, while *LeMADS-MC* governs sepal size and determinate/indeterminate nature of inflorescence. In the spontaneous *rin* mutant, the last exon of *LeMADS-RIN* gene along with probably the regulatory region of *LeMADS-MC* gene is deleted. As a result, fruit ripening, calyx size as well as determinacy of inflorescence are affected in the *rin rin* homozygotes (Vrebalov et al., 2002). Ito et al. (2015) induced knock-out mutations in the *LeMADS-RIN* gene by CRISPR-Cas9, and isolated several mutants, some of which had defective LeMADS-RIN protein. These mutants showed incomplete ripening of fruits, but they had normal sepals and determinate inflorescence; this validates the earlier finding that *LeMADS-RIN* gene governs only fruit ripening.

7.9.7. Virus resistance

Gene editing has been used to generate virus resistance in some plant species, *e.g.*, *Nicotiana benthamiana*, *A. thaliana*. and *Cucumis sativus*. There are two diverse strategies for generating virus resistance: targeting viral genes to disrupt virus multiplication and editing host genes involved in virus infection. The first strategy is used in case of DNA viruses, while the second approach is applicable to RNA viruses. In case of DNA viruses, the CRISPR-Cas9 SDN is designed to edit and, thereby, inactivate an essential gene of the concerned virus to prevent its multiplication and/or systemic spread in the host plant tissues. This strategy has been successfully used in the case of several DNA viruses, such as Beet severe curly top virus, Bean yellow dwarf virus, Tomato yellow leaf curl virus, etc. The viral genes targeted for editing were those encoding coat protein and replication associated protein. In addition, targeting of even intergenic regions of the viral genome was effective. In this strategy, the crop genomes contain the viral gene-specific SDN encoding genes and continuously overexpress them for protection against the concerned viruses. As a result, the virus resistant plants will be subject to regulation as transgenic/GM crops and their consumer acceptance will be uncertain. In addition, continued presence of the SDN in plant cells might induce off-target editing in the plant genomes as well. Thus, this strategy does not seem to be very attractive, although its effectiveness for generating broad spectrum and durable resistance against several DNA viruses has been demonstrated. However, some geminiviruses did evolve variants that were able to evade the CRISPR-Cas9 surveillance of their host plants (Khatodia et al., 2017).

RNA viruses depend on many host proteins, including eukaryotic translation initiation factors eIF4E and eIF(iso)4E, to function and sustain their life cycle. The factors eIF4E and eIF(iso)4E perform redundant functions in the host plants. In cucumber, the genes encoding each of these factors are known to occur in single copy, but in other plant species more than one copy may occur. The protein VPg produced by different potyviruses binds specifically to either eIF4E or eIF(iso)4E; this interaction enables translation of the viral RNA. A mutation in one of these factors that disrupts its interaction with VPg also inhibits the infectivity of those potyviruses that specifically bind to the concerned factor. This specific interaction is the basis for the recessive resistance to potyviruses in several crops. CRISPR-Cas9 SDN system was used to inactivate the gene encoding eIF4E; notably, no off-target editing was detected. Mutant cucumber plants lacking the transgenes were selected and homozygous mutant T_3 plants free from foreign DNA were isolated. The mutant plants were immune to Cucumber Vein Yellowing Virus (Ipomovirus) and resistant to Zucchini Yellow Mosaic Virus and Papaya Ring Spot Mosaic Virus-W (both potyviruses), while the nonmutant parental line was highly susceptible to these viruses. It is important that the virus resistant mutants were comparable to the parental line in morphological features (Chandrasekaran et al., 2016).

Similarly, targeted inactivation of the gene encoding eIF(iso)4E factor in *A. thaliana* by CRISPR-Cas9 produced mutants that were completely resistant to infection by the Turnip Mosaic Virus that infects many vegetable crops and causes substantial losses. In this case also, the mutant plants did not differ from the nonmutant plants for flowering time and dry weight accumulation (Pyott et al., 2016). Cassava yields are markedly reduced by cassava brown streak disease caused by some Potyviruses like Cassava brown streak virus. Cassava has five isoforms of eIF4E; simultaneous inactivation of two of these isoforms (*nCBP*-1 and *nCBP*-2) reduced cassava brown streak disease incidence and severity (Gomez et al., 2018).

7.9.8. Resistance to other biotic stresses

Gene editing has been used to isolate mutants resistant to different biotic stresses like disease causing pathogens (Table 7.5). Pathogens like *Xanthomonas oryzae* pv. *oryzae* (Xoo) activate specific host genes that facilitate host invasion and pathogenesis; these host genes are generally called susceptibility genes. Most Xoo strains either produce the TALE (transcription activator like effector) AvrXa7 or PthXo3. These proteins bind to different effector binding sites in the promoter region of rice gene *Os11N3* (*OsSWEET14*) and activate its expression; this makes sugars available to Xoo and thereby promotes its persistence. Li et al. (2012) designed two pairs of TALENs, one pair specific for each effector binding site, to inactivate either of these sites. They isolated 16 distinct mutations, 6 of which were in homozygous state; most of the mutations were due to small deletions and generally the TATA box (located near the effector binding sites) was not affected. The mutants had normal phenotype, were resistant to the Xoo strains producing the TALE AvrXa7 or PthXo3, which could not induce the expression of gene *Os11N3*. Other rice genes like *Os8N3* (*OsSWEET11*, *OsSWEET13*) also lead to susceptibility to some specific strains of Xoo. The rice *SWEET* genes encode sugar transporters and each of them confers susceptibility to Xoo strains producing specific TALE proteins. Natural variants in the promoters of *OsSWEET13* and *OsSWEET14* confering resistance to the concerned Xoo strains have been identified (Zaka et al., 2019).

Table 7.5. Selected examples of gene editing for abiotic and biotic stress resistance in crop plants

Crop	Locus/Gene	SDN	Type of mutation	Phenotypic change	Remarks
Arabidopsis	*eIF(iso)4E*	CRISPR-Cas9	Inactivation	Virus resistance	Complete resistance to Turnip Mosaic Virus (TuMV)
Cassava	*nCBP-1 and nCBP-2* isoforms of eIF4E	CRISPR-Cas9	Inactivation	Virus resistance	Reduced cassava brown streak disease incidence and severity
Cucumber	*eIF4E*	CRISPR-Cas9	Inactivation	Virus resistance	Resistance broadbased; immune to Cucumber vein yellowing virus; resistance to Zucchini yellow mosaic virus/Papaya ring spot mosaic virus
Flax	*EPSPS*	CRISPR-Cas9	Base substitutions by HDR*	Resistance to herbicide Glyphosate	ssODN used as donor DNA
Maize	*ARGOS8*	CRISPR-Cas9	Promoter replacement or insertion of additional promoter by HDR	Drought resistance	Higher yields under drought stress at flowering time
	NLB18	CRISPR-Cas9	Sequence replacement by HDR	Resistance to Northern leaf blight	Sequence from natural resistance allele used as donor for HDR
Rice	*OsERF922*	CRISPR-Cas9	Inactivation	Resistance to rice blast	Decreased lesion size
	Os11N3	TALEN	Inactivation	Resistance to bacterial leaf blight	Resistance to Xoo strains producing the TALE
	EPSPS	CRISPR-Cas9	Base substitutions by HDR	Resistance to herbicide Glyphosate	Two mutations introduced in the second exon of the gene by intron targeting
Soybean	*Drb2a* and *Drb2b*	CRISPR-Cas9	Inactivation	Drought and salt tolerance	Field trial permitted
Tomato	*SlMlo1*		Inactivation	Powdery mildew resistance	Edited plants are powdery mildew resistant
	J	CRISPR-Cas9	Inactivation	Jointless pedicel	Calyx and pedicel come off the fruits during harvesting
Wheat	*TaMLO-A1, TaMLO-B1, TaMLO-D1*	TALEN	Inactivation	Powdery mildew resistance	Edited plants resistant to powdery mildew; resistance broad-based

*HDR, homology directed repair; ssODN, single-stranded deoxyoligonucleotide; Xoo, *Xanthomonas oryzae* pv. *Oryzae*

In plants, ethylene responsive factors are involved in modulation of responses to several abiotic and biotic stresses. For example, rice ethylene responsive factor encoding genes *OsBIERF1*, *OsBIERF3* and *OsBIERF4* are induced by rice blast pathogen *Magnaporthe oryzae*; in addition, their expression is upregulated by some abiotic stresses like drought, salt, cold, and wounding. Suppression of the expression of ethylene responsive transcription factor encoding gene *OsERF922* by RNAi was shown to improve resistance to rice blast disease. Wang et al. (2016) edited the rice gene *OsERF922* using CRISPR-Cas9 to induce rice blast resistance. Of the 50 T_0 transgenic plants, 21 (42%) plants had indels in the *OsERF922* gene at the targeted site; these mutations were transmitted to the subsequent generations. Foreign DNA-free mutant plants were isolated in T_1 and T_2 generations by selection following segregation. Evaluation of six homozygous mutant T_2 lines showed that these lines had significantly lower number of blast lesions at both seedling and tillering stages. Further, the mutant lines were comparable to the parental line for several agronomic traits, including panicle length, grains per panicle and 1000-seed weight.

The wheat powdery mildew pathogen *Blumeria graminis* f. sp. *tritici* is one of the most damaging plant pathogens and is distributed worldwide. Resistance to this pathogen is not available in wheat germplasm. This fungal pathogen needs Mlo protein function to establish infection. As a result, loss-of-function mutations in *Mlo* genes (*mlo* alleles) are known to generate broad-spectrum resistance to the pathogen. Wheat has three homeologs of *Mlo*, *viz*., *TaMlo-A1*, *TaMlo-B1* and *TaMlo-D1*, which show 98% similarity in their base sequences and 99% similarity in the amino acid sequences of the proteins encoded by them. All the three homeologs of *TaMlo* gene were inactivated by a pair of TALENs specific for the highly conserved region in exon 2. Plants in which all the three *TaMlo* homeologs were inactivated showed broad-spectrum resistance to powdery mildew disease (Wang, Y. et al., 2014a).

Host-pathogen interaction is a very complex process and we have been able to gain only partial insight into it so far. Host resistance to pathogens is generally based on resistance (*R*) genes that encode R proteins. A given R protein typically recognizes a single elicitor produced by the concerned pathogen, and the specificity of this recognition is based on one or few amino acids located in the coiled coil or nucleotide binding domains of the R protein. Therefore, the frequency of resistance breakdown is rather high, and the availability of diverse *R* genes is quite limited. Gene editing can be used to substitute specific amino acids of R proteins to restore their ability to generate resistance. In addition, double amino acid mutations in R proteins tend to enhance cell death during HR. Recombinant DNA and/or gene editing technologies can be used to combine the relevant sequences from different *R* genes conferring resistance to the same pathogen to create *R* genes with novel resistance responses to the pathogens. These technologies can also be used to create synthetic *R* genes capable of specifying resistance to two or more diverse pathogens by combining different pathogen recognition sequences from the concerned *R* genes into a single *R* gene (Andolfo et al., 2016).

In addition, pathogens produce effector proteins that interact with specific host proteins, called effector-target proteins; this interaction leads to suppression of plant resistance to the pathogen. The sequence of a relevant effector-target protein can be suitably altered by gene editing so that the pathogen effector can no longer interact with it. In some cases, change of a single amino acid was able to disrupt this interaction. Further, there is evidence that variation in the effector-target protein sequence can generate quantitative variation in disease reaction of the host (Andolfo et al., 2016).

7.9.9. Herbicide resistance

Resistance of crop varieties to biodegradable herbicides is a very valuable agronomic trait for reducing environmental contamination, persistence of pesticide residues in crop produce and the ease and economy of weed control in crop fields. The herbicide resistant crop varieties in commercial cultivation at present are all transgenic varieties, which are not readily acceptable to all the consumers. The enzyme EPSPS (5-enolpyruvylshikimate-3-phosphate synthase) has a motif that serves as binding site for phosphoenolpyruvate (PEP) and Glyphosate that acts as a competitive inhibitor of the enzyme. It is known that weeds resistant to the biodegradable herbicide Glyphosate have two amino acid substitutions (T102I + P106S) in the enzyme EPSPS, which make the enzyme resistant to glyphosate. Since these mutations are located in the second exon of the gene, Li, J. et al. (2016) designed one sgRNA each for targeting the intron one and intron two flanking this exon; it was ensured that this would not affect the intron sequences involved in splicing. The donor DNA consisted of the genomic sequence located between the two target sites and flanked by the sequences targeted by the two sgRNAs. In addition, the sequence representing the second exon contained the two mutations that make the enzyme EPSPS resistant to Glyphosate. The gene constructs for the two sgRNAs, Cas9 and the donor DNA were introduced into rice cells, and plantlets were regenerated. It is expected that the two sgRNAs would sponsor simultaneous cleavage of the genome and the donor DNA at the same pair of target sites, and this may lead to replacement of the genomic sequence including the exon 2 by the homologous sequence from the donor DNA by NHEJ-mediated DSB repair (Fig. 7.8). This approach of sequence replacement has been termed as intron targeting. Significantly, the two mutations contained in the donor DNA will be introduced into the second exon of the *EPSPS* gene and this is expected to confer Glyphosate resistance to the mutant plants. In eight of the 390 T_0 plants, the exon 2 of the *EPSPS* gene was replaced by that from donor DNA. There was no evidence of editing at putative off-target sites of the mutants. However, deletions or inversions of the genomic fragment, inversion of the donor fragment, etc. were also observed in the T_0 plants. The mutant rice plants expressing the sequence-replaced *EPSPS* gene were resistant to Glyphosate and this trait was heritable. Thus, intron targeting can serve as an alternative approach to HDR-mediated sequence replacements (Li, J. et al., 2016).

The conversion of wild type *EPSPS* gene has been achieved in flax by using CRISPR-Cas9 in combination with a single-stranded oligodeoxynucleotide (ssODN) as donor DNA (Fig. 7.9). The DSB was induced close to and on the downstream side of the sequence to be altered and the ssODN was homologous to the gene sequence surrounding the DSB, except for the desired alteration. About 0.09 and 0.23%, of T_0 plants regenerated from transformed protoplasts showed the desired sequence change in *EPSPS* gene, while the frequency of plants with indels in the gene was rather high (over 19%; Sauer et al., 2016).

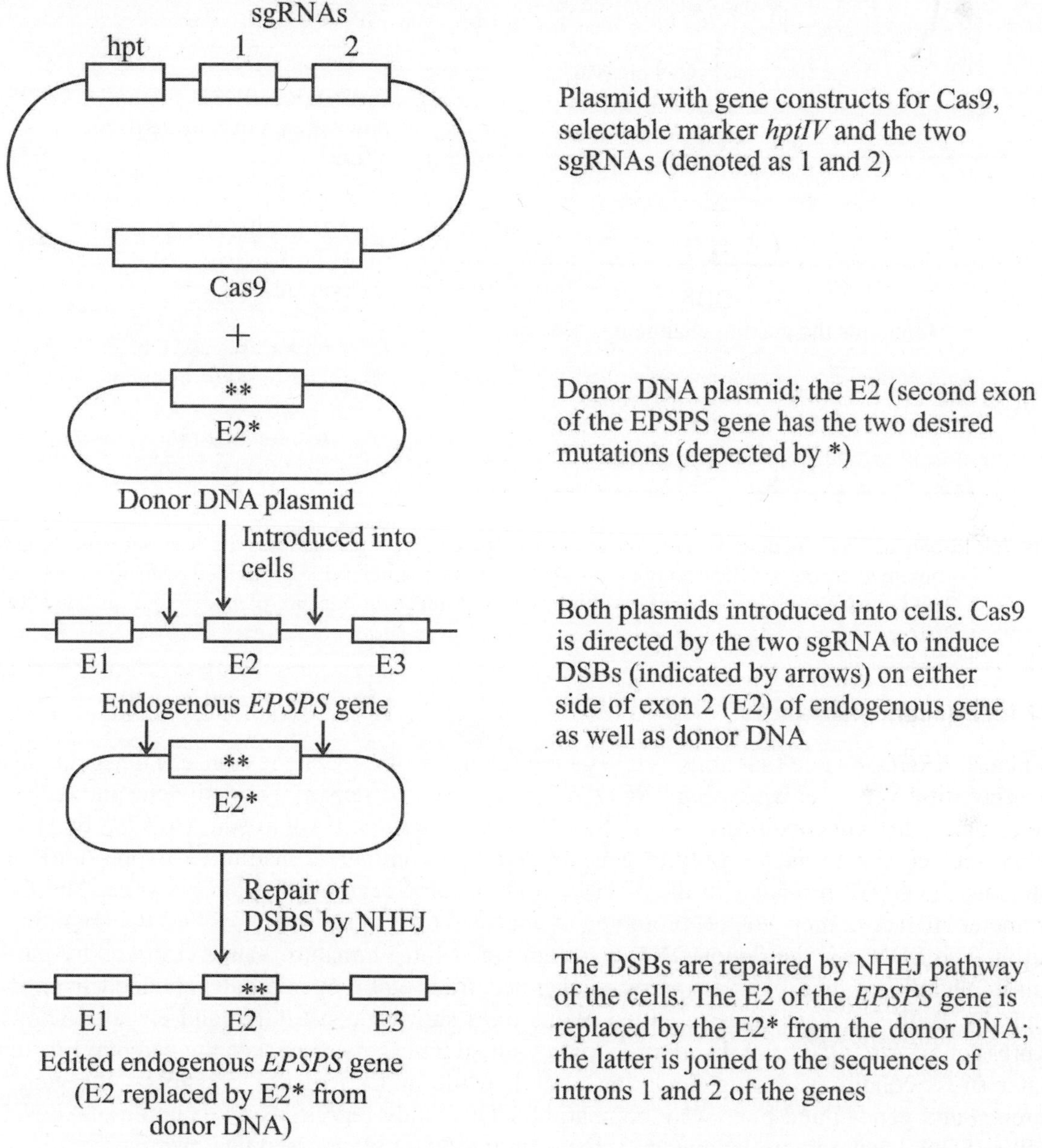

Fig. 7.8. A schematic representation of intron targetting to replace the second exon of the endogenous *EPSPS* gene by the mutant E2* exon from donor DNA to generate Glyphosate resistant rice plants (based on Li et al., 2016).

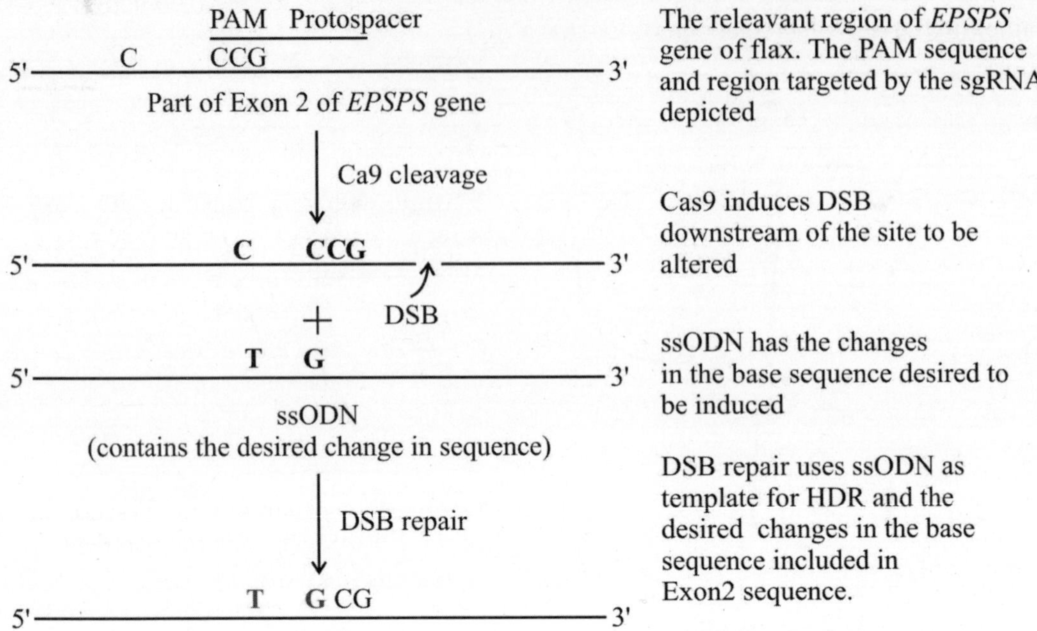

Fig. 7.9. Repair of Cas9 induced DSB in the Exon 2 of flax *EPSPS* gene copies the base sequence changes contained in the ssODN (single-stranded oligodeoxynucleotide) provided with the Cas9 SDN. The edited *EPSPS* genes confers glyphosate resistance in the flax plants (based on Sauer et al., 2016).

7.9.10. Drought resistance

In maize, *ARGOS8* gene functions as a negative regulator of ethylene response. Transgenic maize lines constitutively overexpressing *ARGOS8* display reduced sensitivity to ethylene and enhanced yields under drought conditions. Shi et al. (2017) used CRISPR-Cas9 to edit *ARGOS8* by HDR to either replace the promoter of this gene by *GOS2* promoter (constitutive expression) or to integrate the *GOS2* promoter in the 5' UTR (untranslated region) of *ARGOS8* gene. The *GOS2* promoter in fact comprised the promoter of maize *GOS2* gene and its 5' UTR containing an intron. The SDN and the donor DNA were delivered into immature maize embryos by particle gun bombardment, and plantlets were regenerated from embryogenic calli obtained from these embryos. Both the types of gene-edited maize lines showed constitutive and elevated levels of expression of *ARGOS8* gene. In addition, they yielded 5 bushels more than the parental line under water stress conditions during flowering period, while under irrigated conditions the yields of parental and gene-edited lines were comparable. This study represents an example of SDN-2 as well as SDN-3 technologies applied to achieve an important plant breeding objective.

7.9.11. Nutrition improvement

The nutritional value of crop produce can be enhanced by improving their nutritional status and/or minimizing the contents of antinutritional compounds. During cold storage of potato tubers, a fraction of the starch is converted into glucose and fructose by the enzyme vascular invertase. When cold-stored potato tubers are used to make chips or French fries at high temperatures, the reducing sugars become converted into a dark brown pigment and acrylamide, a strong carcinogen, is also produced. Clasen et al. (2016) used TALEN to edit and inactivate the *VASCULAR INVERTASE* gene. The tubers from the gene-edited potato line produced much lower levels of reducing sugars during cold storage; as a result, the levels of browning and acrylamide were also much lower in the chips and French fries produced from these tubers (Table 7.3).

Phytic acid occurs in seeds of several crops, including maize and pearl millet. Phytic acid serves as storage for phosphorus and is used by the seeds during germination as a source for the same. However, phytate cannot be digested by humans and, in addition, it can sequester nutrients like iron. The maize gene *IPK* encodes an enzyme involved in biosynthesis of phytic acid. The gene *IPK1* was edited using ZFN to generate a maize line that has reduced levels of phytic acid in seeds; this line was undergoing field evaluation (Table 7.3). This gene has also been inactivated using CRISPR-Cas9 and TALEN SDNs (Liang et al., 2014).

7.9.12. Metabolic engineering

In metabolic engineering, the metabolic pathways of the concerned organisms are modified by either silencing or overexpressing endogenous genes or introducing novel transgenes that affect enzymatic, transport and/or regulatory functions in the cells. Some studies have attempted to edit genes that participate in biosynthetic pathways in an effort to modify the pathways and alter the range and/or the types of biochemicals generated by them. For example, Alagoz et al. (2016) attempted to modify the biosynthetic pathway for benzylisoquinoline alkaloid of opium poppy (*Papaver somniferum*) by CRISPR-Cas9 SDN-directed editing of the gene encoding 3'-hydroxy-N-mehtylcoclaurine 4'-O-methyltransferase enzyme. This enzyme converts (S)-3'-hydroxy-N-methylcoclaurine to (S)-reticuline, which is the central intermediate metabolite used for biosynthesis of compounds like morphine, codeine, papaverine etc. The gene-edited plants showed much reduced levels of the alkaloids produced from (S)-reticuline intermediate. In addition, these plants produced a novel uncharacterized alkaloid.

7.10. MULTIPLEXED GENOME EDITING

Simultaneous editing of more than one gene in the same cell is called multiplexed gene editing. Multiplexing can be achieved with ZFNs and TALENs, but it is much simpler and easier to implement with CRISPR-cas9. With this SDN, only the seed or spacer region of the sgRNA needs to be designed for each new target site, while the remaining sequences of sgRNAs and the Cas9 nuclease encoding sequence remain unchanged. Each of the multiple sgRNAs may target a separate gene or all of them may target different sites in a single gene. In the latter case, larger sequences would be deleted from the targeted gene ensuring complete gene inactivation and the frequency of mutations would tend to be higher than when a single sgRNA is used. The multiplexing can be based on a single Cas9 gene construct combined with several individual sgRNA expression cassettes each having its own Pol III promoter and other regulatory sequences (Table 7.6). This scheme limits the number of different sgRNA expression cassettes that can be accommodated in a single plasmid, which limits the number of different genes that can be edited together (Schiml and Puchta, 2016).

Table 7.6. Different strategies for multiplexed gene editing with CRISPR-Cas9 SDN system (based on Cermak et al., 2017)

SgRNA gene construct*	Cas9 gene construct	Remarks
1. *Monocystronic*		
AtU6-sgRNA1-Ter-- AtU6-sgRNA2-Ter-- AtU6-sgRNA3-Ter--AtU6-sgRNA4-ter	Cas9	Each sgRNA is transcribed as separate RNA transcript; only a limited number of sgRNA gene constructs can be included in a single plasmid
2. *Polycystronic*		All sgRNAs are transcribed as a single polycistronic transcript; relatively larger number of sgRNAs can be included in a single plasmid
Pro-Csy4- sgRNA1- Csy4- sgRNA2- Csy4- sgRNA3- Csy4- sgRNA4-Ter	Cas9-Csy4 endoribonuclease fusion protein	Csy4 processes the polycistronic RNA transcript to liberate individual sgRNAs; most efficient in gene editing
Pro-pre-tRNA^{Gly}- sgRNA1- pre-tRNA^{Gly}- sgRNA2- pre-tRNA^{Gly}- sgRNA3- pre-tRNA^{Gly}- sgRNA4-Ter	Cas9	Cellular tRNA processing enzymes process the polycistronic RNA transcript to liberate individual sgRNAs; less efficient in gene editing than the Csy4-based strategy
Pro-Rib- sgRNA1- Rib- sgRNA2- Rib- sgRNA3- Rib- sgRNA4-Ribozyme-Ter	Cas9	The ribozyme present in the polycistronic RNA transcript processes the transcript to liberate individual sgRNAs; least efficient of the three strategies in gene editing

*AtU6, *Arabidopsis thaliana,* U6 promoter; Ter, Transcription terminator; Pro, CmYLCV (Cestrum yellow leaf curling virus) promoter; Csy4, 20 bp Csy4 hairpin; pre-tRNA^{Gly}, 77 bp pre-tRNA^{Gly} gene; Rib, 15 bp ribozyme cleavage site; Ribozyme, 58 bp synthetic ribozyme sequence.

Alternatively, the multiple sgRNAs may be placed in a single polycistronic gene construct and the individual sgRNAs are liberated by processing of the polycistronic RNA transcript by ribonuclease action (Table 7.6). The enzymes used for this processing include CRISPR-associated RNA endoribonuclease Csy4, tRNA processing enzymes present in the plant cells, and ribozymes. Xie et al. (2015) developed an ingenious scheme for a polycistronic gene construct capable of encoding several different sgRNAs. In this gene construct, a tRNA (*e.g.*, 77 bp pre-tRNA^{Gly}) gene is placed between the sgRNA sequences so that processing of the polycistronic RNA transcript by endogenous ribonucleases involved in tRNA processing generates the individual sgRNAs. Xie and co-workers were able to successfully edit up to eight different genes in rice protoplasts using this strategy.

Similarly, Cermak et al. (2017) created polycistronic sgRNA gene construct by placing a 22 bp Csy4 hairpin encoding sequence between neighbouring sgRNA sequences. The Csy4 endonuclease sequence is expressed as Cas9 fusion protein. The polycistronic transcript is processed by the Csy4 endonuclease to release individual sgRNAs. These authors successfully used polycistronic constructs expressing 12 sgRNAs to target six different genes. In the system using ribozyme processing, the neighbouring sgRNAs are separated by a 15 bp ribozyme cleavage site and the 58 bp synthetic ribozyme encoding sequence is placed toward the 3’ end of the polycistronic construct. The synthetic ribozyme processes the polycistronic construct to liberate the individual sgRNAs (Gao and Zhao, 2014). Cermak et al. (2017) compared the three multiplexing systems and concluded that the Csy4 strategy was the most efficient in multiplexed editing, followed by the tRNA scheme, while the ribozyme method was the least efficient of the three procedures.

TRV genome has two single-stranded positive-sense RNA segments, *viz*., RNA1 and RNA2. A positive-sense strand RNA genome can function as mRNA and it can be directly translated. It has been used as an efficient vector for virus-induced gene silencing and for delivering ZFNs and sgRNAs into plant cells. Ali et al. (2015) used TRV to deliver sgRNAs by agroinfiltration into leaf cells of a *N. benthamiana* transgenic line that overexpressed plant codon optimized Cas9. The sgRNA gene construct was integrated into a vector derived from TRV RNA2. *Agrobacterium* cultures carrying the TRV RNA1 were combined with cultures having the TRV RNA2 containing different sgRNAs for agroinfiltration of leaf tissue. In the leaf cells, TRV was reconstituted from the introduced RNA1 and RNA2 segments, and the reconstituted virus spread systemically throughout the plants including the germline cells. The RNA2 segment of this TRV expressed the sgRNA, which associated with the Cas9 produced in the cells of this transgenic tobacco line and carried out efficient gene editing. By combining the *Agorbacterium* cultures carrying RNA2 with those carrying RNA1 containing different sgRNAs, they were able to simultaneously edit two different genes (*PCNA* and *PDS* genes). This SDN delivery system is quite attractive since it obviates the need for tissue culture and plantlet regeneration protocols for gene editing.

7.11. ISOLATION OF TRANSGENE-FREE MUTANT LINES

When the SDN or the base editor is introduced into plant cells as DNA, especially by *Agrobacterium*-mediated gene transfer, the transgene constructs along with some plasmid DNA would become integrated into the plant genome. It is important to remove the foreign DNA from the recovered mutant lines to make them transgene-free in order to minimize biosafety concerns. Mutations are generally detected in T_0 generation, *i.e*., the generation obtained by regeneration from the transformed cells. Many of the T_0 plants would be chimera and would have both mutated and nonmutated sectors; only the mutations occurring in germ line will be transmitted to T_1 generation. The mutations are generally in heterozygous state as the concerned gene is most likely to be edited in only one of the two homologous chromosomes. But a small fraction of mutants may be homozygous for the mutant allele. However, the homozygous/heterozygous state of the mutant allele would not affect elimination of the transgenes so long as transgenes and mutations are located in different chromosomes, and only one of the two homologous chromosomes carries the transgenes. Then the transgenes will segregate independently of the mutation so that a proportion of T_1 plants (progeny of T_0 plants generally obtained by selfing) will carry the mutation in homozygous/heterozygous state but will lack the transgenes. Transgene-free mutant T_1 plants are identified and subjected to selfing followed by selection, if necessary, to establish homozygous foreign DNA-free mutant lines.

It is important to use a sensitive and effective screen for the transgenes during this process. In case of *Agrobacterium*-mediated transformation, the foreign DNA will include the gene encoding the SDN, the selectable marker and the T-DNA sequences. Haun et al. (2014) used three primer pairs to amplify the TALEN SDN (also applicable to ZFN SDNs) encoding genes, the selectable marker, and the T-DNA right border region. In addition, they used the fourth primer pair to amplify the endogenous alcohol dehydrogenase gene to serve as an internal control. It may be expected that in the case of CRISPR-Cas9 system, a pair of primers that amplify the sgRNA encoding DNA would also be included. But Pyott et al. (2016) simply used primer pairs specific for the sequence encoding Cas9 nuclease and for the endogenous house-keeping gene *EF1a* (as internal control) for detecting the presence of transgenes.

The above strategy for detecting the presence of foreign DNA would be useful when the SDN gene constructs are delivered via *Agrobacterium.* This would not serve the purpose if they were delivered as transient expression plasmids by particle gun bombardment or some other delivery route since the plasmid DNA would be cleaved by cellular enzymes in random fragments that may be integrated into the genome. Since the identity of these fragments cannot be predicted, it is not possible to design primers for their amplification and detection. In such cases, the whole genome sequencing of homozygous mutant lines in T_2 or a later generation seems to be the only approach for detecting the presence of foreign DNA sequences in their genomes. However, SDN delivery as proteins, RNA or RNP eliminates the risk of foreign DNA being present in the genomes of the mutant lines. Alternatively, an expression system based on RNA virus vectors may be used to deliver the SDNs. The TRV-based system has been used to deliver ZFNs as well as CRISPR-Cas9 SDNs into plant cells (Puchta, 2017, Jung et al., 2018).

7.12. PROGRAMMABLE GENE REGULATION IN PLANTS

Programmable transcription factors (TFs) are produced by fusing a transcription activation or repression domain with a TALE protein or to dCas9. Fusion of an activator like VP64 yields a transcription activator, while a transcription repressor is created by fusing a repressor, such as SRDX. Programmable TFs have been used for regulation of transcription of specified genes in plants. A programmable TALE-repressor constructed by fusion of TALE protein with SRDX was able to repress transcription of the targeted gene of *Arabidopsis* by over two-fold. Transcription was also blocked by binding of only the TALE protein to the sense strand of the gene. This repression effect was likely the result of physical interference with the transcription process by the TALE protein. Similarly, targeting of the dCas9-VP64 fusion to the methylated promoter of imprinted *AtFIS2* gene in the rosette leaves of *Arabidopsis* resulted in 400-fold higher expression level of the gene. In addition, a CRISPR-dCas9-based toolkit for simultaneous modulation of the expression of multiple genes has been developed. However, there has been only limited work on gene activation in plants using programmable TFs and further refinements in the tools are needed for promoting the application of transcriptional regulation researches in plants (Malzahn et al., 2017).

7.13. REGULATION OF GENE-EDITED CROP VARIETIES

It had been hoped that the products from gene editing would not be treated as genetically modified crops and would, therefore, not be subjected to the same regulatory regime as the GM crop varieties. This hope was based on the fact that gene-edited products are transgene-free and they are modified by deletion, addition or substitution of few base pairs as are the mutants generated by conventional mutation breeding (Lusser et al., 2012; Podevin et al., 2013). More particularly, the mutant lines derived by oligonucleotide directed mutagenesis are indistinguishable from those derived by conventional mutagenesis. In addition, the gene editing procedures generally cause fewer unintended off-target effects than do conventional mutagen treatments. The regulatory procedures are rather involved and intricate, and they require the generation of a dossier on various safety assays that takes considerable time (~ 5-6 years) and is very expensive (approximate cost, US $ 30-50 million). These requirements virtually exclude institutions other than multinational companies from developing and commercializing transgenic crop varieties. The SDN-3 technology can alleviate many of the safety concerns from transgenic crops by ensuring transgene integration into 'safe' genomic sites that enable stable and unhindered transgene expression, and eliminate the risk of unintended activation of undesirable or inactivation of desirable endogenous genes (Bortesi and Fischer, 2015).

In USA, the USDA (United States Department of Agriculture) follows the principle "... to allow innovation when there is no risk present". In its view, the "... methods, such as genome editing, expand traditional plant breeding tools because they can introduce new plant traits more quickly and precisely, potentially saving years or even decades in bringing needed new varieties to farmers" (https://gizmodo.com/c/genetics). Keeping with this, the regulation of plant varieties is based on the features of the varieties under consideration and not on the basis of the procedure followed for their development. Therefore, the varieties developed by gene editing are not considered as GM crops and, consequently, as regulated articles provided they are free from the transgenes and the plasmid sequences that were used for gene editing and the edited gene(s) are not expected to turn them into weeds and make then more persistent. However, it is expected that as and when such a variety is commercialised, necessary approvals, if any, will be obtained from the relevant authorities. In addition, appropriate measures will be taken in case the farming community is concerned about gene flow from fields growing gene-edited varieties to the surrounding fields having conventional varieties of the same crops (https://www.aphis.usda.gov/aphis/ourfocus/biotechnology/am-i-regulated).

The European Union, on the other hand, follows a regulatory framework for GM crop varieties that focuses on the process of their production and not on the features of the varieties themselves. Therefore, a variety produced by gene editing will be regulated differently from another variety that has identical features but was developed by conventional mutation breeding. According to the current practice, a gene-edited variety will be treated as GM crop and will be regulated in the same way as transgenic varieties. There have been representations seeking reversal of this approach and treat gene editing as an extension of the conventional plant breeding techniques, and to exempt the gene-edited varieties from the regulatory burden. However, the European Court of Justice has ruled that the gene-edited varieties should be regulated in the same way as GM crop varieties. The rigorous application of GM regulatory procedures to gene-edited varieties will dampen research and development in the field and confine them to few very big multinational companies. It has been suggested that if "the standards applied to GM crops were applied with equal rigour to the results of classical plant breeding, few new cultivars would survive the process" (Georges and Ray, 2017). The regulation of gene-edited crop varieties is discussed in a greater detail in Section 12.6.

7.14. ENDOGENOUS GENE SUPPRESSION: RNAi vs. GENE EDITING

A gene of interest can be inactivated by RNAi as well as gene editing. The gene constructs used for RNAi-mediated suppression of the targeted genes, like any other transgene, are subject to gene silencing and, as a result, variable levels of gene suppression. Therefore, many transgenic events must be developed and evaluated to identify a line with stable suppression of the targeted gene. In addition, the RNAi-gene suppression lines are, of necessity, transgenic lines, and they must be subjected to costly and time-consuming biosafety evaluation and deregulation procedures before commercial cultivation. In contrast, the lines derived by gene editing show stable gene inactivation, and they are transgene-free (Haun et al. 2014) since the transgenes used for gene editing are eliminated by a suitable breeding scheme like selfing and selection. For these and some other reasons (Section 10.3.4), gene editing is far preferable to RNAi for suppression of endogenous genes.

7.15. GENE EDITING vs. TRANSGENIC TECHNOLOGY

The gene editing technology differs from the transgenic technology with respect to several important features (Table 7.7). Although both the technologies aim at genetic modification of crop plants to make them more useful to humans, the tools and techniques as well as the outcomes of the two are quite different. For example, new DNA sequences, including transgenes and plasmid sequences are introduced into plant cells in both the technologies but they are ultimately eliminated from the mutant lines in the case of gene-edited plants, while they are retained, with the possible exception of marker genes, in the case of transgenic plants. The site of transgene integration is unpredictable due to random integration, but the site of gene editing is pre-decided and highly specific. As a result, there is risk of unintended activation of undesirable endogenous genes or inactivation of desirable genes due to the transgene integration at random genomic sites. In contrast, this risk is non-existent in the case of gene editing and any off-target effects can be detected and eliminated by a suitable breeding procedure like backcrossing. At the same time, the transgene expression is prone to position effects and transgene silencing, which necessitate elaborate evaluation of many independent transgenic lines for selecting lines with the desired level of stable expression. This problem can be avoided by the SDN-3 technology by transgene integration into 'safe' harbour sites. The approval of transgenic crops for commercialization in EU is virtually impossible, while some gene-edited crop varieties may be deregulated in at least some member countries (Section 12.6). Finally, the SDN technology is far cheaper than the transgenic technology: it is estimated that a single transformation event could cost up to 250,000 US dollars, while the cost of genome editing could be as low as US $ 30 (Smyth, 2017).

The uncertainties associated with transgene integration, and the effects of transgenic sequences and their products on human/animal health and the environment raise many safety concerns in the minds of the public at large. This has forced the implementation of a complex, time-taking and costly regulatory procedure that has tended to limit development and commercialization of transgenic crop varieties to giant multinational companies, and has practically driven out smaller companies and public institutions from the scene. It is sometimes felt that the negative inclination towards GM crops may be partly due to the overemphasis of vegetable and fruit breeders on shipping quality of the produce rather than their flavour and palatability. This has led to the development and commercial cultivation of varieties that have been dubbed as 'the "cardboard" strawberry, the flavourless plum, the bouncing tomato' and have been the objects of mockery by consumers. This negative feeling was considerably strengthened by the unfair deal offered to farmers by companies in the transgenic seed business. This negative public perception is likely to adversely affect the consumer response to gene-edited crops as well. Therefore, it is important that the scientist community makes available fact-based correct information about various aspects, more particularly the safety aspects of gene-edited crop varieties to enable consumers make their choice based on correct information rather than misinformation (Georges and Ray, 2017).

Table 7.7. A comparison between transgenic and gene editing technologies

Feature	Transgenic technology	Gene editing technology
Basis of genetic modification	Transgene integration in genome and its expression	Mutations induced by repair of DSBs* induced by SDNs and base editing leading to base substitutions
Type of genetic modification	Addition of new gene functions or suppression of endogenous gene expression	Endogenous gene inactivation or restoration of function by sequence replacement by HDR* or base editing
Transgene source	Transgenes from any source	Typically, donor DNA has sequences from the same species or related species
Gene functions used	The desired transgenes, selectable markers	The selected SDNs*, selectable markers, sgRNAs and, where needed, donor DNA sequences; base deaminases for base editing
Gene functions provided by	Gene constructs delivered as DNA	Gene constructs delivered as DNA or by the concerned RNAs, RNPs* or proteins delivered into cells plus the cellular DNA repair machinery
Fate of plasmid DNA	Present in the transgenic varieties used for commercial cultivation	Varieties used for commercial cultivation are 'clean' as all foreign DNA is removed by segregation and selection
Integration of foreign DNA	Random; at any site in the genome by nonhomologous end joining (NHEJ)	At the specified or targeted genomic site by HDR or even NHEJ
Stability of transgene expression	Transgene expression level may be affected by position effects and/or gene silencing; elaborate evaluation and selection for stable expression required	Since a 'safe' genomic site is chosen, there in no risk of position effects and/or gene silencing; gene expression is stable and not subject to position effects
Risk of unintended activation of undesirable or inactivation of desirable endogenous genes	Present; elaborate evaluation required to rule out such effects	Minimal since only the target gene is affected; the off-target editing can be detected simply by whole genome sequencing of the isolated mutants and eliminated by backcross breeding
GM status of commercial varieties	GM crops	Products of SDN-1 and SDN-2 are similar to varieties developed by conventional mutagenesis; SDN-3 products would be similar to GM crops since a transgene integration will be involved
Biosafety concerns and regulatory process	Several biosafety concerns; data from elaborate safety assays required by the regulatory authorities	USA considers SDN-1 and some SDN-2 varieties as non-regulated, while European Union regards them as GM*

*DSBs, double-stranded breaks; GM, genetically modified; HDR, homology-directed repair; RNPs, ribonucleoproteins; SDNs, site-directed nucleases; SDN-1, SDN-2 and SDN-3, site-directed nuclease 1, 2 and 3, respectively

7.16. MERITS OF GENE EDITING

1. SDN-1 technology enables targeted and stable inactivation of specific endogenous genes. This approach is preferable to the other methods of gene suppression since gene-edited lines are free from transgenes and other foreign DNA sequences.

2. The SDN-2 technology enables replacement of few to several bases of the targeted genes with the desired sequences that either correct a mutant gene or generate a new allele of the gene. This approach allows specified mutations in the targeted genes, and the products will be free from foreign DNA if the donor DNA sequences were based on sequences from the same crop species.
3. The SDN-3 technology enables insertion of transgenes in such genomic regions that support stable expression of transgenes without any risk of unintended activation/inactivation of endogenous genes.
4. The SDN-3 technology also enables trait stacking, which greatly facilitates the use of stacked genes in breeding programmes since all these genes behave as a single locus during segregation (Section 7.5.5., 7.5.6).
5. The varieties developed by SDN-1 and even SDN-2 technologies are free from transgenes and other foreign DNA sequences. Therefore, these varieties are not considered as regulated articles in USA, Argentina and Brazil. It is likely that some other countries might also adopt the same policy for these varieties.
6. Even in case of transgenic varieties, targeted integration of the transgenes in selected genomic regions would greatly reduce the safety evaluation requirements since it would eliminate the risks of gene silencing as well as unintended activation/inactivation of endogenous genes. It would also minimize the need for evaluation of the transgenic lines for the level and stability of transgene expression.

7.17. LIMITATIONS OF GENE EDITING

1. The SDNs are delivered into plant cells/protoplasts and complete plantlets are regenerated from the transformed cells. But in many crop species, dependable protocols for plantlet regeneration are not available. This difficulty can be overcome by using SDN delivery techniques like floral dip in *Agrobacterium* culture, particle gun delivery into shoot tips and agroinfiltration of leaf tissues (in conjunction with the use of TRV vectors) since they do not require plantlet regeneration from cells/protoplasts.
2. SDNs do induce some off-target effects albeit at low or even negligible frequencies. Whenever such effects are produced, additional time, effort and cost will be imposed by the breeding schemes used for eliminating these effects from the mutant lines.
3. The varieties developed by SDN technologies are treated as GM crop varieties in European Union and some other countries (Section 12.6). This imposes considerable additional investment of time, effort and monetary resources for generating the necessary dossier and obtaining the regulatory approvals. This would tend to discourage investments in research and development activities in this potent area and may slow down the progress in crop improvement, which could have adverse impact on food security.

7.18. ACHIEVEMENTS

The gene/genome editing tools and techniques were developed for animal system, especially, cultured human cells. They have been successfully extended to plant species, including many crop species. The SDN technology has been used for targeted gene inactivation, sequence replacement to restore functions of mutated genes and to create new alleles of genes, integration of transgenes at specific genomic sites as well as trait stacking. The mode and the routes of SDN delivery, the types of genetic changes induced and the frequency of off-target editing have been investigated. The SDN technology has been used to analyse and/or validate the functions of genes and regulatory sequences and to induce desired mutations leading to useful phenotypes. Induction of deletion of large targeted genomic segments has been used to delete entire genes, *e.g.*, a gene encoding microRNA in *Arabidopsis*. However, the targeted genes were generally those whose functions are already known, and the mutant alleles generated by editing are generally known to occur in the germplasm/breeding populations. But continued research may enable induction of novel modifications in genes.

Gene editing has been used to develop mutant lines with one or more desirable features in several crops and many of the edited lines are already under field trials. The bulk of these lines have been developed by SDN-1 technology and only few of them have been derived by SDN-2 technology. Some examples of gene-edited lines of mainly food crops permitted by APHIS (Animal and Plant Health Inspection Service), USDA for field trials (https://www.aphis.usda.gov/aphis/ourfocus/biotechnology/am-i-regulated) are briefly described here (Table 7.8). A tomato line has been developed that has a deletion in the gene *J* induced by CRISPR-Cas9. The fruits of this mutant line do not retain the calyx and a part of the pedicel when harvested, which saves considerable labour and time needed for removing these unwanted parts after harvesting the fruits of parental and other nonmutant lines. Although a natural recessive allele *j1* of this gene has been introgressed from the wild relative *Solanum cheesmaniae*, but the homozygous *j1j1* lines produce misshaped fruits. The gene-edited line is free from the undesired phenotypes associated with the natural mutant allele. A mutant line of *Camelina sativa* has three edited genes, namely, *C3008a*, *C3008b* and *C3009* (copies of the *FAD2* gene). The mutant line has enhanced oil yield and improved oil quality (enhanced omega-3 fatty acid content).

Table 7.8. Some gene-edited lines declared as nonregulated and cleared for field trials by APHIS, USDA, U.S.A.

Crop	Edited gene(s)	Trait modified
Camelina sativa	*C3008a, C3008b* and *C3009*	Oil yield and quality
Maize	*NLB18*	Resistance to the Northern leaf blight
	Waxy1	Starch grains exclusively containing amylopectin
Setaria viridis	Homologue of maize gene *ID1*	Late flowering
Soybean	*Drb2a* and *Drb2b*	Drought and salt tolerance
Tomato	*J*	Jointless pedicel
Wheat	*TaMLO-A1, TaMLO-B1, TaMLO-D1*	Resistance to powdery mildew
Button mushroom*	*PPOI* (polyphenol oxidase)	Nonbrowning phenotype

*The first gene edited plant to be declared as nonregulated article by APHIS, USDA, U.S.A.

In maize, the gene *NLB18* (encodes a pattern recognition receptor kinase) is involved in resistance to the Northern leaf blight (NLB) caused by *Setosphaeria turcica.* The susceptibility to NLB is specified by an allele of this gene, while another allele of the gene specifies resistance.

CRISPR-Cas9 was used to induce DSBs and stimulate HDR using as template the appropriate sequence from the resistance specifying allele of *NLB18*. As a result, the NLB-susceptibility allele was converted into NLB-resistance allele. CRISPR-Cas9 SDN was used to inactivate genes *Drb2a* (double-stranded RNA-binding protein 2a) and *Drb2b* to produce drought and salt tolerant soybean mutant lines. The maize gene *Waxy1* encodes granule-bound starch synthase, the enzyme that catalyses amylose synthesis. This gene was inactivated by using CRISPR-Cas9 and the mutant maize lines had starch grains containing exclusively amylopectin. Another gene-edited maize line exempted from regulation has lower levels of phytate, which is an antinutritional compound.

Late flowering mutants of green bristle grass (*Setaria viridis*) were produced by inactivating the gene that is homologue of maize gene *ID1*. Some other examples of gene-edited crop lines are *FAD2*-inactivated soybean (oil quality), *FAD3*-inactivated soybean (oil quality), *MLO* gene-inactivated wheat (powdery mildew resistant line), wheat with enhanced nutritional quality, etc. But the first gene-edited plant to be declared as nonregulated article by APHIS, USDA was a nonbrowning mutant of button mushroom (*Agaricus bisporus*); this mutant was produced by CRISPR-Cas9-directed inactivation of the gene encoding polyphenol oxidase responsible for the browning trait. This mutant mushroom is being marketed in U.S.A.

7.19. FUTURE PERSPECTIVES

One of the main issues that needs to be resolved concerns the creation of plants expressing the appropriate SDNs for gene editing. The SDNs are typically delivered into cultured plant cells/protoplasts, from which plantlets are regenerated. However, dependable protocols for plantlet regeneration from cultured cells and/or protoplasts are not available for many crop plants. Therefore, it is imperative to develop techniques for regeneration from cells/protoplasts and for high frequency transformation of recalcitrant plants like monocot species. The SDNs may be delivered into organized shoot meristems to avoid the need for plantlet regeneration from cell cultures. Further, the use of genes *Babyboom* and *Wuschel* has been reported to improve transformation in poorly responsive monocots (Osakabe and Osakabe, 2014; Malzahn et al., 2017). Another aspect of SDN technology that requires investigation and improvement relates to off-target editing. Although many investigators have reported very low or even negligible frequencies of off-target editing, these data may not be very accurate in view of the relatively small sample size used by them. It might be useful to devise effective techniques for detection and scoring of off-target editing in plants and to collect reliable estimates of these effects. In any case, off-target effects do not pose a serious problem in plants since they can be easily eliminated by segregation and selection after crossing with the parental line, the only caveat being that these effects should be easily and readily identifiable.

One other issue that needs to be resolved is the relatively low frequencies of gene editing in several plant species like *Arabidopsis*. Attempts are being made to enhance the editing efficiency by optimizing various factors of CRISPR-Cas9 system, including sgRNA design, the promoter driving Cas9 (germline-specific promoters are more efficient), the vector used for and the mode of SDN delivery (Table 7.9), and the method used for obtaining the gene-edited plants after floral dip transformation strategy. Durr et al. (2018) designed a novel binary vector, an effective selection procedure for sgRNAs that are highly efficient *in vivo*, and regeneration of plantlets from transformed seeds obtained by floral dip transformation procedure. By combining these refinements, they were able to induce heritable genetic changes, including large deletions. Targeted deletions of gene clusters and regulatory sequences located between genes might be useful for high-throughput functional genomics studies in plants. Similarly, Klimek-Chodacka et al. (2018) evaluated three different versions of Cas9 gene sequence, *viz*., *Arabidopsis* codon-optimized (*AteCas9*), maize codon-optimized (*zCas9*) and plant codon-optimized (*Cas9p*) sequences, for efficiency of gene editing in carrot and found the *AteCas9* to be the most efficient in gene editing.

Table 7.9. Some important relevant features of the CRISPR-Cas9 SDN system used for gene editing in plants

Component/ Procedure	Feature	Description	Remarks
sgRNA	Promoter	U6 snRNA gene promoter from *Arabidopsis*	Active in a wide variety of species; promoters from the same species have also been used
Cas9	Sequence	From *Staphylococcus pyogenes*; generally optimized for expression in plants	Generally, codon-optimized for *Arabidopsis* (for dicots), cereals like rice or for plants in general
	promoter	*CaMV35S* promoter from Cauliflower mosaic virus (CaMV)	Used in a wide variety of species; constitutive expression
		DD45 gene promoter from *Arabidopsis*	Egg cell- and early embryo-specific promoter; enhanced editing efficiency in *Arabidopsis*
		Meristematic tissue-specific promoters	Strong activity in meristematic tissue; highly efficient transmission of mutations to offspring
The SDN gene construct	Vector	Ti plasmid (binary vector)	Generally used; T-DNA integrated into genome
		Geminivirus vectors	SDN gene construct amplified; strong SDN expression; active in both monocots and dicots; the gene construct does integrate into plant genome
		TRV vectors	Effective in the TRV host species; no risk of foreign DNA integration in plant genomes
The SDN delivery	Delivered as	DNA	Risk of foreign DNA being integrated into plant genomes
		RNA, RNP	Preferred route; no risk of foreign DNA being integrated into plant genome
	Delivery method	*Agrobacterium* coculture	Delivers SDN in form of DNA stable expression; of gene construct; DNA integrates into plant genome
		PEG-mediated delivery	Delivered in protoplasts; applicable to only those species for which plantlet regeneration from protoplasts is available
		Particle gun	Applicable to all species; can be used to deliver SDNs to organs; screening of transformants is challenging

Recently, Bhowmik et al. (2018) optimized the protocol for transformation of wheat microspores with CRISPR-Cas9 SDN gene constructs by electroporation. They reported that electroporation of minimum of 75,000 microspores with 10-20 μg DNA and a pulsing voltage of 500V was optimal for transformation of wheat microspores. The electroporated microspores were cultured to regenerate doubled haploid androgenic plantlets. They successfully edited two genes using this technology. It is satisfying that CRISPR-Cas9 SDN is being used for genome editing in woody perennial species that are cross-pollinated and present challenges due to allelic heterozygosity, and sequence polymorphisms. Tools and techniques to handle these difficulties are being devised. It is hoped that the SDN technology will facilitate genetic analyses and breeding activities in these challenging plant species (Tsai and Xue, 2015).

The CRISPR-Cas9 system is amenable to multiplexed gene editing, but the delivery of multiple sgRNAs along with Cas9 gene constructs by a single plasmid presents problem due to the DNA insert size. Three different strategies have been developed to generate multiple sgRNAs by processing of single monocistronic RNA transcripts. These strategies utilize the cellular tRNA processing enzymes, the Csy4 riboendonuclease or a ribozyme for processing of the RNA transcripts. A comprehensive toolkit has been developed for gene editing of both dicot and monocot plants using TALEN and CRISPR-Cas SDNs designed for fast and modular cloning, capable of accommodating a variety of regulatory sequences, and optimized for single or multiplexed gene editing. A web-based tool has been developed to facilitate SDN designing and selection. The use of geminivirus vectors permits HDR-mediated DSB repair and gene editing. The use of Trex2 exonuclease enhances gene editing by 2.5-fold, and Cas9 nickase delivered via geminivirus vectors edits genes at frequencies comparable to that by native Cas9 nuclease (Cermak et al., 2017).

7.20. CONCLUSIONS

Plant breeding aims to improve genotypes of crop plants to make them more useful to humans. This is generally achieved by various breeding schemes that utilize the genetic variation created by spontaneous mutations over very long periods of time. But often the desired genetic variation is not available in the germplasm of concerned crop species. Therefore, scientists have been constantly on the lookout for novel tools for generating the desired genetic variation. The SDN technology is the most recent tool deployed for this purpose. This technology uses SDNs to induce sequence changes at targeted genomic sites. Initially, this tool was developed and optimized for animal cells, but it has now been applied for gene editing in many crop plants. It has been used to inactivate native genes (SDN-1 technology), to carry out sequence replacements for inducing or reversing mutations (SDN-2 technology), and to integrate DNA sequences, including transgenes, at specified genomic sites (SDN-3 technology). The SDN systems have been suitably modified for inducing gene mutations by base editing, in which targeted bases are deaminated leading to their irreversible and permanent conversion into other bases. It is now possible to directly convert A-to-G, and C-to-T, while G-to-A and T-to-C conversions are indirect and based on the conversions of C-to-T and A-to-G, respectively, in the complementary strand of the DNA duplex.

In simple terms, the SDN technology consists of the following operations: (1) designing of the SDN specific for the genomic site to be edited, (2) delivery of this SDN as DNA, RNA, RNP or protein into the cells of concerned crop using a suitable transformation method, (3) regeneration of complete plantlets from the transformed cells/tissues, and (4) detection and isolation of mutant plants/lines. SDNs induce DSBs at the targeted genomic sites, which are repaired by the endogenous DNA repair pathways of the affected cells. This repair is generally based on NHEJ, which typically leads to insertion or deletion of one to few base pairs resulting in inactivation of the targeted genes. Sometimes, DSBs may be repaired by HDR resulting in insertion of the supplied donor DNA sequences into the genomes. Once the mutant plants are identified, they are freed from the foreign DNA sequences by selection following segregation.

The gene editing procedures are being optimized for various plant systems to enhance their specificity and efficiency. Researchers are refining various aspects of the SDN technology, including codon optimization of Cas9 gene construct, the promoter used for Cas9 expression, efficiency of sgRNA, the vector used for SDN delivery, etc. The editing efficiency can be considerably improved by using a combination of some of these factors. Efforts have also been made to devise strategies for efficient multiplexed gene editing using the CRISPR-Cas9 SDN. It appears that the multiplexing strategy based on the Csy4 nuclease is superior to the other two strategies that rely on tRNA or ribozyme processing. Several web-based tools have been developed for designing of sgRNAs to be used in plants. Efforts are being made to extend gene editing to woody plants, which present several challenges including allelic heterozygosity, and sequence polymorphisms.

Gene editing has been used to edit several useful genes in many crops, and many of the mutant lines have been taken to field trials, and some of them have reached the market. For example, a nonbrowning mutant of button mushroom (*A. bisporus*), produced by editing of the gene encoding polyphenol oxidase, is being marketed in U.S.A. This has become possible because APHIS, USDA treats the products of gene editing (SDN-1 and some SDN-2 products) as non-GM and nonregulated articles if they are shown to be free from foreign DNA. However, European Union treats them as GM articles since gene editing involves a transitory transgenic state. The informed opinion is being mobilised to convince EU authorities to relieve gene-edited crop varieties from the regulatory burden. If this does happen, the research and development activities in this potent area will be boosted and many smaller players will enter the arena. It is widely believed that the gene/base editing tools will find many novel applications in plant breeding and their power and prowess is expected to increase with our understanding of the genetic and molecular bases of development of various trait phenotypes, including agronomic performance.

■■■■

8

GENE EDITING IN ANIMAL BREEDING

8.1. INTRODUCTION

Animals are an important source of protein and energy in human diet: in 2009, animal products, *viz*., milk, meat and eggs, provided 28% and 13% of the global human protein and energy needs, respectively. Animal breeding programs have been highly successful in increasing milk, meat and egg production, which has resulted in reduced emission of greenhouse gases per unit of milk/meat produced. However, the global demand for milk and meat is growing continuously due to increase in both global human population as well as affluence especially in countries of the low- and middle-income strata. It will be necessary to enhance the genetic gains from animal breeding programmes to meet the increased demands and this, in turn, might require suitable modifications in breeding strategies. The genetic gain per year depends on the amount of genetic variation present in the relevant population, heritability of the trait being improved, selection intensity and generation interval of the concerned animal species. Any strategy that would increase one or more of the first three variables or decrease the fourth one would improve the rate of genetic gain per year. Alternative modern reproduction technologies like *in vitro* fertilization (IVF) and somatic cell nuclear transfer (SCNT), along with elaborate breeding designs, such as, genomic selection are being regularly used in concert to accelerate the rate of genetic gain and to reduce generation interval in the cattle breeding programmes (Tait-Burkard et al., 2018; Van Eenennaam, 2019).

In addition, animals serve as human disease models, which are invaluable for studies on the mechanisms and progression of various human diseases, and the development of novel therapeutic agents and treatment procedures. The classical disease models have been genetically modified nematodes, fruit flies, zebrafish and rodents, which have been extremely useful in enhancing our understanding of many human diseases. But these models are often inadequate since the relevant biological features of these organisms are not comparable to those of humans. Therefore, there is demand for the use of large animals like nonhuman primates (NHPs), pigs, dogs, etc. as disease models since they are more closely related to humans in evolution and/or are more like humans in their physiological features during both health and disease. It is expected that they would serve as excellent models particularly for complex disorders like neurodegenerative diseases, cancer and cardiovascular diseases. Further, the use of these models may be expected to enhance the rate of translation of pre-clinical findings with animal models to clinical trials. At the same time, the large animals could serve as organ donors for xenotransplantation into humans. However, the genetic modifications necessary for these applications have been very slow especially in pigs and dogs mainly due to the lack of embryonic stem cell (ESC) technology, and the very low efficiency of homology-directed repair (HDR; Tait-Burkard et al., 2018; Van Eenennaam, 2019; Zhao et al., 2019).

The SDN (site-directed nuclease) technology (Section 1.8) offers excellent opportunities for gene editing to produce genetically altered large animals for improvement of agricultural traits as well as for creating specific disease models, organ donors and animal bioreactors for production of pharmaceutically valuable proteins. Animal genome editing is a rapidly developing area and several reviews covering various aspects of the discipline have been published; some of the reviews are, Wang and Zhou (2003); Tait-Burkard et al. (20018), Yang and Wu (2018), Yum et al. (2018), Van Eenennaam (2019), and Zhao et al. (2019). Gene editing has also been used to induce specified genetic alterations in human zygotes and the so-called 'CRISPR-babies' have already been borne (Cyranoski, 2019b). This development has shaken all concerned and has invited vocal criticisms from various quarters.

8.2. ANIMAL BREEDING

The livestock have been subjected to one or the other form of selection since they were brought under domestication. They have been subjected to systematic controlled hybridization and selection especially during the twentieth century leading to the development of many distinct breeds that exhibit remarkable improvements in performance (Table 8.1). The conventional breeding schemes utilize natural genetic variation present in the relevant populations and selection is based on trait phenotype. In addition, selection for desirable traits is generally associated with reduction in genetic variation, increase in inbreeding depression and sometimes expression of detrimental alleles of genes. Continued genetic improvement requires the presence of genetic variation in the traits of interest, and creation of new genetic variation without disturbing the characteristic features of the concerned breeds is a major challenge. At the same time, breeding programmes in livestock are generally costly and the progress is rather slow. Therefore, suitable innovations in breeding programmes are necessary to address these limitations. It is satisfying that modern reproduction technologies like IVF and SCNT, and advanced breeding schemes, such as genomic selection are being used to enhance the rate of genetic gain and to reduce the generation interval in cattle breeding programmes (Van Eenennaam, 2019).

Table 8.1. Improvement in certain livestock production traits achieved by conventional breeding from 1960 to 2005 (based on Tait-Burkard et al., 2018)

Livestock	Trait	Improvement
Pigs	Litter size	50% increase
	Lean pork meat	37% increase
	Lean pork meat per kg of feed intake	100% increase
Chicken	Days to acquire mass of 2 kg	Reduced to 40 days from 100 days (60% reduction)
	Breast meat	Increased from 12 to 20% (66.7% increase)
	Feed conversion ratio	Reduced by 50%
	Eggs per year	30% increase
	Eggs per tonne of feed	80% increase
Cattle	Milk production	67% increase

In simple terms, the selected females are induced to super-ovulate, oocytes are collected, fertilized *in vitro* and cultured for few days. Healthy young embryos are selected and transferred into the uteri of recipient females synchronized for the purpose. After some time, foetuses are collected, and fibroblast cell lines are established from each foetus. The cell lines are genotyped for many markers densely distributed over the entire genome and used for genomic selection. In genomic selection, the contribution of each marker to the breeding value of the trait(s) under selection is estimated based on historical data, *i.e.*, data from earlier breeding populations, on trait phenotype(s) and marker genotypes. These data are analysed according to an appropriate genomic selection model to estimate the contribution of each marker to the concerned trait. The marker genotype of each cell line is used to compute its genomic estimated breeding value (GEBV) and fibroblast cell lines with superior GEBV are selected. The selected cell lines are used for SCNT, in which enucleated fertilized oocytes are fused with cells from the selected cell lines. The fusion products are cultured, and young embryos are transferred into recipient females to obtain improved progeny as a result of the genomic selection (Fig. 8.1; Kasinathan et al., 2015: Yudin et al., 2016).

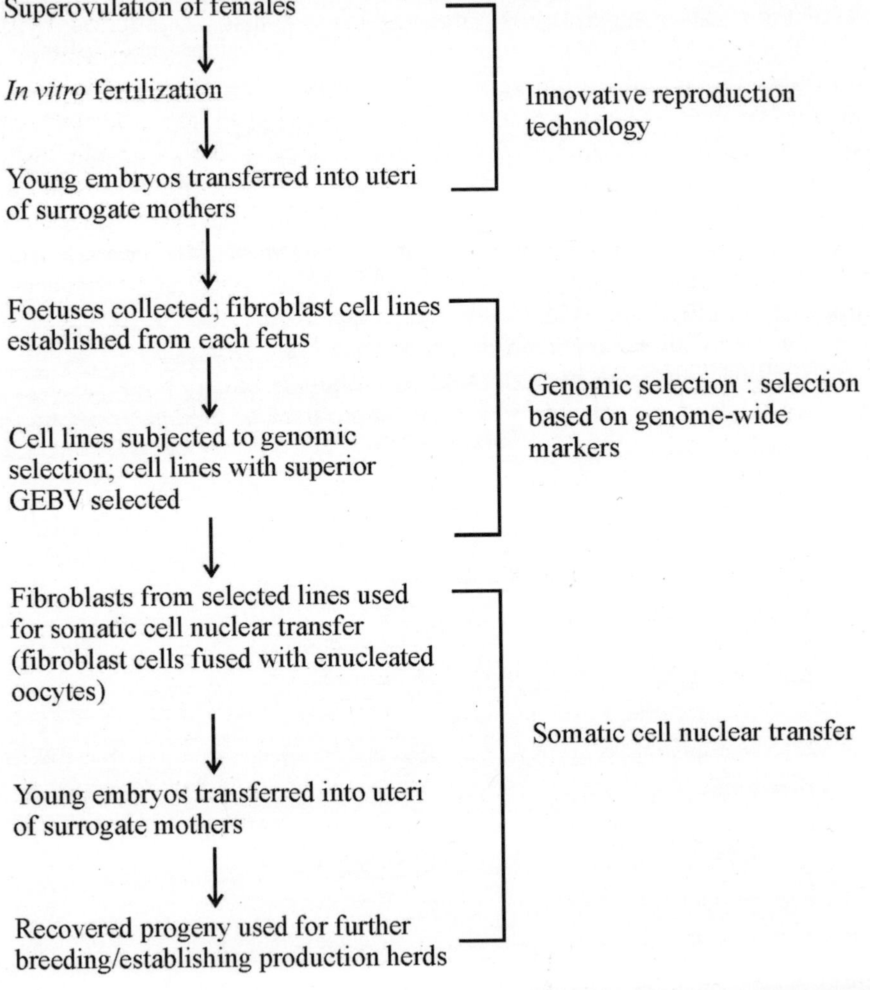

Fig. 8.1. Seamless integration of innovative reproduction technology, genomic selection and somatic cell nuclear transfer in modern cattle breeding programmes (based on Kasinathan et al., 2015).

Recombinant DNA technology has been used to produce transgenic animals in the hope that this technology will enable novel genetic modifications that would be useful in livestock improvement. Transgenic animals have been produced in most livestock species, but they have either served as proof-of-concept or some other specific purpose like production of some pharmaceutical proteins, and none of them has been instrumental in improvements relevant to the conventional uses of these animals. The transgenic technology has been used to create many disease models in rodents and rabbits, and recently efforts have been directed at large animals as well. The genome editing technology based on SDNs provides opportunities for conservation and utilization of beneficial alleles, introduction of novel targeted genetic variation and shortening of the timeframe for generating desired mutant animal lines. Since the mutant strains developed by this technology are free from foreign DNA, they may be expected to be exempt from the GMO regulations. At the same time, they are likely to be acceptable to the consumers as human food and animal feed. The SDN technology has been used to create several different strains of many large animals, including NHPs, that have been designed to serve specific objectives (Zhao et al., 2019).

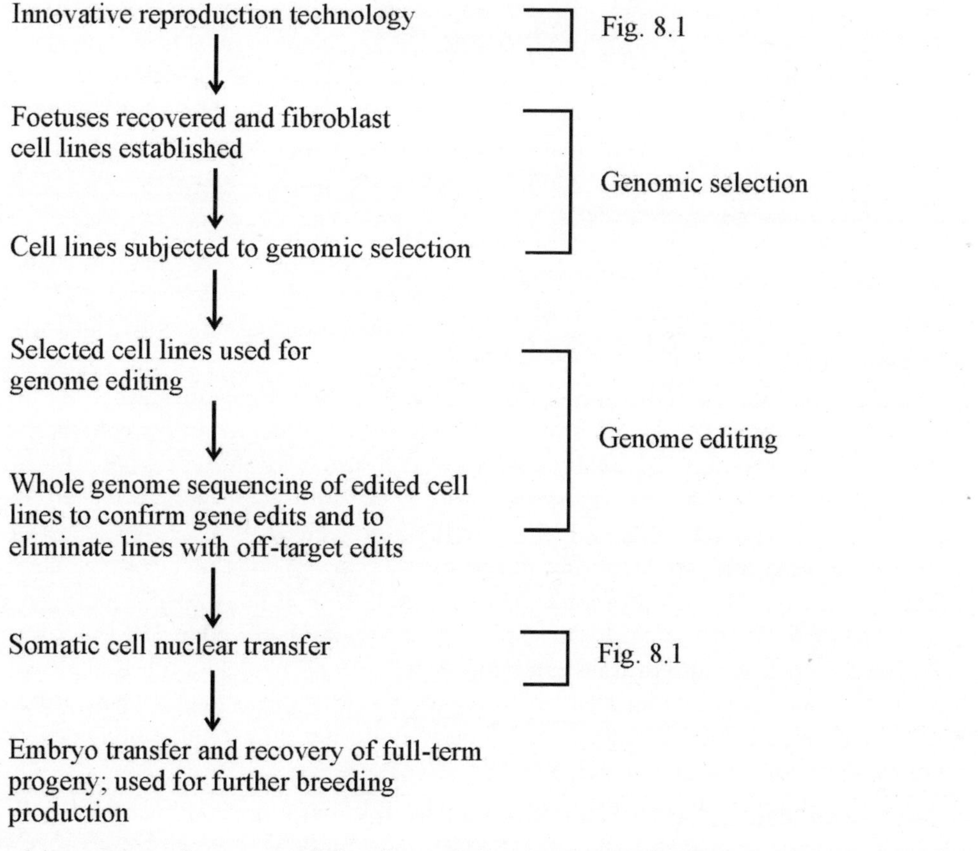

Fig. 8.2. Proposed integration of genome editing in a modern cattle breeding scheme based on modern reproduction technologies, genomic selection and somatic cell nuclear transfer (based on Van Enennaam, 2017).

It has been proposed that genome editing can be seamlessly integrated into the modern livestock breeding schemes based on superovulation, IVF, embryo transfer and SCNT technologies, and the genomic selection procedure (Fig. 8.1). Briefly, the fibroblast cell lines established from individual foetuses are subjected to genomic selection, and the cell lines with superior GEBVs are used for genome editing generally by CRISPR-Cas9. The edited cells are subjected to whole genome sequencing, and the sequences are analysed to select such cell lines that have the intended editing of the targeted genes and are free from off-target edits. The selected cell lines are used for SCNT followed by embryo transfer to recover full-term calves with improved genetic merit (Fig. 8.2; Van Eenennaam, 2017).

8.3. GENERAL PROCEDURE FOR GENOME EDITING IN ANIMALS

The production of gene-edited animals involves *in vitro* editing of single cell embryos or cultured somatic cells or embryonic stem cells (ESCs), SCNT (when somatic cells or ESCs are edited), *in vitro* culture of the zygotes or reconstituted zygotes, followed by transfer of the developing embryos into recipient females to recover full-term calves. Thus, gene editing can be achieved by microinjecting the SDNs into embryos or treating the cultured cells with SDNs. The SDN is generally delivered as protein, ribonucleoprotein (in case of CRISPR-Cas9) or mRNA (Cas9 mRNA and sgRNA in case of CRISPR-Cas9). Genome editing can be used for basic investigations like determination of gene function or creation of useful lines of livestock, *e.g.*, disease resistant animals, animals with improved agricultural traits, and strains producing designer milk and/or pharmaceutical proteins. ZFNs, TALENs and CRISPR-Cas9 have been used to generate numerous gene-edited animals, including cattle, pigs, sheep and goats. But CRISPR-Cas9 is the current SDN of choice.

8.3.1. Embryonic stem cell technology

Embryonic stem cells are derived by culturing the inner cell mass of blastocyst embryos. The ESCs are pluripotent so that they can differentiate into a variety of cell types and organs, can be maintained for long periods, and retain their pluripotent state provided they are cultured under specified conditions. The ESCs have been extensively used to produce transgenic mice, and the phrase ESC technology typically describes this application. In simple terms, the ESCs are transfected with the desired transgene construct, following which the transgenic ESCs are selected and cloned. The selected ESCs are then introduced into the blastocoel of blastocyst stage embryos isolated and prepared for the purpose. The ESCs can be introduced into blastocysts either by injection or coculture with the blastocyst stage embryos. Alternatively, the ESCs can be fused with enucleated oocytes to obtain zygote like cells that develop into embryos. The embryos/reconstituted zygotes are cultured *in vitro* to allow selection of healthy embryos that are transferred into recipient females prepared for the purpose and full-term progeny are obtained. The mice obtained in this manner are chimeric, but pure transgenic animals can be obtained from them by a suitable breeding scheme (Singh, 2012).

The availability of germline transmitted pluripotent embryonic stem cells is the main reason for rapid advances in creation of transgenic mice models for many human diseases. However, such ESCs are not available in cattle. In mice and humans, induced pluripotent stem cells (iPSCs) can be produced from somatic cells by expressing five embryonic transcription factors, *viz.*, Oct4, Sox2, cMyc, Klf4, and Nanog. iPSCs have been produced in cattle, but the technology needs to be further refined to enable its use for production of gene-edited animals. In case of mice, ESCs are subjected to gene editing, and the cells are cultured to generate single cell clones. The clones are screened to identify gene-edited clones, which are then introduced into blastocyst stage embryos to produce chimeric animals (Yum et al., 2018)

8.3.2. Microinjection into oocytes

The embryos for microinjection are generated by *in vitro* fertilization of mature oocytes recovered according to standard procedures (Fig. 8.3). Generally, the females used for oocyte collection are super-ovulated by subjecting them to a suitable hormone regime. Typically, cryopreserved semen is used for IVF of these oocytes. The injection buffer with the desired SDN is microinjected into either the cytoplasm or the pronucleus of zygotes. The microinjection into pronuclei is rarely practised now in view of its low efficiency. The SDN is generally microinjected as mRNA or sgRNA and protein (in case of CRISPR-Cas9); it can also be electroporated into the oocytes. The zygotes are cultured *in vitro* for about 6 days, after which healthy blastocyst stage embryos are transferred into recipient females by laparoscopy-assisted minimal surgery procedure. Standard ovulation control procedures are used to synchronize the recipient females to be on the day 6 of their oestrous cycle when the blastocysts are placed inside their uteri. Typically, two blastocysts are transferred into each recipient. The editing efficiency and off-target activity can be determined by whole genome sequencing of a proportion of the morula and blastocyst stage embryos. Skin and/or limb muscle samples may be collected from the off-spring about one week after birth for DNA isolation and various analyses, including T7E1 assay, and DNA sequencing for confirmation of gene editing and detection of off-target editing (Carlson et al., 2012; Proudfoot et al., 2015; Crispo et al., 2015; Yang and Wu, 2018).

At least in the case of pigs, this procedure is not efficient for production of gene knockins. But young embryo microinjection was very efficient in sheep for production of knockout animals. In addition, eGFP (enhanced green fluorescent protein) mRNA can be microinjected along with SDN mRNA into the zygotes to enable selection of gene-edited embryos based on eGFP fluorescence (Carlson et al., 2012; Proudfoot et al., 2015; Crispo et al., 2015; Yang and Wu, 2018). Some workers prefer a strategy that combines microinjection of SDNs into fertilized oocytes with SCNT technology for cloning of the gene-edited animals. In this approach, fertilized oocytes are subjected to gene editing to produce several gene-edited animals, which are characterized for their phenotypes. The animals with the correct phenotypic expression are identified and fibroblast cell cultures are established from them. The cultured cells are used for SCNT to generate clones of the selected animals. The cloning step generates genetically identical animals, which would reduce the number of animals needed for each treatment in various biomedical experiments. This approach was used for cloning of macaque monkeys gene-edited to disable the gene *BMAL1*, which is crucial to their sleep-wake cycle. But the cloning of monkeys was quite inefficient (only five monkeys were obtained from 325 embryos transferred into 65 recipient females) and highly expensive. Further, the findings from clones with specific genotypes may not precisely represent the response of human population that would comprise individuals with different genotypes (Cyranoski, 2019a).

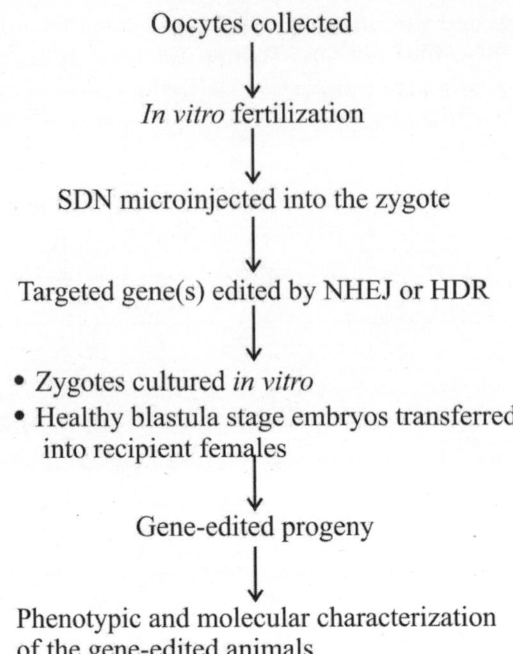

Fig. 8.3. Young embryo microinjection for gene editing. For HDR, donor DNA is also microinjected with the SDN. For gene knockin, the donor DNA has the complete transgene construct. The progeny may show mosaicism. SDN, site-directed nuclese; HDR, homology-directed repair; NHEJ, non-homologous end-joining (based on Yum et al., 2018)

The young embryo microinjection strategy is far simpler and more efficient than SCNT in animals like cattle and sheep especially when it is combined with oocyte collection and IVF. But this method often generates individuals with mosaic genotypes so that several cycles of breeding and selection are needed to develop homozygous mutant animals. Further, in case a chimeric individual has wild type germ cells, it will not produce progeny with the mutant allele. In comparison, the SCNT procedure (Section 8.3.3) yields pure mutant individuals and there is no risk of mosaicism. Further, gene editing can be confirmed and off-target action can be detected before the cells are used for SCNT. Embryo microinjection shows high pregnancy rate, healthy off-spring and no genome instability. In contrast, the SCNT procedure shows low pregnancy rate and high rate of abortion or sudden death of offspring. In addition, there may be genome instability and the epigenetic reprogramming may be incomplete (Carlson et al., 2012; Proudfoot et al., 2015; Crispo et al., 2015; Yang and Wu, 2018; Yum et al., 2018).

8.3.3. Editing of foetal fibroblasts followed by somatic cell nuclear transfer

In this approach, cultured foetal fibroblast cells are subjected to gene editing. The edited cells are identified, used for SCNT, and the young blastocyst embryos obtained from *in vitro* culture of the SCNT products are transferred into recipient females. Typically, fibroblast cell cultures are initiated from young foetuses since they have the capability of extensive proliferation. The cultured cells are exposed to the appropriate SDN for gene editing (Fig. 8.4). The edited cells are cloned, and each clone can be screened for the intended mutational change(s) in the targeted gene(s). Clones with the intended mutation may be subjected to whole genome sequencing to select those clones that are free from off-target edits. All the above operations are carried out according to optimized procedures. Generally, the gene-edited cells have lower cloning efficiency than the wild type cells, and their SCNT products show reduced embryonic development. Therefore, many laboratories prefer embryo microinjection to SCNT procedure at least in the case of pigs. In cattle, long-term somatic cell cultures exhibit low efficiency in SCNT (Yang and Wu, 2018; Yum et al., 2018).

Fibroblast cells cultured *in vitro*

↓

- Cells subjected to gene editing using the selected SDN
- Gene editing based on NHEJ (for gene knockout) or HDR (for sequence replacement or gene knockin)

↓

- Gene-edited cells selected and multiplied
- SCNT : cells fused with enucleated zygotes produced by *in vitro* fertilization

↓

Gene-edited progeny

↓

Phenotypic and molecular characterization of the gene-edited animals

Fig. 8.4. Gene editing of fibroplast cells, followed by SCNT. For HDR, donor DNA must be provided along with the SDN. For gene knock-in, the complete transgene construct is included in the donor DNA. SDN, site-directed nuclease; HDR, homology-directed repair; NHEJ, nonhomogous end-joining; SCNT, somatic cell nuclear transfer (based on Yum et al., 2018).

The first report of successful use of SCNT procedure was in cattle by Sims and First (1994) who produced calves by transferring nuclei from cell cultures of the inner cell mass or embryoblast of a primordial embryo. But the first report of production of a gene-edited large animal by the SCNT procedure was that of lambs from gene-edited cultured fibroblast cells (McCreath et al., 2000). The somatic cells used in SCNT are generally foetal fibroblast cells cultured and multiplied *in vitro*. The recipient oocytes are collected and enucleated following an appropriate procedure. The edited fibroblast cells can be either injected into the enucleated oocyte or they can be fused with the latter by electrical pulses. The cells produced by these operations are comparable to zygotes and are often called reconstituted zygotes; they are stimulated for further development by a suitable physical or chemical stimulus. In pigs, activation of the reconstituted zygotes 3 hr after the nuclear transfer is reported to improve blastocyst formation. The developing embryos are cultured *in vitro* till blastula stage and then transferred into surrogate females prepared for the purpose. Generally, more than one embryo is transferred into each female, but in pigs a minimum of four embryos are needed for establishing pregnancy (Lai and Prather, 2003). A transposon co-selection strategy can be used for the enrichment of gene-edited cells. Thus, the SCNT procedure allows isolation and cloning of gene-edited cells prior to their introduction into enucleated oocytes and transfer of the resulting blastocysts into recipient females; this reduces the risk of needless use of too many surrogate females that are diverted from production (Carlson et al., 2012). This strategy has been successfully used for producing gene-edited monkeys, cattle, pigs, sheep, etc.

Both gene knockout and gene knockin animals can be generated by the transgenic technology based on ESCs; the SDN technology can achieve the same by utilizing ESCs, young embryos as well as somatic cells for gene editing. At the same time, the SDN technology is much faster and generally more efficient than the ES cell technology. Further, it is applicable to even such species, in which ESCs are not available. The animals produced by SDN technology generally show high germline transmission of the genetic changes. Therefore, the phenotypic evaluation of mutants could be initiated in the G1 generation itself. In contrast, pure transgenic animals must be generated by breeding of the chimeric animals produced by the ESC technology before initiation of phenotypic evaluation. The effectiveness of ESC technology is strain-specific, while that of SDN technology is not; therefore, the latter can be applied to any inbred line. However, there is risk of off-target editing in case of the SDN technology. The off-target edits, if present, must be removed by repeated backcrossing to the parental strain. Alternatively, the presence of off-target edits can be ruled out by comparing the phenotypes of two or more independent mutant lines (Mashimo, 2014), and by analysis of the whole genome sequences of these lines along with that of the parental line.

8.4. APPLICATIONS OF GENOME EDITING

Genome editing has been used for the development of livestock strains to serve a variety of needs ranging from animal models for human diseases, large animal donors of organs for xenotransplantation in humans, animal bioreactors for the production of valuable pharmaceutical proteins, and agricultural breeding (improvement of traits related to performance, product quality, disease resistance, animal/human welfare, etc.; Zhao et al., 2019). Selected examples of each of the above applications of genome editing are briefly described in the following sections.

8.4.1. Animal models of human diseases

Animal models of human diseases are essential for understanding the mechanisms and progression of various human diseases, and for the development of novel therapeutic agents and procedures for their treatment. The classical model organisms like nematodes, fruit flies, zebrafish and rodents have long been used to create disease models, which have been extensively used in biomedical research. Experimental data accumulated from these models have provided novel insights into human biology and disease pathogenesis. Animal models are used for screening of novel therapeutic compounds for developing new drugs and for the development of new treatment procedures. It is, therefore, necessary to select the correct animal model so that the findings from the biomedical studies based on them are transferred to humans with some degree of confidence. It has been observed that for many diseases the findings from smaller animal models, including rodent models, do not precisely reflect the situation prevailing in humans. For example, the mouse lipoprotein metabolism and profile differ from those in humans, while pigs are more comparable to humans in these aspects.

In view of the above, there is need to use animal species that are evolutionarily more closely related to humans for creation of human disease models. Such animals may be expected to more closely represent the human physiological features than the classical animal models. The use of such models for preclinical studies, therefore, may be expected to enhance the translation of their findings to the clinical trials. Large animals, particularly NHPs, pigs and dogs, are remarkably like humans in anatomy, organ size, physiology, and metabolism. Therefore, these animals can serve as donors of organs and as animal models for specifically such diseases as neurodegenerative disorders and cardiovascular diseases. The large animal models would also be extremely useful for studies in pathogenesis, researches in regenerative medicine, and discovery and development of novel drugs (Zhao et al., 2019).

Gene editing has been used to create human disease models in cultured human cells/organoids as well as in several animal species including rodents (Section 9.2). Several large animals like monkeys, cynomolgus monkeys (NHPs), pigs and dogs have also been gene-edited to generate several disease models (Table 8.2). The modelled diseases include RETT syndrome, microcephaly, adrenal hypoplasia congenital and hypogonadotropic hypogonadism (AHC-HH), Duchenne muscular dystrophy, atherosclerosis, Von Willebrand disease, Parkinson's disease, Huntington disease, cancer, etc. These models have been typically produced by gene inactivation, but HDR was exploited to create the precise nucleotide substitutions in *p53* gene in the monkey models designed for investigation of tumorigenesis. In addition, *HTT* gene knockin pigs were produced by using CRISPR-Cas9 to replace the pig *HTT* gene exon 1 having 18 CAG repeats with the human *HTT* gene exon 1 with 150 CAG repeats. The gene-edited pigs showed movement and behavioural anomalies along with perfect recapitulation of selective neurodegeneration observed in patients with Huntington disease. Some of the large animal models were clinically closer than rodent models to the concerned human patients. For example, the clinical features of human patients with AHC-HH were more closely represented by *DAX1*-deficient monkey models than *DAX1*-deficient mouse models. Similarly, the cynomolgus monkey models with inactivated *microcephalin 1* (*MCPH1*) gene duplicate most of the clinical features found in human patients suffering from microcephaly, including smaller head circumference, behavioural abnormalities, etc. (Zhao et al., 2019).

Gene editing has been used to create models for different cancers. A pig model that expresses Cas9 protein in response to induction by Cre recombinase permits both *ex vivo* and *in vivo* gene editing when Cas9 expression is induced and the desired sgRNA is provided. The feasibility of lung cancer induction in this pig model was studied by intranasal delivery of the sgRNAs specific for tumour suppressor genes *TP53*, *PTEN*, *APC*, *BRCA1*, and *BRCA2*, and the oncogene *KRAS*. Three months later, the pigs showed signs of pneumonopathy and pathological features like human adenocarcinoma in their lung tissues. Pigs with inactivated *p53* and *RUNX3* genes by TALENs and CRISPR-Cas9, respectively, have been produced, but their carcinogenic phenotypes are yet to be investigated. Three pig models of Parkinson's disease (PD) have been developed by knocking out one, two or three genes, *viz.*, gene *DJ-1* alone, genes *PARK2* and *PINK1* or genes *Parkin*, *DJ-1* and *PINK1*. These genes were inactivated by either TALENs or

CRISPR-Cas9 SDN system. The triple knockout Bama miniature pigs were produced by multiplexed gene editing using Cas9 mRNA; these pig models have been used for studies on PD pathology as well as for the development of therapeutic intervention for PD (Yang and Wu, 2018; Zhao et.al., 2019).

Table 8.2. A list of some gene-edited animal models of human diseases (based on Yang and Wu, 2018; Zhao et al., 2019)

Animal	Edited gene	SDN[1]	Model for the human disease	Remarks
Rhesus and cynomolgus monkeys	*Methyl-CpG binding protein 2 (MECP2)*	TALENs	Rett syndrome (RTT)	Clinically similar to human patients
Cynomolgus monkeys	*Microcephalin 1 (MCPH1)*	TALENs	Microcephaly (smaller head circumference and other symptoms)	Recapitulated most of the symptoms of human patients
	DAX1 (dosage-sensitive sex reversal, adrenal hypoplasia critical region, on chromosome X, gene 1)	CRISPR-Cas9	AHC-HH (adrenal hypoplasia congenital and hypogonadotropic hypogonadism)	Deletion of the *DAX1* gene
	BMAL1 (*Brain and Muscle ARNT-Like 1*; crucial to sleep–wake cycle)	CRISPR-Cas9	Circadian disorders	Strong circadian disorder phenotypes, including reduced sleep time, increased anxiety and depression, schizophrenia-like behaviour, etc.
Monkey	*p53**	CRISPR-Cas9	Tumorigenesis	First model produced by HDR
	dystrophin	CRISPR-Cas9	Duchenne muscular dystrophy	Deletion of the *dystrophin* gene
Pigs**	*PPAR-ϒ*	ZFN	Myocardial events	First report of gene inactivation by gene editing in a large animal
	low density lipoprotein receptor (LDLR)	TALENs	Familial hyper-cholesterolemia and atherosclerosis	Considerably elevated levels in total and LDL cholesterol
	Npc1l1 (Niemann-Pick C1-Like 1)	CRISPR-Cas9	Cardiovascular and metabolic diseases	Gene *Npc1l1* deleted
	apolipoprotein E (ApoE) and *LDLR*	CRISPR-Cas9	Atherosclerosis	Both genes inactivated; atherosclerosis related abnormal lipid metabolism
	MYH7	TALENs	Hypertrophic cardiomyopathy (HCM)	Induced, point mutation R723G; mimicked HCM phenotype
	vWF (*von Willebrand Factor*)	CRISPR-Cas9	Von Willebrand disease (vWD)	Mimicked vWD phenotype: reduced factor VIII and severe bleeding

Animal	Edited gene	SDN[1]	Model for the human disease	Remarks
	Adenomatous polyposis coli (APC)	TALENs	Colon cancer	A model for colon cancer
	PKD1 (*polycystic kidney disease 1*)	ZFNs	Polycystic kidney disease	Good model for studying renal cystogenesis
	DJ-1; *PARK2* and *PINK1*; *Parkin*, *DJ-1* and *PINK1*	TALENs and CRISPR-Cas9	Parkinson's disease	Being used for pathology studies and therapeutic development
	Huntingtin (*HTT*)	CRISPR-Cas9	Huntington disease	*HTT* knockin (transgene integration); perfect recapitulation of neurodegeneration of Huntington disease
	MITF	CRISPR-Cas9	Waardenburg syndrome	White coat colour and other clinical manifestations
	Hoxc13	CRISPR-Cas9	Ectodermal dysplasia-9 (ED-9)	The gene deleted; mimicked symptoms of ED-9
	TP53	TALENs, CRISPR-Cas9	Cancer	Various tumour phenotypes
	RUNX3	CRISPR-Cas9	Cancer	Disease phenotype yet to be evaluated
	Tumour suppressor genes TP53, PTEN, APC, BRCA1, BRCA2, and oncogene *KRAS*	CRISPR-Cas9	Lung cancer	Pneumonopathy: pathological features similar to the human adenocarcinoma
	IL2RG	ZFN	X-linked SCID	T and NK cells were absent; an essential preclinical model for evaluation of stem cell therapy
	RAG1/2	TALENs	SCID phenotype	T and B cells were absent
	IgM heavy chain gene	CRISPR-Cas9	Human B cell deficiency	Both B cell and antibody depletion
	DAZL	TALENs, CRISPR-Cas9	Infertility (azoospermia)	HDR- or NHEJ-mediated editing; oligonucleotide template for HDR
	APC	TALENs, CRISPR-Cas9	Colon cancer	HDR-mediated editing using oligonucleotide template
Dogs	*MSTN* (myostatin)	CRISPR-Cas9	Double-muscle phenotype of thighs	Phenotype observable in 4 months

* Mutant allele produced by HDR (homology-directed repair) of DSB

** Fewer ethical concerns than with nonhuman primates (NHPs)

[1] SDN, Site-directed nuclease; DSB, double strand break; SCID, severe combined immune deficiency

Animal models can be produced by microinjection of SDNs in oocytes/young embryos or by gene-editing of cultured fibroblast cells and using the edited cells for SCNT followed by embryo transfer. The SCNT procedure is preferable since it allows recovery of genetically uniform animal models as well as weeding out of any edited cells carrying off-target edits and/or unintended on-target edits (Zhao et al., 2019). Custom made CRISPR-Cas9 gene-edited (gene knockout/gene knockin) mouse and rat models are commercially available, *e.g.*, at Cyagen Biosciences (apac.cyagen.com/services.html?). The breeding of large animals is expensive and the production of various animal models in sufficient numbers needed for biomedical studies may be much more difficult than in the case of smaller animals. In addition, ESCs are not available, and the SCNT and cloning procedures are relatively inefficient in most of the large animals than in smaller animals. These observations are particularly applicable to NHPs, which are also regarded to have a higher moral status than other animals since they possess high levels of cognition. As a result, many find the use of NHPs to model human diseases rather objectionable. It is also argued that for studies on many human diseases, mice models are equally appropriate, far cheaper, and more acceptable to the public than NHP models. But most scientists consider mice models inadequate for investigations on disorders of brain and higher cognitive functions, and of cardiovascular diseases (Cyranoski, 2019a; Zhao et al., 2019).

8.4.2. Xenotransplantation

There is acute shortage of human organs for transplantation in patients suffering from end-stage organ failures. Xenotransplantation is regarded as a promising option to meet this shortage. In xenotransplantation, organs, tissues and/or cells from one species are transplanted into another species. During 1963 and 1964, 13 kidney and one liver transplant from chimpanzee were made in humans, and one person with kidney transplant survived for 9 months. But xenotransplantation from NHPs suffers from limitations of limited organ supply, risk of organ infection and difficulties in breeding of these species. Pigs seem to be promising organ donors in view of their anatomical, physiological and organ size similarities with humans and their systematic breeding and cost-effective multiplication. But there are two major difficulties, *viz.*, graft rejection and cross-species infections in xenotransplantation of pig organs. The phenomena responsible for graft rejection include hyperacute rejection (HAR), delayed xenograft rejection, acute cellular rejection, and chronic rejection; each of these processes involves different proteins/sets of proteins. Gene editing technology is being used to target the genes involved in these processes to create potential organ donor animals that would present minimum or, ideally, no risk of xenograft rejection (Table 8.3). The term xenogeneration has been coined to describe the production of humanized organs in gene-edited pigs or pigs produced by blastocyst complementation for transplantation in human patients (Yang and Wu, 2018; Zhao et al., 2019).

Pig to human xenotransplantation leads to HAR of the transplants, which develops immediately after transplantation. HAR involves antigen-antibody interaction, complement activation and most likely blood-mediated inflammatory reaction. The antigen antibody interaction is primarily due to the binding of antibodies already present in human plasma with the Galα(1,3)Gal residues of glycoproteins located on the surface of pig cells. The Galα(1,3)Gal epitope is synthesized by the enzyme α-1,3-galactosyltransferase (encoded by the gene *GGTA1*) this enzyme is not found in humans. Therefore, it may be expected that the inactivation of gene *CGTA1* by gene editing would reduce HAR. Gene knockout strategy has validated this expectation and has identified two additional genes (*CMAH* and *β4GalNT2*) that contribute to HAR. CRISPR-Cas9 was used to inactivate the pig genes *GGTA1*, *CMAH* and *β4GalNT2*; the peripheral blood mononuclear cells (PBMCs) from these multiple gene-edited pigs showed reduced binding to human IgM and IgG *in vitro* in comparison to the PBMCs from pigs with edited *GGTA1* and *CMAH* genes (Tseng et al., 2005; Lutz et al., 2013; Butler et al., 2016; Yang and Wu, 2018; Zhao et al., 2019).

Table 8.3. Some of the genes relevant in xenotransplant rejection and the progress made in making large animals a likely source of organs for humans (based on Zhao et al., 2019 and other sources)

Animal	Gene	Method	Organ transplanted	Remarks
Pig	*GGTA1*	HR-mediated disruption	Heart	Survived 6 months in baboons
	CMAH and *β-4GalNT2*	ZFN-mediated inactivation	Cells used for antibody-mediated complement-dependent cytotoxicity assays	Lower humoral barrier to xenotransplantation than in case of *GGTA1* knockout
	GGTA1, *CMAH* and *β4GalNT2*	CRISPR-Cas9-mediated inactivation	*In vitro* assay using PBMCs	Less human immunoglobulin (Ig)M and IgG binding *in vitro* to their PBMCs than in *GGTA1*$^-$ and *CMAH*$^-$ inactivated pigs
	PERV	CRISPR-Cas9-mediated inactivation	–	Greatly (~1000-fold) reduced transmission of PERV to human cells/tissues
	CD46, *CD55* and *CD59*	Transgenic animals expressing human genes	Cytotoxicity assays; xenografts, *e.g.*, of heart	Xenotransplants better tolerated in cynomolgus monkeys and baboons
Pig and cattle	Interspecies chimera	hESCs or hiPSCs grafted into pig blastocysts	Gene(s) for development of the desired organ inactivated in the embryo donor	Blastocyst complementation; organs developed from hESCs or hiPSCs available from chimeras; experimental stages

HR, homologous recombination; SDN, site-directed nuclease; PBMCs, peripheral blood mononuclear cells; PERV, Porcine Endogenous Retrovirus; hESCs, human embryonic stem cells; hiPSCs, human induced pluripotent stem cells.

Complement activation is a potent factor involved in HAR. Many proteins involved in complement regulation, *e.g.*, CD46, CD55 and CD59, have been identified as potent targets for inactivation for reducing complement activity during xenotransplantation. Xenotransplants from transgenic pigs expressing human *CD55*, and *CD46* and *CD59* genes were better tolerated in cynomolgus monkeys and baboons, respectively, while those from pigs expressing human *CD59* gene alone were protected from complement attack. The blood-mediated inflammatory reaction is triggered by physiological incompatibilities between the host and the xenograft, and some factors like CD39 have been identified for modification. In addition, xenogenic cellular response also affects xenograft survival. This response results from differences between pig and human leucocyte antigens. Thus, it seems likely that gene editing would be able to create pig strains lacking such gene functions that contribute to xenograft rejection. In addition, anti-inflammatory and antiapoptotic genes may be introduced into pigs to minimize xenograft rejection (Zhao et al., 2019).

The PERV (Porcine Endogenous Retrovirus) occurs in many copies that are distributed in the pig genome. PERV is harmless to pigs, but it could become activated in xenotransplanted organs and cause harm to the recipient humans. CRISPR-Cas9 was used to inactivate all the 25 copies of PERV present in the genome of a pig cell line. The gene-edited cells were used to generate PERV-inactivated pigs by SCNT. These pigs would provide organs with greatly reduced risk of PERV transmission into the human recipients of the xenotransplants. Overall, there has been remarkable progress in creation of gene-edited pigs that offer organs for xenotransplantation, which are far more compatible with the human immune system as compared to those from their wild type counterparts. It may be hoped that further refinements in the genetic makeup of donor pigs would enable xenotransplantation of their organs in human patients (Yang et al., 2015; Niu et al., 2017; Zhao et al., 2019).

Functional organs amenable to xenotransplantation can also be generated by creating chimeric embryos, *i.e.*, young embryos of the donor species containing human embryonic stem cells (hESCs) or human induced pluripotent stem cells (hiPSCs). Genes controlling the development of individual organs have been identified in many model animals. Disruption of gene(s) governing the development of a specific organ leads to the development of embryos or foetuses lacking that organ. Such embryos contain empty niches that can be filled with hESCs or hiPSCs, which would give rise to the concerned organ/tissue in the animals developed from the chimeric embryos/foetuses so obtained. For example, disruption of gene *pdx1* leads to the absence of pancreas in mouse foetuses/neonates. This approach for production of chimeric embryos and animals is referred to as blastocyst complementation. The first interspecies chimera was produced in 2017, when rat pluripotent stem cells (PSCs) were injected into *pdx1*-deficient mouse blastocysts; as expected, the resulting chimeric mice had rat pancreas. In the same year, Wu et al. (2017) showed that human pluripotent stem cells vigorously established themselves in preimplantation blastocysts of both cattle and pig to generate chimeric human-pig and human-cattle embryos. These findings raise the hope that soon chimeric large animals would be created for providing human organs for transplantation (Yang and Wu, 2018; Zhao et al., 2019).

8.4.3. Animal bioreactors

A large number of human proteins are used as therapeutic agents and they are in great demand. Recombinant DNA technology has enabled the production of many of these proteins in a variety of organisms, such as bacteria, yeast, fungi, animal cells, transgenic animals as well as transgenic plants. The transgenic animals used for production of such proteins are generally referred to as animal bioreactors or simply bioreactors. The animal bioreactors offer certain advantages over other systems in terms of correct post-translational processing, folding and glycosylation of the proteins, and lower production costs than cultured animal cells. Three recombinant pharmaceutically valuable proteins, *viz*., ATryn®, Ruconest® and Kanuma® have been approved By FDA (Food and Drug Administration), U.S.A. for clinical use. ATryn® is human antithrombin-α obtained from transgenic goats; it is the first drug produced from a transgenic animal to be approved for clinical use. ATryn® is used for the treatment of hereditary antithrombin deficiency. Ruconest® is C1 esterase inhibitor produced from transgenic rabbits and is used for the treatment of hereditary angioedema. Kanuma® is sebelipase-alfa produced from transgenic chickens; this recombinant protein is used for the treatment of lysosomal acid lipase deficiency. The transgenic animals are produced by random integration of the concerned transgenes. As a result, it is not possible to predict the transgene integration site, its expression level and the yield of recombinant protein. The SDN technology can be used to ensure transgene integration by HDR at a specified genomic site like the whey acidic protein locus that would support high expression level and generate little or no detrimental effect on the transgenic animals. For example, the transgene for lysostaphin was integrated within the cattle β-casein locus using ZFNs; the transgenic cows showed high expression level of the transgene and production of large volumes of recombinant lysostaphin (Yum et al., 2018).

Table 8.4. A list of some gene-edited animals developed to serve as bioreactors (based on Zhao et al., 2019)

Animal	Gene expressed	SDN	Modification	Product	Remarks
Goat	*human lactoferrin* (*hLF*)	TALENs	*hLF* integrated into the endogenous *β-lactoglobulin* (*BLG*) gene	hLF protein	High level hLF expression; absence of BLG in milk
Cattle	*human serum albumin* (*HSA*)	TALE nickase	Transgene integrated within endogenous *BLG* gene	Human serum albumin	High level of expression of human serum albumin
	Lysostaphin	ZFNs	Transgene integrated into *β-casein* locus	Lysostaphin	High level of expression and large volume production
Pig	*HSA*	CRISPR-Cas9	cDNA of *HSA* integrated into the endogenous *albumin* gene	Human serum albumin	HSA recovered from blood

Gene editing has been used to create human gene knockin animals for recovery of the concerned pharmaceutical proteins (Table 8.4). The recombinant proteins are recovered from either milk or blood of the gene-edited animals. For example, human serum albumin (HSA) is in great demand as it is used for the management of several severe diseases like liver failure and traumatic shock. The HSA derived from human blood is in short supply and there is risk of infection being present in the source human blood. Transgenic pigs were produced by knocking-in the *HSA* gene cDNA within the endogenous pig albumin locus by CRISPR-Cas9. The gene-edited piglets contained HSA protein in their blood. *HSA* gene has also been integrated in the *BLG* (*β-lactoglobulin*) gene of cattle using TALE nickases. The gene-edited cows showed high levels of HSA expression in their milk. The *human lactoferrin* (*hLF*) gene was integrated in the endogenous *BLG* gene of goats, and the transgenic goats showed large-scale hLF expression in their milk. At the same time, both goats and cows produced milk devoid of BLG, an allergen found in normal cow and goat milks (Peng et al., 2015; Zhao et al., 2019).

8.4.4. Gene editing for animal improvement (agricultural breeding)

Breeding of animals for traits relevant to commercial raising of herds/flocks and utilization of animal products, including product quantity and quality is generally referred to as agricultural breeding. Conventional breeding programmes utilized existing natural genetic variation by various mating schemes combined with selection. The modern versions of these schemes incorporate improved reproduction technologies, genomic selection, SCNT, and embryo transfer to enhance the genetic gain per year. The transgenic technology has also been deployed for trait modification or introduction of novel traits, but the regulatory approval of transgenic animals for food has been highly discouraging. The SDN-based gene editing technologies offer transgene or foreign DNA-free products, which may be more acceptable to consumers. This technology has been deployed to create many animal strains with desirable agricultural traits, such as milk quality, meat yield and/or quality, thermoregulation, disease resistance, etc. (Table 8.5). The proposed breeding schemes incorporating gene or genome editing have been termed as '*in vitro* breeding' or 'breeding by design'.

Table 8.5. A list of some gene-edited animals developed for enhanced production, produce quality, disease resistance, etc. (based on Tait-Burkard et al., 2018; Zhao et al., 2019, and other sources)

Animal	Gene	SDN	Trait modified	Remarks
Cattle	*β-lactoglobulin (BLG)*	ZFNs and TALENs	Allergenic property of BLG	Milk from TALEN-edited cattle free from mature BLG molecules
	PRNP (prion protein)	TALENs, CRISPR-Cas9	Prion-free cattle	One PRNP mutant pregnancy by SCNT; characterization of the calf not known
Pig	*MSTN (myostatin)*	ZFN	Increased lean meat	In Chinese Meishan pigs, increase in muscle by 100% and decrease in fat mass; in other Chinese breeds, abnormal legs; even piglet mortality
	FBXO40	CRISPR-Cas9	Muscle hypertrophy phenotype	Normal survival; no detectable pathology changes in major organs
	mouse adiponectin-UCP1	CRISPR-Cas9	Improved thermoregulation under cold exposure	Pigs lack UCP1 protein; *UCP1*-knockin pigs showed improved thermoregulation and reduced fat deposition
	CD163	CRISPR-Cas9	PRRSV (Porcine Reproductive and Respiratory Syndrome Virus) resistance	Receptor for PRRSV; no clinical signs observable in *CD163* knockout pigs when exposed to PRRSV
Goat	*MSTN*	CRISPR-Cas9	Improved meat quality	Enhanced body weight and muscle fibre size
	fat-1 gene	CRISPR-Cas9	*fat-1* gene integrated within the endogenous *MSTN* gene	Improved muscle growth and reduced n-6 PUFA* to n-3 PUFA ratio
Cattle	*MSTN*	TALEN	Increased lean meat	Enhanced body weight

Animal	Gene	SDN	Trait modified	Remarks
	mouse SP110 gene	TALE nickase	Resistance to bovine tuberculosis (*Mycobacterium bovis*)	*Mouse SP110* knockin cattle showed increased resistance to tuberculosis
	Human lysozyme gene	ZFN	Mastitis resistance	Transgene integrated within the *β-casein* locus
	POLLED	TALEN	Hornless cattle	*POLLED* allele knockin using the SDN
Sheep	*MSTN*	CRISPR-Cas9	Increased lean meat	Increased body mass and muscle growth

* PUFA, polyunsaturated fatty acid.

8.4.4.1. *Milk quality*

Human milk is typically free from BLG, which is the main protein present in whey from ruminants, including goat and cow. BLG is known to cause mild to severe allergy symptoms that can be life-threatening. The allergenic potential of BLG is generally reduced by either heat processing or enzymatic degradation, but these procedures add to the cost and might adversely affect the nutritive value of whey. Gene editing based on ZFNs and TALENs was used to inactivate the BLG encoding gene of cattle. The milk from gene-edited cows was free from BLG. This application of gene editing improves milk quality by eliminating an allergenic protein that would otherwise necessitate costly heat or enzymatic processing of whey.

8.4.4.2. *Meat yield and quality*

The primary function of myostatin (MSTN) protein is negative regulation of muscle growth. Natural mutations of *MSTN* gene were first described in cattle and later discovered in sheep, dogs and humans; these mutations lead to increased muscle growth and generate the 'double muscle phenotype'. The natural mutations in *MSTN* gene are caused by one of the following changes at the DNA level: an 11-bp deletion (in Belgian Blue cattle), a SNP (single-nucleotide polymorphism; in Piedmontese cattle) and a mutation that creates target sites for two microRNAs, which downregulate the *MSTN* mRNA (in Texel breed of sheep). Thus, *MSTN* gene is an attractive target for gene editing to increase lean meat yield from livestock. This gene has been edited in cattle, sheep, pig, goat, and Channel Catfish; in general, the gene-edited animals showed increased body mass and muscle growth. But pigs homozygous for the *MSTN* knockout allele suffered from some abnormalities including leg weakness syndrome, inability to stand and walk, and even death. However, a pig breed that accumulates high levels of fat is not reported to show abnormalities due to *MSTN* disruption. Thus, it would be desirable to subject homozygous *MSTN* knockout animals to rigorous genome sequencing for ruling out off-target editing, and to detailed phenotypic evaluation for determining their usefulness for meat production (Tait-Burkard et al., 2018; Zhao et al., 2019).

8.4.4.3. *Disease resistance*

Gene editing can be used to generate resistance to certain important diseases of livestock. For example, porcine reproductive and respiratory syndrome (PRRS) is caused by a virus and it is the most important disease of pigs worldwide. PRRS is estimated to cause economic losses of 2.5

billion US dollars annually in the USA and Europe. Vaccines against this virus provide limited protection mainly due to the existence of genetic diversity in the pathogen. The pig gene encoding protein CD163 serves as the receptor for PRRS virus; therefore, it is the obvious target for inactivation to block infection by this virus. CRISPR-Cas9 was used to induce the following four types of changes in the *CD163* gene of pigs: deletion of the gene, deletion of exon 7 of the gene (this exon encodes the domain SRCR5 of CD163 protein), generating a stop codon by NHEJ in exon 7, and replacing the exon 7 of pig gene with the orthologous sequence from human gene (domain swapping). These modifications conferred complete resistance to PRRS virus, except in the case of domain swapping, which generated complete resistance to Type 1 but not to Type 2 PRRSV. Thus, inactivation of SRCR5 domain of *CD163* gene or null mutation of this gene generates complete resistance to PRRSV (Tait-Burkard et al., 2018; Yang and Wu, 2018; Zhao et al., 2019). Further research and development efforts are being made to commercialize the gene-edited PRRS resistant pigs.

The bacterial pathogen *Mycobacterium bovis* is the causal organism of tuberculosis in cattle. This disease is developing into a serious threat to global cattle industry and represents a serious hazard to human health since the pathogen can be transmitted from cattle to humans. The gene encoding the natural resistance-associated macrophage protein-1 (NRAMP1) generates innate resistance to several intracellular pathogens like *Mycobacterium*, *Leishmania*, *Salmonella* and *Brucella*. Transgenic cattle expressing the *NRAMP1* gene show resistance to bovine tuberculosis. But integration of the resilient allele of *NRAMP1* gene into cattle genome by means of CRISPR–Cas9 resulted in reduced growth of *M. bovis* in peripheral blood monocytes and lower interferon response in the gene-edited animals. The resilient or tolerant animals do not develop symptoms of the disease and maintain an acceptable level of productivity when they are infected by the concerned disease. The mouse gene *SP110* has been integrated into cattle genome using TALEN, and the transgenic cattle showed greatly reduced growth of *Mycobacterium* and transmission of tuberculosis to their penmates. This gene generates its effects by limiting growth of the bacterium in infected macrophages and causing apoptosis of the infected cells (Tait-Burkard et al., 2018; Zhao et al., 2019).

The anaerobic Gram-negative bacterial pathogen *Mannheimia haemolytica* causes shipping fever, enzootic pneumonia and peritonitis in calves, sheep and lambs. It produces a leukotoxin that binds to the signal peptide of protein CD18 present on the surface of animal leucocytes. The mature CD18 proteins of mice and humans do not have the signal peptide and these species are not affected by the leukotoxin. ZFNs were used to mutate the CD18 protein of cattle by replacement of one amino acid. The leucocytes of the gene-edited foetuses did not show the symptoms of cytotoxicity caused by *M. haemolytica* leukotoxin (Tait-Burkard et al., 2018).

The epidemiological theory predicts that only a proportion of the herd needs to be gene-edited to avoid spreading of a disease in the entire local population. Since a variety of factors can influence such estimates, refined models need to be developed for a reliable prediction of the proportion of animals that must be gene-edited for resistance to a given disease in each herd. However, disease resistance generated by gene editing can be overcome by new pathogen strains produced by mutation and/or recombination, and mutation rates are particularly high in some pathogens like retroviruses. On the other hand, disease resilience or tolerance imposes much less selection pressure on the pathogen for evolution of novel virulence. But tolerance does not prevent infection of the resilient animals and the disease would spread to the other animals in the herd. Therefore, the presence of tolerant individuals in a herd is likely to increase the incidence and prevalence of the concerned disease in a herd (Tait-Burkard et al., 2018).

8.4.4.4. *Thermoregulation*

Livestock generally possess the ability to regulate their body temperatures so that they can avoid ill-effects of cold stress. Pigs do not have a functional *UCP1* gene, as a result of which they lack brown adipose tissue (BAT) that mediates adaptive non-shivering thermogenesis. Therefore, piglets are sensitive to low temperatures and may die under cold stress. The *mouse adiponectin-UCP1* gene was integrated into the pig genome using CRISPR-Cas9. These gene-edited pigs showed improved thermoregulation and reduced fat deposition, which is an important pig breeding objective. Further, they were able to maintain body temperature when exposed to extreme cold. These gene-edited pigs can serve as a valuable genetic resource for pig breeding and production (Zheng et al., 2017; Yang and Wu, 2018).

8.4.4.5. *Welfare traits*

In cattle, traits like polled or hornless are called welfare traits since they are important for protection of both humans as well as the animals from injury. A vast majority of producers of dairy cattle resort to dehorning, which is costly, unkind and open to criticism. It is, therefore, highly desirable to increase the frequency of polled allele in the popular breeds. The *polled* allele is prevalent in beef cattle, but its frequency is very low in dairy cattle, *e.g.*, <0.01 in the breeds of U.S.A. The genetic basis of polled phenotype is not clear as it has been reported to be associated with an indel, a duplication and an SNP at the *POLLED* locus. Fibroblast cell lines from young embryos were used for editing the *POLLED* locus with TALENs. The gene-edited cells were used for SCNT followed by embryo transfer to obtain two surviving polled calves (Tait-Burkard et al., 2018; Zhao et al., 2019).

A simulation study was carried out to compare gene editing with conventional breeding for eliminating *horned* allele from the Jersey population. It was concluded that the decline in frequency of *horned* allele was faster in case of gene editing than conventional breeding. In addition, the inbreeding was at acceptable level and the genetic gain was maintained when gene editing was used (Mueller et al., 2018).

8.4.4.6. *Quantitative traits*

The usefulness of gene editing for improvement of quantitative traits has been investigated *in silico*. A computer simulation study compared the effectiveness of genome editing plus genomic selection with that of genomic selection alone. The quantitative trait used in simulation had 10,000 quantitative trait nucleotides (QTNs), and the number of edits per sire ranged from 0-100, while that per generation varied from zero to 500. The response to genome editing plus genomic selection ranged between 1.08 and 4.12 times higher than that for genomic selection only after 20 generations of selection; the response was higher with more edits per sire and greater number of sires edited. It was predicted that 1–20 beneficial edits per quantitative trait could maintain or even accelerate the genetic gain. It was concluded that genome editing can contribute significantly to livestock breeding programmes, but the level of inbreeding needs to be controlled. This approach of breeding has been termed as promotion of alleles by genome editing (Jenko et al., 2015).

Another computer simulation study showed that gene editing in even a small proportion of the select animals effectively reduces the frequency of harmful alleles in dairy cattle. Further, the efficiency of gene editing technology had more pronounced effect on allele frequency than the proportion of select animals subjected to gene editing. It was suggested that gene editing should focus on loci with large, positive effects on animal health, while the recessive alleles having deleterious effects on quantitative traits of animals can be better handled by suitable mating schemes than by gene editing (Cole, 2017).

8.4.5. Implementation of modern breeding schemes

Genome editing can be used to facilitate implementation of modern breeding schemes in such geographical areas and/or animals like chicken, in which cryopreservation of semen and/or egg/young embryo is either inefficient or not feasible at all. The modern breeding schemes use techniques like artificial insemination with semen from high breeding value males, IVF and embryo transfer techniques, which rely on cryopreservation of semen and/or egg/young embryo. One answer to this problem is to create sterile animals, into which germ cells from individuals with high breeding value can be transplanted. Sterile animals can be created by targeted gene editing to inactivate gene(s) concerned with germ-cell development. Gene editing has been used to produce sterile pigs (knockout of *NANOS2* gene) and chickens. The early diploid germ cells from chicken embryos can be successfully cryopreserved. When these germ cells were transplanted into proxy or surrogate individuals, they produced functional gametes. Thus, the cryopreserved germ cells can be transplanted into the sterile chickens to reconstitute pure poultry flocks. Further, the animals with transplanted germplasm can be distributed to such areas that lack facilities for cryopreservation and artificial insemination for use in breeding programmes (Tait-Burkard et al., 2018; Yang and Wu, 2018).

8.5. CRISPR BABIES

On November 28, 2018, the Chinese scientist He Jiankui described his work on gene editing of human embryos at a gene-editing conference in Hong Kong. His work had resulted in the birth of gene-edited twin girls early in November 2018, and the establishment of a second pregnancy with a gene-edited embryo. He was severely criticized at the meeting, and his findings remain unpublished. Subsequently, He was 'censured by the health ministry in Guangdong, where he worked, and fired from his university.' He carried out human germline editing when many questions, including the implications of inactivation of the targeted gene and the off-target effects of gene editing technology are yet to be resolved. In addition, He 'forged ethics-review documents during recruitment of participants and swapped blood samples to skirt laws against allowing people with HIV (human immunodeficiency virus) to use assisted reproductive technologies' (Editor, 2019; Cyranoski, 2019b). It has been suggested that since there has been no independent confirmation of He's claims, it could well be a very bold fraud similar to the claim of cloning human embryos by Hwang Woo-Suk in 2004-05.

He Jiankui had edited the human embryos to inactivate *CCR5* gene by induction of the *CCR5*-Δ32 (pronounced as "CCR5-delta-32"), which occurs naturally in human population and makes the affected individuals resistant to HIV. But there is no definitive evidence that *CCR5* gene was in fact inactivated in the embryos edited by He. The CCR5 protein is thought to contribute to protection of people from other infections like flu and West Nile Virus. Further, homozygotes for inactivated *CCR5* gene may have shortened lifespan. On the other hand, there is some evidence that *CCR5*-Δ32 mutation increases recovery from stroke. Thus, it seems that the full spectrum of effects generated by *CCR5*-Δ32 mutation is not yet known. For some years now, many scientists have been editing human embryos for basic research. Whether this type of research would receive a backlash due to the He episode is presently uncertain, but a call for international moratorium on all forms of human embryo gene editing has already been issued. On the other hand, some well-known scientists support human embryo gene editing for implantation on the grounds that it provides a useful alternative to preimplantation genetic diagnosis (PGD) and for some couples this may be the only viable option to produce healthy babies. PGD involves screening of multiple embryos generated by *in vitro* fertilization for specific mutations to select a healthy one for implantation. Meanwhile, a Russian scientist has announced his plans to proceed with human germline editing for inactivation of the *CCR5* gene (Editor, 2019; Cyranoski, 2019b).

There are several limitations of gene editing by CRISPR-Cas9, including off-target editing and unintended on-target editing (Section 8.6). Unintended on-target editing seems to be more challenging to manage since there is only limited information on the preferred mechanisms of DSB repair in human embryos. The 'prime editing' technique is reported to induce high frequency of HDR and avoid unintended on-target editing (Section 3.10.5). It may be expected that refined reagents and procedures might soon make gene editing and base editing much safer or even completely safe. But the question remains as to what kinds of sequence changes in the human germline are safe. The public opinion tends to favour human germline editing for treating/preventing incurable diseases, but it does not support editing for enhancing performance. There is, however, some degree of disagreement among scientists as to which diseases merit gene editing, but it is hoped that an acceptable list of ailments would soon be developed (Ledford, 2019a). The US National Academies of Sciences, Engineering, and Medicine headed an international initiative that outlined the prerequisites for human germline editing. One of these requirements is that the sequences generated by gene editing must be common in the human gene pool and there must not be any known risk of disease (NASEM, 2017).

8.6. OFF-TARGET AND UNINTENDED ON-TARGET EDITING

Generally, the action of CRISPR-Cas9 (and other SDNs) is investigated in ~1 kb genomic region surrounding the targeted site and *in silico* predicted off-target sites. These analyses preclude the detection of changes in regions distal to the DSBs, and have led to the conclusion that the CRISPR-Cas9 SDN is reasonably specific. However, it is known to induce variable frequencies of off-target editing, which is affected by several factors, including the targeted genomic site. Even the high specificity variants of Cas9 are known to induce some off-target edits (Section 5.9). It is generally believed that NHJ-mediated DSB repair induces indels of less than 20 bp, and larger deletions are rare. But recent detailed investigations reveal that Cas9 used with a single sgRNA induced several different types of unintended on-target edits, including several kilobases long deletions, asymmetric deletions, inversions, other more complex genomic rearrangements and even crossover events in different cell types. Further, alterations were also observed at genomic sites distal to the targeted sites. When pairs of sgRNAs are used to induce localised deletions, complex genomic aberrations like longer than intended deletions, inversions, and insertion of both endogenous and exogenous DNA are also produced. Further, deletion homozygotes may be produced, many of which may have two different deletion alleles differing in the size of deleted sequences. It seems that the intervening regions between the two DSBs tend to become inverted when the intervening sequences are large. Finally, simultaneous use of several sgRNAs targeting a locus almost exclusively leads to large deletions that include the entire genomic region demarcated by the two outermost sgRNAs. *It is, therefore, desirable to use long-range PCR to amplify up to 14 kb genomic regions around the DSBs and combine it with long-read sequencing to detect unintended on-target edits* (Kosicki et al., 2018; Shin, H.Y. et al., 2017).

There is little information so far on the occurrence of off-target gene editing and unintended editing at the on-target sites of gene-edited animals. The whole genome sequence data of the polled calves posted online by Recombinetics were analysed by FDA, USA. It was discovered that both the calves had the intended edit in one of the alleles of the *polled* locus, while their second allele contained the intended edit as well as 'the plasmid and a second copy of the repair template sequence at the target site.' The plasmid sequence included two antibiotic (neomycin/kanamycin and ampicillin) resistance genes. These claves were produced by using TALENs to introduce the *Celtic polled* allele (*Pc*; produced by a 212-bp duplication in the *polled* locus) by HDR using a donor DNA plasmid as template. The authors concluded that in earlier reports integration of template plasmid in the genomes of gene-edited individuals was either 'under reported or overlooked.' They further concluded that the standard methods of screening for detection of unintended genome sequence changes in gene-edited individuals are inadequate and some modifications for the 'detection of plasmid integration and integration of multiple template copies' were suggested (Norris et al., 2019).

8.7. ETHICAL ISSUES

Gene editing and cloning of livestock is being increasingly used especially for furthering basic research. These research applications are generally accepted by the biologists and, to a large extent, by the public at large. But the various applications of gene-editing technology are generally viewed differently, and the products from gene editing for agricultural traits must be cleared by regulatory authorities in USA and EU. The use of gene-edited animals as disease models for various biomedical researches raises certain bioethical concerns, which largely depend on the animal species used for the purpose. Bioethicists place considerable emphasis on assigning a moral status to various animal species. However, the animal moral status is a subject of continuing debate among the bioethicists. The moral status depends mainly on the sentience and self-consciousness of the concerned species. Sentience is 'the ability to feel and to suffer', while self-consciousness describes 'the ability to reason, think, remember, and know'. One of the main issues debated by animal bioethicists is whether human beings have a special status among the diverse animal species.

The use of moral status as the criterion for selection of an animal species to model human diseases presents the following paradox: the closer is the evolutionary relationship between the animal species and humans, the better it would serve as disease model, and the more bioethical questions are likely to be raised against such use. For example, mouse and rat models have been created and used for biomedical research for several decades without any serious bioethical objection. But the use of NHPs for the same purpose is questioned since these animals are assigned a much higher moral status than other animal species. On the other hand, scientists argue that NHPs are matchless as animal models for brain research, and biomedical studies on neurological and cardiac disorders. A possible reason for objections to the development and use of novel tools for genetic modifications in animals like NHPs is the fear that sooner or later the same tools might be used in human beings. It is important that this fear is justified by our past experiences. It is also argued that humans tend to consider other animal species as commodities and feel that they can do whatever they wish with them; such a viewpoint may result in other animal species being treated inappropriately (Ogden, 2019).

In addition to the bioethical issues, gene-edited lines of farm animals must find regulatory approval and consumer acceptance since the products from these lines must be marketed and consumed by the public at large. The national policies and public opinion about gene-edited animals have not been able to keep pace with the developments in the field of biotechnology. As a result, the policies of different countries are often quite divergent. For example, cattle, goats, and pigs developed by the SCNT procedure are accepted as food in the USA, but the European Union does not accept these animals as food. Similarly, the gene-edited (SDN-1 and some SDN-2) plant varieties are treated as non-GM (non-genetically modified) in USA, while gene-edited animals are regulated as new drugs (Section 8.8). The farm animals edited to function as bioreactors are likely to be treated in the same way as the animal bioreactors produced by transgenic technology (Yum et al., 2018; Van Eenennaam et al., 2019).

8.8. REGULATORY PERSPECTIVES

The enormous present-day animal breeds and plant varieties have been developed by breeders using breeding schemes that utilize genetic variation represented as alleles of different genes by selecting for the desirable genotypes/phenotypes. The various alleles represent changes in DNA sequences that arise either naturally or in response to treatments with chemical or physical mutagens. The existing variation in DNA sequences of different livestock species is phenomenal. For example, the U.S. 1000 Bull Genomes Project identified 86.5 million variants (84 million SNPs, 2.5 million indels) in the whole genome sequences of 2,703 cattle. According to a recent study, every animal is expected to carry an average of 60 new SNPs and 5 new indels. Thus, everyday animal and plant products containing myriads of DNA variations are consumed as food seemingly without any ill-effect on consumers. In short, 'genetic variation *per se* does not pose a unique hazard as it relates to food safety' (Van Eenennaam et al., 2019).

The animal breeds and plant varieties developed by conventional breeding methods, including chemical and physical mutagenesis, are neither analysed for unintended changes at the molecular level nor subjected to health and environmental safety evaluations before commercialization. Gene editing uses SDNs to bring about gene inactivation by inducing indels or SNPs, and gene activation by sequence replacement or gene knockin; all these changes are induced at the specified genomic sites in a highly specific manner. The USDA (US Department of Agriculture) announced in 2018 that it would not regulate gene-edited crop varieties 'so long as the crop is not a plant pest or developed using plant pests.' But in 2017, FDA, USA released its draft proposal entitled "Regulation of Intentionally Altered Genomic DNA in Animals". This proposal treats the gene-edited food animals as drugs and intends to regulate them accordingly (Van Eenennaam, 2018; Van Eenennaam et al., 2019).

It is unreasonable or even misleading to treat an intentionally induced change in DNA sequence as drug even when the same or similar change exists in the gene pool of the concerned species. The FDA proposal intends to use the same approach for gene-edited animals that has been applied to transgenic animals. Thus, FDA regulates transgenic as well as gene-edited animals based on the method of their production in the place of characteristics of the animals themselves. Significantly, no food from a transgenic animal had reached the US market till the beginning of 2019. Although the transgenic salmon earned FDA approval in 2015, it cannot be marketed in USA in the absence of a defined product labelling policy. But this salmon can be marketed in Canada. The cost of obtaining regulatory approval is prohibitive and it requires considerable time and effort as well. This situation has limited or even prevented the use of transgenic technology by public sector institutions and small companies for the development of animals with novel traits (Van Eenennaam, 2018; Van Eenennaam et al., 2019).

It makes little sense to regulate polled cattle developed by gene editing differently from the naturally occurring polled individuals having the same allele. Further, regulatory oversight should be rigid for products that pose high risk, while it should be relaxed for products with little or no risk. The proposed US regulation requires genotypic and phenotypic evaluation over multiple generations, and inheritance studies for at least two generations. These and other requirements are rather exacting and would discourage the use of gene editing in livestock breeding programmes. In addition, it would force research as well as commercialization activities related to gene-edited livestock to be shifted to countries with more favourable regulatory regimes. It may be pointed out that Argentina proposes to regulate gene-edited plants and animals in the same way, and both of them will not require regulatory approval if they do not contain "novel combination of genetic material", *i.e.*, foreign DNA and new combinations of genetic material. Canada and Brazil follow similar approaches. The 2019 Plant and Animal Genome meeting has petitioned that plant and animal lines developed by modern methods like gene editing, which can be developed by traditional breeding methods as well should not be subjected to regulation (Van Eenennaam, 2018; Van Eenennaam et al., 2019).

8.9. ADVANTAGES AND LIMITATIONS

Gene editing can be used to generate the desired mutation, including correction of the mutant allele by sequence replacement in a highly specific manner. This technology is particularly suited for mutation induction in oligogenic or qualitative traits, but it can also be used to modify quantitative traits. Since it is expected to affect only the targeted loci, the technology can be gainfully used for modification of specific traits in well-established livestock breeds without affecting their characteristic features. The gene knockouts generated by gene editing are free from foreign DNA, which is an advantage over the transgenic technology. Further, the sequence changes produced by gene editing involve indel production and intended sequence replacement for mutant correction/induction, and these sequence variations are present in the gene pool of the species. Thus, these changes are not entirely new, and they are comparable to the natural sequence variation.

Gene-editing enables targeted genetic changes in livestock breeds/lines more rapidly, efficiently and precisely in comparison to the conventional breeding methods. This technology can be used to generate in domesticated animals the beneficial genetic changes that exist in natural populations. For example, resilience to ASFV (African swine fever virus) is found in the wild warthog population but not in domesticated pigs. This allele has been created in pigs by gene editing, and the gene-edited pigs are expected to be resilient to the pathogen. Gene editing can also create such alleles of endogenous genes that are not known in a species and its relatives. For example, resistance to PRRSV is not known in either pigs or its relatives. The exon that encodes the domain SRCR5 of CD163 protein (receptor of the virus), which is involved in interaction with PRRSV, was deleted to create a 'designer allele' of *CD163* gene; this allele confers complete resistance to the virus. *These applications of gene editing provide examples that cannot be achieved by any other breeding scheme, including genomic selection* (Tait-Burkard et al., 2018).

The SDN technology can be used to create gene knockins; this strategy is superior to the transgenic technology since it ensures transgene integration at the predetermined genomic sites. This eliminates the risk of position effects on transgene expression and the unpredictable consequences to the transgenic animals due to unintended inactivation of an essential gene or activation of a detrimental gene as a result of the transgene integration. Therefore, gene knockin livestock produced by SDN-3 technology would be more desirable as bioreactors, organ donors, etc., than those obtained by the transgenic technology.

The SDNs used for gene editing also cause, albeit in low frequencies, off-target gene edits and unintended on-target edits. So far, the focus of different investigators has been to demonstrate successful editing of the targeted genes leading to the expected phenotypes in gene-edited lines. There is little information so far on the occurrence of off-target gene editing and unintended editing at the on-target sites. However, an unexpected and disturbing finding concerns the detection of unintended donor plasmid sequences in the genomes of gene-edited hornless calves that were earlier claimed to be free from off-target edits and foreign DNA sequences (Norris et al., 2019). It was expected that gene-edited animals would not be regulated in the same way as transgenic animals, but this does not seem to be the likely approach at least in U.S.A. (Section 8.8). This is one of the major limitations of gene editing technology as it has created a dearth of public funding for research projects in this area. It is feared that getting funds from the industry too might become increasingly difficult if the gene-edited livestock products are tightly regulated irrespective of the characteristics of the editing products.

A tight regulatory regime might cause migration of research and, especially, commercialization activities to other countries with more favourable regulatory regimes. For example, some of the companies that have developed gene-edited animals with valuable traits are already seeking to commercialize them in other countries in the place of seeking FDA approval in USA. The company AquaBounty Technologies, which had developed the fast-growing transgenic salmon AquqAdvantage, is trying to market its gene-edited fast-growing tilapia fish in Argentina, which has exempted it from regulation. Similarly, Recombinetics, St. Paul, Minnesota is attempting to commercialise its gene-edited polled cattle in Brazil after FDA did not grant approval (Ledford, 2019c). However, the genomes of Recombinetics polled calves contained the donor plasmid, including included two antibiotic resistance genes as well a second copy of the donor DNA sequence at the target site (Norris et al., 2019).

8.10. ACHIEVEMENTS

Gene editing has been successfully used for genetic modification of livestock species to achieve the following objectives: (i) creation of animal models of human diseases, (ii) development of livestock strains suitable as donors of organs for xenotransplantation in humans, (iii) generating animal bioreactors for production of pharmaceutically valuable proteins, and (iv) improvement of livestock breeds for agricultural traits. Several animal models have been developed in NHPs, pigs and dogs (Table 8.2). It is expected that these models will be particularly useful for biomedical researches on complex human diseases concerned with brain and heart functions. The pig models have been used for investigating PD pathology and to develop therapeutic interventions for the disease. Pigs appear to be the most promising as organ donors. Many of the genes involved in xenotransplant rejection have been modified in pigs leading to increased xenotransplant survival in NHPs (Table 8.3). Further fine-tuned genetic alterations in pigs may ultimately enable their use or organ donors for human patients with terminal stage organ failure. Transgenic strains of cattle, goat and pig have been developed to serve as bioreactors for production of human lactoferrin, HSA, and lysostaphin. These strains show high levels of production of the concerned proteins, but they are yet to achieve commercial status.

Several different strains of gene-edited cattle, pig, goat and sheep have been developed for improved agricultural traits like milk/meat production, produce quality, disease resistance, etc. (Table 8.5). Some of the gene-edited animal strains on the verge of commercialization are, gene-edited fast-growing fish tilapia developed by AquaBounty Technologies, and the gene-edited polled cattle generated by Recombinetics, U.S.A. Efforts are also being made to commercialize the PRRS resistant gene-edited pig strain.

8.11. FUTURE PERSPECTIVES

Considerable progress has been made in gene editing of large animals, including farm animals. However, the genome editing technology needs to be further refined to make it more efficient and, particularly, more specific in view of the recent reports of unintended on-target genomic alterations. It might be necessary to introduce some modifications in the DNA repair pathways to be able to attain some control on the type of genomic reorganizations induced by SDNs since the role of SDNs is expected to be limited to induction of DSBs. Further, zygote microinjection generally produces chimeras; therefore, a strategy for rapid single-step establishment of pure gene-edited animals is highly desirable. Some strategies for reducing chimera production are being developed (Zhao et al. 2019). The efficiency of SCNT needs to be further improved especially when adult fibroblast cell cultures are used for gene editing. In addition, SCNT procedure needs to be refined to minimize the risk of undesirable phenotypes being present in the SCNT-derived progeny.

Organ xenotransplantation from gene-edited animals raises safety as well as ethical issues. Considerable genetic refinements are still required in the gene-edited animals to make their xenotransplantats to be safe enough for clinical use in humans. The blastocyst complementation strategy seems to be a promising approach for generating specific organs for transplantation in human patients. This approach generates interspecies chimeras, which might serve as donors of human organs. One of the major concerns related to this method is the risk of integration of human cells into the brain and/or the germline of some of the chimera individuals leading to unexpected and possibly unacceptable consequences. It is, therefore, extremely important that the blastocyst complementation technology is refined and streamlined to such an extent that these possibilities are virtually ruled out.

So far, only traits controlled by single oligogenes with large effects have been subjected to gene editing. Even for many oligogenic traits, the specific genes governing them need to be conclusively identified, and the types of sequence changes needed to generate the desired phenotypic effects need to be reliably established. At the same time, many traits of interest, including milk/meat/egg production, etc. are governed by several genes with small individual effects. The polygenes are mapped with the help of molecular markers as quantitative trait loci (QTLs). Each QTL may cover relatively large genomic region and would usually contain more than one gene. It would be essential to establish the specific gene located in a QTL that is involved in the expression of the concerned quantitative trait. Further, the specific sequence change in the given gene that would generate the desired phenotype of the concerned trait must be determined before gene editing can be applied for improvement of such traits. Integrated and co-ordinated international research efforts to identify target genes and the desirable sequence changes could result in huge benefits to livestock breeding programmes and, ultimately, the livestock industry. Further, it would be necessary to edit multiple QTLs for modification of each quantitative trait; therefore, techniques for multiplexed gene editing must be refined so that they can be integrated as a routine component of animal breeding programmes. Computer simulations reveal that genome editing in combination with genomic selection may be useful for increasing frequencies of positive alleles of polygenes and reducing the frequencies of deleterious alleles (Tait-Burkard et al., 2018; Zhao et al. 2019).

According to the epidemiological theory, only a proportion of the herd needs to be gene-edited for resistance to a disease in order to avoid spreading of the disease in the entire herd/flock. It is well known that a variety of factors can influence the estimates of proportion of local animal population that must be gene-edited. Therefore, the prediction models need to be further refined for obtaining dependable estimates (Tait-Burkard et al., 2018). Finally, the regulatory regime applied to a new technology determines the time, effort and money that must be invested in seeking approval for commercialization of its products and may also affect their consumer acceptance. The proposed FDA draft regulations for gene-edited animals are generally regarded as strict since they are more severe than those applied to the gene-edited crop plants. The future course of research and development activities in gene editing of livestock will be determined, to a large extent, by the regulatory regime applied to products of this technology particularly in U.S.A. and EU.

8.12. CONCLUSIONS

Farmers, industry, consumers as well as the environment have all benefited from conventional animal breeding schemes with or without integrated modern reproduction technologies and genomic selection. However, these breeding strategies can only utilize the genetic variation already present in the livestock species. The modern genomic resources have enabled the development of gene editing technology that has been applied for the improvement of agriculturally relevant traits of livestock. In addition, it has been used to generate livestock strains suitable for various nonconventional uses. For example, gene-edited animals have been developed as models of many human diseases, organ donors for xenotransplantation in humans, and bioreactors to produce pharmaceutically important proteins. It is hoped that the livestock disease models would greatly facilitate biomedical research, including development of novel therapeutic agents and procedures, including those in regenerative medicine.

ZFNs, TALENs and CRISPR-Cas9 have been used for gene editing in large animals, but the CRISPR-Cas9 system is the current SDN of choice. Gene-edited animals are either produced by microinjection of the SDN with or without donor DNA plasmid preferably into the cytoplasm of zygotes/early embryos. Alternatively, cultured fibroblast cells are gene-edited, cloned and characterized before they are used for SCNT procedure. The embryo microinjection procedure is more efficient, but it generally yields mosaic progeny, while fibroblast cell-based method yields pure progeny, but the procedure has low efficiency. Animal models of many diseases have been developed in large animals like NHPs (monkeys), pigs and dogs. It is believed that these models, especially the NHP models, will be better suited for biomedical researches and therapy development particularly for complex diseases like those concerning brain and heart functions. But breeding of large animals is problematic and costly, and their use for biomedical research may be ethically objectionable particularly in the case of NHPs. Overall, pig models seem to be more useful in biomedical research than mouse and rabbit models, and they are likely to raise few, if any, ethical objections.

Pigs appear to be a potential source of organs for xenotransplantation in humans. But several pig proteins contribute to acute, delayed and chronic xenotransplant rejection reactions. Gene editing has been used either to inactivate some of the relevant genes or to introduce their human counterparts to generate pigs, xenotransplants from which show improved survival in NHPs. But considerable work is still required to make the gene-edited pig strains suitable for xenotransplantation in humans. The emerging technique of blastocyst complementation seems to hold considerable promise for producing human organs in livestock. But the technique is still in developing stages and some important safety issues like integration of the hiPSCs in brains and/or germline of the resulting chimeras need to be addressed. Animal bioreactors have been created to produce at least three therapeutic human proteins, *viz*., hLF, HSA and Lysostaphin, but they are yet to achieve commercial status. Animal strains with improved traits like enhanced meat production, increased disease resistance, improved milk quality (allergen-free), etc. have been developed and at least three such strains are in the process of being commercialized. So far, gene editing has been confined to oligogenic traits with single gene control, but computer simulation studies indicate that this technology can be gainfully used for improvement of quantitative traits as well.

The SDNs are known to induce off-target editing as well as unintended on-target edits. However, there is little data on these aspects in gene-edited animals possibly because of inadequacies of the detection techniques used and/or even understatement of such changes. In addition, the bioethical aspects of many livestock gene editing applications, especially those concerning biomedical researches based on NHPs, need to be deliberated to develop a widely accepted guideline. Meanwhile, the birth of twin 'CRISPR babies' in China has surprised and shocked the biological scientists and the public at large, and the pros and cons of this development are being hotly debated. In addition, the regulation of gene-edited animals for human use is not clear at least in USA and EU, and the proposed US regulations seem to be somewhat unfair and rather harsh. It is feared that a strict regulatory regime would discourage research funding for gene editing in livestock, which will slow down research and development activities in this important emerging area.

■■■■

9

HUMAN HEALTH

"Genome editing — a suite of methods for creating changes in DNA more accurately and flexibly than previous approaches — was hailed as the 2011 Method of the Year by Nature Methods, and the CRISPR/Cas9 system of genome editing was named the 2015 Breakthrough of the Year by Science. The technology has excited interest across the globe because of the insights it may offer into fundamental biological processes and the advances it may bring to human health."

— NASEM (2017)

9.1. INTRODUCTION

The motivation for research efforts to develop site-directed nucleases (SDNs) was provided by the observation that double-strand breaks (DSBs) in genomic DNA promote transgene integration by homologous recombination or, more appropriately, homology-directed repair (HDR) of the DSBs. This observation suggested that targeted DSB induction by SDNs could enable the correction of specific gene mutations causing genetic diseases by HDR-mediated sequence replacement in cells *in vitro*, and the gene-corrected cells could be used for transplantation into patients for gene therapy. The zinc finger nucleases (ZFNs) were the first programmable nucleases successfully used for targeted gene editing. The ZFNs triggered a burst of research activity in this field that soon led to the development of several SDN systems. Of the various SDN systems, ZFNs, transcription activator like effector nucleases (TALENs) and CRISPR-Cas9 (clustered regularly interspaced short palindromic repeat-CRISPR-associated 9) systems have emerged as efficient gene editing tools. They have been used for site-specific induction of DSBs and a variety of genetic changes, including base substitutions, HDR-mediated sequence replacement and targeted transgene integration. The CRISPR-Cas9 system has rapidly overtaken the other two SDNs to become the most widely used gene editing tool due to its simplicity, versatility, low cost and ease of implementation. But ZFNs continue to find therapeutic applications and many of the gene therapy protocols are based on them.

The possible valuable contributions of SDNs were predicted in the early years of their development by Chandrasegaran and Smith (1999) who wrote, "The availability of chimeric nucleases, a new type of molecular scissors that target a specific site within the human genome, will likely contribute and greatly aid the feasibility of genome engineering and, in particular, *ex vivo* gene therapy using stem cells". The CRISPR-Cas system has emerged as a powerful experimental tool that has enabled the development of additional useful and effective gene editing tools like base editors and transcription activators/repressors, etc. Together these tools enable a variety of modifications in gene structure and/or function that would have extremely valuable therapeutic applications (Goodman et al., 2017).

Table 9.1. Selected examples of various CRISPR/Cas9 approaches and their potential clinical applications (Based on Goodman et al., 2017, and other sources)

CRISPR-Cas9 approach	Target gene/disease	Target cells**	Objective
Gene inactivation	*CCR5**	HSPCs, T cells	Resistance to R5-tropic infection by HIV-1
	CXCR4	HSPCs, T cells	Resistance to X4-tropic infection by HIV-1
Mutation correction	*ADA* (ADA-deficiency SCID)	HSPCs	Correction of SCID causing mutations in *ADA* gene
	SNPs associated with asthma	HSPCs, T cells	Changing the SNPs associated with asthma to safe SNPs
Gene replacement/ insertion	Defective *IL2RG* (causing X-linked SCID)	HSPCs	Complete cDNA of normal allele of *IL2RG* gene integrated at the mutant *IL2RG* locus; alternatively, the normal allele expression cassette integrated into a 'safe harbour' site of the genome
	CFTR locus	Primary adult intestinal stem cells	Mutation corrected by HDR-mediated sequence replacement
Gene activation repression	Genes for transcription factors and cytokine mediators (associated with Th1 phenotype)	HSPCs, epidermal stem cells, T cells, tissue-specific cells (*e.g.*, macrophages, dendritic cells, etc.)	Activation of expression of the transcription factors and cytokine mediators associated with Th1 response (*e.g.*, T-bet, IFN-ϒ and or IL-2) or Treg response (*e.g.*, FoxP3, IL-10, and/or TGF-β) to shift the phenotype toward Th1 or Treg phenotype and away from the atopic Th2 phenotype
	HIV-1 provirus	Cells with latent HIV-1 infection	Activation of provirus leading to accumulation of viral proteins and death of the cells with latent infection
Gene repression	Genes for transcription factors and cytokine mediators (associated with Th2 phenotype)	HSPCs, epidermal stem cells, T cells, tissue-specific cells (*e.g.*, macrophages, dendritic cells, etc.)	Repression of the expression of transcription factors and cytokine mediators associated with Th2 response (*e.g.*, GATA-3, IL-4, IL-5 and/or IL-13) to shift the phenotype toward Th1 or Treg and away from the atopic Th2 phenotype

* CCR5, C–C chemokine receptor type 5 (coreceptor for HIV-1); R5-tropic infection, HIV-1 enters host CD4 cells by binding to CCR5; ADA, adenosine deaminase; SCID, severe combined immunodeficiency; SNP, single-nucleotide polymorphism; IL2RG, interleukin 2 receptor subunit gamma; X-SCID, X-linked severe combined immunodeficiency; Treg: regulatory T cell; Th1, Type 1 T helper cell; Th2, Type 2 T helper cell (associated with atopic diseases like allergy); GATA-3, GATA-binding protein 3; FoxP3, forkhead box p3; T-bet, T-box transcription factor TBX21; HSPCs, hematopoietic stem/progenitor cells; IFN-ϒ, interferon gamma; IL-2,4,5, etc., interleukin 2,4,5, etc.; TGF-β, transforming growth factor β.

** Obtained from the patients to be treated with gene-edited cells.

There are several excellent and comprehensive reviews devoted to the varied applications of SDN technologies to combat genetic and/or pathogenic diseases and to promote human health; some of these reviews are, Porteus (2016), Savic and Schwank (2016), Tsai and Joung (2016), Chandrasegaran (2017) Goodman et al. (2017), Pulecio et al. (2017), Moreno and Mali (2017), Shen et al. (2017), Chen et al. (2018), Dunbar et al. (2018) and Broeders et al. (2020). Some selected examples of the various CRISPR/Cas approaches and their potential clinical applications are presented in Table 9.1. and many of them are discussed in this chapter.

9.2. SDNS IN HUMAN HEALTH

The SDN systems, especially the CRISPR-Cas system, have a variety of potential applications in the field of human health. They provide a relatively simple and effective tool for genome-wide screens for gene functions, which enable dissection of complex and intricate biological processes, including disease onset and development. Such studies would unveil the host factors essential for infection by pathogens and replication and propagation of viruses; the genes encoding these factors could serve as potential targets for gene editing for preventing and/or treating/curing the concerned diseases. For example, a genome-wide screen using cultured human cells has identified several host proteins that may be indispensable for HIV-1 (Human Immunodeficiency Virus-1) replication; the human genes encoding these proteins offer themselves as potential targets for therapeutic gene editing. These studies would also suggest targets for the development of novel drugs for treatment of various diseases. Genetic interaction maps might enable the identification of such gene pairs that would provide the target for 'synthetic lethal' therapy already being used for treatment of some cancers (Section 9.13.2). The CRISPR-Cas9 technology also provides a robust approach for studies on the role of genomic rearrangements in the initiation and progression of human diseases including cancers (Chandrasegaran and Carroll, 2016).

The SDN technology can be used to develop human disease models in animals that greatly facilitate both basic and applied researches in human health. For example, the ZFN technology was used to knockout the interleukin-2 receptor gamma chain (IL2RG) encoding gene in rat for investigations of the X-linked severe combined immunodeficiency (X-SCID) disease. Subsequently, the ZFN technology was used to develop rat models deficient in both *Prkdc* and *Il2rg* gene functions. These rat models can serve as hosts for evaluation of human stem cell/tissue xenotransplantats. Genome-wide association studies indicate that human genome has many SNPs (single nucleotide polymorphisms) that appear to be associated with important diseases. Gene editing technology can be used to create rat models with SNPs that are equivalent to specific putative disease-associated human SNPs in order to validate the findings from the association studies; these models can be used for HDR-mediated sequence replacement in the concerned genomic regions to analyse the effects of these sequence changes (Mashimo, 2014).

The creation of animal models of human diseases using gene editing tools offers several advantages over the traditional gene knockout technology. The most important advantage is that gene editing is fast: it generates knockout rats in 4-6 months as compared to 12-18 months needed for the embryonic stem (ES) cell-based technology. The second advantage is that, unlike the ES technology, gene editing method is not strain-dependent. Thirdly, the gene editing tools can be used to induce several different types of genetic alterations, which cannot be done by the traditional methods. Since mouse models fail to mimic human phenotypes of many diseases, gene editing may be used to generate pig models of human diseases. Gene editing would allow creation of novel disease models that precisely recapitulate the genetic changes known to occur in patient populations. In contrast, the traditional disease models are merely phenocopies of the concerned human diseases, and they do not reveal the roles of actual genetic changes involved in the concerned diseases (Hsu et al., 2014; Mashimo, 2014; Chandrasegaran and Carroll, 2016).

The SDNs can be used to either induce or correct specific gene mutations in isogenic ES cells and induced pluripotent stem cells (iPSCs) to generate cellular disease models. The multiplexing capacity of the CRISPR-Cas system can be exploited to simultaneously mutate several different relevant genes to model polygenic diseases like diabetes, heart afflictions, etc. The available techniques permit the experimenter to make the iPSC cultures to develop as organoid-like systems that are being used as models of intestines, brain, kidney, etc. The CRISPR-Cas technology has been used for gene editing of these *in vitro* organoids to create sophisticated disease models, *e.g*., mouse mammary tissue organoid model for breast cancer, organoid model of human colon adenocarcinoma, etc. The organoid disease models have also been used to evaluate the use of CRIPR-Cas9 for therapeutic correction of the concerned diseases. For example, CRIPR-Cas9 was used for correcting the mutation in cystic fibrosis transmembrane conductor receptor (CFTR) protein encoding gene in the primary intestinal organoids derived from the stem cells of cystic fibrosis patients. It was observed that the gene-edited organoids did not exhibit the cystic fibrosis phenotype. The *in vitro* developed organoids have been successfully transplanted into mice and rat models. This suggests that the *in vitro* grown organoids derived from stem cells may be gene-edited *in vitro* and then transplanted back into patients to replace the diseased tissue/organ with gene-corrected tissue/organ (Katrekar et al., 2017).

9.3. GENETIC INTERACTION (GI) MAPS

Genetic interaction (GI) or epistasis is a phenomenon, in which the functions of alleles of one interacting gene are affected by those of the other gene and *vice-versa*. The presence of GI between two or more genes signifies that these genes perform related functions due to, for example, the proteins encoded by them being components of the same protein complex or their participation in the same metabolic pathway. The phenotypic effects of simultaneous mutations of interacting genes cannot be predicted on the basis of the effects generated by the mutant alleles of individual genes. Mapping of GI is based on analysis of the functional relationships among the concerned genes and involves systematic collection of data on phenotypic effects of mutations in individual genes as well as in all possible combinations of two or more of the genes being investigated. So far, most of the data on GI in eukaryotes have been obtained from studies in yeast and *Drosophila*. Studies on GI in mammalian cells have been based on large-scale RNAi (RNA interference). The CRISPR-Cas system has recently been used for high throughput analyses of GIs in human cell cultures (Gilbert et al., 2014; Du et al., 2017; Shen et al., 2017)

Shen et al. (2017) developed a high throughput approach for GI mapping in human cells using the CRISPR-Cas9 SDN system. They inactivated 73 cancer genes both individually as well as in all possible pairs and recorded the phenotypic effects of these genetic alterations on cell growth. Du et al. (2017), on the other hand, used CRISPRi (CRISPR-interference; based on CRISPR-dCas9-transcription repressors) to knockdown 107 factors involved in chromatin regulation in human cells both singly and in pairs. The data collected in both the studies were subjected to appropriate analyses to identify pairs of interacting genes. It may be pointed out that

CRISPRa (CRISPR-activation; based on CRISPR-dCas9-transcription activators) can be used to activate the expressions of targeted genes, and the data so obtained may be combined with CRISPRi data to analyse the phenotypic effects of these alterations in gene expression patterns and to generate GI maps (Gilbert et al., 2014).

GI mapping is expected to unveil the gene interaction networks that affect various biological processes, including disease development, *e.g.*, tumorigenesis. It is also expected to help identify more precise therapeutic targets and the development of precision drugs for treatment of various ailments. Since different tumours are likely to exhibit diverse genetic interactions, samples from many patients need to be analysed for GI mapping. In addition, the specificity and the efficiency of gene editing/activation/repression by the CRISPR-Cas system should be further improved and the off-target effects should be minimized to increase the reliability and usefulness of the GI maps generated from these studies. In basic studies like this, the off-target gene editing could confound the interpretation of phenotypic effects produced by the on-target gene editing. There are experimental approaches to virtually rule out the confounding effects of off-target editing, *e.g.*, the use of multiple sgRNAs to induce mutations in the same gene, reversal of the mutational event to verify if the wild type phenotype is restored, and to carry out complementation test for the mutants (Tsai and Joung, 2016; Shen et al., 2017). These control experiments are powerful tools, but they are rather demanding and may not be practicable on a scale needed for GI mapping.

In future, it might become feasible to generate comprehensive GI maps for different diseases by combining findings from CRISPR-Cas gene editing, gene expression inhibition and activation, as well as data from proteomic and chemogenetic analyses. GI maps could unravel the manner, in which expression levels of different genes, either singly or in groups, and the levels of the proteins encoded by them affect the various cellular functions and phenotypes. GI maps also have therapeutic implications: for example, simultaneous inactivation of two specific gene functions might lead to cell death, while inactivation of only one of them may be involved in certain types of cancers. Such interacting gene pairs provide a potential route for 'synthetic-lethal' therapy for the concerned cancers (Section 9.13.2; Gilbert et al., 2014; Shen et al., 2017).

9.4. TARGET SITE SELECTION

Therapeutic gene editing aims to change the genotype at a relevant genomic site to either cure a disease or generate resistance to a pathogen. The locus to be targeted can be selected based on the vast amount of available human genetics data, which would also help determine the type of DNA sequence change required to achieve the desired objective. In simple words, genetic studies provide information on the mode of inheritance (monogenic or polygenic) of the genetic diseases as well as identify the genes involved in these conditions. Linkage studies and molecular genetics investigations provide information on the genomic locations of the concerned genes and genomics data give an idea of the gene sequence, its putative functions and the genetic variation present in the locus in human populations. The genetic polymorphism at the concerned locus can be compared with the phenotypes of the concerned individuals for the relevant diseases to identify the DNA sequences common in the individuals free from the concerned diseases. Even if a single healthy individual with a particular DNA sequence at the relevant locus is known to be resistant to infection by a pathogen or to have been cured of a disease, recreation of the concerned

DNA sequence by gene editing in other individuals can be expected to be safe as well as therapeutically useful (Porteus, 2016).

An example of the above is provided by the discovery of resistance to HIV-1 infection conferred by the *Δ32* deletion in the *CCR5* (C–C chemokine receptor type 5) gene. This information emerged from the observation that an AIDS patient suffering from leukaemia was virtually cured of AIDS after receiving a bone marrow transplant from a donor homozygous for the *CCR5Δ32* mutation. This observation encouraged gene editing to recreate the (C–C chemokine receptor type 5) mutation in an effort to generate HIV-1 resistance. Similarly, it is known that patients suffering from sickle cell disease are homozygous for a mutation in the *HBB* gene (the gene encoding the β-chain of haemoglobin). In contrast, individuals heterozygous for this mutation do not have the sickle cell disease. This clearly suggests that correcting the concerned mutation in the *HBB* gene in even one of the two mutant alleles in sickle cell patients will relieve them from the disease (Porteus, 2016). Therapeutic targets are also indicated by genome-wide association studies. For example, a genome-wide association study suggested that deletion of the enhancer of *BCL11A* gene will block its expression in erythroid precursor cells without affecting its expression in non-erythroid cell lineages; this might be useful in the treatment of diseases involving β-haemoglobin. However, such leads must be validated in suitable models of the concerned diseases. Both animal and organoid/cell models of many human diseases are available, and new models are being continuously developed (Section 9.2).

9.5. THERAPEUTIC GENETIC CHANGES

The types of genetic alterations produced by SDN treatment will depend on the type of DNA repair pathway, *viz*., NHEJ- or HDR-mediated repair, used by the affected cells for repair of the site-specific DSBs induced by the treatment (Table 9.2). It may be pointed out that all these types of alterations can be therapeutically useful depending on the genomic loci involved. The NHEJ pathway can lead to short indels that either inactivate functional genes or restore activities of nonfunctional mutant genes by compensating for the frameshift mutations that inactivated them. In addition, it can produce targeted deletions of predecided genomic sequences. SDN-induced indels have been used to inactivate genes that are involved in certain diseases, *e.g*., inactivation of the *CCR5* and *CXCR4* genes to generate resistance to HIV-1 infection (Section 9.13), gene *PCSK9* knockout to lower the low-density lipoprotein (LDL) cholesterol levels in mice, etc. Many diseases are the result of frameshift mutations in genes. For example, about 60% of the mutant alleles of *dystrophin* gene that cause Duchenne muscular dystrophy (DMD) disease are produced by frameshifts due to large indels in this extremely large (2.6 Mb, megabase pair; 79 exons) gene. When SDNs were used to induce indels on the upstream side of the frameshift mutations, about one-third of the indels restored the correct reading frame in the cells collected from DMD patients (Ousterout et al., 2015; Porteus, 2016).

Deletions of selected genomic segments can be induced by using SDNs to cleave the genomic DNA at two appropriate sites. In general, the deletion frequency decreases with increasing size of the deleted fragment, but other factors also affect this frequency. Deletions within the selected genes can be induced to increase inactivation frequencies of targeted genes. It can also be used to remove genomic segments containing disease causing mutations/exon(s) from relevant genes or even complete genes, repeat expansions of triplets associated with specific diseases, or regulatory sequences to alter the expression pattern of a therapeutic protein. This strategy has also been used to delete virus genomes integrated as provirus into the host genomes. Simultaneous induction of two DSBs can lead to inversion of the intervening genomic sequence in about 1% of the cases. This could be used to correct disease-causing inversions in genes, such as the inversion in Factor VIII encoding gene that causes haemophilia A in most of the patients (Porteus, 2016).

Table 9.2. A summary of the types genetic changes produced by SDN treatments and their potential therapeutic applications (based on Porteus, 2016 and other sources)

Genetic change	Therapeutic application	Remarks
Nonhomologous end joining (NHEJ)		
Indels	Gene inactivation	*CCR5* inactivation for resistance to HIV-1 infection
	Correction of frameshift mutations	Duchenne muscular dystrophy (DMD) disease caused by frameshift mutant alleles of *dystrophin* gene
Deletions (two DSBs induced at the borders of the genomic region to be deleted)	Deletion of a small sequence from a gene	Complete inactivation of the gene, *e.g.*, deletion of a portion of *CCR5* gene (for HIV-1 resistance)
	Removal of an exon or exons, or complete genes	Deletion of Exon 51 of the *dystrophin* gene involved in Duchenne muscular dystrophy; deletion of the gene for α or β chain of the endogenous T cell receptor to create 'universal' T cells
	Deletion of disease-causing repeat expansions	The expanded GAA repeat excision in case of Friedreich's ataxia relieves molecular symptoms in the edited cells
	Deletion of regulatory sequences	The deletion of enhancer of *BCL11A* gene will block its expression in erythroid precursor cells without affecting its expression in non-erythroid lineages; may be useful in diseases involving β-haemoglobin
Inversion (two DSBs induced at the borders of the genomic region to be inverted)	Correction of a disease-causing inversion	Proposed to be used for correction of the inversion in gene for Factor VIII, the most common cause of Haemophilia A; inversion is 100-fold less frequent than deletion of the fragment located between the two DSBs
Homology-directed repair (HDR)		
Gene correction	Replacement of the mutant sequence by the wild type sequence	In case of several genes involved in genetic diseases, such as sickle cell anaemia, Duchenne muscular dystrophy, Xeroderma pigmentosum, Fanconi anaemia, beta-thalassemia, cystic fibrosis, etc.
Functional gene correction	Integration of full-length wild-type cDNA at the start codon of the endogenous gene	Provides wild-type gene function without correcting the mutational changes in the endogenous mutant gene; the gene subject to the endogenous regulatory sequences; in cases of β-thalassemia, sickle cell disease, SCID-X1
Gene addition in 'safe' harbours	Integration of transgene expression cassette in 'safe' harbours	Can be used when precise regulation of the transgene expression is not essential, or a higher level of expression is necessary

Genetic change	Therapeutic application	Remarks
Targeted addition of transgenes	Transgene integrated within an endogenous gene	The transgene expression is regulated by the regulatory sequences of the endogenous gene
Other therapeutic reagents based on CRISPR-Cas9		
Targeted correction by base substitution at the mutated site	Base editors used to chemically change the base at the mutation site to the base present in the wild-type allele of the gene	The available base editors can convert C into T and A into G; the conversion of T-to-C and G-to-A can be achieved by editing the A-to-G and C-to-T, respectively, in the complementary strand
Programmable epigenetic effectors	Gene transcription activation or suppression	Example: activation of HIV-1 provirus to induce death of the cells carrying latent HIV-1 infection

On the other hand, HDR-mediated DSB repair replaces the sequence at DSB site with that of the donor DNA used as template during HDR. The DSB should be <30 bp from the intended site of sequence replacement and the donor DNA provided with the SDN must contain the sequence to be copied at the targeted site. The therapeutic alterations based on HDR have been divided into the following four groups by Porteus (2016): (i) gene correction, (ii) functional gene correction, (iii) gene addition in a 'safe harbour' and (iv) targeted addition of transgenes. Gene correction strategy is used in cases of such monogenic genetic diseases, in which the mutational change affects a single site in the concerned gene, *e.g.*, sickle cell anaemia, Xeroderma pigmentosum, Fanconi anaemia, etc. In cases of such diseases, SDN targeting a single site in the affected gene along with a donor DNA for correction of the mutational change is delivered into the cells. HDR of the DSB corrects the defect in the gene; this has been demonstrated in the cases of the diseases listed above and some other diseases (Porteus, 2016).

In case of most human genetic diseases, mutations are located at many different sites in the concerned genes of different patients. In such cases, a single and different SDN and donor DNA set may have to be designed for each patient. This can be avoided by producing full-length cDNA (complementary DNA) of the functional wild-type allele of the concerned gene and target its SDN-directed integration at the start codon of the mutant endogenous gene by HDR. The expression of this cDNA will be under control of the regulatory sequences of the endogenous gene and would be able to correct all the different disease causing mutations located downstream of the start codon in the genes of different patients. This strategy is termed as functional gene correction since it provides the wild-type gene function without correcting the mutational changes in the genes. This strategy has been used in cases of sickle cell anaemia, β-thalassemia and SCID-X1. If required, the level of gene expression may be modulated by addition of an intron into the cDNA and modifying the codon usage (Porteus, 2016).

In case a precise regulation of the therapeutic transgene is not essential or when a higher level of transgene expression is necessary, the transgene along with appropriate regulatory sequences can be integrated into a 'safe' harbour (Section 9.6) by HDR-mediated repair of the DSBs induced by SDNs. But sometimes it may be desirable to use the regulatory sequences of an endogenous locus to modulate the transgene expression to achieve a high expression level and/or the expression pattern of the selected endogenous locus. In such cases, the transgene coding sequence is integrated by HDR into the locus of interest (Porteus, 2016). For examples, the transgenes encoding Factor IX, iduronate-2-sulfatase, and α-L-iduronidase have been inserted into the albumin locus of hepatocytes for treatment of Haemophilia B, mucopolysaccharidosis type II (Hunter's syndrome) and mucopolysaccharidosis I, respectively; these *in vivo* gene therapies are in clinical trials (Table 9.3).

Table 9.3. The *in vivo* gene therapies approved for therapeutic use or in clinical trials (based on Dunbar et al., 2018 and other sources)

Disease	Vector*	Transgene*	Approval*	Remarks
Childhood-onset spinal muscular atrophy	AAV9	*SMN*	FDA (2016); EMA (2017)	Caused by mutant *SMN* gene; target: central nervous system cells
Haemophilia B	AAV8	*Factor IX*	FDA (2014); EMA (2017)	Target: liver cells
	AAV100	*Factor IX Padua***	FDA (2016); EMA (2017)	Target: liver cells
	AAV5	*Factor IX*	FDA (2017); EMA (2017)	Target: liver cells
	AAV2/6	*Factor IX* and ZFNs	FDA (2017)	Target: liver cells; AAV2/6 is known to be liver tropic
Haemophilia A	AAV5	*Factor VIII*	EMA (2017)	Target: liver cells
Lipoprotein lipase (LPL) deficiency	AAV1	*LPL*	EMA (2012)[#]	Target: muscle cells
Inherited retinal dystrophy (inherited blindness, Leber's congenital amaurosis)	AAV2	*RPE65*	FDA (2017)	Caused by an autosomal recessive mutation; target: retina[1]
Haemophilia A	AAV2/6	B domain-deleted Factor VIII and ZFNs	In clinical trials	Target cell: liver; AAV2/6 exhibits liver tropism
Mucopolysaccharidosis type II (Hunter's syndrome)	AAV2/6	IDA and ZFNs	In clinical trials	Targeted therapeutic gene insertion into the albumin locus of hepatocytes mediated by ZFNs[##]
Haemophilia B	AAV-derived vector	Factor IX and ZFNs; intravenous delivery	In clinical trials	
Mucopolysaccharidosis I	AAV-derived vector	α-L-iduronidase and ZFNs; intravenous delivery	In clinical trials	

* AAV, adeno-associated virus; EMA, European Medicines Agency; FDA, U.S. Food and Drug Administration; ZFNs, zinc finger nucleases; IDA, iduronate-2-sulfatase; RPE65, retinal pigment epithelium-specific 65 kDa protein; SMN, survival motor neuron 1; CAR, chimeric antigen receptor.

** Factor IX Padua, a natural variant of factor IX having high specific activity.

[#] License not renewed as of 2017.

[##] The albumin locus ensures high level of liver-specific transgene expression.

[1] Marketed as LUXTURNA™ (manufactured by Spark Therapeutics, Inc.)

9.6. 'SAFE' HARBOUR

In simple terms, a 'safe-harbour' site is a genomic locus that is not likely to generate undesirable effects because of integration of a transgene expression cassette within this locus. Integration of transgene cassettes at random locations in the genome can produce undesirable effects due to unintended activation/inactivation of genes located at or near the sites of integration. Another important consideration is that the transgene integration sites should support sustained transgene expression or at least should not cause transgene silencing. In physical terms, a 'safe' harbour should not be located within the coding region of a gene and should be situated at a certain distance from known oncogenes (Sadelain et al., 2012). But a variety of functional genetic elements other than the coding regions of genes, *e.g.*, sequences encoding noncoding RNAs, enhancer sequences and sequences yet to be discovered and described, are scattered all over the human genome. It is not feasible to develop a set of physical criteria based on all the known and putative, let alone 'yet to be discovered', functional elements present in human genome. Therefore, whether a putative 'safe' harbour can be really considered as 'safe' will depend on the effects of transgene integrations into this locus on the phenotypes and behaviour of the transgenic cells (Porteus, 2016).

The *AAVS1* site is the preferred site for integration of adeno-associated virus. This site is in an intron of the gene *PPP1R12C* and is considered as a 'safe' harbour. Therapeutic genes for α-thalassemia and chronic granulomatous disease have been integrated into this site. But transgene integration into this site can interrupt the expression of *PPP1R12C* gene, which encodes a protein phosphatase. Further, the long-term effects of transgene integration into the *AAVS1* site are not known. The *CCR5* gene is regarded as another 'safe' harbour, and disruption of this gene leads to HIV-1 resistance. *CCR5* has been used for integration of many therapeutic genes for HIV-1 resistance and for overexpression of therapeutic proteins like erythropoietin. But recent findings suggest that disruption of *CCR5* can lead to susceptibility to some other viruses. Continued search for 'safe' harbours in the human genome may be able to identify safer and more useful sites for specific clinical applications. It has been suggested that 'safety switches' may be integrated into therapeutic strategies so that the gene-edited cells can be readily eliminated after they have served their curative purpose or whenever they show pathological tendencies (Porteus, 2016).

9.7. GENE THERAPY

Gene therapy involves favourably changing the base sequence or expression pattern of an endogenous gene so that the diseased phenotype is reversed to the normal or near normal phenotypic state. Gene therapies are of the following two types: (i) *in vivo* gene therapies and (ii) *ex vivo* gene therapies. In case of *in vivo* gene therapy, the therapeutic reagents, including a therapeutic vector carrying the necessary gene constructs are either delivered locally into the concerned tissue or administered systemically, *i.e.*, introduced into the blood stream, so that they are carried throughout the body. Following administration of the therapeutics, monitoring is done for assessing the efficacy and safety of the therapy, and degree of improvement in the disease phenotype. The therapeutics may be re-administered if and as required for a sustained amelioration in the symptoms of the disease (Fig. 9.1). A safe and efficient delivery system is essential for gene therapies, and the therapeutics must satisfy the requirements for safety, efficacy and quality control. At present, there are several guidelines issued by FDA (Food and Drug Administration, U.S.A.) and EMA (European Medicines Agency), European Union, *e.g.*, 'FDA guidance for human somatic cell therapy and gene therapy' (1998), EMA guidelines 'Quality, preclinical and clinical aspects of gene transfer medicinal products (2001)', etc., that are relevant to the various aspects of gene therapies. The delivery systems based on cationic lipids or polymers are safer than viral delivery systems, and they can be modified for transfection of specific target cell types (Shim et al., 2017; Dunbar et al., 2018).

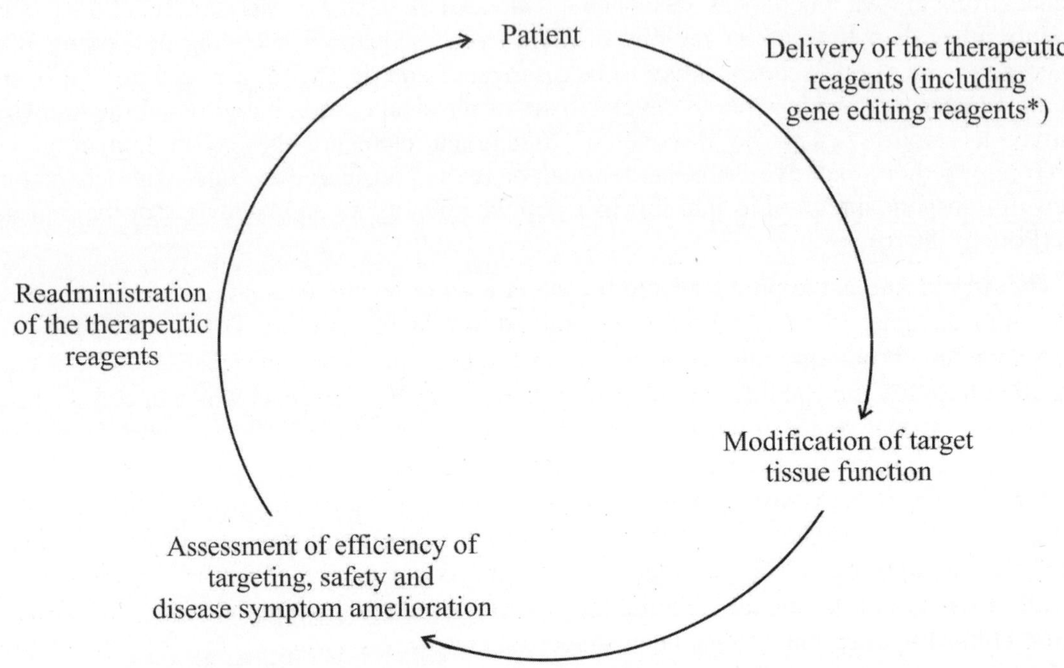

Fig. 9.1. A simplified schematic representation of *in vivo* gene therapy.*, CRISPRa/i seem to hold greater promise among the different gene editing technologies (based on Moreno and Mali, 2017).

In the case of *ex vivo* gene therapy, on the other hand, appropriate cells are isolated from the body of concerned patient, subjected to the desired genetic modification, including gene editing, *in vitro* and then introduced back into the patient. In general, nonviral delivery systems are preferred for *ex vivo* transfection of cells, and electroporation is the most widely used technique. In addition, lentiviral and γ-retroviral vectors have been used in almost comparable number of approved cases (Table 9.4). In contrast, the adeno-associated virus vectors have been exclusively used in all the cases of both approved and under clinical trial *in vivo* gene therapies. The genetically altered cells are selected and purified before they are transferred back into the body of patient who donated them. It is desirable to use pluripotent stem cells for gene therapy/gene editing. Autologous (derived from the concerned person/patient) induced pluripotent stem cells (iPSCs) can now be induced by introduction of stem cell transcription factors and can be used for gene therapy (Section 9.7). The transgenes for stem cell transcription factors are removed from the cell genomes after iPSCs have been induced from somatic cells (Fig. 9.2; Shim et al., 2017; Dunbar et al., 2018).

Table 9.4. The *ex vivo* gene therapies approved for therapeutic use or in clinical trials (based on Dunbar et al., 2018 and other sources)

Disease	Vector	Gene	Approval	Remarks
Adult ALL*	γ-RV	CD19 (CD28) CAR-T**	FDA (2014)	Target cells: T cells; express second-generation CAR with improved antitumor activity
Refractory paediatric ALL	LV	CD19 (4-1BB) CAR-T	FDA (2017), EMA (2016)	CAR T therapy; targets CD19 receptor protein on the surface of cancer cells[1]
Diffuse large B cell lymphoma	γ-RV	CD19 (CD28) CAR-T	FDA (2014)	Target cells: T cells; express second-generation CAR with improved antitumor activity
	γ-RV	CD19 (CD28) CAR	FDA (2015), EMA (2016)	Target cells: T cells; express second-generation CAR with improved antitumor activity
	LV	CD19 (4-1BB) CAR-T	FDA (2016), EMA (2016)	Target cells: T cells
Relapsed or refractory large B-cell lymphoma	CAR-T cells	CD19 CAR-T	FDA (2017)	CAR T therapy; targets CD19 receptor protein on the surface of cancer cells[2]
Synovial sarcoma	LV	NY-ESO-TCR#	FDA (2016), EMA (2016)	Target cells: T cells
β-Thalassemia	LV	Anti-sickling	FDA (2015), EMA (2016)	Target cells: HSPCs
Adenosine deaminase deficiency	γ-RV	ADA	EMA (2016)	Target cells: HSPCs
	LV	ADA	FDA (2015)	Target cells: HSPCs
Metachromatic leukodystrophy	LV	ARSA	EU Orphan Drug (2007)	Target cells: HSPCs
Acquired immunodeficiency syndrome	Electroporation	CCR5	In clinical trials: T cell-based therapy, Phase I/II completed; HSPC-based therapy, Phase I ongoing	The first SDN-based therapy to reach clinical trials; ZFN electroporated to inactivate CCR5; generates resistance to infection by HIV-1; target cells: T cells and HSPCs
Acute myeloid leukaemia	CD123 CAR T	CD123	In early trials	Universal T cells generated by TALEN-mediated gene editing
Blastic plasmacytoid dendritic cell neoplasms	CD123 CAR T	CD123	In early trials	–

* ALL, acute lymphoblastic leukaemia; ARSA, arylsulfatase A; CCR5, C-C motif chemokine receptor type 5; EMA, European Medicines Agency; FDA, U.S. Food and Drug Administration; γ-RV, murine γ-retrovirus; HIV, Human immunodeficiency virus; HSPCs, hematopoietic stem and progenitor cells; LV, lentivirus; ZFN, zinc finger nuclease.

** T cells expressing second generation CAR, *viz.,* CD19RCD28 and CD19(4-1BB); CD28 and 4-1BB are co-stimulatory molecules. These T cells show enhanced persistence and greater antitumor effect.

[#] T cell receptor modified to bind to the NY-ESO1 protein, which is expressed by many tumour cells.

[1] Marketed as KYMRIAH™ (manufactured by Novartis Pharmaceuticals Corporation)

[2] Marketed as YESCARTA™ (manufactured by Kite Pharma, Inc.)

Gene therapy can also be classified based on the strategies employed for gene therapy. The traditional gene therapy usually delivers the therapeutic factors or genes via viral vectors that are efficient in gene transfer, incapable of replication and, nontoxic to human somatic cells. The transgenes delivered into the cells may either integrate into the cell genome and support long-term expression or they may remain in an extrachromosomal state and lead to medium- or short-term expression. The gene editing-based gene therapy, on the other hand, uses specific SDNs to achieve one of the following three objectives: inactivation of an endogenous gene, mutation correction of a defective gene, and integration of a therapeutic gene into a 'safe harbour' site of the genome or within the targeted endogenous gene. Several traditional gene therapies and some gene therapies based on gene-edited T cells ('universal' T cells) have already been approved for use as drugs for disease treatment and more than one dozen of these therapies have been designated as "breakthrough therapy" (Table 9.3; Dunbar et al., 2018).

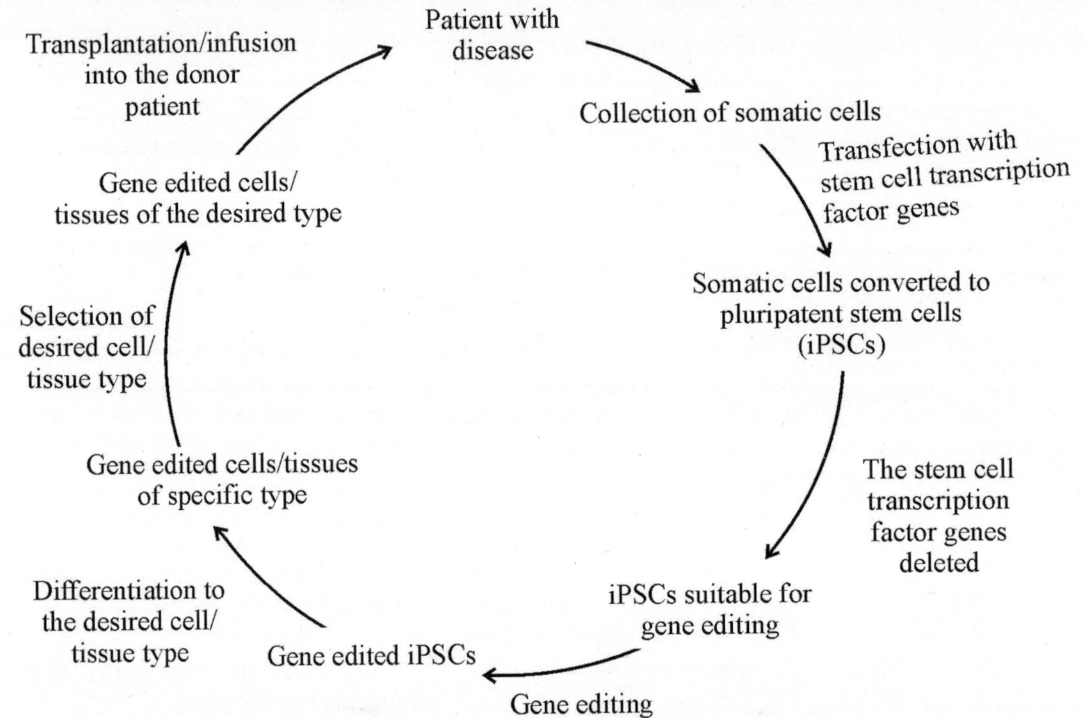

Fig. 9.2. A simplified schematic representation of *ex vivo* gene therapy. The stem cell transcription factor expression casette is flanked by *loxP* sequences. This casette is deleted from iPSCs (induced pluripotent stem cells) by a Cre recombinase treatment (based on Moreno and Mali, 2017)

9.8. VECTORS FOR MAMMALIAN CELLS

The SDN gene constructs have been delivered into human cells using both viral and nonviral delivery system (Table 9.5). The viral delivery systems use the following four types of vectors derived from virus genomes: (i) retrovirus vectors, (ii) adenovirus (AV) vectors, (iii) lentivirus (LV) vectors and (iv) adeno-associated virus (AAV) vectors. These viruses have evolved in nature to deliver their genetic material into mammalian cells with high efficiency. Therefore, viral vectors generally transduce mammalian cells with high efficiency. But these vectors are complex, problematic-to-produce, and necessitate additional regulatory clearances. The retroviral and lentiviral genomes integrate into mammalian genomes, while the genomes of other two viruses (AV and AAV) persist in extrachromosomal state in the host cell cytoplasm/nuclei. The retroviral vectors are the most suitable for long-term expression of their therapeutic payloads, but their random integration into the genome can cause insertional mutagenesis. The gamma-retroviral vectors (γ-RV) were developed during 1980s and early 1990s and several *ex vivo* gene therapies based on them have been approved as drugs for therapeutic use. The AV vectors are suitable for carrying all types of CRISPR-Cas systems since they can accommodate large DNA inserts. The recombinant AV genome carrying the SDNs can be easily packaged into virus particles using several different engineered packaging host cell lines. However, the process of producing recombinant AV is costly and labour intensive. At the same time, the AV shows strong immunogenic effect in human cells; but the second and third generation AV vectors developed by genetic engineering do not suffer from this limitation (Moreno and Mali, 2017; Chen et al., 2018).

Lentiviruses (LV) were originally derived from HIV-1. The LV vectors are highly potent with large (~9.7 kb) packaging capacity and have been widely used to deliver CRISPR-Cas SDNs into both dividing and nondividing human cells. However, these vectors cannot deliver payloads into quiescent G_0 cells. The LV vectors can be made capable of infecting a given cell type by inclusion of an envelope protein specific for the cell surface receptor of that cell type. But LV vectors can become integrated into human genome with the associated high risk of tumorigenesis/carcinogenesis. The new generation of IDLVs (integrase-deficient lentivirus vectors) have mutated *integrase* and LTR (long terminal repeat) sequences. As a result, they present greatly reduced, but not zero, risk of insertional mutagenesis. LV integration into the genome will also insert the Cas nuclease expression cassette, but the consequences of such insertions are not well understood. In addition, the IDLVs do not seem to be well-suited for the delivery of TALEN sequences that are highly repetitive. The LV vectors lead to long-term expression of the therapeutic genes. They have been extensively used for *ex vivo* gene therapies based on hematopoietic stem cells (HSCs) and T cells. Several *ex vivo* gene therapies based on LV vectors have been approved for treatment of patients. Currently, the LV vectors are the most preferred for payload delivery into HSCs, but γ-RVs are still in use for certain T cell-based procedures. The LVs and γ-RVs used in most of the current clinical trials have a 'self-inactivating' SIN design (Gaj et al., 2013; Moreno and Mali, 2017; Chen et al., 2018; Dunbar et al., 2018).

The AAV vectors are highly versatile and widely used in clinical research due to their overall safety, low toxicity to host cells, large range of serotype specificities and relatively low risk of carcinogenesis following integration into host genome. In addition, it exhibits high-infection efficiency, permits medium-term transgene expression, and shows low immunogenic potential in human cells. But AAV vectors can cause immune response on re-administration and the efficiency of delivery also tends to decline on repeat administration. The AAVs are derived by deleting all the coding sequences from the virus genome, and the expression cassette for the gene of interest is inserted in their place; this recombinant DNA is maintained in an episomal state in the cells. The AAV vectors can carry relatively small DNA inserts so that they can accommodate only one TALEN monomer driven by a minimal promoter. An AAV vector cannot accommodate both SpCas9 and sgRNA expression cassettes. But when smaller Cas9 orthologs like SaCas9 and St1Cas9 (both ~3.3 kb) or Cas12a are used, the expression cassettes for both Cas and sgRNA can be included in a single AAV vector. The smaller Cas9 orthologs require more complex PAM sequences, which limits the genomic sites they can target. But this limitation can be resolved by engineering mutants with simpler PAM requirements. In addition, all the smaller Cas9 nucleases are not as robust as SpCas9. The SaCas9-sgRNA delivered through AAV vector was used to excise HIV-1 provirus genome from host genome *in vivo*. Some of the AAV vectors (AAV8 and AAV9) have been used *in vivo* in mouse models for Duchenne muscular dystrophy for correcting the gene mutations causing DMD. Therapies based on these vectors are already in clinical trials. Continued efforts for serotype engineering of AAV and discovery of novel Cas9 orthologs might enhance the usefulness of AAV vectors in gene therapy (Gaj et al., 2013; Moreno and Mali, 2017; Chen et al., 2018).

Table 9.5. A summary of the salient features of the various vectors used for delivery of SDNs into target cells

Vector	Expression	Risk of integration in cell genome	Immunogenicity	Remarks
Retroviruses	Long-term	Yes (oncogenicity observed)	Low	Random integration into host genome can cause insertional mutagenesis
Lentiviruses	Long-term	Yes (low oncogenicity)	Low	Adapted from HIV-1; being used in clinical trials; large (9.7 kb) cargo capacity; infect both dividing and nondividing cells; can cause insertional mutations**
Adenoviruses (AV)	Medium-term	Low	High	Immunogenicity
Adeno-associated viruses (AAV)	Medium-term	Low	Low	Preferred vector for gene therapy; high infection efficiency; ~4.7 kb packaging capacity*
Mini-circle Plasmids	Short-term	Low	Low	–
Conventional Plasmids	Transient	Low	Low	–
Proteins	Transient	No	Being evaluated	–
RNA	Transient	No	Low	–
Nanoparticles/ biomaterials	Transient	Low	Being evaluated	–
Liposomes	Transient	Low	Being evaluated	–

Note: 1. Integration of viral vectors into the genome can lead to oncogenicity.

2. Immunogenicity is particularly relevant in case of repeated administration of the therapeutic reagents via the vectors.

3. The efficiency of genome targeting is affected by the cell type and, particularly, the mitotic state of cells as well as the efficiency with which the gene editing tools are delivered into the cells.

4. Immunogenicity of the vector system is also affected by the delivery payload.

5. Transfection of primary cells with naked DNA initiates interferon response that leads to cell death and reduced editing efficiency

* Cannot accommodate SpCas9 with sgRNA.

** Integrase deficient lentivirus vectors (IDLV) greatly reduce, but not completely eliminate, the risk of insertional mutagenesis.

The full length of SpCas9 encoding gene is ~4.2 kb; both SpCas9 and sgRNA expression cassettes can be accommodated in a single AV or LV vector, but not AAV vector. Two AAV vectors, one carrying the SpCas9 and the other bearing the sgRNA expression cassettes, have been used to overcome this limitation, but this strategy reduces the efficiency of delivery unless higher viral titters are used. The split-Cas9 (Section 5.7) strategy was developed to enable the delivery of SpCas9 and sgRNA by AAV vectors. The two parts of SpCas9 are delivered by two separate AAV vectors into the same cells. Inside the cell, the two parts are produced, and they reassemble to form the complete functional SpCas9 molecule. However, this strategy also suffers from the same limitation as the use of two AAV vectors one each for SpCas9 and sgRNA genes (Moreno and Mali, 2017; Chen et al., 2018).

The nonviral delivery vehicles for plasmid DNA include nanoparticles like cationic nanocarriers, liposomes, dendrimers and polymeric materials. Among the nonviral delivery vehicles, liposomes are the most extensively used as they are highly biocompatible and have high transfection efficiency. These systems have overall safety, low immunogenicity, large loading capacity, and ease of production. Further, they could be suitably modified for specifically targeting the desired cells and tissues. For example, they might be coated with such ligands that are specific to the receptors present on the surfaces of target cells to enable the delivery of therapeutic agents specifically into these cells. But these methods must overcome many cellular physical barriers to efficient delivery of the payload into cells. Both Cas9 and sgRNA have net negative charge, *i.e.*, are anionic; therefore, they are readily integrated into cationic liposomes and polymers. Cationic lipids and bioreducible lipid nanoparticles have been used to successfully deliver Cas9 and sgRNA complex into target cells *in vitro* and *in vivo*, respectively. Since these vehicles deliver Cas9 and sgRNA as ribonucleoprotein (RNP), the period of Cas9 action is short and the risk of off-target effects is greatly reduced (Moreno and Mali, 2017). ZFNs have been delivered into cultured mammalian cells as purified protein for targeted gene inactivation. This delivery method reduces off-target activity due to short period of ZFN action, is cell type-independent, and avoids many safety and regulatory issues (Gaj et al., 2013).

9.9. DELIVERY METHODS

The method of delivery of SDNs for gene therapy must be efficient and above all safe; the development of such delivery procedures is one of the major challenges of this field. In many research programmes, it is adequate to introduce the genome-editing reagents as plasmid DNA, but this method is not suitable for several other research applications and for *in vivo* therapeutic gene editing. The genes or mRNAs encoding the gene-editing proteins or the proteins and sgRNAs can be delivered by several methods like electroporation, lipid/cationic lipid-mediated delivery, viral vectors, etc. These methods are satisfactory for many cell types, but they do not work well with certain other cell types like HSCs and some primary cells, *i.e.*, cells obtained directly from normal body tissues (Komor et al., 2017b).

The CRISPR-Cas and other SDN systems can be delivered into human cells as DNA, RNA (Cas mRNA:sgRNA), protein, or RNP (Cas protein:sgRNA). Transient expression of SDNs is highly desirable; it can be achieved by their delivery either as RNA, protein or RNP. Electroporation and microinjection are the preferred physical methods of delivery of SDNs since these methods can be applied to many cell types *ex vivo* and to several tissue types *in vivo*. Microinjection is generally used for transfection of animal zygotes/embryos for production of gene-edited animals, while electroporation is usually applied to cultured animal cells. Electroporation is the best method for SDN delivery, but it requires highly technical electroporation procedure and very large number of primary cells in view of cell mortality caused by the high voltage used for electroporation. At the same time, the production of Cas protein (produced as recombinant protein in *E. coli*) or mRNA and sgRNA (both obtained by *in vitro* transcription) requires considerable amount of technically demanding work and financial resources (Hsu et al. 2014; Ain et al., 2015; Moreno and Mali, 2017).

The Cas9 and sgRNA gene constructs can be delivered using AV, LV and AAV vectors. Each of these viral vectors has some advantages and some disadvantages as discussed earlier in Section 9.8. Transfection of primary cells with naked DNA initiates interferon response that leads to cell death and reduced editing efficiency. The genome-editing efficiency can be enhanced by culturing the cells, into which nuclease has been delivered, at 30ºC for 16–24 h in the place of normal 37ºC (Porteus, 2016). The plasmid DNA, mRNA or protein can be delivered by cationic lipid-based reagents as well as electroporation, but both these transfection methods can be toxic to cells and are limited to certain cell types. Another method commonly used to deliver naked DNA or siRNA (small interfering RNA) is hydrodynamic injections, *viz*., high volume (8–10% of mouse body weight) injections delivered in 5-7 seconds into the vascular system. In several studies targeting liver, CRISPRCas9 has been delivered *in vivo* by hydrodynamic injections in tail veins of rodents. But this method is not suitable for the delivery of therapeutic reagents since it might damage liver and heart (Gaj et al., 2013; Moreno and Mali, 2017; Chen et al., 2018).

The reagents for gene editing and gene therapy are generally delivered *in vivo* by means of injections. These reagents may be injected into the musculature or into the circulatory system by intravenous injection. AAV vectors are used as delivery vehicles and some of these vectors are known to target specific cell types. For example, the AAV vector 2/6 is known to deliver its cargo into liver cells. Commercially available cationic lipid reagents used for delivery of nucleic acids have been used to deliver RNP particles comprising Cas9 and sgRNA both *in vivo* (in mice) and in cultured cells. Cas9 mRNA and sgRNA can be delivered directly into animal oocytes or zygotes by microinjection. This delivery method has been used for inactivation of specific genes in several animals, including frogs, mice, pigs etc. and for HDR-mediated sequence replacement in mice and zebrafish (Komor et al., 2017b; Moreno and Mali, 2017; Dunbar et al., 2018).

9.10. TARGET CELLS FOR GENE EDITING

The target cells for gene therapy are selected on the basis of their usefulness in therapy of the concerned disease. In general, somatic cells are obtained from the patient to be treated and subjected to gene editing before being transplanted into the patient. An alternative approach is to use pluripotent stem cells (PSCs) for gene editing and then differentiate the desired cell type from the gene-edited PSCs. The human pluripotent stem cells (hPSCs) are capable of continued cell division and can give rise to all the three primary germ layers, namely, ectoderm, mesoderm and endoderm, that can develop into all the cell types, tissues and organs of the body. They can maintain their pluripotent state and genome stability even after extensive handling as *in vitro* cell cultures. Therefore, they are particularly well-suited for gene editing using SDNs. It is now feasible to develop hiPSCs (human induced pluripotent stem cells) as per the need of experimenter. For gene therapy, they can be obtained from somatic cells of the patients themselves and these autologous hiPSCs can be corrected for the genetic defect by gene editing before being transplanted back into the patient. This approach has been projected as the future regenerative medicine (Chandrasegaran and Carroll, 2016).

The stem cell transcription factor genes are conventionally integrated into the genomes of somatic cells at random sites. But the use of SDNs enables their integration at selected 'safe' or even useful genomic sites that could make the hiPSCs capable of some additional functions. For example, the transcription factor genes may be integrated within the *CCR5* gene, which will make the hiPSCs resistant to invasion by HIV-1. The ZFN, TALEN and CRISPR-Cas9 SDNs have been used for targeted insertions of the stem cell transcription factor genes, *viz*., *Oct4, Sox2, Klf4, Lin28* and *Nanog* (*OSKLN*). The expression cassette containing these five genes is flanked by the loxP sequences; this cassette can be represented as loxP-*OSKLN* expression cassette-loxP. After the stem cell transcription factors convert the somatic cells into hiPSCs, the cells are treated with the Cre recombinase; this deletes the loxP-*OSKLN*-loxP expression cassette leaving the loxP site within *CCR5* gene, which inactivates *CCR5*. The sickle cell disease has been used by many workers to demonstrate the feasibility of mutation correction by gene editing in hiPSCs (Hanna et al., 2007; Chandrasegaran and Carroll, 2016).

The likely advantages of using hiPSCs for therapeutic gene editing are as follows. Since hiPSCs can be multiplied *in vitro* for prolonged periods, single cell clones of gene-edited hiPSCs can be subjected to comprehensive phenotypic characterisation and whole genome deep sequencing to select clones free from off-target effects. The hiPSCs are capable of differentiation into any cell type found in body so that they can serve as the source of gene-edited cells of such cell types that ordinarily cannot be obtained from the concerned patient. The main limitation to the use of gene-edited hiPSCs for gene therapy relates to production of the large numbers of highly purified differentiated cells of the type required for therapeutic transplantations. But the cell types derived from hiPSCs using the current techniques have embryonic phenotype in place of the desired adult phenotype, which may not be clinically desirable. It is encouraging that gene-edited iPSCs were successfully used for gene therapy in a mouse model of sickle cell disease (Porteus, 2016).

9.11. MINIMUM EDITING EFFICIENCY NEEDED FOR THERAPY

The therapeutic effectiveness of gene editing depends mainly on the frequency of edited cells and the level of expression of therapeutic gene(s) in relation to the threshold level of gene product needed for the reversal of disease pathology. For some diseases like SCID-X1, few or possibly even one, edited stem or precursor cell may be sufficient for therapy since the 'corrected' cells are expected to have a selective advantage over the endogenous 'uncorrected' cells. A similar, albeit smaller, advantage occurs for the gene-edited cells in diseases like sickle cell disease and β-thalassemia; in such cases, ~2-10% gene-edited stem cells may be enough for curing the diseases. The frequency of gene-edited cells must be high for obtaining therapeutic effectiveness if the edited cells produce the therapeutic proteins at the normal endogenous levels. But relatively fewer cells will produce effective amount of the therapeutic protein if each edited cell produced and secreted high (supratherapeutic) levels of the concerned therapeutic protein(s) (Porteus, 2016). For example, high expression levels of therapeutic genes encoding Factor IX, iduronate-2-sulfatase, and α-L-iduronidase have been achieved by inserting them into the albumin locus of hepatocytes.

9.12. CANCER IMMUNOTHERAPY

Cancer is a disease characterised by abnormal or uncontrolled cell growth with the potential for invasion of other body tissues and organs. The cancer cells may express tumour antigens that are generally recognized by the human immune system leading to an immune response and elimination of these cells. But some tumour cells accumulate mutations so that they are able to evade the immune surveillance and continue to proliferate. In addition, cancer cells also develop several other mechanisms, including reduced expression of MHC-I on their surface and induction of apoptosis of T lymphocytes to escape the immune response. In view of this, development of strategies to enable T cells to recognize cancer cells and to mount an immune response against them is being continuously pursued as an important objective of cancer immunotherapy.

The T cells can be modified to act against tumours by introducing in their genomes a gene encoding a tumour-specific chimeric antigen receptor (CAR); such T cells are generally referred to as CAR-T cells. The first-generation CARs had the following components: a single chain variable region from a monoclonal antibody specific for the targeted tumour antigen, a CD8 hinge domain, the transmembrane domain from T cell receptor and the intracellular signalling domain of the CD3-zeta chain (Fig. 9.3A). The CAR design has been continuously refined to improve the performance as well persistence of CAR-T cells, and the fourth generation of CARs is now available. The CARs must be designed for each tumour-specific antigen, but efforts are being made to design and develop 'universal' CAR. The current research focus is to develop split, universal, and programmable (SUPRA) CARs as they are expected to provide further flexibility and control in the therapeutic applications of CAR-T cells (Zhao et al., 2018).

The CAR-T cells against a given tumour-specific antigen need to be produced *de novo* for each patient from the T cells donated by the concerned patients to achieve prolonged therapeutic efficacy. This constraint has been removed by creating 'universal' or 'off-the-shelf' third-party T cells by TALEN, ZFN and CRISPR-Cas SDN-mediated gene editing. The SDNs are used to delete the endogenous T cell gene encoding either α- or β-chain of the T cell receptor (TCR). These gene-edited T cells do not respond to TCR stimulation and are much less likely to lead to graft-versus-host disease. An improved version of 'universal' CAR-T cells that affords a novel collection of tools and techniques for cancer immunotherapy was undergoing clinical trials during 2018 (Dunbar et al., 2018; Zhao et al., 2018).

CD19 is considered as the tumour-specific antigen for most acute as well as chronic malignancies of B cells since it is constitutively expressed and displayed on the surfaces of these cells. Therefore, anti-CD19 CAR (or simply CD19 CAR)-T cells, *i.e.*, T cells expressing the CD19-specific CAR, were developed to treat B cell malignancies expressing CD19. The persistence and antitumor activity of CD19-CAR-T cells was improved by refining the CAR architecture to generate the second-generation CARs designated as CD19RCD28 CAR and CD19(4-1BB) CAR. The new versions of anti-CD19 CAR provide additional functions due to the inclusion of costimulatory domains of CD28 and 4-1BB (CD137) costimulatory factors placed between the transmembrane domain of T cell receptors and the signal transduction domain of TCR CD3-zeta chain to yield the CD19RCD28 CAR and CD19(4-1BB) CAR, respectively (Fig. 9.3B). The T cells expressing CD19(4-1BB) CAR provide sustained and efficacious antileukemic action for more than 6 months. The *ex vivo* gene therapies based on the improved versions of

CARs have been approved for use as drugs to treat acute lymphoblastic leukaemia and diffuse large B cell lymphoma. In addition, T cell receptor modified to recognize and bind to the NY-ESO1 protein expressed by many tumour cells has been approved for the treatment of synovial sarcoma (Table 9.4; Dunbar et al., 2018; Zhao et al., 2018).

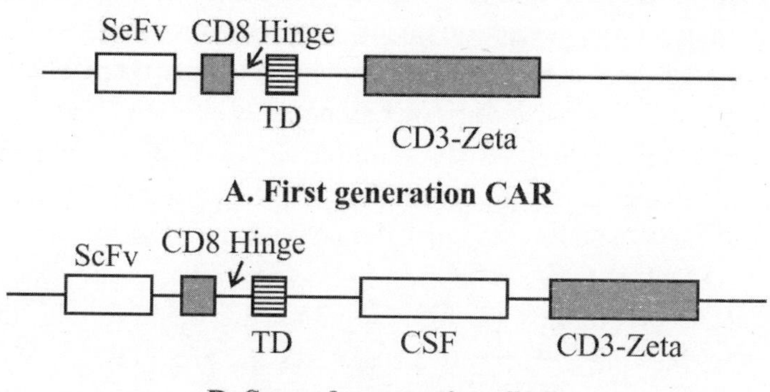

Fig. 9.3. Architecture of chimeric antigen receptors (CAR). **A.** First generation receptors have a ScFv, a TD and CD3-Zeta signalling domain. **B.** The second generation CARs have a costimulatory factor (CD28 or 4-1 BB) in addition to the domains present in the first generation CARs. ScFv, single N-chain variable region from a tumour antigen-specific monocloal antibody; TD, trans-membrane domain of T cell receptor; CD3-zeta, intracellular signalling domain of CD3-zeta chain; CSF, costimulatory domain of factor CD28 or 4-1 BB (based on Zhao et al., 2018).

The transgene CD19 CAR was introduced into and expressed in the 'universal' T cells. In 2017, these T cells were used for the treatment of two patients suffering from refractory B cell acute leukaemia on sympathetic considerations. These patients responded to the treatment; this *ex vivo* gene therapy was in Phase I clinical trials in 2018. In addition, early *ex vivo* gene therapy trials have been initiated for the use of TALEN-generated universal T cells expressing anti-CD123 CAR for treatment of acute myeloid leukaemia and blastic plasmacytoid dendritic cell neoplasms (Dunbar et al., 2018; Zhao et al., 2018).

In October 2016, the first clinical trial using CRISPR-Cas9 SDN system was initiated for treating a patient suffering from an aggressive form of lung cancer. The gene for PD-1 (programmed cell death protein 1) was inactivated in the peripheral lymphocytes collected from this patient by gene editing with CRISPR-Cas9, and the gene-edited cells were purified and transferred back into the patient. Similar clinical trials for PD-1 inactivation-based treatment of bladder cancer, prostate cancer and renal cell carcinoma have been approved. Another clinical trial for the use of CRISPR/Cas9 SDN system to engineer T cells for targeting several types of cancer cells has been approved in U.S.A. In this therapy, the CRISPR/Cas9 is used to delete the PD-1 receptor and the endogenous TCR encoding genes, and to insert a gene for a CAR designed to target the cancer cells. In addition, at least nine clinical trials for the use of CRISPR-Cas SDN system mainly to inactivate/delete the gene encoding the PD-1 receptor in T cells designed for tumour-targeting have been approved by the Chinese regulatory agencies (Katrekar et al. 2017; Dunbar et al., 2018).

9.13. VIRUS RESISTANCE

The processes of viral infection and disease development involve complex, dynamic and persistent virus-host interactions based on many components (Table 9.6). Viruses use many specific host factors to enter host cells and to sustain and multiply themselves once they are inside the cells. On the other hand, hosts use various components of their innate and adaptive immune systems as well as general restriction factors to inhibit virus infection, replication and spread or to clear viral infections. The viruses counteract the host defence strategies by employing appropriate escape tactics to avoid the host immune surveillance systems and the restriction factors (Table 9.7). The CRISPR-Cas SDN system can be used to treat viral diseases by targeting either the viral genomes (RNA or DNA) or the host factors, including the positive and negative regulators as well as interferons, cytokines and interferon-stimulated gene products.

Table 9.6. A generalised summary of the components of host-virus (DNA and RNA) interaction and the CRISPR-Cas systems that can be used to target them (based on Chen et al., 2018)

Component	CRISPR-Cas system	Remarks
DNA genome (virus, provirus in case of retroviruses)	SpCa9, SaCas9, Cas12a, C2c1/3 (Cas13)	The CRISPR-Cas systems are used for gene editing; CRISPRi is used for CRISPR-dCas-based transcriptional inactivation of genes
RNA genome (virus)	FnCas9, Cas13b	Cleavage of RNA genomes of viruses
Protein factors:		
Virus receptor/coreceptor Negative regulators Dependent molecules	SpCa9, SaCas9, Cas12a, C2c1/3 (Cas13) CRISPRi	Gene editing to modify the proteins so as to interfere with the virus infection/function; CRISPRi used for transcriptional repression of genes encoding the factors
Nucleic acid sensors Antiviral factors Positive regulators Signalling adaptors Transcription factors Interferons Cytokines Interferon-stimulated gene products Genomic DNA (virus)	CRISPRa	Transcriptional activation of the genes encoding protein factors

Table 9.7. Selected examples of restriction factors produced by human cells to combat viral pathogens

Restriction factor	Action	Viral response	Remarks
Tripartite motif 5-alpha (TRIM5α)	Interacts with viral capsid; blocks replication	-	-
APOBEC3G*	Converts C-to-U	Counteracted by HIV-1 protein Vif	D128K mutation in the restriction factor interferes with the inhibition by Vif factor of HIV-1
Tetherin	Prevents replication by inhibiting release of viral particles	Counteracted by Vpu protein of HIV-1	Interferon-induced antiviral protein; amino acid substitution T45I makes Tetherin resistant to inhibition by Vpu
SAMHD-1*	Inhibits lentivirus infection in myeloid cells and HIV-1 infection in resting $CD4^+$ T-cells	Protein Vpx counteracts it by inducing its proteasomal degradation	It hydrolyses dNTPs and reduces intracellular dNTP levels to levels insufficient for HIV-1 reverse transcription
Myxovirus resistance protein 2 (Mx2)	Prevents nuclear import of subviral complexes during late stages of infection	-	Interferon-induced inhibitor of HIV-1 infection

* APOBEC3G, Apolipoprotein B mRNA-editing enzyme, catalytic polypeptide 3G; SAMHD-1, Sterile Alpha Motif- and HD-domain containing protein 1.

The viral genomes and some of the host factors can be the targets of gene editing, while host factors and even viral DNA genomes can be subjected to CRISPRi or CRISPRa. CRISPR-Cas systems have been used to develop tools for combating the human diseases caused by HIV (Human Immunodeficiency Virus), HBV (Hepatitis B Virus), HPV (Human Papillomavirus), HCV (Hepatitis C Virus), EBV (Epstein–Barr Virus), dengue virus, Ebola virus, etc. The salient features of the findings from these studies are summarized here with special reference to HIV. The CRISPR-Cas strategies can target either the processes of viral infection, replication and propagation or the viral genomes present as provirus in the genomes of host cells with latent viral infection (Table 9.1; Chen et al., 2018).

9.13.1. Targeting viral infection, replication and propagation

HIV-1 is the causal agent of human acquired immunodeficiency syndrome (AIDS), which occurs worldwide and there is no effective vaccine against this dreaded pathogen. AIDS treatment is based on HIV-1 antiretroviral therapy (ART), which keeps the HIV-1 reservoir at a low level; this therapy is very expensive, has to be taken life-long, and can damage the hepatic or cardiovascular system. In 2009, it was reported that the *CCR5Delta32* mutation confers resistance to infection by HIV-1 virus. This discovery originated from the observation that an American patient suffering from AIDS and leukaemia lived free from HIV-1 infection after he received a bone-marrow transplant from a donor who was homozygous for the *CCR5Delta32* mutation. As a result, this patient did not require ART after the transplant. Thus, the gene encoding CCR5 presented itself as an attractive target for gene editing for resistance to HIV-1 infection. HIV-1 uses CD4 (cluster of differentiation 4) as the primary receptor for entering target cells. In addition, protein CCR5 serves as an essential coreceptor for HIV-1 during primary infection (Porteus, 2016; Chen et al., 2018).

The first report of successful use of CRISPR-SpCas9 to edit hiPSCs was by Ye et al. (2014). They induced the naturally occurring *CCR5Delta32* mutation in the hiPSCs and showed that the monocytes/macrophages differentiated from the edited hiPSCs were resistant to HIV-1. Subsequent studies validated the observation that inactivation of *CCR5* resulted in resistance to HIV-1 infection in other cell types like human primary $CD4^+$ T cells, and primary $CD34^+$ hematopoietic stem and progenitor cells. The R5-tropic (infection using CCR5 as receptor) HIV-1 strains also use another coreceptor, called C-X-C chemokine receptor type 4 (CXCR4), in the later stages of AIDS for entering new target cells. The disruption of CXCR4 in primary human $CD4^+$ T cells using CRISPR-SpCas9 generated resistance to HIV-1 infection. Further, simultaneous disruption of CCR5 and CXCR4 conferred on $CD4^+$ cells resistance to dual tropic (CCR5-tropic and CXCR4-tropic) infection by HIV-1 (Chen et al., 2018).

An alternative strategy for blocking HIV-1 infection is to knockout the genes encoding host proteins like LEDGF, PSIP1, TNPO3, Nup358, etc. that are essential for for HIV-1 replication, packaging and budding. Inactivation of the gene encoding LEDGF or TNPO3 by CRISPR-Cas9 delivered as RNP into primary $CD4^+$ T cells resulted in reduced tropism-independent HIV-1 infection. In addition, human cells have developed virus defence system based on proteins, called restriction factors, that interfere with key steps in virus propagation. CRISPRa can be used to activate the transcription of genes encoding these factors and thereby boost the human defence against HIV-1. For example, factors TRIM5α and Tetherin block virus replication and protein Mx2 interferes with nuclear import of subviral complexes of HIV-1. There is some evidence that a combination of CCR5 disruption with overexpression of selected restriction factors tends to provide significant protection against both R5-tropic and X4-tropic HIV-1 infection (Chen et al., 2018).

Genome-wide screens based on CRISPR-Cas9 system have enabled the identification of mechanisms used by EBV to evade the host tumour suppressor responses, and three host proteins essential for cell-to-cell transmission of HCV as well as for direct entry into cells during fresh infections. CRISPR-Cas9 has been used for genome editing of herpesviruses like HSV-1 and Human cytomegalovirus. The Cas13b and possibly FnCas9 can be used to target the genomes of RNA viruses and the host mRNAs encoding proteins essential for viral multiplication and/or propagation. The FnCas9 seems to have potential for RNA genome editing since it is reported to target an endogenous mRNA. FnCas9 has been used to target the 5'- and the 3'-untranslated regions of the positive-sense ssRNA (single-stranded RNA) genome of HCV in human hepatocellular carcinoma cells. The FnCas9 did not cleave the RNA genome but it inhibited its translation and replication by physically interfering with these processes (Chen et al., 2018).

It is noteworthy that frequent mutations during replication of viral genomes allows them to escape the host immune surveillance. The high mutation rate also allows them to evolve rapidly and to use alternative host proteins including receptors for infection, replication and propagation. Therefore, it is important that multiple sgRNAs be used to target viral genomes and to target more than one host factor for CRISPRi/a to afford efficient gene therapy against viral diseases (Chen et al., 2018).

9.13.2. Targeting the provirus in genomes of cells with latent infection

The HIV treatment based on combined antiretroviral therapy is a "functional cure" and suffers from resurgence of HIV-1 after withdrawal of the therapy. The therapy is unable to eliminate the virus from latently infected cells as the virus integrated into the host genome is transcriptionally inactive. "Shock and kill", the latest strategy of HIV treatment, uses drugs to activate the expression of viral genes in latently infected cells leading to death of these cells by viral cytotoxicity and/or due to the host immune system. This is followed by pulsed antiretroviral treatment to prevent infection of new cells by the virus. But the "shock and kill" strategy suffers from less than satisfactory effectiveness, non-specificity as it affects noninfected cells as well, resistance of cells to apoptosis, drug toxicity, and the need for repeated drug administration (Zhang et al., 2015).

The CRISPR-Cas system can be used to excise the HIV-1 genome integrated as provirus into the human genome. This has been achieved in cultured progenitor/neural stem cells from transgenic mice and in C11 latent cell lines. Zhang et al. (2015) used the dCas9-SAM system (Section 10.3.2.2) for highly specific, very potent and persistent activation of HIV-1 virus present in latently infected cells. Two sequences surrounding the enhancer region in the LTR region of viral genome served as excellent targets for the dCas9-SAM system. The virus reactivation resulted in production and accumulation of viral proteins leading to death of the concerned cells. Significantly, this did not affect HIV-negative cells in the neighbourhood. The effectiveness of this strategy is not likely to be affected by the site of HIV-1 integration and/or the epigenetic state of the genomic regions surrounding the integration sites. The efficiency of viral reactivation can be increased by deploying multiple msgRNAs targeting different sites in the viral genome. This strategy needs to be further refined to increase the efficiency of delivery of dCas9-SAM system into the target cells, and to establish that the activation system does not produce off-target effects before it can be advanced to clinical trials (Zhang et al., 2015).

The CRISPR-Cas9-based transcription repressor can be deployed to create a 'synthetic lethal' lesion in an essential cellular pathway that would cause death of the diseased cells and tissues. For example, many cancers are produced due to the loss-of-function mutations in specific genes. A GI map of the concerned pathways would allow identification of such pairs of proteins that complement each other so that a cell can survive loss-of-function of any one but not both members of the pair. In case inactivation of one member of such a pair is involved in the development of a disease, *e.g.*, a type of cancer, the other member of the pair will be an excellent target for inactivation for the treatment of the concerned cancer since this will be lethal to the cancerous cells. The normal cells would survive the inactivation of the targeted gene since the functional allele of the other member of the gene pair will complement this loss-of-function. This approach is termed as 'synthetic lethal' therapy. It is being used for treatment of some cancers using specifically designed drugs to inhibit the targeted proteins. Similarly, CRISPR-Cas9-based transcription repressors can be used to block transcription of the targeted genes and thereby create 'synthetic lethals'. The recently approved drug Olaparib, is an excellent example of this strategy: this drug inhibits PARP1/2 function and is used for the treatment of tumours caused by inactivation of either *BRCA1* or *BRCA2* gene. The inhibition of both PARP1/2 and BRCA1/BRCA2 functions is lethal to the cells. As a result, Olaparib is lethal to tumour cells lacking BRCA1/BRCA2 function but is not detrimental to the normal cells that have intact BRCA1 and BRCA2 functions (Gilbert et al., 2014; Shen et al., 2017).

9.14. CHOLESTEROL LEVEL

The level of LDL cholesterol is markedly affected by gene *PCSK9*. This gene encodes Proprotein convertase subtilisin/kexin type 9 (PCSK9) protein, which is secreted by hepatocytes into plasma. Protein PCSK9 functions as an antagonist of the receptor for LDL cholesterol, and this ultimately reduces LDL cholesterol degradation. It is well known that individuals with spontaneous loss-of-function mutation in the *PCSK9* gene have lower levels of LDL cholesterol than the $PCSK9^{+}$ $PCSK9^{+}$ homozygotes. The heterozygotes ($PCSK9^{+}PCSK9^{-}$) have lower LDL levels than the wild-type homozygotes, while persons homozygous for the mutant allele ($PCSK9^{-}$ $PCSK9^{-}$) have still lower LDL levels. At the same time, both heterozygous and homozygous mutant individuals are known to be completely healthy (Porteus, 2016; Savic and Schwank, 2016). The CRISPR-Cas9 SDN system was delivered into mouse liver cells by Ding et al. (2014) using adenoviral vectors. The SDN induced loss-of-function mutation of *PCSK9* gene in up to 50% of the cells, and biallelic mutations (mutation in both the alleles of the gene in a cell) were also produced. In the gene-edited mice, the hepatic LDL receptor levels increased and, ultimately, the plasma LDL levels declined as compared to unedited control mice. This proof-of-concept study may lead to the development of a therapeutic procedure for lowering the LDL levels in humans.

9.15. DUCHENNE MUSCULAR DYSTROPHY

Duchenne muscular dystrophy is a X-linked recessive trait that affects about one in 5,000 male babies. The symptoms of DMD become manifest by the age of 4 yrs and patients have average life expectancy of 26 yrs. It is caused by loss-of-function mutations in the extremely large *dystrophin* gene; about one-third of the cases of DMD are due to new mutations. The protein Dystrophin is important for maintenance of cell membranes of muscle fibres. The absence of this protein disrupts the normal structure of muscle fibres, which may be replaced by fat bodies in the muscle tissues. The muscle loss occurs first in thighs so that patients have trouble standing up. The muscle loss progresses to arms and, ultimately, respiratory and cardiac muscles are also affected. The mutational defects in *dystrophin* gene have been corrected by using three distinct SDN-based approaches, *viz*., indel induction, HDR-mediated exon addition, and large sequence deletion by multiplexed editing. Ousterout et al. (2015) used TALENs to target exon 51 of the *dystrophin* gene to induce indels since this exon carried a frameshift mutation. About one-third of the indel events produced by TALENs restored the reading frame of the gene and production of the dystrophin protein in the edited cells (Fig. 9.4). Gene editing was carried out in skeletal myoblasts and dermal fibroblasts obtained from DMD patients after these cells were converted by MyoD, a myogenic regulatory factor, into cells of the myogenic lineage.

Li et al. (2015) used HDR-mediated insertion of Exon 44 in the *dystrophin* gene of a DMD patient with missing Exon 44. They used patient-derived fibroblasts to generate iPSCs and introduced into these cells CRISPR-Cas9 and a donor template that included Exon 44. HDR-mediated correction of the *dystrophin* gene resulted in insertion of the missing Exon 44 into the gene (Fig. 9.5). The cells with the corrected gene were selected, subjected to comprehensive analysis for off-target effects and differentiated towards skeletal muscle cells. These cells expressed full-length wild-type Dystrophin protein. Subsequently, Ousterout et al. (2015) used multiplexed CRISPR-Cas9 to delete the Exons 45-55, which are mutation 'hot-spots' and contain over 60% of the mutations found in DMD patients. This large deletion restored the expression of dystrophin protein in DMD patient-derived immortalized myoblasts (Fig. 9.6). The gene-edited cells continued to produce the Dystrophin protein even after they were transplanted into mutant mice.

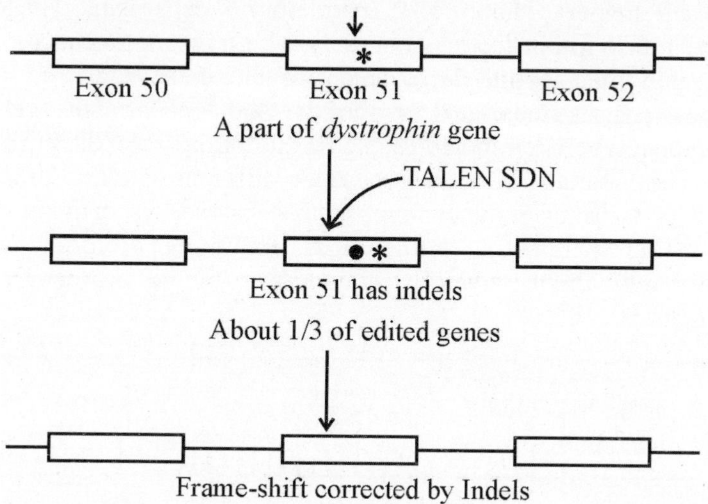

Fig. 9.4. Correction of frame-shift mutations in *dystrophin* gene. **A.** Exon 51 has frame-shift mutation, mutation site indicated by *. The arrow indicates the site targeted by TALEN SDNs. **B.** TALEN SDN induces DSB and indels at a site upstream of the mutation site (indicated by a solid circle). **C.** In about one-third of the gene-edited cells, the frame-shift mutation is corrected and the function of *dystrophui* gene is restored.

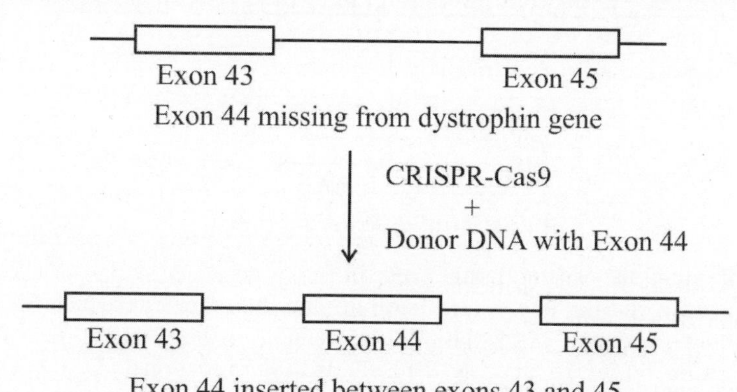

Fig. 9.5. In some mutant *dystrophin* genes, the Exon 44 is missing. In such cases, mutation correction is based on HDR-mediated insertion of exon 44 between exons 43 and 45.

9.16. ALLERGIC AND IMMUNOLOGIC DISEASES

In simple terms, allergy is an abnormal immune response to common environmental substances that are harmless to most people. Some persons are genetically predisposed to develop allergic diseases; this condition is called atopy. The Type 2 T helper (Th2) cells that are involved in response against extracellular parasites are also associated with atopic diseases like asthma, atopic dermatitis, allergic rhinoconjunctivitis, etc. The T helper (Th) cells are $CD4^+$ that develop as either Th1 (Type 1 T helper), Th2 or Treg (regulatory T cell) phenotypes depending on the expression of specific transcription factors (Fig. 9.7). The transcription factor GATA-3 (GATA binding protein 3) is important for the development of Th2 cells, while the transcription factor FoxP3 (Forkhead box p3) is important for the development of Treg cells. Similarly, the development of Th1 phenotype is promoted by the transcription factor T-bet (T-box transcription factor TBX21). The Th1, Th2 and Treg cells produce different cytokine mediators that lead to Th1 response, Th2 response and immune suppression response, respectively. The Treg cells, in addition, are involved in maintaining homeostasis, inhibiting proliferation of T cells and production of cytokines by them; they also play a key role in prevention of autoimmunity (Goodman et al., 2017).

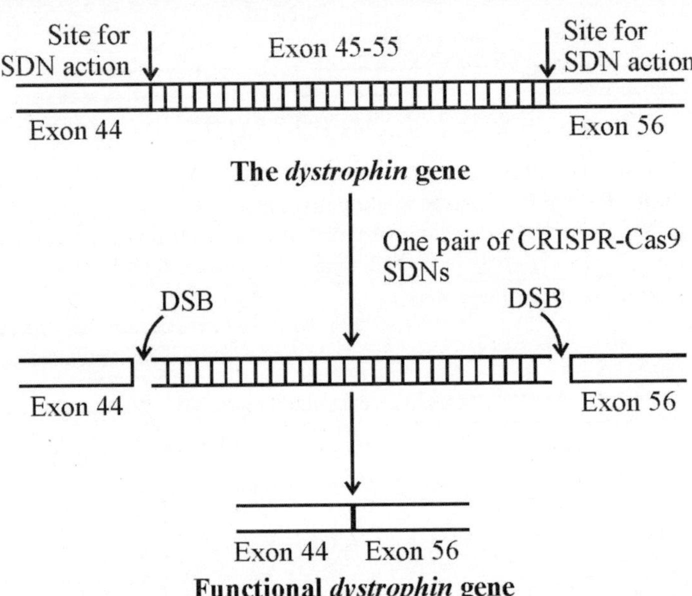

Fig. 9.6. About 60% of mutations causing DMD occur in Exons 45 to 55. A pair of CRISPR-Cas9 SDN is used to induce two DSBs, one before Exon 45 and the other after Exon 55, to delete the gene segment containing Exons 45-55. The shorter version of the genes is freed from the mutations located in Exons 45 to 55 and, as a result, is fully functional. DMD, Duchenne muscular dystrophy; DSB, double-strand break.

The CRISPRa/i can be used to activate/suppress the expression of relevant transcription factors/cytokine mediators to shift the balance from Th2 response to Th1 and/or Treg response to manage/treat atopic diseases. This could be achieved by activating the expression of transcription factors T-bet and/or cytokine mediators IFN-γ (interferon-gamma) and IL-2 (interleukin 2) for promoting Th1 response and transcription factor FoxP3 and/or cytokine mediators CTLA-4 (cytotoxic T lymphocyte–associated protein 4), IL-10 and TGF-β (transforming growth factor β) for increasing the Treg response. Alternatively, the expression of transcription factor GATA-3 and/or cytokine mediators IL-4, IL-5 and IL-13 could be targeted by CRISPRi since their repression would shift the response away from Th2 response. It may be pointed out that the allergic diseases have a complex control and their development may involve interaction among two or more genetic components, some epigenetic changes, and environmental factors. Therefore, it has been suggested that multiple genes may need to be simultaneously targeted for a proper treatment of atopic diseases. The multiplexed CRISPRa/i can be implemented by using gene transcription regulators (Section 10.3.2.2; Goodman et al., 2017).

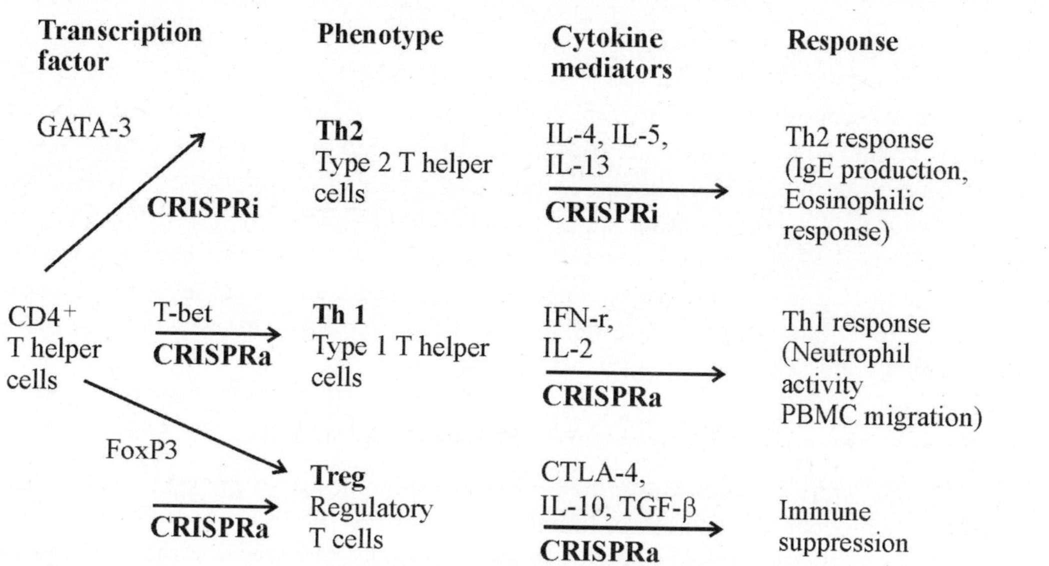

Fig. 9.7. Schematic representation of the different T helper cell responses and the possible targets for CRISPR-Cas action. CD4, cluster of differentiation 4; GATA-3, GATA bindng protein 3; FoxP3, forkhead box p3; T-bet, T-box transcription factor TBX21; IL-2, 4, 5, 10, 13, Interleukin 2, 4, 5, 10, 13 respectively; IFN-γ, Interferon γ; CTLA-4, cytotoxic T lymphocyte-associated protein 4; TGF-β, Transforming growth factor β; PBMC, Peripheral blood mononuclear cell; CRISPRi, CRISPR-interference; CRISPRa, CRISPR-activation.

The CRISPR-Cas technology has made significant contributions to both basic and clinical research efforts in immunology. For example, CRISPR-Cas9 has been used to induce class-switching recombination in the heavy chain encoding gene of antibodies to obtain the desired subclass of antibodies or even Fab (fragment antigen-binding) fragments. These efforts have greatly facilitated a more careful and detailed analysis of antibody subclasses and provided novel methods for easy generation of Fab fragments both for research as well as therapeutic applications. Gene editing has enabled the development of an *in vitro* model of blockage in T cell development due to deficiency in Janus kinase 3 (JAK3), which has permitted identification of the block in T-cell development (Goodman et al., 2017).

One of the most notable contributions of the CRISPR-Cas technology relates to the strategies that may be useful for the treatment of AIDS, *e.g.*, inactivation or deletion of the gene encoding

CCR5 or CXCR4 coreceptors for HIV-1 virus (Section 9.13.1). In addition, the provirus present in the genomes of cells with latent HIV-1 infection can be either deleted or the cells themselves may be eliminated by activating the provirus (Section 9.13.2). JAK3 deficiency is a form of SCID caused by a recessive loss-of-function mutation in the autosomal gene *JAK3*. The loss of JAK3 function leads to the absence of T cells and natural killer cells as well as poor function of B cells. The CRISPR/Cas9 was used to correct the mutational change in the *JAK3* gene of the *in vitro* model, which restored the JAK3 function and permitted normal T-cell development (Chang et al., 2015).

Gene editing can be used to correct genetic defects responsible for primary immunodeficiencies like ADA-SCID (ADA deficiency), X-SCID, asthma, etc. The ADA-SCID is a form of SCID caused of loss-of-function mutation in the autosomal ADA (adenosine deaminase) encoding gene. Similarly, X-linked SCID is caused by a mutant X-linked gene *IL2RG* that leads to near absence to T cells and natural killer cells. The mutation in *ADA* and *IL2RG* genes can be corrected by gene editing and the gene-edited HSPCs may be used for gene therapy. In case of asthma, some SNPs are known to be associated with the ailment. The bases of these SNPs can be subjected to base editing to convert them to other bases that are not associated with asthmatic condition. But as in the case of allergy, a multiplexed strategy would be required to produce a demonstrable effect of gene therapy (Goodman et al., 2017).

9.17. FUNCTIONAL TOXICITY ASSAYS FOR GENE EDITING

The evaluation of safety of gene editing-based therapies should be on the same lines as those for other therapeutic procedures. Therefore, functional toxicity of the therapy should be evaluated in addition to the specificity and off-target activities of the SDNs used for the therapy. In these assessments, a distinction must be made among the following three types of toxic effects of gene editing protocols: (i) initial transient inhibition of cell behaviour that disappears with time, (ii) a permanent injury to a cell that reduces its function as compared to injury-free cells, and (iii) alterations that have no adverse effects initially but negative effects become manifest sometime later. An example of the last type of change would be such alterations induced by gene editing that permit the cells to appear normal initially but predispose them to transform into cancerous cells over time. There are four major assays for measuring functional toxicity of gene editing, *viz*., (*a*) cell cycle dynamics analysis, (*b*) determination of short-term cell survival and proliferation, (*c*) analysis of clonal dynamics in large, heterogeneous populations, and (*d*) assessment of long-term cell behaviour (Porteus, 2016).

The cell cycle analysis aims to detect and quantify any disturbance in progression of cell cycle caused by the SDN treatment. The cell survival or proliferation assay estimates the proportion of cells that survive SDN treatment and/or measures their ability to proliferate. The clonogenicity assay attempts to determine if the clonal composition of the large cell population generated from SDN-treated cells is affected by the treatment. Generally, barcodes are used to classify cells into clones derived from different individual cells subjected to SDN treatment. The lineage reconstitution assay involves analysis of the ability of gene-edited cells to give rise to appropriate cell types after they are transplanted into an immunodeficient mouse, and determination of whether they would undergo cell transformation (Porteus, 2016).

9.18. IDENTIFICATION OF DRUGS FOR VENOMS

Gene inactivation by CRISPR-Cas9 SDN can be used to identify genes that are involved in sensitivity to a toxin and use this information to develop drugs for alleviating or eliminating the effects of the toxin. For example, four genes involved in cholesterol regulation pathway were found to be important for production of toxic effects by transparent box jellyfish venom. The drugs used for the regulation of concerned cholesterol pathway were found to prevent the venom from acting on human cells in culture.

9.19. REGULATORY CONCERNS

The regulatory concerns pertaining to gene editing-based therapeutics relate mainly to safety, efficacy, and quality control. The efficacy and quality control aspects may be assessed following the existing guidelines for conventional *in vivo* and *ex vivo* gene/cell therapies, *e.g.*, 'FDA guidance for human somatic cell therapy and gene therapy' (1998), EMA guidelines 'Quality, preclinical and clinical aspects of gene transfer medicinal products (2001)', etc. But the use of gene-editing tools introduces new safety issues related to their delivery systems and the SDNs themselves. Generally, the delivery of gene editing tools into *ex vivo* cells poses lower risk than their introduction into cells *in vivo*. In general, the use of viral vectors for delivery of the therapeutics poses greater safety concerns than physical and chemical delivery agents. The SDNs used in gene therapy might cause off-target genotoxicity, *i.e.*, toxic side effects produced by DSB induction at non-target genomic sites; this risk is one of the most significant safety concerns. The CRISPR/Cas9 SDN system is likely to generate higher off-target effects than TALENs and ZFNs. The initial assessment of off-target effects in gene-edited cells may be done by functional analysis of the edited cells, followed by detection of large chromosomal alterations and small indels. This may be followed by sequencing of the exome and whole genome, the latter being necessary for detecting the insertion of vector DNA into the cell genomes and donor DNA integration at off-target sites (Shim et al., 2017).

The assessment of immunogenicity of gene-editing reagents is required even when viral vectors are not used for their delivery. It is important that the processing of proteins used as gene-editing reagents and their presentation to the immune system has not been investigated so far. The host cell impurities remaining in the viral vector preparations can be immunogenic, and even the AAV vectors are reported to cause immune response. The non-viral vectors like polymer/lipid nanoparticles are regarded to be much less immunogenic than viral vectors. It is important to evaluate the distribution and concentrations of therapeutic reagents in the relevant cells/tissues, and the duration for which they persist in them. In addition, the frequency and persistence of gene-edited cells should also be monitored. Finally, in case of *in vivo* gene therapies based on injection of viral vectors, the presence of SDNs and the introduced DNA levels in the germ cells must be assayed (Shim et al., 2017).

Another aspect that needs careful evaluation is the tumorigenicity of gene-edited cells, which may be caused by factors like insertion of the viral vectors in the vicinity of proto-oncogenes. The tumorigenic potential of gene-edited cells can be screened by using Onco-chip to monitor changes in the expression of oncogenic proteins. In any case, tumorigenicity of the cells must be ruled out using appropriate animal models. The safety issues in case of viral vectors are immunogenicity, integration into the cell genome, duration of expression, sexual transmission to partners and germline transmission. In case of non-viral nanoparticles, the safety issues relate to toxicity and inflammation caused by them. Similarly, infection, inflammation, and tissue damage caused by the therapeutic delivery equipment and/or procedure are also important. The safety evaluation of gene editing therapeutics should be done in 'humanized' animal models, which have genomes that mimic the human genome. It would be desirable to conduct *in silico* analysis of the target genomic sites of animal models and the human genomic sites being targeted by the therapeutics to determine the degree of homology between them (Shim et al., 2017).

The evaluation of efficacy of gene-editing therapeutics should include dose selection, determination of the relevant animal models, gene editing efficiency, and the duration of their efficacy. The therapeutic dose needs to be carefully determined as it would be affected by several factors, including the form and the method of SDN delivery. The efficiency and the duration of gene editing should be carefully determined since prolonged SDN expression would lead to off-target action; this aspect is particularly relevant in case of *in vivo* therapies. The chief issues pertaining to quality control include purity, and freedom from microbial contamination and endotoxins. In case of *ex vivo* gene-editing, phenotypic characteristics, viability and the proportion of gene-edited cells are important quality parameters. Finally, the manufacturing processes should ensure purity, sterility and consistency of the gene-editing reagents (Shim et al., 2017).

9.20. ADVANTAGES OF GENE EDITING-BASED GENE THERAPIES OVER CONVENTIONAL GENE THERAPIES

Gene editing therapies are preferable to traditional gene therapies since they involve correction of the mutational defects of the concerned endogenous genes, while the latter either aims at temporary reduction in disease symptoms or random integration into the genome of therapeutic factors, usually, the normal alleles of the concerned genes. Gene therapy is capable of only addition of gene function, while gene editing-based therapy can add novel gene functions, correct mutated genes and restore normal gene function, inactivate endogenous genes, and modulate gene expression levels and patterns. The therapeutic usefulness of gene editing is influenced by many factors, including efficiency of gene editing, fitness of the gene-edited cells, and efficiency of delivery of the edited cells into the target tissues *in vivo*. Many of these issues limit the effectiveness of therapeutic gene editing; therefore, the tools and procedures need to be further refined for enhancing the effectiveness of gene therapy. Since most immunologic diseases target hematopoietic cells, autologous HSCs can be subjected to corrective gene editing *ex vivo* and used for transplant to completely or partially replace the defective endogenous cells (Goodman et al., 2017).

Most diseases result from inactivation of genes by base substitutions or base deletions/ insertions. In such cases, it would be necessary to correct the relevant mutations by HDR-mediated sequence replacement in the concerned mutant loci. In case the deletion is large, it may become necessary to integrate the complete gene into the genome. Random integration of transgenes is undesirable as it may activate some harmful genes like oncogenes as was observed in the case of X-SCID gene therapy trials. The SDNs can be used to integrate complete genes, including their 5' and 3' regulatory sequences, in 'safe harbor' sites of human genome (Section 9.6). But if these genes are integrated within the concerned endogenous mutant loci whose missing functions they are designed to provide, they will be subjected to the same controls as their normal endogenous counterparts; this situation is highly desirable (Carroll, 2014).

Some monogenically controlled genetic diseases are produced by duplications of specific genomic sequences. For example, diseases like Fragile X syndrome (the most common form of inherited mental retardation) are the result of expansion of trinucleotide repeats. Such disorders may be corrected by deleting the expanded repeat sequences using suitably designed SDNs to induce a pair of DSBs at the two ends of the sequence to be deleted. This approach will be more likely to produce the desired result in case of diseases like Friedreich's ataxia, in which the duplicated sequence is located in the noncoding region of the concerned gene since there will be little risk of NHEJ-mediated gene inactivation. It has been suggested that the relative ease in designing and implementing CRISPR-Cas9 SDNs might facilitate gene therapy for rare genetic diseases at an affordable cost (Hsu et al., 2014).

9.21. LIMITATIONS OF GENE EDITING FOR GENE THERAPY

All SDN systems, including CRISPR-Cas9, are known to produce variable frequencies of off-target effects; off-target sites differing for 2-3 bp from the target site are prone to CRISPR-Cas9 action. For therapeutic applications, it is highly desirable to eliminate the risk of detrimental effects due to off-target action of the SDNs used especially for *in vivo* therapies. It has been argued that it may not be possible to eliminate the off-target action of SDNs since their target DNA sequence recognition and binding domains have evolved in nature to tolerate some degree of mismatch. Therefore, it would be highly desirable to reach a consensus on the acceptable level of off-target action for therapeutic applications of SDNs. In this context, the nature and the functions of the off-target sites would be far more important than their frequency. The method used for off-target detection should be unbiased and highly sensitive, and the whole genome must be covered. Although several different methods for detection of off-target effects are available, none of them is comprehensive and capable of detecting <0.1% off-target action. Therefore, it is important to develop more sensitive methods for the detection of off-target actions (Tsai and Joung, 2016; Chen et al., 2018).

Many human diseases cannot be treated by *ex vivo* gene therapy and diseases like cystic fibrosis affect several organs and tissues. In such cases, the delivery of recommended doses of the therapeutic agents into all the relevant tissues of patients presents many challenges. Further, a precise and targeted correction of mutated genes requires sequence replacement based on homology-directed repair (HDR). But the frequency of HDR remains quite low as compared to NHEJ. At the same time, HDR requires a donor DNA, the delivery of which presents difficulties that must be resolved. It is encouraging that the helper-dependent adenoviral vectors can carry the genes for both the members of a TALEN pair and the donor DNA. These vectors were used to deliver the cDNA of *CFTR* gene along with the TALENs into the cultured lung airway epithelial cells. In about 5% of the cells, there was targeted integration the *CFTR* cDNA into the *AAVS1* locus. This strategy is expected to ameliorate the deleterious effects of all the different mutations in *CFTR* gene that lead to cystic fibrosis. In addition, the donor DNA itself can become integrated at random sites in the genome and, thereby, lead to off-target effects independent of the SDN action. In this context, novel adenoviral donor vectors with chemically modified ends may reduce the risk of random insertion of the donor DNAs. The 'prime editing' procedure is believed to overcome many of the limitations related to HDR and donor DNA (Section 3.10.5; Carroll, 2015; Tasan and Zhao, 2017; Dunbar et al., 2018; Xia et al., 2019).

Gene editing is being explored as a tool for prevention and/or treatment of viral infections, and the procedure for *CCR5* inactivation for HIV-1 resistance is in clinical trials. It may be pointed out that virus genomes mutate at relatively high rates during replication to escape human immune surveillance. Similarly, they can evolve to use alternative receptors for entering the host cells. To overcome these limitations, multiple sgRNAs could be used to target viral genomes and the host genes encoding receptors and coreceptors. Since the sequences of viral genomes present in individual patients can be determined, the development of personalized set of sgRNAs seems to be an attainable goal (Chen et al., 2018).

It is disturbing that the action of CRISPR-Cas9 is more efficient in cultured human cells with impaired p53 function than in those with wild-type p53. The p53 tumour suppressor protein coordinates the monitoring of DNA damage in cells. The above finding suggests that in a cell population exposed to CRISPR-Cas9 action, cells with impaired p53 function would increase in frequency and would be more likely to undergo gene editing than those with intact p53 since they are preferentially killed by the SDN treatment. As a result, when such cells are reinfused into patients, they would be vulnerable to tumorigenic genetic changes that would make them cancerous. Similarity, the efficiency of genome editing in hPSCs with a normally functional *p53* gene is much lower than that in mouse embryonic stem cells. It appears that protein p53 triggers apoptosis in hPSCs, in which DSB is induced by CRISPR-Cas9. This DSB-induced cell death is a potent challenge to high-throughput CRISPR-Cas9-mediated gene editing and screening in hPSCs. It is, therefore, imperative that the function of p53 protein is carefully monitored during the gene editing of cells for therapeutic applications (Haapaniemi et al., 2018; Ihry et al., 2018).

The gene editing technology raises another concern that is common to other 'humane and patient-centric medical care' for patients suffering from genetic diseases. These medical interventions allow even those individuals to reproduce who would have otherwise died before adulthood. It is feared that these measures would lead to an increase in the genetic load of future human generations. This concern is a vexing issue that has been discussed for decades without any consensus on what could be a proper response (Porteus, 2016).

9.22. CLINICAL TRIALS

The efficacy and safety of new therapeutic reagents and procedures are ultimately evaluated on human participants in a series of clinical trials. According to the U.S. government official website, 'A clinical study involves research using human volunteers (also called participants) that is intended to add to medical knowledge… In a clinical trial, participants receive specific interventions according to the research plan or protocol created by the investigators. These interventions may be medical products, such as drugs or devices; procedures; or changes to participants' behavior, such as diet. Clinical trials may compare a new medical approach to a standard one that is already available, to a placebo that contains no active ingredients, or to no intervention'(https://clinicaltrials.gov/ct2/about-studies/learn#Considerations).

The first clinical trial for SDN-mediated gene inactivation and use of gene-edited cells for gene therapy was based on ZFN-directed inactivation of *CCR5* gene that encodes coreceptor for HIV-1 virus. The ZFNs targeting the *CCR5* gene were delivered via an AV vector into T cells obtained from HIV-positive patients. The *CCR5* gene was inactivated by NHEJ-mediated repair of the DSBs induced by the ZFNs. The gene-edited T cells were reinfused into the respective donors with encouraging results. This gene therapy had completed the Phase I/II clinical trials by 2018. Subsequently, ZFNs were used to target and inactivate the *CCR5* gene in HSCs; the gene-edited HSCs were undergoing Phase I clinical trials in 2018. It is noteworthy that this SDN-based *ex vivo* gene therapy progressed from the proof-of-concept experiment in laboratory to Phase I of clinical trials in less than five years. It may be expected that as the therapeutic gene editing becomes more well-known to the regulators as well as public at large, the period required for progression from proof-of concept to clinical trials may become significantly shorter (Urnov et al., 2010; Carroll, 2014; Porteus, 2016; Dunbar et al., 2018).

SDNs have been used for creation of 'universal' CAR T cells that can be infused into any patient; this has obviated the need for construction of autologous CAR T cells for each patient. CAR T cells express a chimeric antigen receptor designed to be specific to the concerned tumour-specific antigen. CD19 is regarded as tumour-specific antigen for many malignancies of B-cells. 'Universal' T cells expressing the CAR specific for CD19, *i.e.*, CD19 CAR, were used in 2017 for treatment of two patients suffering from refractory B cell acute leukaemia on sympathetic considerations. These patients responded to the administration of anti-CD19 CAR T cells, and this *ex vivo* gene therapy was in Phase I clinical trials in 2018. In addition, early *ex vivo* gene therapy trials have been initiated for the use of TALEN-generated universal T cells expressing anti-CD123 CAR for the treatment of acute myeloid leukaemia and blastic plasmacytoid dendritic cell neoplasms (Dunbar et al., 2018; Zhao et al., 2018). Two CD19-directed CAR T therapies, Kymriah and Yescarta, have recently been approved for clinical use (Table 9.8)

In October 2016, the first clinical trial using CRISPR-Cas9 SDN system was initiated for treating a patient suffering from an aggressive form of lung cancer. The SDN was used to inactivate the gene encoding PD-1 in the peripheral lymphocytes and the gene-edited cells were transferred back into the patient. Similar trials for PD-1 inactivation-based treatment of bladder cancer, prostate cancer and renal cell carcinoma have been approved. Another clinical trial in U.S.A. is based on CRISPR/Cas9-mediated deletion of the PD-1 receptor and the endogenous TCR genes, and for targeted insertion of the gene for a CAR against the cancer cells. In addition, at least nine clinical trials based on CRISPR-Cas (mainly used for inactivation/deletion of the PD-1 receptor gene in T cells) for tumour-targeting have been approved by the Chinese regulatory agencies (Katrekar et al. 2017; Dunbar et al., 2018).

In addition, clinical trials for at least four *in vivo* gene therapies based on SDNs had been initiated/approved for initiation till 2018 (Table 9.2). These trials are for the treatment of haemophilia A by insertion of the transgene encoding B domain-deleted Factor VIII into the genomes of liver cells. Haemophilia B is to be treated by integration of the transgene for Factor IX into the genome. Further, the therapies of mucopolysaccharidosis type II (Hunter's syndrome) and Mucopolysaccharidosis type I diseases are based on insertion of transgenes encoding the enzymes iduronate-2-sulfatase and α-L-iduronidase, respectively, into the genomes of hepatocytes. In all these cases, ZFNs are being employed for targeted integration of the transgenes into the albumin locus of the targeted cells. The therapeutic sequences including the SDNs will be administered by intravenous injections (Dunbar et al., 2018).

9.23. ACHIEVEMENTS

SDNs are being extensively used for basic investigations, including genome-wide screens for gene functions, analysis of gene interactions and development of GI maps, and creation of disease models. These applications are greatly facilitating a clearer and deeper understanding of the various biological phenomena including disease initiation and progression and enabling identification of novel and effective targets for drug development and gene therapy. In addition, gene editing has been used to develop gene therapies for several diseases, some of which are in different stages of clinical trials (Tables 9.3, 9.4, Sections 9.22) and many others are at various stages of preclinical development. The 'universal' T cells, a product of gene editing that supports the development of gene therapies against cancers, have been approved in U.S.A. for clinical use in cancer immunotherapy. Some CAR-T cell gene therapies based on the 'universal' T cell have been approved for clinical use (Table 9.8). Other noteworthy developments are the *CCR5*-inactivated T cells for resistance to HIV-1, inactivation of *PD*-1 in the peripheral lymphocytes or deletion of the gene encoding PD-1 receptor for treatment of various types of cancers, and the targeted transgene insertion in the albumin encoding gene of liver cells for treatment of Haemophilia A/B and Mucopolysaccharidosis type I/II (Section 9.22).

Table 9.8. The gene editing-based gene therapies approved by the FDA for clinical use (till mid-2019)

Therapy	Clinical use	Company	Cost
Kymriah	CD19-directed CAR-T cell therapy for persons of up to 25 years age with refractory B-cell acute lymphoblastic leukaemia	Novartis, Paris (France)	US $475,000 per treatment
Yescarta	CD19-directed CAR-T therapy for adult patients with relapsed or refractory large B-cell lymphoma	Kite Pharma, California (USA)	US $373,000

The ZFNs, TALENs and CRISPR-Cas systems have been repurposed as nucleic acid (RNA and/or DNA) detection tools. For example, Cas12a has been used to develop a DNA detection tool called DETECTR (DNA endonuclease targeted CRISPR *trans*-reporter), while Cas13a has been used to devise the tool SHERLOCK (Specific High-sensitivity Enzymatic Reporter unLOCKing). These tools have attomolar sensitivities and can be used in a variety of nucleic acid detection and/or quantification applications, including detection of specific strains of viruses, distinguishing pathogenic bacteria, genotyping human DNA, etc. (Gootenberg et al. 2017; Batista and Pacheco, 2018).

9.24. FUTURE PERSPECTIVES

The applications of gene editing technologies to resolve various issues concerning human health have made rapid progress during the past few years, especially with the development of the CRISPR-Cas system. Many scientists feel that it is not possible to predict, with any degree of confidence, the future course of this technology. The following discussion, therefore, focusses on the current limitations that need to be addressed for making this technology safer, more efficient and capable of contributing to promotion of human health. One of the main challenges is to develop an efficient and safe method of SDN delivery specifically to the targeted tissues. Another concern relates to efficient gene editing, particularly HDR-based sequence replacement, especially in hiPSCs. The protocols for differentiation of different types of stem cells like hematopoietic stem cells from hiPSCs and the isolation and purification of these stem cells free from hiPSCs before transplantation need to be further refined and standardized (Chandrasegaran and Carroll, 2016; Parekh et al., 2017).

The efficiency of HDR is significantly improved with concomitant suppression of NHEJ since the latter would lead to inactivation of the target gene even after its sequence has been corrected by HDR. However, adequate resolution of these limitations remains a major challenge; the 'prime editing' may address the issues related to HDR. The HDR is greatly reduced in post-mitotic cells (G_1 and G_0 stages) and the human body is largely composed of such cells; this poses a serious limitation for *in vivo* gene correction by HDR. It has been suggested that the deployment of transposases or recombinases for gene editing might alleviate this constraint. It is hoped that as we gain clearer and deeper insights into the mechanisms of these pathways of DNA repair, we may be able to favourably modify them to serve our objectives. For example, it was discovered that HR suppression in G_1 cells involves inhibition of resection of the DNA ends at DSBs and the recruitment of the BRCA2 to the DSB sites. HDR was promoted in G_1 cells when these two limitations were rectified. Therefore, it may be expected that further insights into the mechanisms and the regulations of HDR and NHEJ pathways of DNA repair might provide strategies for enhancing and/or supressing HDR and/or NHEJ.

The asymmetrical dissociation of Cas9 from the target DNA after cleavage has also been exploited to promote HDR by providing ssDNA donors complementary to the DNA strand to be first released by Cas9. In addition, the human Geminin protein could be linked to the N-terminus of Cas9 and thereby target it for ubiquitin-mediated destruction during G_1 and limit its action to $S/G_2/M$ phases of the cell cycle. The issue of HDR and NHEJ can be completely bypassed by using base editors to correct mutations since dCas9 does not induce breaks in DNA strands (Moreno and Mali, 2017). But base editors can correct only such mutations that are caused by base substitutions, and base editors also cause off-target edits and bystander mutations (Section 6.8, 6.9).

In view of the potential off-target action of SDNs, decisions regarding therapeutic applications of gene editing would have to be based on "risk versus benefit analysis and informed consent" of the patients. For this, it is important to define the "acceptable" or "minimum" level of risk. It has been proposed to "define the minimum risk profile as a list of NHEJ mutations in regions of the human genome that, if they result from treatment, would not adversely affect the well-being of the patient". Once the whole genome profile of 'safe' NHEJ mutations is developed, the decision on safety of a gene therapy procedure would become largely rational and objective. In addition, the methods for detecting the off-target effects need to be further refined to increase their sensitivity and reliability. A set of guidelines with 'minimum information' requirements for reporting the research findings on specificity of CRISPR-Cas9 should be implemented as has been done for quantitative real-time PCR and several other fields of investigation. It has been suggested that targeted deep sequencing should be adopted as the standard procedure for confirming off-target effects since none of the unbiased methods is 'yet perfectly sensitive and reliable'. However, its experimental procedure and the computational analyses need to be standardized to enable comparisons across studies. The factors that need standardization include the number of PCR cycles (should be reduced), the DNA polymerase used (should have proof-reading capacity), the reference genome used for alignment of the sequence reads, the bioinformatics tools used for sequence data analysis, etc. CRISPR-GA and CRISPResso are two open access robust indel analysis bioinformatics tools (Tycko et al., 2016; Chandrasegaran and Carrol 2016).

The vectors used for delivering therapeutic reagents should have adequate packaging capacity and high transduction efficiency, and should not trigger an immune response against either themselves or their payloads upon being readministered. There is evidence that Cas9 induces humoral as well as cellular immune responses, which would reduce the efficacy of CRISPR-Cas9 system on repeated administrations. In addition, the immune response to *in vivo* gene therapy procedures evaluated in mice and even larger mammals like dogs tends to underestimate the human 'pathologic immune response' to these protocols (Porteus, 2016; Moreno and Mali, 2017).

There are two disturbing issues, one concerning the consequences of gene therapy itself and the other is related to the procedure of gene editing. Gene editing seems to be more successful in cells deficient in p53 function and such cells are more likely to undergo transformation to cancerous cells after being introduced into patients. Therefore efforts must be made to develop such gene therapy procdures that select and use gene edited cells with intact p53 function (Haapaniemi et al., 2018; Ihry et al., 2018). The gene therapy strategies with or without gene editing are likely to increase the genetic load of future human generations. This concern has been a frustrating issue that awaits an appropriate response (Porteus, 2016).

9.25. CONCLUSIONS

The interest in SDN development was triggered by the desire to use them for targeted transgene integration in cells to be used for gene therapy. This technology has found a variety of other applications in addition to targeted transgene integration. Gene editing has greatly facilitated genome-wide analysis of gene functions, development of GI maps, identification of novel drug targets and suggested 'synthetic lethal' routes for treatment of some types of cancers. Gene editing is being used to create more precise animal models of human diseases, and cell and organoid models of diseases that have greatly facilitated biomedical researches directed at analysis and treatment of the concerned diseases. It has also enabled the development of several therapeutic strategies for many genetic and certain infectious diseases, many of which have gone beyond the proof-of-concept stage into the clinical trial stage and some of them have been approved for clinical use. Gene editing enabled the development of 'universal' T cells that have made an immense contribution to cancer immunotherapy since these cells eliminate the need for development of cancer cell-targeting T cells for each individual patient. These T cells are already being used for cancer immunotherapy.

SDN-mediated inactivation of gene *CCR5* in primary $CD4^+$ T cells, and primary $CD34^+$ hematopoietic stem and progenitor cells generates resistance to HIV-1 infection. Gene therapy based on this strategy has completed Phase I clinical trials. Many other gene therapy strategies based on gene editing had been approved for clinical trials till 2018. For many diseases like Duchenne muscular dystrophy, LDL cholesterol level, cystic fibrosis etc., proof-of-concept is already available. Further, putative therapeutic Targets for gene editing have been identified for even complex diseases like allergy and we may soon have information on the therapeutic usefulness of these targets. The cells used for gene editing are primary cells, hematopoietic stem cells or induced pluripotent stem cells; in the cases of the latter two cell types, the edited cells are selected, characterised and differentiated into the desired cell type, purified free from the stem cells and transfused into the patients. The frequency of gene editing in stem cells, especially iPSCs, is relatively low. In addition, gene editing seems to more efficient in cells with defective p53 function, which predisposes such cells to cancerous cell transformation. The gene therapy strategies based on gene editing should consider this aspect of gene editing to avoid complications to the patients receiving the therapy. In addition, the gene editing procedures should be subjected to functional toxicity assays before being used as therapeutic strategy. The fear that gene therapy strategies with or without gene editing may increase the genetic load of future human generations needs to be discussed and resolved.

The SDNs can be delivered into human cells as DNA (gene constructs encoding SDNs), mRNA, protein or RNP (Cas protein complexed with sgRNA). For transient expression, the SDNs should be delivered as either RNA or RNP. The SDN gene constructs have been delivered into human cells using both viral and nonviral delivery system. The viral delivery systems use the following four types of vectors derived from virus genomes: (i) γ-retrovirus (γ-RV) vectors, (ii) adenovirus (AV) vectors, (iii) lentivirus (LV) vectors and (iv) adeno-associated virus (AAV) vectors. The nonviral delivery vehicles include nanoparticles like cationic nanocarriers, liposomes and polymeric materials. The nonviral systems have the advantage of overall safety, low immunogenicity, large loading capacity, and ease of production. The LV and γ-RV vectors used in most of the current clinical trials have a 'self-inactivating' SIN design. The *in vivo* gene therapies in clinical trials are based almost exclusively on AAV vectors. The SDNs can be delivered by several nonviral methods like electroporation, lipid/cationic lipid-mediated delivery, etc. These methods are satisfactory for many cell types, but they do not work well with certain other cell types like HSCs and some primary cells, *i.e.*, cells obtained directly from normal body tissues.

Gene editing is expected to provide effective therapies for an increasing number of diseases caused by mutant genes and pathogens. It may provide avenues for overcoming dreaded diseases like cancer and AIDS, and even diseases that have perplexing and complex genetic bases. Gene therapies based on induced pluripotent stem cells have been projected as the future regenerative medicine. However, several issues like off-target action, high gene editing frequency, effective SDN delivery, iPSCs differentiation to different cell types with adult phenotypes and purification of these differentiated types etc. need to be resolved to make these therapeutics safe, reliable and dependable.

■■■■

10

MODULATION OF GENE EXPRESSION

10.1. INTRODUCTION

The concepts and designs of two of the initially used site-directed nucleases (SDNs), *viz.*, zinc-finger nucleases (ZFNs) and transcription activator-like effector nucleases (TALENs) emerged from the organization and function of endogenous transcription factors, namely, zinc finger proteins of eukaryotes and transcription activator-like effectors produced by certain pathogenic bacteria. The ZFN and TALEN SDNs were created by replacing the transcription activation domains of the concerned transcription factors by the nuclease domain of FokI. In addition, their DNA binding domains were custom designed by protein engineering to enable them to bind to the targeted DNA sequences at the specified genomic sites. These programmable SDNs, therefore, can be designed to induce double-strand breaks (DSBs) or nicks and, thereby, generate various genetic changes at almost any genomic site of interest to the experimenter. The SDN concept was soon extended to design programmable or site-directed effectors other than nucleases like transcription activators, transcription repressors, etc.

In 1998, Beerli et al. (1998) redesigned zinc finger protein to regulate the expression of *erbB-2/HER-2* gene. The programmable effector proteins are created simply by replacing the nuclease domain of the SDNs by the desired effector domain, such as transcription activation or repression domain, chromatin remodelling enzyme domain, etc. As a result, these sequence-specified chimeric effectors comprise the desired effector domain and a programmable DNA binding site, and they can be designed to carry out the specified modification at any genomic site selected by the researcher. Thus, the story of site-directed nuclease development has now completed the full circle beginning from designing SDNs based on the structure and function of transcription factors and ending with the development of programmable/site-directed transcription factors by applying the information gained from SDN design and function. However, designing of ZFN- and TALEN-based regulator proteins specific for each gene target is a challenging task and in case of genome-wide studies it becomes virtually unmanageable.

The CRISPR-Cas9 SDN system is completely different from ZFN and TALEN SDNs both in design and function, but it achieves the same results, and serves equally well or even better when used to modify transcription and other cellular functions. In this system, the Cas9 protein functions as carrier of the desired effector domain to the targeted genomic site: Cas9 identifies this site in association with its appropriate sgRNA that has base sequence complementary to the targeted site. Since for each new target site only the spacer sequence of 21 or so nucleotides (nt) of its sgRNA component must be redesigned, CRISPR-Cas system offers an easy and versatile tool for various genome-wide applications.

The programmable chimeric effectors have been deployed to achieve a variety of objectives, including base editing (Chapter 6), transcription regulation, chromatin structure modification, labelling of specific genomic loci, interrogation of functions of genes and cis-regulatory sequences, etc., which are considered in the following sections in some detail. There are several recent reviews that together provide an excellent overview of these aspects (Doudna and Charpentier, 2014; Gilbert et al., 2014; Tanenbaum et al., 2014; Katrekar et al., 2017; Pulecio et al., 2017).

10.2. REGULATION OF GENE ACTION IN EUKARYOTES

Regulation of gene action in eukaryotes is much more complex than in prokaryotes as the eukaryotic DNA is associated with proteins and RNA, and organized in form of chromatin. Eukaryotic gene expression can be regulated at several levels ranging from chromatin organization through transcription initiation to post-translational modification of proteins. The chief mode of regulation of gene expression, however, is based on chromatin modification and transcription initiation. The first step in expression of a gene is conversion of chromatin to an 'active' state. The genes present in an 'active' chromatin region are available for transcription initiation, which is mediated by many transcription factors.

10.2.1. Chromatin activation

In eukaryotes, DNA, proteins (both histone and nonhistone proteins) and RNA form a complex structure called chromatin that is organized into chromatin fibres and ultimately typical chromosomes. The changes in chromatin organization are mediated by chromatin remodelling complexes that are recruited to the chromatin by activator/repressor proteins since these complexes themselves lack DNA binding function. Chromatin remodelling involves both histone and DNA modifications, which are generally referred to as epigenetic changes. The chief epigenetic modifications that affect transcription regulation in mammals are DNA methylation, histones alterations and participation of non-coding RNAs (ncRNAs). Histone modifications involve acetylation, deacetylation, methylation, demethylation, phosphorylation and dephosphorylation of specific amino acid residues of certain histone proteins; these changes together with DNA methylation comprise epigenetic marks of chromatin. These modifications indicate the active/inactive state of chromatin, and they together constitute the histone code (Table 10.1). Typically, histone acetylation is associated with chromatin activation, while deacetylation is indicative of inactive chromatin. Histone acetylation is catalysed by histone acetyltransferases (HATs), while deacetylation function is performed by histone deacetylases (HDACs). In simple words, the epigenetic regulation comprises two sets of proteins, *viz*., the effectors that write or erase the epigenetic marks, and the reader proteins that detect, deposit and remove specific chromatin modifications. Generally, it is very difficult to ascertain the causal relationships between distinct epigenetic marks at specific genomic loci and expression of the genes in the concerned region (Pulecio et al., 2017).

Table 10.1. The effects of different histone modifications on transcription

Histone	Amino acid*	Modification	Effect
H3	Lys-4	Methylation	Transcription activation; *demarcates euchromatin*
H3	Lys-9**	Methylation	Chromatin condensation and inactivation; required for DNA methylation; *di-methylated Lys-9 is a mark for heterochromatin*
H3	Lys-9	Acetylation	Transcription activation
H3	Ser-10	Phosphorylation	Transcription activation; prevents di-methylation of Lys-9 and promotes acetylation of Lys-14
H3	Lys-14	Acetylation	Prevents methylation of Lys-9; transcription activation
H3	Lys-27	Methylation	Transcription activation
H4	Lys-12	Acetylation	Promotes H3 Ser-10 phosphorylation
H4	Lys-16	Acetylation	Nucleosome assembly; *Drosophila* X activation

* The numbers associated with amino acid symbols denote the position of this amino acid residue in the polypeptide.

** Methylated Lys-9 attracts protein HP1, which in turn recruits DNA methyltransferases to carry out DNA methylation.

Histone and DNA methylations, similarly, are associated with inactive chromatin/repressed gene transcription, while their demethylation occurs during chromatin activation. But methylation of certain amino acids, *e.g.*, Lys at position 4 (Lys-4) of H3, is associated with activation of chromatin and ultimately transcription. Methylation of histone lysine residues is catalysed by lysine methyltransferases, while their demethylation is the function of lysine demethylases. DNA methylation is performed by a group of enzymes called DNA methyltransferases or simply DNA methylases. But removal of methyl groups is indirect and involves both active (including base excision and repair, and chemical modification) and passive (DNA methylation rate being slower than cell proliferation rate) routes. DNA methylation, especially of C residues in CpG doublets, promotes heterochromatinization, and histone and DNA methylations tend to reinforce each other.

10.2.2. Transcription regulator proteins

The regulator proteins either activate (activator proteins) or repress (repressor proteins) transcription initiation. True transcription activators of eukaryotes bind to DNA and exert their effect by making physical contact with the basal transcription apparatus assembled at the promoter; this contact is either direct or mediated by coactivator proteins. Another group of activators, called antirepressors, bind to enhancers and recruit chromatin modifying proteins that change the chromatin from inactive to active state. The eukaryotic repressors can act by either binding to the nuclear transport signals (blocks nuclear transport) or activation domains of activator proteins (prevents transcription activation), or by competing with them for the same DNA binding sites in the promoter regions of the concerned genes. Transcription repression can also be achieved by chromatin remodelling leading to conversion of chromatin from an open or active to a closed or inactive state that does not support transcription initiation.

Typically, the transcription activators and repressors comprise a DNA binding domain and an effector (activator/repressor) domain that are linked together by a flexible connector. As a result, the effector and DNA binding domains from two separate activators/repressors can be linked to create a chimeric regulator protein that performs the function specified by its effector domain at the site for which its DNA binding domain is specific. Programmable transcription activation and repression of targeted genes have been achieved by deploying site-directed effectors derived from native effector proteins involved in transcription activation/repression as well as chromatin remodelling. A brief discussion on these programmable transcription regulators and chromatin remodelling effectors is provided in Sections 10.3 and 10.4, respectively.

10.3. SITE-DIRECTED TRANSCRIPTION FACTORS

Controlled activation or repression of specific genes is necessary for a variety of investigations, including interrogation of gene functions and construction of genetic interaction maps. The conventional approach for achieving controlled expression of a gene is to create a transgenic, in which an inducible or repressible promoter is used to drive this gene. As a result, the expression of the gene can now be regulated externally by providing the appropriate chemical or physical stimulus. This strategy requires the use of a new promoter for each gene of interest, and application of the appropriate stimulus to activate/repress each of the genes. In addition, a single stimulus may activate several endogenous genes that share the concerned mechanism of regulation, as a result of which the activation of individual target genes is virtually impossible. The site-directed transcription activators and repressors provide an effective and highly specific tool for regulating individual gene expression since they bind with high specificity to their target sequences in the genomic DNA.

A chimeric transcription activator was created by Sadowski et al. (1988) by fusing the DNA-binding domain from GAL4, a yeast transcriptions activator, with the highly acidic transcription activator domain of protein VP16 from Herpes Simplex Virus. This chimeric transcription activator was able to activate transcription of mammalian genes with unusual efficiency on binding either close to or even at large distances from the concerned genes. But site-directed or programmable transcription factors were designed and developed only after the SDNs were designed and used for genome editing. The target DNA recognition function of these site-directed transcription factors is provided by engineered zinc-finger or TALE protein DNA binding domains, or by specifically designed sgRNAs of the CRISPR-Cas9 system. The nuclease activity of Cas9 is abolished to generate 'dead' Cas9 (dCas9), which is used to construct the programmable transcription activators/repressors. These site-directed transcription factors can precisely and predictably activate/repress transcription of the targeted genes directly by recruiting/blocking the proteins involved in the transcription process. Alternatively, they can exert their infleunce indirectly by triggering appropriate modifications in chromatin structure by altering the status of histone acetylation/methylation or the state of DNA methylation (Doudna and Charpentier, 2014; Gilbert et al., 2014; Puchta, 2017).

10.3.1. Chimeric TALE effectors

A TALE protein consists of a translocation domain at the N-terminus, a DNA-binding domain in the middle, and a nuclear localization signal (NLS) and an acidic domain for transcription activation at the C-terminus. The TALENs and chimeric TALE effectors for transcription activation (TALE activator, TALEa)/repression (TALE repressor, TALEr) are typically created by replacing the NLS and the activation domain by a suitable NLS for the concerned organism and the desired nuclease or transcription activation/repressor domain, repectively. The central

DNA binding domain of TALE proteins comprises a variable number of imperfect repeat units arranged in tandem. In each repeat unit, the residues at positions 12 and 13, called repeat variable di-residues (RVD), recognise a single specific DNA base of the target DNA. The DNA binding domain of TALENs and TALE effectors is designed to bind the target DNA sequence. In case of transcription activation/repression, the target DNA is typically the promoter region of the desired gene. The TALEr binds to a specific region in the promoter of targeted gene and its repressor domain then represses the transcription of this gene. But TALEs can also repress transcription of a targeted gene by simply binding to the sense strand in the exon region of the gene. It is noteworthy that this kind of transcription repression does not depend on the effector domain fused at the C-terminus of the TALEr protein. Thus, it appears that binding of the TALE protein to the sense strand interferes with the transcription process leading to repression of transcription of the concerned gene (Lin et al., 2016).

One of the major limitations in the wide use of TALE effectors is that the synthesis of novel TALE proteins is a challenging task as it involves protein engineering and elaborate evaluation procedures. This is mainly due to the repetitive nature of their DNA binding domains, which makes the construction of novel TALEs by PCR-based gene assembly or serial ligation of DNA segments quite a difficult task. Although commercial services are available for the synthesis of novel TALEs, they are costly particularly when large numbers of new TALEs are to be designed for genome-wide applications. Zhang et al. (2011) developed a reliable strategy for high throughput synthesis of novel TALEs based on hierarchical DNA ligation. In this strategy, the similarity among DNA sequences encoding the DNA-binding sequence of TALE repeat units is minimized without affecting their amino acid sequences. This is achieved by exploiting the degeneracy of genetic code. The sequences encoding the amino acids Gly-Leu representing the junctions linking the repeat monomers are modified using a different codon for these two amino acids to design four different 4 bp sticky-end ligation adapters. These ligation adapters allow simultaneous hierarchical ligation to create a sequence encoding a TALE protein with 12 tandem repeat units. This strategy allows construction of TALE encoding DNA sequences at a fraction of the cost and in a considerably shorter period than that required for their creation via commercial DNA synthesis.

The level of transcription activation by TALE activators seems to be affected by some factors other than TALEa design; these factors include the effects of context on DNA binding strength of the individual repeat units, and components of the complex transcription processes in mammals and other eukaryotes. Further, the first 48 amino acids from the N-terminus of Hax3 TALE protein can be safely deleted without an adverse effect on the level of transcription activation/repression. But the 68 amino acids at the C-terminus of Hax3 TALE protein must be retained to ensure the highest level of activity since this region contains an element important for DNA binding by TALE proteins (Zhang et al., 2011).

10.3.2. CRISPR-mediated transcription activation/repression

The CRISPR-Cas9 SDN system can be repurposed for targeted genome-wide modulation of gene expression. Qi et al. (2013) observed that when dCas9 is targeted to the coding region of a gene, it can suppress its transcription; this suppression is dramatic in bacteria but only moderate in mammalian cells. In this case, suppression results from binding of dCas9 to the target gene, which interferes with the transcription process. The CRISPR-dCas9-based transcription activators were first constructed by Mali et al. (2013) for regulation of gene expression in human cells. These programmable transcription activators consisted of dCas9, sgRNA and the VP64 transcription activator. VP64 was tethered to either the C-terminus of dCas9 or to the sgRNA that contained two copies of the RNA stem-loop that specifically binds to the coat protein of MS2 bacteriophage. In the latter case, VP64 was expressed as fusion of VP64 with the MS2 phage coat protein, and it was attracted to the two copies of RNA stem-loop linked to the 3' end of sgRNA (Fig. 10.1). Both the types of CRISPR-dCas9 activators produced sequence-specific

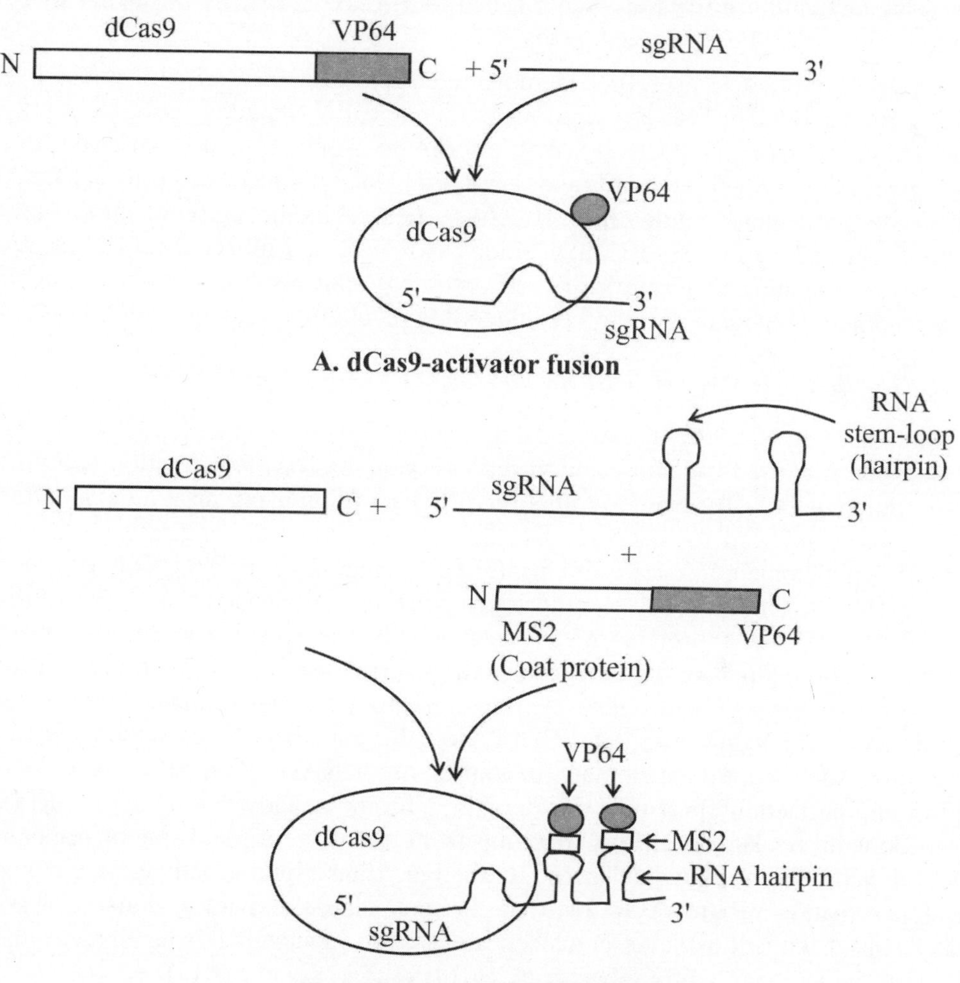

Fig. 10.1. A schematic representation of the CRISPR-dCas9-based transcription activator developed by Mali et al. (2013). **A.** Activator VP64 fused to the C-terminus of dCas9 (sgRNA is not modified). **B.** VP64 (expressed as fusion protein with MSZ phage coat protein) is attached to the 3' end of modified sgRNA. The RNA hairpin (two copies) specific for MS2 protein is added to the 3' end of the sgRNA. The MS2-VP64 fusion protein binds to these RNA stem-loops (hairpins).

robust gene activation when the sgRNA was specific to a sequence in the promoter region of the targeted gene. The dCas9-activator fusion showed 1.5 to 30-fold higher activation than the sgRNA-mediated activator recruitment. But tethering of the activator to the sgRNA can, in principle, allow multiplexing using a single dCas9 in combination with multiple sgRNAs. Each of these sgRNAs contained a different RNA hairpin sequence and was expressed in combination with a fusion protein consisting of the protein that specifically interacts with the concerned RNA hairpin and the desired activator domain. Transcription was activated by a single sgRNA, but the activation was many-fold higher when multiple sgRNAs were targeted to the promoter region of each gene. The activation level was higher when the sgRNAs targeted sequences closer to the transcription start site of the concerned genes.

Subsequently, Gilbert et al. (2013) demonstrated that dCas9 fused with transcription activator and suppressor domains can serve as programmable RNA-guided tools for highly specific activation and suppression, respectively, of transcription in human cells, yeast and, in principle, other biological systems. The activator and suppressor domains were fused to the C-terminus of dCas9. This system of gene regulation is effective when promoter region or introns are targeted. This method repressed gene expression as stably as RNAi, had minimal impact on nontargeted genes, and was amenable to multiplexing. The strategy can also be used for high throughput functional mapping of enhancers and other noncoding regulatory elements (Section 10.14).

10.3.2.1. *CRISPR-interference (CRISPRi) and CRISPR-activation (CRISPRa)*

Gene regulation using CRISPR-dCas9-based activators/repressors is called CRISPR-interference (CRISPRi) when it is used for transcription repression and CRISPR-activation (CRISPRa) when applied for transcription activation. Modulation of transcription using programmable transcription regulators is often referred to as epigenome editing or epigenome engineering. Therefore, the tools used for epigenome editing are often referred to epigenome editing tools or programmable epigenome effectors. The epigenome comprises a collection of various chemical modifications of genomic DNA and histone proteins of an organism. In the epigenome editing tools based on CRISPR-dCas9, the regulatory domains of activator or repressor proteins are linked to dCas9 to construct chimeric dCas9 with gene activation/repression capabilities. Some examples of activator domains are VP64, p65, Rta, EDLL domain, and the activator domains from TALE proteins, while some repressor domain examples are KRAB (Kruppel-associated box), SID domain (mSin interaction domain), CS domain (chromo-shadow domain of HP1a), WRPW (repressor domain of Hes1), SRDX domain from the transcription factor ERF, and EAR (repressor domain from plants) (Table 10.2). For transcription activation, four copies of transcription activator VP16 (VP64) or a single copy of the activation domain of protein p65 (p65AD) is fused with dCas9; these fusions are able to achieve high level of gene activation (Doudna and Charpentier, 2014; Gilbert et al., 2014; Katrekar et al., 2017).

Similarly, the KRAB domain is fused with dCas9 for high level of transcription repression.. The dCas9-mediated repression of transcription needs to be refined further since the repression is only partial even when helper functional domains are added to the repressor domains. CRISPRi can prevent transcriptional elongation, binding of RNA polymerase to promoters, or association of transcription factors to their binding sites in the promoters. The quantum of effect generated by CRISPRi depends on the particular sequence, to which the concerned sgRNA is specific. Both CRISPRi and CRISPRa can be used for simultaneous reversible repression and activation, respectively, of single or multiple genes. These strategies can regulate the transcription of endogenous genes to obtain a wide range of transcript levels. Further, properly designed sgRNAs are highly specific and effective (typically 90%–99% repression) and the off-target effects are minimal. The combined range of modulation of gene expression levels is ~1000-fold. The sgRNAs targeting the region between ~400 bp to ~50 bp upstream of the transcription start site showed high CRISPRa activities, while the peak CRISPRi activities were exhibited by sgRNAs specific for the region between ~50 and ~300 bp downstream of the start site (Doudna and Charpentier, 2014; Gilbert et al., 2014; Katrekar et al., 2017).

Table 10.2. A selection of transcriptional activator and repressor domains used in programmable transcription regulators

Effector domain	Source	Effect on transcription	Remarks
VP16	Herpes Simplex Virus	Activation	Activator domain, acidic; forms complex with transcription factors Oct-1 and HCF
VP64*	VP16	Better activation than VP16	Consists of four copies of VP16 activator domain
p65	Transactivation domain of human transcription factor NF-κB	Activation	Involved in formation of NF-κB heterodimer, its nuclear translocation and transcription activation
Rta	Epstein-Barr virus transcription activation domain	Activation	Interacts with TBP and TFIIB (and some other proteins) to activate transcription
HSF1	Human	Activation	Activation domain of heat-shock transcription factor 1; highly conserved in eukaryotes
EDLL motif	Plants: AtERF98/TDR1 transcription regulators	Activation	Strong activation domain; has a unique arrangement of acidic amino acid and hydrophobic leucine residues
Acetyltransferase p300	Human	Activation	Acetylates Lys-27 of H3 histone
TET1CD	Human	Activation	DNA demethylation
KRAB	-	Repression	Trimethylation of Lys-9 of histone H3
DNA methyltransferase (DNMT3A)	Human	Repression	DNA methylation
mSin3 interaction domain (SID)	Mad proteins	Repression	Ultimately, activates histone deacetylase leading to histone deacetylation
4xSID (Enhanced SID repressor)	The SID domain	Better repression than SID	Four copies of SID used

Effector domain	Source	Effect on transcription	Remarks
CS (chromo-shadow domain)	HP1a protein (heterochromatin-associated protein 1; the N-terminal region)	Repression	Distantly related to the chromo-domain; causes chromatin condensation that represses transcription
WRPW repressor domain	Mammalian Hes1 protein	Repression	WRPW motif is a functional transcriptional repression domain
SRDX domain (EAR-repression domain)	Plants; transcription factor ERF (ERF-associated amphiphilic repression, EAR)	Repression	Can be used to detect protein–protein interactions; repression partly overcome by EDLL motif
LSD1	Human	Repression	Demethylation of dimethylated Lys-4 residues of H3 histone

* dCas9-VP64 is superior to dCas9-p65 and dCas9-Rta.

CRISPRi/a has been used for a variety of purposes, including validation of gene functions, identification and/or analysis of promoter and enhancer regions, conversion of differentiated cells to pluripotency, and even to develop methods with therapeutic potential. Gene activation enables identification of limiting components of biochemical processes and the molecular targets for drugs, and activation of the key rate-limiting steps of metabolic pathways. It might permit analysis and understanding of cellular pathways that are difficult to analyse by gene knock-down strategies due to redundancy. CRISPRa may enable activation of otherwise inactive cellular pathways and reveal interactions among complex cellular networks. Finally, CRISPRi and CRISPRa can be used to construct genetic interaction (GI) maps (Section 10.7; Gilbert et al., 2014; Pulecio et al., 2017). Liu et al. (2015) developed a bioinformatics tool called CRISPRERA that supports comprehensive design of CRISPR-mediated gene editing, as well as transcription repression and activation; this tool is publicly available.

10.3.2.2. *Modulation of transcription level*

The level of transcription activation can be modulated by using different activator domains, targeting different sites in the promoter region of the gene, simultaneously targeting two or more sites for activator attachment in the promoter region (Fig. 10.2A), and recruiting multiple copies of activators at a single site (Figs. 10.3, 10.4). Further, two or more distinct activator domains can be linked to the same dCas9 protein to create a more potent activator (Table 10.3). For example, the dCas9-VP64-p65-Rta (dCas9-VPR) transcription activator was formed by modifying the dCas9-VP64 activator (Fig. 10.2B). The dCas9-VP64 activator itself was created by fusing the VP64 transcription activator to the C-terminus of dCas9. The transcription activators p65 and Rta were then linked to the C-terminus of dCas9-VP64 to construct the dCas9-VP64-p65-Rta activator (Fig. 10.2B). This transcription activator, therefore, places all the three activation domains at the targeted site and leads to a much higher level of transcription activation than the individual activator domains. It has been concluded that the VPR (***V***P64-***p***65-***R***ta), synergistic activation mediator (SAM), and SUperNova or SunTag activator systems work best in combination with dCas9.

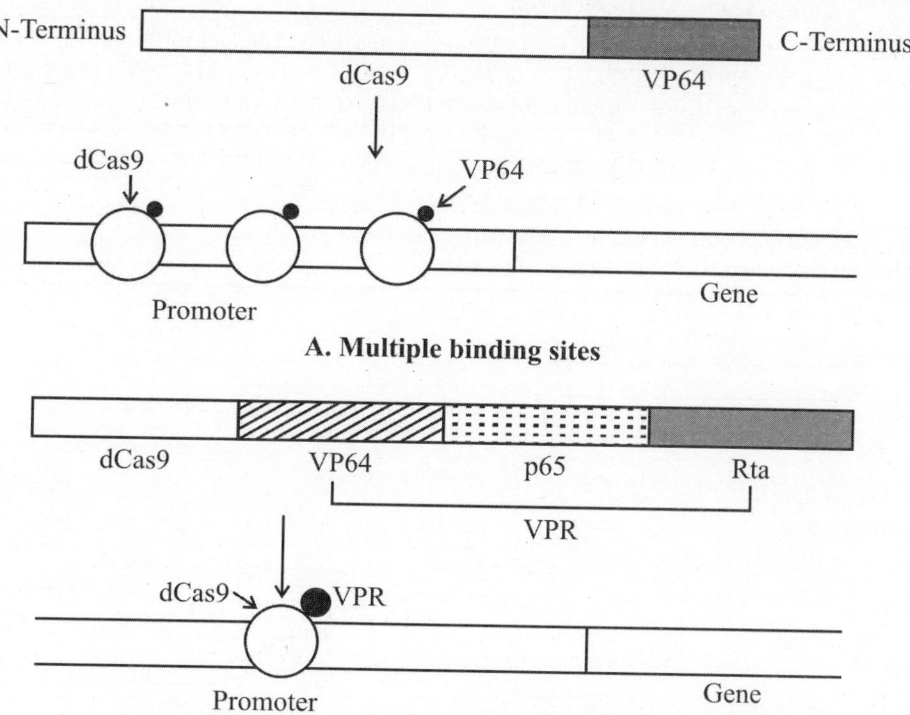

Fig. 10.2. Two strategies for placing multiple activators in the promoter region of a gene. **A.** dCas9-VP64 activator binds to multiple sites. **B.** dCas9-VPR activator having three different activators binds to a single site. dCas9, dead Cas9; VPR, VP64, p65 and Rta.

Table 10.3. The strategies used for recruiting multiple activator domains to the promoter region of a gene

Strategy	Components	Description
Targeting multiple sites in promoter region	dCas9-VP64 and multiple sgRNAs	Each sgRNA targets a different site in the promoter region and recruits one dCas9-VP64 to that site of the promoter
Linking multiple activators to dCas9	dCas9-VPR (dCas9-VP64-p65-Rta)	The activators VP64, p65 and RTa are linked to dCas9 C-terminus in the above optimal order; enables assembly of the three activators at a single site, to which the dCas9-sgRNA complex binds
dCas9-SAM system	dCas9-VP64, a modified sgRNA and MS2-activator fusion proteins	sgRNA contains multiple copies of RNA hairpin specific for MS2 coat protein; activators expressed as fusion of MS2 coat protein, *e.g.*, MS2-p65-HSF1; each hairpin attaches to two MS2-activator fusion molecules
SunCas9 or SunTag activator system	dCas9-SunTag, sgRNA, activator-antibody fusion proteins	dCas9 is linked to SunTag comprising repeat units; each SunTag unit is recognized by an antibody; up to 24 copies of activator-antibody fusion molecules can bind to one SunTag

Strategy	Components	Description
CRISPRainbow*	dCas9, a modified sgRNA and MS2-fluorophore fusion proteins	Based on SAM approach; sgRNA contains different RNA hairpins, *e.g.*, MS2, PP7 or boxB hairpins, specific for MS2 , PP7 or boxB proteins; different fluorescent proteins fused with these proteins, *e.g.*, GFP-MS2, BFP-PP7 and RFP-boxB; the fusion proteins bind to the related RNA hairpins; enables simultaneous labelling of seven different loci or of a single locus in six different colours

* Strategy used for multicolour imaging. The other strategies listed in this table have also been used for enhancing fluorescence intensity to enable robust imaging.

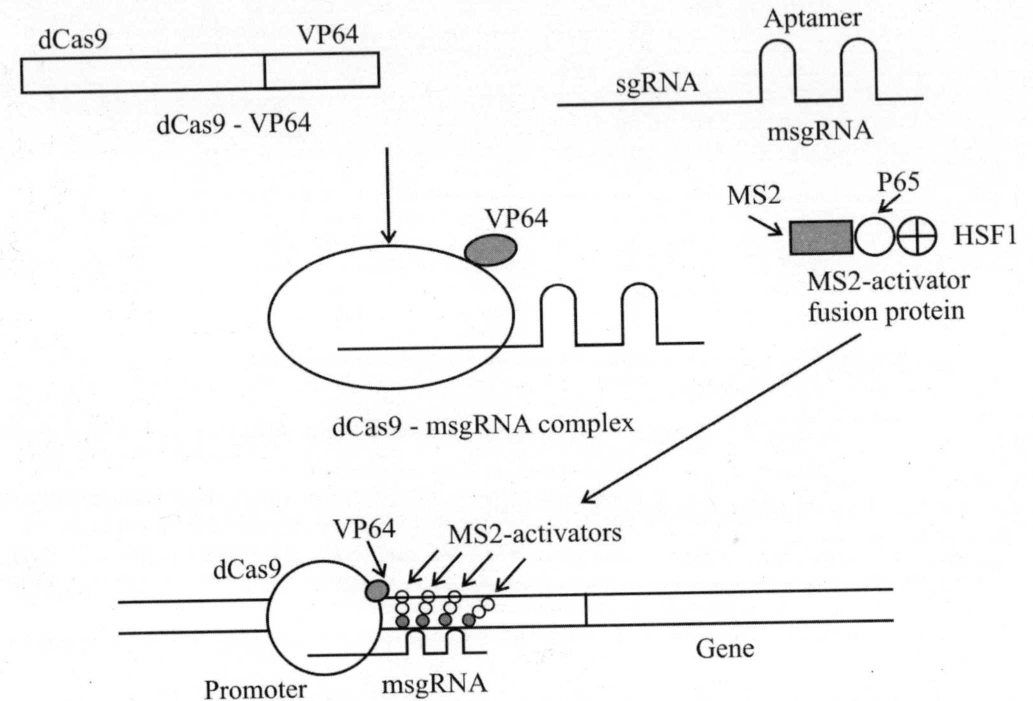

Fig. 10.3. A schematic representation of the dCas9-SAM (synergistic activation mediator) system of transcription activation. The aptamer is an RNA hairpin, to which MS2 bacteriophage coat protein binds. MS2, coat protein of MS2 bacteriophage; VP64, p65 and HSF1, transcription activators. The msgRNA recruits MS2-p65-HSF1 activators to the dCas9-msgRNA complex bound to the targeted site.

The dCas9-synergistic activation mediator (dCas9-SAM) system enables association of multiple transcription factors with a single dCas9 molecule. In simple words, the dCas9-SAM system comprises dCas9-VP64, a modified sgRNA (msgRNA), and transcription activators fused with the coat protein of MS2 bacteriophage (Konermann et al., 2015). The sgRNA is modified by integration of RNA hairpin aptamers that specifically bind to MS2 bacteriophage coat protein. These aptamers are inserted at the tetra loop and the stem loop 2 of sgRNA. The RNA hairpins are exposed at the surface of dCas9-sgRNA complex and are easily accessible to MS2 coat protein. The transcription factors p65 and HSF1 are fused with MS2 coat protein and expressed as

MS2-p65-HSF1 fusion protein. The dCas9-VP64 activator is used in combination with the msgRNA and the MS2-p65-HSF1 fusion protein. The msgRNA not only guides the dCas9-VP64 activator to the target site, it also recruits the MS2-p65-HSF1 fusion protein (two molecules per RNA hairpin) to this site due to specific interaction of MS2 protein with the msgRNA hairpin aptamers. Thus, the dCas9-SAM system enables accumulation of multiple transcription activators at the genomic site recognized by CRISPR-dCas9-msgRNA complex and leads to robust high level expression of the targeted gene (Fig. 10.3).

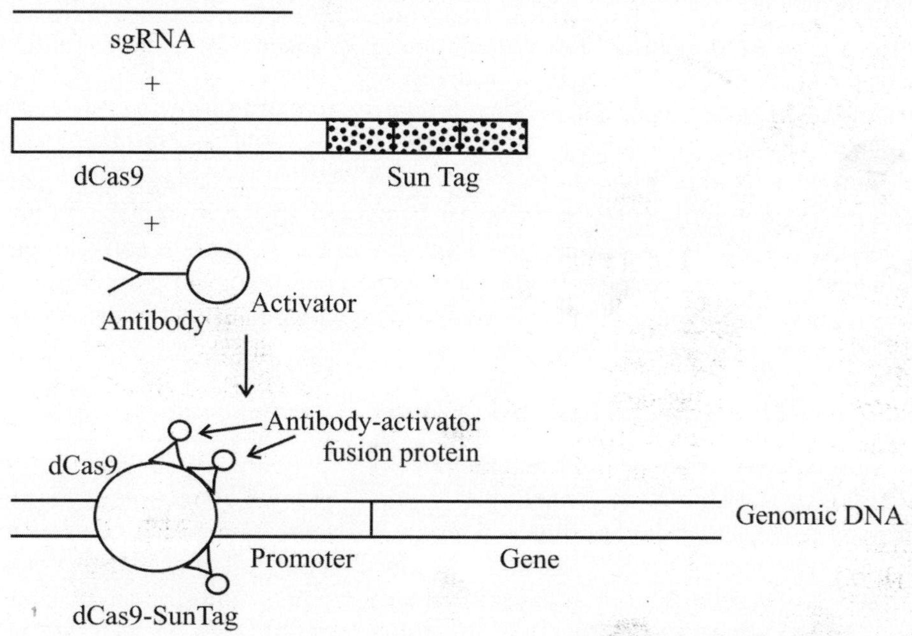

Fig. 10.4. A simplified schematic representation of dCas9-SunTag activation system. sgRNA is unmodified, dCas9 is linked to SunTag and the activator is attached to antibody specific for the oligopeptide repeat unit making up the SunTag. Here, only three repeats are shown, while a single SunTag polypeptide can recruit 24 antibody-activator fusion molecules.

The SundCas9 or SunTag activator system is an improved CRISPR-dCas9 activation method that enables robust transcription activation by a single sgRNA targeted at a single site in the promoter region of the desired gene (Tanenbaum et al., 2014). This activation system has three components, namely, modified dCas9, unmodified sgRNA, and transcription activators fused with antibodies to which NLS had been attached (Fig. 10.4). The dCas9 is modified by linking to it a designer polypeptide sequence called SunTag. SunTag comprises an array of oligopeptide repeats that specifically interact with and bind to antibodies. The antibodies used in the SunTag method are capable of binding tightly to short peptide epitopes that can be distinct from the endogenous epitopes encoded by the concerned cell genome. Single-chain variable fragment (scFv) antibodies are created by fusing the epitope-binding regions of the light and heavy chains to yield a single polypeptide. The scFv version of antibody GCN4 (scFv-GCN4) was fused with the superfolder-GFP (sfGFP), and the small solubility tag GB1 was also added (scFv-GCN4-

sfGFP-GB1) to avoid aggregation of the antibody-GFP fusion molecules. This fusion protein was used for robust imaging of genomic loci. For transcription activation, the scFv-GCN4 was fused with the activator VP64 to generate the scFv-GCN4-VP64 fusion protein. This strategy allowed recruitment of multiple copies of VP64 activator to the targeted site which supported robust transcription activation. The neighbouring antibody-binding epitopes in the SunTag are separated by spacer sequences of five amino acids to allow unhindered binding of the antibody-GFP molecules to all the epitopes in SunTag. Of the two versions of SunTag (v1 and v4), the v4 version under the control of a weak promoter was more suitable than v1 for many *in vivo* imaging applications (Tanenbaum et al., 2014).

The sgRNA guides the modified dCas9 to the targeted genomic site and the antibodies fused with the activator domains specifically bind to the SunTag repeat units. As a result, multiple (up to 24) copies of the single activator accumulate at the targeted promoter, and this may lead to a 5 to 25-fold higher activation of the targeted gene (Fig. 10.4). This strategy can also be used to accumulate multiple activators at the targeted promoter; this can be achieved by using a SunTag that has different repeat oligopeptides each repeat being recognized by a different antibody, and attaching a distinct activator to each antibody. This system can be used to activate inactive genes present in a latent virus, which may have diagnostic and even therapeutic applications. Any one of the above systems can be used to achieve a robust epigenetic modification since they multiply the action of programmable epigenetic effectors.

10.3.2.3. *Using several site-directed epigenetic effectors in the same cell*

The use of multiple programmable epigenetic effectors in a single cell enables simultaneous modification of several different epigenetic marks, which permits dissection of the relationships between various marks and/or a thorough alteration in the transcriptional state of the targeted gene(s). This can be achieved in the following two ways: (i) utilization of multiple dCas9-sgRNA orthologs or SpdCas9 variants having distinct PAM requirements, and (ii) use of several different sgRNAs each with a different RNA hairpin in its scaffold region. The different PAM requirements and use of distinct RNA hairpins ensures the delivery of the specific effector to the specified genomic site. dCas9 orthologs like SpdCas9, StdCas9, NmdCas9 and SadCas9 have different PAM requirements (Table 5.2). Each dCas9 ortholog can be fused with a different epigenetic effector. The sgRNAs for each of these orthologs can be designed to target a different site; this would ensure the recruitment of the desired effectors to the site specified by the respective sgRNAs. The same end is achieved by using variants of SpCas9 having different PAM requirements (Table 5.3). In this case also, each SpdCas9 variant is fused with a different effector, and targeted to the desired genomic site by designing the sgRNA specific for the concerned site. Confident implementation of these strategies would require information on the occurrence of different PAM sequences in the concerned genome, and the targeting efficiencies of different Cas9 orthologs/SpCas9 variants (Pulecio et al., 2017).

The second approach uses dSpCas9 in combination with multiple sgRNAs, each of which carries a distinct effector to the genomic site targeted by it. Each of these sgRNAs is modified by including in its scaffold region a different RNA hairpin that binds to a specific viral protein. Each epigenetic effector is expressed as a fusion protein with a distinct viral protein that interacts with the respective RNA hairpin integrated into the relevant sgRNA (Pulecio et al., 2017). This strategy is an extension of the dCas9-SAM method, and has been used in the CRISPRainbow system. In addition, the dCas9-SunTag system as well can be adapted for this purpose (Section 10.3.2.2).

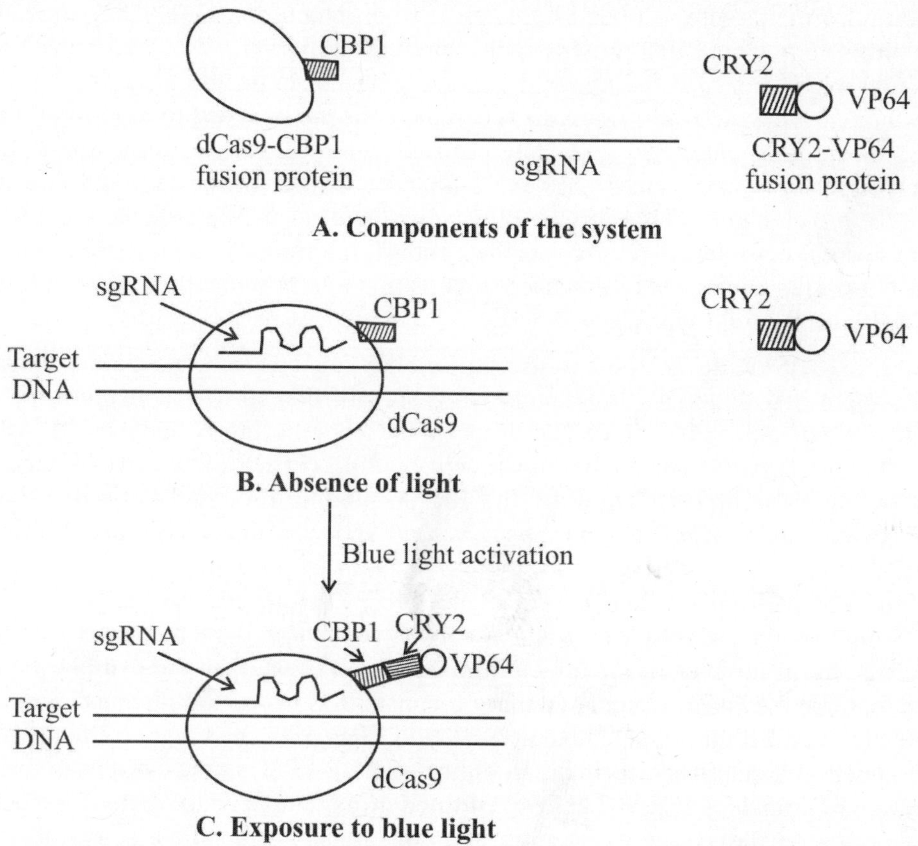

Fig. 10.5. The light inducible CRISPR-dCas9 activator system. **A.** The components of the light-inducible system. **B.** In the absence of light, dCas9-CBP1-sgRNA complex binds the target site, but the CRY2-VP64 fusion does not join this complex and there is no transcriptional activation. **C.** But on exposure to blue light the CRY2-VP64 fusion protein binds to CBP1 fused with dCas9 and there is gene activation as a result. CRY2, light-sensitive cryptochrome 2, CBP1, CRY2 binding protein 1; VP64, transcription activator. This strategy is also used to construct a light-inducible Cas9 nuclease by using Cas9 in the place of dCas9 and ommitting the VP64 from CRY2-VP2 fusion protein.

10.3.2.4. *Temporal and/or spatial regulation of CRISPRa/i*

It is well recognized that many biological processes involve highly specific gene expression patterns that are regulated by a variety of exogenous as well as endogenous stimuli. In contrast, the standard programmable transcription activation/repression tools based on CRISPR-dCas9 or TALEs do not offer any spatial or temporal control on gene expression patterns. Several different strategies have been developed for temporal and spatial regulation of CRISPR-Cas9 nuclease action (Section 5.9.2.2); some of these strategies can be adapted for the development of such

CRISPR-dCas9-based activators/repressors that would afford spatial and/or temporal control on transcription regulation. For example, a light-activated version of CRISPR-dCas transcription activation system developed by Nihongaki et al. (2015) comprises dCas9-CIB1 fusion protein, sgRNA specific for the targeted site and light-sensitive cryptochrome 2 (CRY2)-activator (VP64 or p65) fusion protein. The CIB1 protein is the binding partner of CRY2 protein. The dCas9-CIB1-sgRNA complex binds to the targeted genomic site but is unable to activate transcription since it does not carry the activator domain. When the cells are exposed to blue light, the CRY2-activator fusion protein interacts with CIB1 fused with dCas9 to form the CIB1/CRY2 heterodimer (Fig. 10.5). This places the CRY2-linked activator at the targeted site leading to activation of the targeted gene. This system allows rapid and reversible targeted mammalian gene activation in response to light. It also enables robust multiplexed gene activation by using multiple sgRNAs. This strategy was also used to construct a light inducible Cas9 nuclease system by fusing CIB1 with the wild type Cas9.

Similarly, a CRISPR-dCas-VP64 transcription activator has been developed based on the split dCas9 system that dimerizes in response to a specific ligand to reconstitute the complete functional dCas9 molecule (Fig. 10.6). The dCas9 was first split into N-terminal [dCas9(N)] and C-terminal [dCas9(C)] parts. The FKBP rapamycin binding (FRB) domain was linked at the C-terminus of dCas9(N) fragment. Similarly, the FK506 binding protein 12 (FKBP) domain was fused to the N-terminus of dCas9(C) and the activator VP64 was appended to its C-terminus. The FKBP and FRB domains are derived from the mammalian target of rapamycin. In addition, the dCas9(N) fragment had two nuclear export signal (NES) sequences that sequestered it in the cytoplasm. Similarly, the dCas9(C) part had two NLS sequences, which sponsored its transport into the nucleus. In the absence of rapamycin, the dCas9(N) remains in the cytoplasm, as a result of which the complete dCas9 molecule cannot be constituted inside the nucleus (Fig. 10.6). But when rapamycin is added into the cells, the FKBP and FRB domains linked to the dCas9(N) and dCas9(C) fragments, respectively, form heterodimers (FKBP-FRB) in the cytoplasm itself leading to the formation of complete dCas9. The reconstituted dCas9 moves into nucleus, associates with sgRNA and binds to the targeted genomic site; this places the VP64 activator fused with dCas9(C) at the target site, which activates transcription (Zetsche et al., 2015b).

The above inducible systems can be used to develop inducible programmable repressor systems as well. Thus, these light/small molecule inducible programmable epigenome editing tools would allow dynamic spatial and temporal control of CRISPRa/i.

10.3.3. Factors affecting epigenome editing

Epigenome editing is affected by several factors related to the programmable epigenetic effectors as well as other conditions of the experiment, including the genomic site targeted for epigenetic modification. In general, the programmable effectors constructed using catalytic domain of an effector are more efficient than those comprising the complete effector protein. In addition, the amino acid sequence and the length of the linkers used to join the effector domains to dCas9 have a profound effect on the robustness of chromatin modifications induced by the resulting programmable effectors. Nearly 100% nuclear localization of the site-directed effectors is achieved when two NLS sequences are used, one NLS being located at the N-terminus of dCas9 and the other being integrated either at the N- or the C-terminus of the effector domain fused with dCas9 (Pulecio et al., 2017).

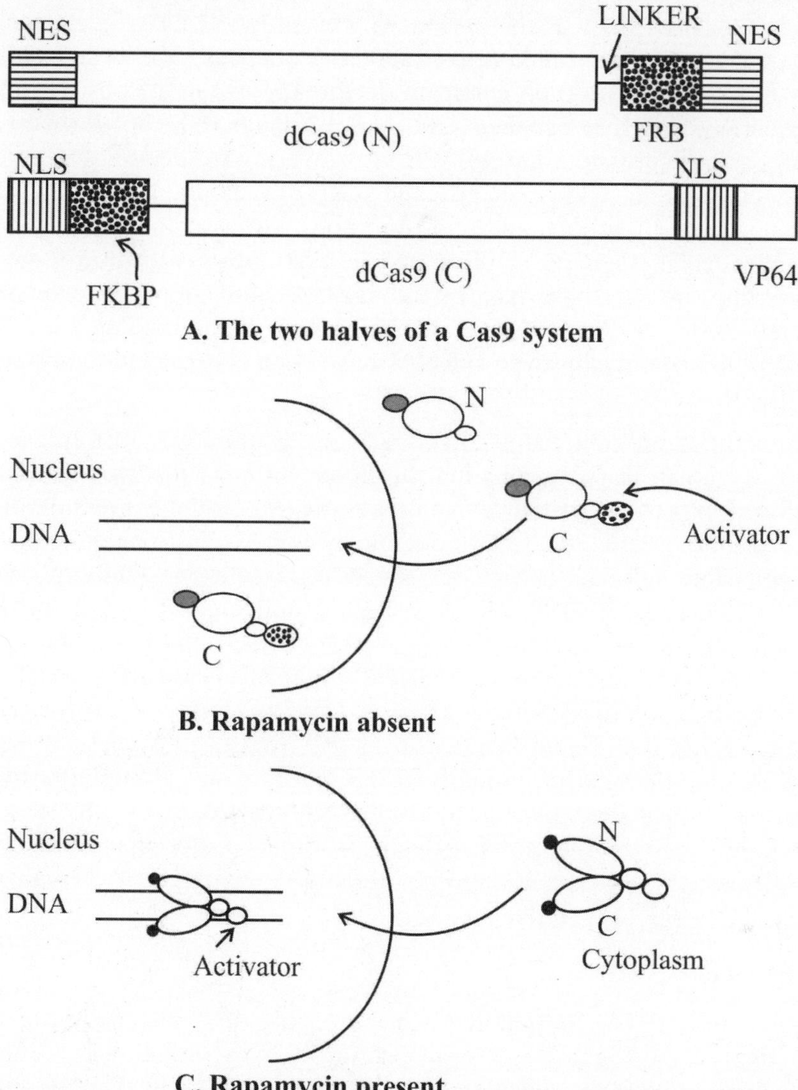

Fig. 10.6. Schematic representation of inducible split Cas9 activation system. **A.** Organization of the N-terminal and C-terminal parts of Cas9. **B.** In the absence of rapamycin, the N-terminal part is translocated into nucleus, but can not bind the target DNA. **C.** But when rapamycin is present, the N-terminal and C-terminal parts are brought together by rapamycin induced heterodimer formation by the FKBP and FRB domains. The reconstituted Cas9 associates with sgRNA, moves into nucleus and binds the target site causing transcriptional activation. dCas9(N) and dCas9(C), the N-terminal and C-terminal parts of dCas9; FRB, FKBP Rapamyein binding protein; NES, nuclear export signal sequence, NLS, nuclear localization signal sequence; FKBP, FK506 binding protein.

The genomic region targeted for gene expression modulation should be carefully selected based on tested databases like FANTOM5, which allow accurate determination of TSS (transcription start site) and the concerned promoter sequences. The best results are obtained when the region from -200 bp (200 bp upstream) to the TSS is targeted by multiple sgRNAs for transcription activation, and that between -200 and +800 bp (800 bp downstream to TSS) is targeted for transcription repression. Several sgRNAs covering the desired genomic region should be evaluated for identification of the most efficient sgRNA(s) having little or no off-target effects. Increased expression level of sgRNA and improved ability to associate with dCas9 may enhance the targeting efficiency. The sgRNA stability and its association with dCas9 are reported to be increased by inserting an A-U base pair in the sgRNA stem loop and extending the hairpin structure that binds to dCas9. The presence of nucleosome at the binding site is likely to affect dCas9 binding, and libraries of improved sgRNAs have been constructed to mitigate this problem (Pulecio et al., 2017).

Stable expression of programmable effectors is necessary for robust epigenome alterations specially in case of human primary cells and pluripotent stem cells. Several inducible systems have been developed to provide spatial and temporal control on the expression of programmable effectors; of the various systems, the light-inducible systems seem to be quite promising. In case of mammalian cells, the sgRNAs can be introduced into cells via lentiviral vectors, but their direct introduction into cells as RNA seems to be a promising delivery method as it has high efficacy and is amenable to multiplexing. In general, genome-wide RNA-Seq data do not show off-target effects, but some authors have reported slight effect on the overall profile of gene expression produced by even dCas9-effector fusion proteins alone, *i.e.*, without associating with any sgRNA. The intensity of fluorescence during *in vivo* imaging declines with time due to photo-bleaching of the fluorescent proteins (FPs); as a result, long-term imaging is quite problematic. But fluorescent proteins attached to sgRNA scaffolds (*e.g.*, in cases of dCas9-SAM and CRISPRrainbow) recover fluorescence about 40-times faster than those fused to the C-terminus of dCas9 protein. As a result, chromatin could be continuously visualized for about 26 hr using FPs attached to sgRNAs (Pulecio et al., 2017).

10.3.4. CRISPRi vs. RNAi

CRISPRi is preferable to RNAi for genome-wide screening of gene functions as it can be easily reprogrammed simply by designing sgRNAs specific for new targets, while implementation of RNAi for each new gene involves considerably greater effort and cost. Further, eukaryotes do not have anything remotely like CRISPR-Cas system, while they have effective endogenous RNAi pathways. As a result, there is no possibility for CRISPRi to compete with an endogenous system, which could complicate the execution of certain experiments and interpretation of the experimental findings. This difficulty is experienced with RNAi when, for example, it is used to suppress the expression of endogenous genes that participate in the endogenous RNAi pathway. Thus, RNAi-mediated suppression of genes encoding Dicer, Argonaute, etc. can yield results that are not easily interpreted. Further, there is the risk that RNA sequences used for RNAi-mediated suppression of specific genes may interfere with gene regulation by endogenous RNAs; this possibility is virtually non-existent in case of CRISPRi. The RNAi technology is capable of only silencing the targeted genes, while CRISPRa/i can be used for both upregulation and silencing of the selected genes as per need of the investigation. Despite continued refinements in RNAi technology, gene silencing can sometimes be incomplete and there may be unpredictable off-target gene silencing; off-target effects are particularly relevant in genome-wide screens. In contrast, the effects of CRISPRi on gene expression are efficient, highly specific and stable, and these effects occur presumably only at the target sites (Doudna and Charpentier, 2014; Komor et al., 2017b).

The available results indicate that dCas9 can serve as a versatile modular DNA-binding tool capable of recruiting various proteins to the specified genomic sequences and is expected to be useful in experiments focussing on genome-wide modulation and screening of gene functions. CRISPRi has helped identify genes concerned with immune response, cancer development, drug resistance, and many other traits related to health and disease (Doudna and Charpentier, 2014; Komor et al., 2017b). However, RNAi offers a unique advantage of tissue-specific suppression of endogenous genes, which is not feasible with gene editing.

10.4. MODULATION OF CHROMATIN ORGANIZATION

Transcription activation and repression can also be modulated by deploying enzymes concerned with chromatin remodelling by modifying histones and/or DNA (Table 10.4). For example, human acetyltransferase p300 catalyses acetylation of H3 histone at Lys-27. The catalytic domain of p300 was fused with dCas9; the resulting hybrid effector catalysed acetylation of Lys-27 residues of H3 histone within a region of up to thousands of base pairs around the site targeted by the sgRNA leading to robust transcriptional activation of genes in the region. Modifications of DNA methylation state has been exploited for chromatin activation. The mammalian TET (Ten-eleven translocation) proteins like TET dioxygenase1 (TET1CD) convert 5-methylcyosine to bases like 5hmC (5-hydroxymehtylcytosine) ultimately leading to demethylation of cytosine and chromatin activation. The fusion of catalytic domain of TET1CD with dCas9 yields a programmable DNA demethylation and chromatin activation tool. This tool can demethylate >90% of the methylated CpG islands up to about 200-bp around its target site. In case of plants, DNA demethylation involves the ROS1 glycosylase-lyase family of enzymes. These enzymes could be similarly harnessed for chromatin activation in plants. The histone demethylase LSD1 catalyses demethylation of dimethylated Lys-4 residues of H3 histone. LSD1 has been fused to the C-terminus of dCas9 to yield a programmable histone demethylase that catalyses demethylation of dimethylated Lys-4 of H3 up to >350 bp from the targeted site leading to inactivation of chromatin and transcription repression in the region (Komor et al., 2017b).

On the other hand, the fusion of catalytic domain of human DNA methyltransferase (DNMT3A) to dCas9 enables targeted methylation of CpG islands within ~100 bp around the targeted site, which promotes heterochromatinization and transcription repression. Similarly, Kruppel-associated box (KRAB) repressor induces trimethylation of Lys-9 residue of histone H3, which leads to heterochromatinization of the concerned genomic region and, as a result, transcription repression. The KRAB domain has been fused with dCas9 and the hybrid effector has been used for programmable silencing of targeted genes. The dCas9-based methylation is more efficient, affects larger genomic regions and causes greater transcription repression than TALE-effector fusions. A transient expression-based methylation/demethylation strategy may require additional chromatin modifications for sustained effect on gene expression. Further, the inclusion of cofactors like TDG (thymine DNA glycosylase; involved in repair of G/T mismatches) for demethylation and DNMT3L for *de novo* methylation can enhance the effectiveness of DNA methylation/demethylation. The SID interacts with the mSin3 domain present in repressors like mSin3A. The dCas9-SID fusion protein recruits repressors like mSin3A to chromatin, which interact with and activate histone deacetylase HDAC1/2. As a result, there is histone deacetylation leading to conversion of active chromatin to inactive state and transcription repression (Doudna and Charpentier, 2014; Pulecio et al., 2017).

Table 10.4. Some examples of chromatin remodellers used for programmable transcription activation and repression

Modification agent	Action	Effect of modification	Remarks
Chromatin activation			
Histone acetyltrans-ferase p300	Acetylation of H3 histone at Lys-27	Robust transcription activation of the targeted genes	A transcription coactivator; plays essential role in regulation of cell division and growth
Ten Eleven Translo-cation Dioxygenase1 (TET1CD)	DNA demethylation	Activation of chromatin leading to transcription	Converts 5-methylcytosine to 5-hydroxymehtylcytosine leading to demethylation
Chromatin inactivation			
DNA methyltransferase (DNMT3A)	Methylation of C residues in CpG doublets	Promotes heterochromatin-ization and transcription repression	Needed for genome-wide *de novo* methylation
Kruppel-associated box (KRAB) repressor	Trimethylation of Lys-9 residue of histone H3	Promotes H3 histone Lys-9 trimethylation, local chromatin compaction and transcription repression	It recruits KAP1/TRIM28 and HP1 proteins, which hinder RNA polymerase II positioning; 50% to 99% transcription repression
mSin3 interaction domain (SID)	Recruits repressors with mSin3 domain, *e.g.*, mSin3A corepressor	Histone deacetylation by recruiting histone deacetylases leading to transcription repression	mSin3A activates histone deacetylase HDAC1/2; 50% to 99% transcription repression
LSD1	Demethylation of dimethylated Lys-4 residues of H3 histone	Inactivation of chromatin leading to transcription repression	Methylated Lys-4 is a mark of euchromatin

Generally, chromatin modifications are transient but they would become persistent if reinforced by endogenous epigenetic machinery. It may be pointed out that programmable transcription activators and repressors may produce off-target effects similar to those of SDN systems. In addition, there could be cross-talk between programmable effector domains and the endogenous effectors with related functions. These possibilities need to be investigated in comprehensive detail before these programmable tools are applied for modification of gene expression *in vivo* particularly in human subjects for therapeutic purposes (Doudna and Charpentier, 2014; Pulecio et al., 2017).

Some epigenetic marks like trimethylated Lys-4 and Lys-27 of H3 histone are generally associated with transcription activation and repression, respectively. But the precise effects of histone modifications depend on the cellular and chromatin context. For example, the above two trimethylated Lys residues co-occur in certain genomic regions and are thought to allow low levels of expression of several developmental genes; they also enable rapid deployment of factors for either inducing or repressing transcription depending on the cellular context. The CRISPRi/a is expected to unravel many aspects of epigenetic modifications, including the precise role of each enzyme involved in histone modification, the nature of interactions among histone modifiers, and the effects of these histone alterations on chromatin organization and regulation of gene expression (Pulecio et al., 2017).

10.5. TARGETED RELOCATION OF NONCODING RNAs

The noncoding RNAs (ncRNAs) do not encode polypeptides and participate in gene regulation by affecting the processes of chromatin remodelling, transcription and translation. Some lncRNAs (long noncoding RNAs) and microRNAs (<30 nt long) are involved in chromatin modifications, including heterochromatinization. The CRISPR-dCas9 system could be repurposed for targeted placement of specific ncRNAs at the desired genomic sites to achieve diverse objectives. This could be done by integration of the selected ncRNA into the sgRNA in a way that the normal sgRNA function remains unaffected and the ncRNA is able to exert the desired effect. ncRNAs like NoRC-binding pRNA stem-loop, the Xist-A repeat and the HOTTIP lncRNA have been placed at specific genomic loci by using the above approach. This ncRNA relocation strategy is expected to provide insights into the roles of ncRNAs in regulation of gene expression, and to establish and maintain epigenetic modifications of DNA and histones (Pulecio et al., 2017).

10.6. OFF-TARGET ACTIVATION/REPRESSION

The programmable nucleases as well as base editors are known to induce off-target gene editing and base editing, respectively, in low to almost negligible frequencies (Sections 5.9 and 6.9). In case of CRISPR-Cas9 system, the frequencies of off-target effects are affected by several factors, including sgRNA design, Cas9 variant, targeted genomic locus and even the organism. These realizations have prompted the development of bioinformatics tools for genome-wide CRISPR-Cas designing and optimization, and identification and validation of CRISPR/Cas off-target sites. However, a precise determination of off-target effects in a comprehensive, robust and highly sensitive manner remains a great challenge especially in the cases of base editing and epigenetic modifications (transcription activation and repression). Many studies on off-target activities of CRISPR-Cas9 genome editing suggest it to be highly specific, and unbiased profiling assays tend to validate the above inference. It is argued that the frequencies of off-target effects are likely to be lower *in vivo* than those estimated from *in vitro* assays as a result of the epigenetic protections *in vivo* (Zhang et al., 2018).

It has been suggested that the off-target effects of CRISPR-dCas9-based activators/ repressors can be expected to be much lower than those for gene editing since activation/ repression would require much longer association of the activators/repressors with the off-target site than that necessary for the nuclease action. Off-target effects of CRISPR-dCas9-based activators/repressors can be assayed by comprehensive mRNA-Seq analyses. Zhang et al. (2018) used the dCas9-SAM technology to activate expression of HIV-1 in latently infected cells to ultimately lead to death of the infected cells. They carried out deep sequencing of the entire transcriptome and subjected the mRNA and lncRNA sequence data to comprehensive analyses for assessing the off-target effects of stable expression of dCas9-SAM-sgRNA (specific for HV-I) on the frequencies of different transcripts, alternative splicing patterns and induction of genetic mutations, including SNPs (single-nucleotide polymorphisms) and indels (insertions and deletions).

It was observed that only two off-target genes were significantly upregulated, while 30 others showed significant downregulation. Similarly, a small number of lncRNAs were also downregulated. The downregulation of mRNAs and lncRNAs is unexpected and cannot be explained. The authors concluded that these effects on transcription level may not be the direct result of dCas9-SAM action. In addition, there was no pronounced effect of dCas9-SAM on the genome-wide alternative splicing pattern. Finally, the exposure to dCas9-SAM did not generate any significant effect on SNP and indel frequencies; this RNA-Seq data-based inference should be supplemented with genome sequence data since only a fraction of the genome is transcribed. Although these findings indicate little off-target effects of dCas9-SAM targeted for activation of HIV in latently infected human cells, further investigations using human primary cells and animal models, and a larger number of replicates should be undertaken for generating dependable comprehensive information on off-target effects of epigenome editing (Zhang et al., 2018).

10.7. GENETIC INTERACTION MAPS

Genetic interaction between two or more genes signifies that these genes perform related functions due to, for example, the proteins encoded by them being components of the same protein complex or participating in the same metabolic pathway. The phenotypic effects of simultaneous mutations in such genes cannot be predicted based on the effects generated by mutant alleles of individual genes. Mapping of GI or functional relationships among various genes involves systematic collection of data on phenotypic effects produced by simultaneous mutations in all possible combinations of two or more genes affecting the trait of interest. It is, therefore, important that a very large number of selected genes are targeted both individually as well as in different combinations, and their phenotypic effects are evaluated. CRISPR-Cas9 system provides a high throughput approach for targeting genes and gene combinations for GI mapping. This system can be harnessed to either inactivate (by gene mutation) or repress/activate the expressions of selected genes. The GI mapping is expected to reveal the gene interaction networks that affect disease onset and development, *e.g.*, tumorigenesis. It is also expected to help in identification of more precise therapeutic targets and the development of precision drugs for treatments of various ailments. The data used for GI mapping should be based on large number of samples to include the genetic diversity present in human population. The gene editing/repression/activation data should be combined with those from proteomic and chemogenetic analyses to develop comprehensive GI maps that would be more reliable and useful than those generated from only one type of data (Gilbert et al., 2014; Shen et al., 2017).

GI maps could unravel the manner in which expression levels of different genes, either singly or in groups, affect the various cellular functions and phenotypes, including diseases. GI maps could also suggest therapeutic targets for precision drug development. For example, it could indicate gene pairs whose simultaneous inactivation would lead to cell death and, thereby, provide a 'synthetic-lethal' route for treatment of diseases like cancer (Shen et al., 2017).

10.8. IMAGING OF SPECIFIC GENOMIC LOCI

In vivo imaging of organelles and other intracellular structures, including biomolecules is based on fluorescent tags, including fluorescent proteins like, dTomato, GFP and its mutants, etc. Imaging is greatly facilitated by fluorescence signal amplification by recruiting multiple copies of the selected FP to the imaging target. Signal amplification by accumulating multiple copies of proteins is common in cells, *e.g.*, accumulation of transcription factors during transcription initiation, binding of multiple protein molecules for translation regulation, and the inclusion of more than one NLS sequences in several proteins. This principle has been used for imaging of proteins by fusing three copies of GFP, but any further increase in the number GFP copies is challenging. RNA localization has been investigated by inserting in the target RNA molecules as many as 24 copies of the RNA hairpin specific for MS2 coat protein, which recruits multiple copies of the MS2-FP fusion molecules (Fig. 10.7). The above RNA hairpin-MS2 system can also be used to image and follow a protein by artificially linking to it multiple copies of the above RNA hairpin (Tanenbaum et al., 2014).

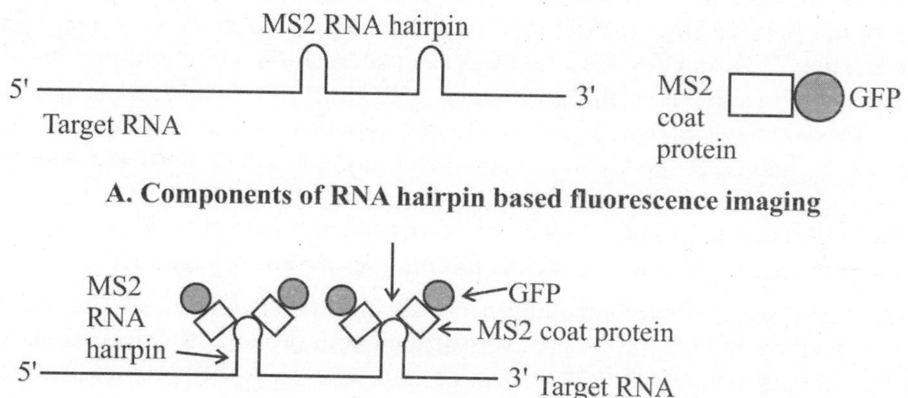

Fig. 10.7. Fluorescence imaging of RNA based on RNA hairpin-viral protein interaction. **A.** Components of the imaging system. **B.** MS2 coat protein-GFP fusion protein attaches to the RNA hairpins integrated into the target RNA. Multiple (up to 24) copies of the RNA hairpin have been inserted into the target RNA for amplifying the fluorescence signal to enable robust imaging.

The labelling of genomic sites can be achieved by linking a FP, such as GFP, to a DNA binding protein like LacI repressor that specifically binds to *lacZ* operator. In this case, the fluorescence signal can be amplified by integrating multiple copies of *lacZ* operator at the targeted genomic site, which will recruit many copies of the GFP-LacI fusion molecules but this method is cumbersome and time-consuming. A simple and versatile strategy for mapping of specific genomic sites is to repurpose CRISPR-Cas9 system to carry and position fluorescent proteins to the targeted genomic sites in living cells. However, multiple copies of the FP must be brought to the targeted genomic region to permit dependable imaging of the locus. The genomic regions like telomeres that have repetitive DNA can be targeted by a single sgRNA, which will bind at multiple sites in the repetitive region and, thereby, amplify the fluorescence signal and allow imaging of the genomic locus. But imaging of nonrepetitive genomic regions would require designing at least 36 different sgRNAs specific to nonoverlapping sequences in the selected region. These sgRNAs are simultaneously used in combination with dCas9-eGFP fusion to achieve strong fluorescence and robust imaging of the labelled genomic site (Doudna and Charpentier, 2014; Tanenbaum et al., 2014).

The CRISPR-dCas9-based imaging tool may facilitate studies on native chromosome conformational dynamics in living cells, especially when multiple colour imaging tools will be used. One possible approach for orthogonal multicolour imaging would be to link separate fluorescent molecules/proteins to sgRNAs and Cas9 proteins. The genomic sites targeted by dCas9-SunTag system (based on SunTag polypeptide and antibody-GFP fusion protein) show much brighter labelling than when they are labelled by dCas9-GFP fusion protein. In theory, it would be possible to use a much smaller number of different sgRNAs in combination with the dCas9-SunTag system for imaging of genomic loci with nonrepetitive DNA sequences (Doudna and Charpentier, 2014; Tanenbaum et al., 2014).

At present, chromatin organization is generally studied by methods like Hi-C and fluorescence *in situ* hybridization (FISH) that require sample fixation. Fluorescently tagged DNA binding proteins like ZFPs and TALEs, and dCas9 in combination with multiple (26–36) sgRNAs have been used for imaging of genomic regions with repetitive DNA and single copy DNA, respectively. An advantage of the dCas9-FP fusion proteins is that they do not displace the proteins bound to telomeres and centromeres, and do not affect telomere movement. The CRISPR-dCas9 system allows monitoring of the timing of appearance and persistence, and the range of various chromatin configurations; it also permits the detection of correlations between these configurations and the variations in cells morphology and/or behaviour.

The SunTag system allows imaging at relatively lower levels of fluorescent protein expression and, thereby, avoids the problems associated with protein overexpression. At the same time, lower light intensities are needed for imaging; this minimizes the phototoxic effects of long-term imaging. The system enables imaging in both cytoplasm and nucleus of even single molecules and the bright fluorescence may help their automated tracking. The SunTag polypeptide fused with dCas9 enables amplified gene activation (Section 10.3.2.2), but its large size (1,400 kDa for SunTag bound to 24 antibody-GFP molecules) can interfere with the activity or the half-life of the protein, to which it is fused (Tanenbaum et al., 2014; Pulecio et al., 2017).

Fluorescence imaging based on CRISPR-dCas9 permits visualization of chromosome dynamics, including chromosome packaging, structure and organization, elongation and disruption of telomeres, and localization of various genomic loci within the nuclei of living cells. The CRISPRainbow system permits tracing of DNA movement in living cells and real time studies of genome organization by using a combination of different FPs that generate fluorescence of disparate colours (Ma, H. et al., 2016). This technique has been used for simultaneous tagging of seven different genomic loci, and for concurrent imaging in six different colours. CRISPRainbow system uses the SAM approach for recruiting different FPs to the targeted sites. Briefly, different RNA hairpins, namely, MS2, PP7 and boxB hairpins, are integrated in the sgRNA scaffolds; these hairpins specifically bind to the MS2, PP7 and BoxB proteins, respectively. The MS2, PP7 and BoxB proteins are fused with different FPs, *viz*., GFP, BFP (blue fluorescent protein) and RFP (red fluorescent protein). The sgRNAs with RNA hairpins are used in combination with dCas9 and the FP-hairpin binding domain fusion proteins. The different FPs are recruited to cognate RNA hairpins leading to multicolour labelling of the genomic site (Fig. 10.8).

In fluorescence resonance energy transfer (FRET), a donor fluorophore is excited by radiation and it transfers its excitation energy to an acceptor chromophore located nearby, as a result of which the latter generates fluorescence. Thus, FRET imaging is based on specific pairs of donor and acceptor fluorophores. The two members of a FRET pair can be linked to two different molecules, *e.g.*, sgRNA and target DNA, and FRET fluorescence imaging can be used to follow the dynamics of interaction between dCas9-sgRNA complex and the target DNA in real-time. Such imaging would provide clearer understanding of the kinetics of target recognition and rejection by dCas9-sgRNA complex, and the effects of sgRNA-target DNA mismatches on these processes.

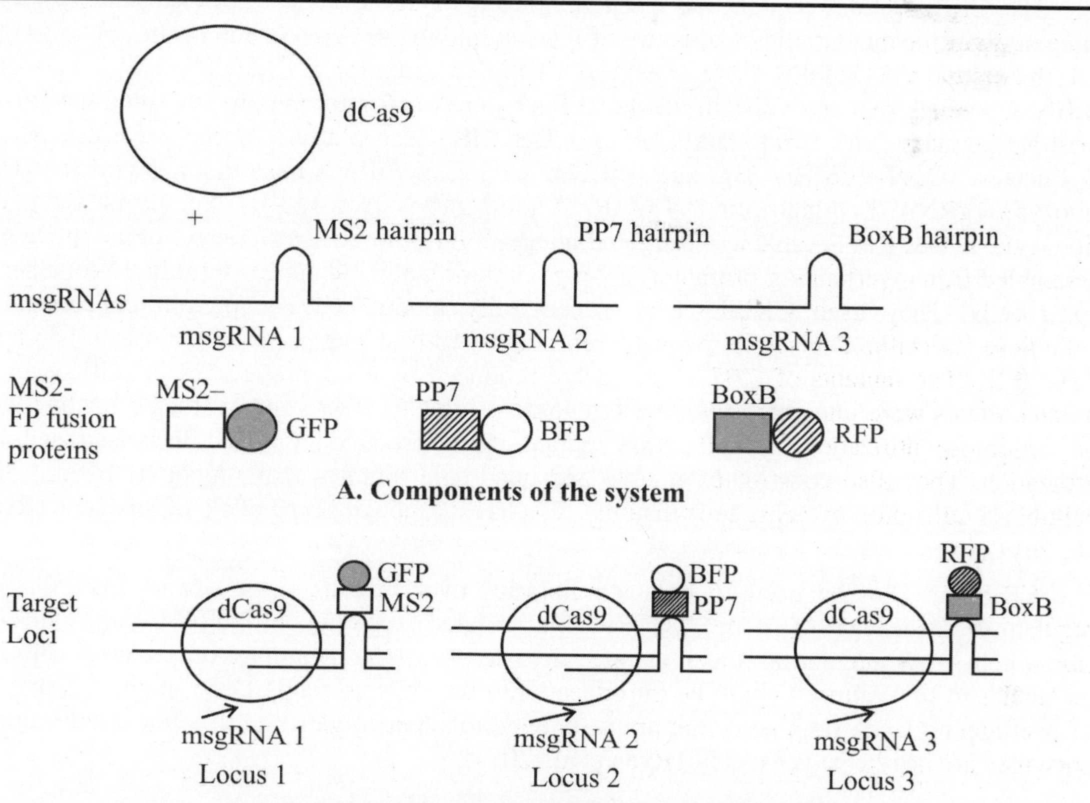

Fig. 10.8. A highly simplified schematic representation of the CRISPR rainbow system of simultaneous imaging of multiple genomic loci in different colours. **A.** Components of the system. **B.** Different viral protein-FP fusions attached to their cognate RNA hairpins integrated into separate msgRNAs bound to the different target loci. BFP, GFP and RFP, blue, green, and red fluorescent proteins, respectively; msgRNA, modified (by RNA hairpin integration) signle guide RNAs; MS2, PP7 and BoxB, proteins specific for the MS2, PP7 and BoxB RNA hairpins, respectively.

10.9. DIRECTED EVOLUTION OF PROTEINS

Directed evolution of proteins involves engineering mutant proteins in the laboratory and selection of the protein(s) with desired functions from among the mutants so generated. In simple terms, this comprises creation of numerous mutant alleles of the gene encoding the selected protein followed by evaluation of the mutant proteins in a suitable test system to enable selection of the allele(s) that encode(s) the most promising protein(s). For evaluation, the mutant alleles are inserted in suitable plasmid or bacteriophage vectors and expressed in appropriate host cells. This procedure is often plagued by several complications, including uncontrolled variation in plasmid copy number per cell and the need for a specific host strain. The directed evolution process is a powerful tool for refining and enhancing protein functions, and even to generate novel functions (Ryan et al., 2014).

The CRISPR-Cas9 system has been optimized by Ryan et al. (2014) to enable targeted integration of the mutant alleles of genes of interest into the selected genomic sites of yeast. They call this optimized CRISPR-Cas9 system as CRISPRm (multiplex CRISPR) since it can induce DSBs at a single genomic site in all the chromosomes (*cis*-multiplexing) or simultaneously at multiple genomic sites (*trans*-mutliplexing). The CRISPRm platform comprises Cas9 with the architecture Cas9-NLS-His tag and sgRNA having 5' tRNA-hepatitis delta virus (HDV) ribozyme-sgRNA 3' organization. The tRNA molecule is removed by the self-cleaving HDV ribozyme. CRISPRm is capable of targeted integration (70 to 80% efficiency) of complete genes assembled from overlapping promoter, coding and terminator sequences introduced together into yeast cells. They used CRISPRm to successfully modify CDT-1 protein concerned with cellobiose utilization; this improvement resulted in 10-fold enhanced cellobiose utilization by yeast cells. The mutants of *CDT-1* gene were produced by error-prone PCR amplification. The mutant alleles were integrated into yeast genome using CRISPRm and evaluated for their effect on cellobiose utilization by cells; one mutant allele produced 2.6-fold increased cellobiose utilization. They also constructed double and quadruple mutants that supported 10-fold higher cellobiose utilization by yeast cells than the cells having the wild type allele of this gene (Ryan et al., 2014).

CRISPRm can be used to evaluate libraries of *cis*-regulatory elements like 5' and 3' regulatory sequences of genes, UTRs (untranslated regions) of mRNA molecules, and riboswitches by integrating them at specific sites into the genomes of selected cells. The molecules to be evaluated could be introduced into the cells as single DNA molecules or as sets of overlapping DNA fragments that are assembled together to generate the functional sequences once they are inside the yeast cells (Ryan et al., 2014).

10.10. SITE-DIRECTED RECOMBINASES AND TRANSPOSASES

The gene editing technology based on SDNs suffers from several limitation like unpredictable off-target effects that are difficult to monitor, dependence on cellular DNA repair mechanisms that may not be available in some cell types, etc. These considerations have prompted the creation and evaluation of other types of programmable enzymes like site-specific recombinases and transposases that could be used for gene editing and other genome engineering applications. Recombinases and transposases can cleave and religate DNA autonomously and their action is not dependent on cellular DNA repair mechanisms. These enzymes can lead to DNA fragment excision, integration and inversion, and there is little or no likelihood of accumulation of DSBs that are toxic to cells. In addition, targeted addition of genes catalysed by these enzymes is noticeable by integration of donor DNA into the genome, which enables easy monitoring of off-target effects (Gaj et al., 2013).

The chief limitation of recombinases and transposases is that considerable refinements are required to improve their performance and, especially, flexibility to make them truly programmable and generally applicable. The catalytic domains of recombinases contribute to the sequence specificities of the native enzymes. Therefore, considerable protein engineering and directed evolution would be needed to make these domains act on a user-defined target sequence. The transposases exhibit high activity at targeted genomic sequences, but they also show substantial off-target activity (Gaj et al., 2013).

10.11. MOLECULAR DETECTION METHODS

Efforts are being made to develop the CRISPR-Cas9 system into an efficient sensor of single-stranded RNA (ssRNA) and single-stranded DNA (ssDNA) based on the toehold mediated strand displacement technique. In this technique, a sequence is inserted into the scaffold region of sgRNA molecule, as a result of which the secondary structure of this region is changed and the sgRNA is unable to interact with and activate the Cas9 nuclease. When a ssRNA or ssDNA molecule that is complementary to the sequence inserted into the sgRNA is provided, this ssRNA/ssDNA base-pairs with the inserted sequence. This base-pairing generates a change in the secondary structure of sgRNA scaffold region so that it can now interact with and activate the Cas9 nuclease. This method has been demonstrated *in vitro* and needs to be evaluated in *in vivo* systems (Jakimo et al., 2018).

The sgRNA containing the inserted sequence is called switchable guide RNA (swigRNA) since it becomes active in response to a trigger, in this case, ssRNA or ssDNA complementary to the inserted sequence. One type of swigRNA can associate with Cas9 only when it is activated by an interaction with the oligonucleotide trigger (a ssRNA or ssDNA) called trigger RNA (trigRNA) or trigger DNA (trigDNA). Therefore, this switchable CRISPR-Cas9 system can cleave DNA only after the swigRNA interacts with the specific trigRNA of trigDNA; these swigRNAs serve as off-to-on switches. Another type of swigRNAs prevents cleavage by Cas9 in response to specific trigRNA/trigDNA; these swigRNAs are termed as on-to-off switches. The swigRNAs were computationally designed by predicting the dependence of guide RNA secondary structure and function, *i.e.*, association with and activation of Cas9 nuclease, on hybridization with trigRNAs/trigDNAs. This application of CRISPR-Cas9 system is in initial stages of development. It is expected to allow profiling of RNAs and DNAs that can help in the classification of a variety of biological conditions, including diseases, metabolism, tissue and cell types, etc. (Jakimo et al., 2018).

The swigRNA has been combined with appropriate Cas9 variants to create 'signal conductors' and 'logic gates' that can be used to modulate transcription of targeted endogenous genes. The time and duration of this modulation can be controlled by providing the exogenous signal in form of trigRNA/trigDNA. This strategy has been used to modify the endogenous signalling networks in cells by creating interaction links between such metabolic pathways that do not otherwise interact with each other (Katrekar et al., 2017).

10.12. ISOLATION OF SPECIFIC ENDOGENOUS mRNAs

The CRISPR-Cas9 system can be used in association with specially designed PAMmers (PAM presenting oligonucleotides) to selectively isolate specific native mRNAs from cells without any kind of tagging. A PAMmer is an oligodeoxynucleotide that has PAM sequence along with a short 3' and an 18 nucleotide-long 5' extension; the latter region of PAMmer base-pairs specifically with a sequence of the desired mRNA. The selective isolation of the desired mRNA is achieved by mixing the total cellular RNA extract with biotinylated Cas9-guide RNA complex and PAMmer specific for the mRNA to be isolated. The PAMmer pairs with the targeted region of mRNA and generates a double-stranded region around the PAM sequence of the PAMmer (Fig. 5.13). The Cas9-guide RNA complex recognizes the PAM sequence present in the PAMmer and associates with it. It then begins to unwind the double-stranded target region and the guide RNA pairs with the targeted mRNA sequence. The Cas9-guide RNA-PAMmer-mRNA complex is then pulled out using streptavidin-coated magnetic beads and the mRNA is separated by an appropriate treatment. The PAMmers used in this procedure are chemically modified, *e.g.*, by 2'-O-Methylation, to prevent activation of the Cas9 HNH nuclease domain and, as a result, cleavage of the associated mRNA. This approach of specific mRNA isolation is robust and simpler than other approaches used for this purpose (O'Connell et al., 2014). The need for chemical modification of the PAMmers can be avoided by using dCas9 in the place of active Cas9 nuclease.

10.13. TARGETED mRNA TRACKING

The CRISPR-Cas9 can be lured to associate with ssRNA like mRNA in a sequence-specific manner by using PAMmers (Section 5.11); this modification has been termed as RNA-targeted Cas9 (RCas9). The PAMmers are chemically modified to protect the mRNA-PAMmer hybrid regions from RNase H digestion. In addition, the nuclease action deficient dCas9 was used to protect the mRNA from cleavage by the activated HNH domain of Cas9. RCas9 can be modified for labelling and tracking the targeted mRNAs in living cells. For this purpose, the Cas9 is linked to a fluorescent protein (FP) like eGFP and localized in nuclei. The dCas9-FP fusion associates with sgRNA and recognizes and associates with the PAM sequence of the PAMmer paired with the target mRNA. The sgRNA is specific for the PAMmer sequence so that the sgRNA pairs with the PAMmer-mRNA complex and remains associated with the target mRNA during export and localization of the latter. The FP attached to dCas9 enables visualization of the mRNA transport and localization and analysis of their dynamics in living cells. This approach avoids the need for tagging of the targeted mRNAs with genetically encoded tags or by molecular beacons injected into the cells (Nelles et al., 2016).

PAMmer-mediated association of a single Cas9-eGP with each RNA transcript enabled mRNA tracking. Similar association of Cas9-eGP in the UTR region of mRNA does not seem to affect its stability or translation. It might be possible to visualize even single endogenous RNA molecules by using more sensitive detection methods, targeting multiple Cas9-eGP molecules to the mRNA molecule or the use of Cas9 either bearing multiple fluorophores or capable of recruiting them, *e.g.*, according to the dCas9-SAM or SundCas9 scheme (Section 10.3.2.2). In future, the RCas9 method could be repurposed to develop programmable effectors that could control various RNA processing steps; this development might be useful in synthetic biology, disease modelling and even disease treatment (Nelles et al., 2016).

10.14. IDENTIFICATION AND CHARACTERIZATION OF *CIS*-REGULATORY REGIONS

In eukaryotes, there are four types of regulatory regions: (i) promoters, (ii) enhancers, (iii) silencers, and (iv) insulators. Enhancers and silencers stimulate and inhibit, respectively, transcription by interacting with promoters. Insulators, on the other hand, protect genes from regulation by inappropriate regulatory sequences by preventing interactions between inappropriate enhancers/silencers and promoters. The epigenetic states of these regulatory sequences are correlated with the status of transcription and cell phenotype. The CRISPR-Cas9-based genome and epigenome editing tools have been used for intensive investigation of promoter and enhancer sequences. Diagnostic sequences like TATA box and their location close to the transcription start site enable relatively easy identification of promoters. In contrast, enhancers lack such identifiable sequences and are located at megabase distances on either side of the promoters they affect; as a result, the methods currently used for identifying enhancers have false discovery rates of 40-73% (Pulecio et al., 2017).

CRISPR-Cas9 has been used to abolish the functions of promoters and enhancers by inducing indel mutations in them and, thereby, disrupting protein binding and/or protein-DNA interactions. However, large sgRNA libraries are required to systematically interrogate the noncoding regions of a genome. For example, ~18,000 different sgRNAs will be required for mutating merely 715 kb of the non-coding sequences on one side of a gene. Therefore, previous knowledge, such as binding sites for the relevant transcription factors, interactions with gene promoters, etc., is used to narrow down the genomic area to be searched for promoters and, particularly, enhancers. Alternatively, large deletions can be induced using pairs of sgRNAs specific to the sites defining ends of the sequences to be deleted; this allows interrogation of a large genomic region included in a deletion. The CRISPR-Cas9-based approach is easily scalable and permits high resolution definition of enhancers within their native context. This method has permitted the discovery of novel enhancers and super-enhancers, and the unveiling of hitherto unappreciated features of enhancer organization and function. For example, only a limited number of enhancer domains is needed for the enhancer function (Pulecio et al., 2017).

The CRISPR-dCas9-based chromatin remodelling tools have been used in combination with sgRNA libraries to identify new regulatory regions. An advantage of these dCas9-based tools is that they typically induce alterations in chromatin architecture over relatively larger genomic regions, which is expected to reduce the number of different sgRNAs needed for interrogation of a given genomic region for regulatory sequences. The identification and characterisation of regulatory regions is expected to provide targets for epigenome editing to modulate transcription of targeted genes and for gaining insights into the mechanisms of gene expression regulation (Pulecio et al., 2017).

10.15. THERAPEUTIC APPLICATIONS

The programmable transcription activators and repressors can be used for activating and repressing, respectively, specific genes for treatment of certain diseases. For example, the dCas9-SAM system (Section 10.3.2.2) can be used to activate HIV-1 virus present as provirus in the genomes of latently infected human cells. The commonly used combined antiretroviral therapy for the treatment of HIV infections is effective against active HIV-1 but fails to act against the latent virus. As a result, there is resurgence of HIV-1 after withdrawal of the therapy. Therefore, the recent approach, called "Shock and Kill", is to activate the latent HIV-1 by using specific drug activators leading to death of the cells with latent infection due to the cytotoxicity of viral protein accumulation and/or action of the host immune system. But this therapy is nonspecific as it adversely affects normal cells and its effectiveness is less than satisfactory. The dCas9-SAM system was used for highly specific and high level activation of HIV-1 virus present in cells with latent infection using msgRNAs specific for two sequences surrounding the enhancer region in the long terminal repeat region of the viral genome. The reactivation persists over a period and is not likely to be affected by the site of HIV-1 integration and/or the epigenetic state of the site. The risk of off-target action needs to be assessed and the delivery procedure needs to be optimized for therapeutic applications (Zhang et al., 2015).

The CRISPR-Cas9-based transcription repressor systems may also find therapeutic applications. It can be used to create a "synthetic lethal" that may have therapeutic value in some diseases. Some types of cancers are produced due to loss-of-function mutations in specific genes. In such cases, GI maps of various pathways may identify pairs of such genes that cause cell death when both the genes are inactivated, but loss-of-function mutation in only one of them leads to the concerned disease. Once the gene pair is identified, the nonmutated member of the pair can be repressed using dCas9-repressor fusion protein in combination with the sgRNA specific for the gene. This repression will create a 'synthetic lethal' gene that will kill the targeted cancer cells but is not likely to affect normal cells. This strategy is being used for treatment of some cancers using specifically designed drugs to inhibit the targeted protein.

10.16. SPECIFICITY ASSAY FOR PROGRAMMABLE DNA BINDING SYSTEMS

Mali et al. (2013) developed an assay based on CRISPR-dCas9-based transcription activator system for monitoring off-target activity of CRISPR-dCas9/TALE proteins. This approach of off-target monitoring is expected to be less sensitive than the strategies based on scoring of DSBs, but it avoids complications due to the toxic effects of DSBs and the mutagenic effects of DSB repair. In addition, it can permit high throughput monitoring as well as multiplexing. This assay strategy, in very general terms, is as follows. A library of reporter gene constructs comprising the following sequences is created: two invariant sequences, namely, the sequence encoding dTomato fluorescent protein and a minimal promoter to drive this gene, and three variable sequences, *viz.*, two binding sites for CRISPR-dCas9/TALE transcription activators and a barcode. The dTomato protein produces bright red fluorescence that is six-fold brighter than the enhanced green fluorescent protein (eGFP); this makes dTomato an ideal fluorescent protein for imaging of living cells. Each reporter gene construct has a variable 24 bp (A/C/G) barcode downstream of the minimal promoter and two binding sites for programmable transcription activators located upstream of the minimal promoter (Fig. 10.9). One binding site, the *common site*, is constant in all the reporter constructs and binds to the transcription activator used as control. The second binding site, termed as *target site*, is for test transcription factors; it comprises a large number of sequences that differ for one to several base pairs from the target sequences, for which the test transcription factors are designed. The barcode sequences are so diverse that the chances of a barcode to be associated with two target sequences is rare; therefore, it allows reliable identification of each distinct target site.

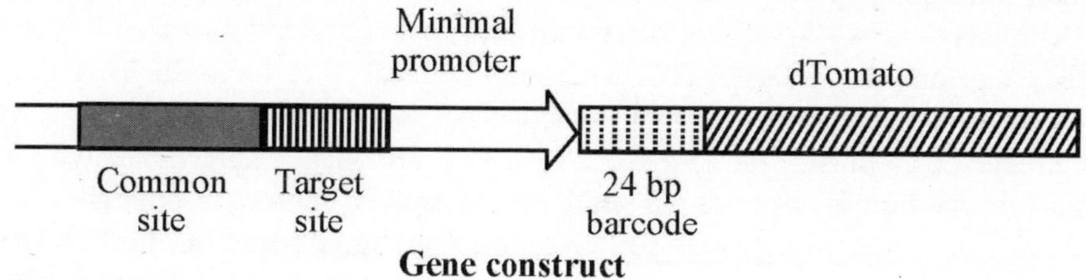

Fig. 10.9. The gene construct designed for off-target activation assay by CRISPR-Cas9 or TALE activators. dTomato, encodes a highly fluorecent protein; 24 bp barcode, variable sequence for identification of transcripts from different constructs; common site, binding site shared by all constructs for binding of control transcription factor; target site, binding site for test transcription factor : sequence varies among different constructs.

The reporter gene construct library is first activated by the control transcription factor, and gene expression is readily detectable as bright red fluorescence of dTomato protein. The cells are harvested, and the RNA transcripts are sequenced and classified based on the barcode. The library is then activated by the test transcription factor and the RNA transcripts are sequenced and classified. It is expected that the control transcription factor will equally activate all the gene constructs in the library. In contrast, the test transcription factor will activate most strongly that gene construct, which has the sequence at its target site that matches the sequence recognized by the TF, and the level of activation will decline with the increasing divergence of the target site from the TF-targeted sequence. The frequencies of transcripts from the control transcription factor experiment are used to normalize those obtained from activation by the test transcription factors. The relative frequencies of the transcripts from different reporter constructs of the library are then subjected to appropriate analyses that provide an insight into the effects of the mismatches for different numbers of bases located at diverse positions in the target binding site. Based on this assay, the authors concluded that CRISPR-dCas9 based transcription activators and 18-mer TALEs could tolerate 1-3 and 1-2 mismatches, respectively (Mali et al., 2013).

10.17. CRISPR BARCODES

In simple words, a DNA barcode is a short DNA sequence that enables unambiguous identification of cells/individuals/species carrying it from other similar entities. Similarly, a CRISPR barcode is a DNA sequence that acquires sequence variation due to the action of CRISPR-Cas9 and can be used to follow cell lineage during development and differentiation. A CRISPR barcode consists of a unique DNA sequence introduced into the genome along with CRISPR-Cas9, which induces mutations and, as a result, changes in the barcode sequence. Since these sequence changes are expected to differ among cells, the barcode sequences can be used to determine cell lineage relationships among cells produced by division of these cells. Typically, each barcode is long enough for CRISPR-Cas9 action and several barcodes may be linked in an array to provide large sequence diversity among the barcodes of the cell population after mutation induction.

Kalhor et al. (2016) used the homing CRISPR system to develop an evolving CRISPR barcode. In this system, a homing gRNA (hgRNA) encoding locus is used as one barcode, and each barcode can store about 5 bits (= 2^5 = 32) of information, *i.e.*, generate 32 different alleles of the barcode. A hgRNA is a type of sgRNA that can target its own locus for Cas9 action. In contrast, a sgRNA is unable to target its own locus because this locus lacks a PAM sequence immediately on the downstream (3') side of the sgRNA spacer region. Therefore, a hgRNA locus for SpCas9 is created by altering its sequence so that it encodes a sgRNA in which the 'GUU' sequence located immediately after the 3' end of the spacer region is changed to GGG. This generates the PAM sequence at the locus so that it can now be targeted by the hgRNA it encodes. The Cas9 action may change the hgRNA spacer encoding sequence of this locus, but this change will be faithfully transferred into the sequence of hgRNA encoded by the mutated locus. As a result, the mutated hgRNA will target the mutant locus and the sequence of the hgRNA locus will continue to evolve with time.

After the desired period of growth and development, the sequences of the CRISPR barcodes are determined by *in situ* single cell DNA sequencing. The barcode sequence information is then used to construct cell lineage dendrogram, which traces the descent history of analysed cells to the common parent cell. It has been estimated that merely seven distinct hgRNA loci can provide unique barcode genotypes for over 12 billion cells (a rough estimate of cell number in a mouse). The evolving CRISPR barcode is superior to simple CRISPR barcode in respect of the following. (i) It requires only one locus, the hgRNA locus, for each barcode, while two (sgRNA and barcode) loci are required for each CRISPR barcode. (ii) It can tolerate mutations in the spacer regions of hgRNAs, while the simple CRISPR barcode system cannot. The evolving CRISPR barcode system, however, suffers from the following three limitations: (i) the hgRNAs soon become inactive due to deletion in their spacer regions, (ii) only limited sequence diversity is generated in each barcode, and (iii) the effects of integration of an array of DSB induction sites (in the form of barcodes) into the genome of an organisms on its growth and development are not yet known (Kalhor et al., 2016).

10.18. FUTURE PERSPECTIVES

The programmable epigenome editing tools, particularly those based on dCas9, have revolutionized epigenetic research as they allow direct modification of specific regulatory sequences as well as of epigenetic marks for assessing their roles in gene expression and cell function. The amenability of CRISPR-dCas9-based epigenome editing tools to multiplexing enables simultaneous investigation of several genomic loci and multiple epigenetic marks, which facilitates dissection of even complex networks of epigenetic regulation. In addition, the delivery of these tools into live mammalian cells is relatively easy. It may be expected that in near future epigenome editing might be applied to live whole organisms. When this happens, it would greatly facilitate investigations about the roles of various epigenetic alterations in growth, development, and disease onset and progression (Pulecio et al., 2017).

The tools and the procedures of epigenome editing need to be further refined to enable their application in various fields of enquiry, especially therapeutics. For example, comprehensive set of rules for designing the most efficient and specific set of gRNAs need to be developed. At the same time, the experimental designs and procedures must be refined and streamlined to account for the background noise generated by various factors of the experimental system other than the variables being investigated, including the programmable effectors used in the study. In addition, robust techniques and tools need to be developed that would enable spatial as well as temporal regulation of transcription activation and/or repression. This is very important since the expression of a large fraction of native genes is endogenously regulated either in time or space for normal growth and development of organisms. Similarly, the frequencies of off-target effects generated by different epigenome editing tools need to be catalogued; this would require standardization of the procedures used for analysis of specificity and off-target effects (Pulecio et al., 2017).

Pulecio et al. (2017) have suggested the following five levels of measurements for analysis of specificity and off-target action of epigenome editing tools: (i) the association of dCas9-based tools with the targeted loci, (ii) appearance of the planned epigenetic marks at the targeted loci, (iii) assessment of the altered expression patterns/levels of the targeted genes, (iv) genome-wide analysis of the targeted chromatin remodelling by techniques like ChIP-seq (ChIP sequencing), bisulfite sequencing, and RNA-seq (RNA sequencing), and (v) monitoring of cell phenotype and behaviour to determine if the targeted epigenetic modifications have been able to generate the desired cellular phenotypes.

10.19. CONCLUSIONS

Transcriptome analyses have revealed the number of transcribed loci to be much larger than that of protein encoding ORFs (open reading frames). The RNA transcripts include a variety of noncoding RNAs like upstream antisense RNAs, enhancer RNAs and other RNAs encoded by intergenic regions. The functions of many of the genes and most of the noncoding RNAs are not well known. This deficiency can now be effectively addressed by ingenious use of programmable gene activators and repressors. The site-directed recombinases, transposases and transcription factors together represent a very promising collection of tools that could be refined for various applications in genome engineering (Doudna and Charpentier, 2014; Gilbert et al., 2014; Puchta, 2017).

The concept and design of SDNs has been used to develop programmable site-directed effectors, *e.g.*, transcription activators and repressors, to modulate gene expression. These effectors are best described as programmable epigenetic effectors, the modification catalysed by them are called epigenetic modifications, and this process is described as epigenome editing. These effectors are generated by replacing the nuclease domain of a TALEN with the desired effector domain. In case of CRISPR-Cas9 system, the effector domain is either linked to the C-terminus of dCas9 or is attached to the sgRNA by RNA hairpin-viral protein (*e.g.*, MS2 coat protein) interaction, the effector domain being fused with the viral protein. In general, a single epigenetic effector induces only partial repression or activation, and it becomes necessary to amplify their effects. The amplification can be achieved in one of the following ways: (i) fusion of multiple effector domains, *e.g.*, VP64, p65 and Rta, to a single dCas9 molecule to yield dCas9-VPR, (ii) targeting multiple sgRNAs in the genomic region, *e.g.*, promoter region, of interest, (iii) integrating multiple copies of a RNA hairpin to which multiple viral protein-effector fusion molecules become attached (dCas9-SAM system is a good example), and (iv) linking multiple copies of effector-antibody fusions to the SunTag peptide fused with dCas9 by antibody-epitope interaction.

The CRISPR-dCas9 system has been modified for imaging of targeted loci *in vivo* by either fusing a fluorescent protein to dCas9 or attaching it to sgRNA via RNA hairpin-viral protein interaction. The fluorescence signal must be multiplied for robust imaging using one of the approaches listed at (ii) to (iv) above. The imaging procedure causes photobleaching of the FP; the recovery from photobleaching effect is faster when the FP is linked to sgRNA than to dCas9. dCas9-based epigenetic tools that respond to light or a ligand have been developed; these tools permit temporal regulation of the epigenetic modifications induced by them. The PAMmer technique allows Cas9 to bind to and cleave single-stranded RNA. This technique has also been adapted to enable imaging and tracking of single RNA molecules, and for isolation of a specific mRNA from the total cellular RNA.

The dCas9-based epigenome editing tools have been applied to activate the expression of viral genomes in latently infected cells and, thereby, causing death of these cells. CRISPR-Cas9 SDN has been used to develop a homing gRNA-based evolving DNA barcode that can be used to determine cell lineage relationships among very large cell populations. It has been used for several other diverse purposes, including identification and characterisation of *cis*-regulatory elements like promoters and enhancers, directed evolution of proteins, and specificity assay for programmable DNA binding systems. It is pertinent to note that epigenome editing studies have so far been confined to animal/human cells and microbes like yeast, while there are few publications with plants. It is difficult to prophesy about what and how many other applications these tools will be put to in the future, but the range and the novelty of these deployments is likely to touch the horizons of human imagination.

An example of potential therapeutic application of programmable transcription factors was provided by Wang et al. (2020) who created human brown-like adipocytes by activating gene *UCP1* (Uncoupling Protein 1) in human white preadipocytes by CRISPR-SAM-gRNA. Transplantation of these cells in mice enabled them to avoid weight gain and diabetes when they were fed high fat diet.

■■■■

11

GENETIC MODIFICATION OF NATURAL POPULATIONS

11.1. INTRODUCTION

Many wild species cause considerable human suffering and economic losses by causing disease, transmitting pathogens and parasites, damaging materials and structures, reducing crop yields, etc. Humans have devoted substantial resources for devising effective measures for their control and several measures, including chemicals, management practices and biological agents have been used for this purpose. These control strategies have met with variable success, but most of the problem causing species continue to impose heavy costs. It has been suggested that genes could be deployed in the noxious natural populations to either eliminate or suppress them to an acceptable level or mitigate their harmful effects. The idea of deploying gene drives based on genetic elements that exhibit non-Mendelian transmission for genetic modification of natural, particularly, mosquito populations was first suggested during the early 1960s and was discussed extensively (Burt, 2003; Esvelt et al., 2014). There are some naturally occurring site-specific selfish elements like homing endonuclease genes (HEGs), group II introns and some site-specific LINES-like transposable elements. Burt (2003) concluded that HEGs were the most suitable for engineering to create gene drives that can invade the targeted population and either suppress, eradicate or merely make it less noxious.

In 2011, gene drives based on HE genes were successfully constructed (Fig. 11.1) and shown to work in mosquitoes (Windbichler et al., 2011) and *Drosophila melanogaster* (Chan, 2011). The concept of gene drive construction based on CRISPR-Cas9 was presented and analyzed in detail by Esvelt and coworkers in 2014; one year later, they were created and evaluated in yeast (Dicarlo et al., 2015) and *D. melanogaster* (Gantz and Bier 2015). In addition, some other types of gene drives have been developed and tested for achieving specific objectives (Section 11.4). However, gene drives based on the CRISPR-Cas9 system are the most promising and are being almost exclusively studied and refined to make them very safe, and highly effective and efficient for various applications.

There are several excellent reviews (listed in Rode et al., 2019) covering various aspects of gene drives, including their effects on natural populations; some of these reviews are Champer et al. (2016), Esvelt et al. (2014), Esvelt and Gemmell (2017); Zentner and Wade (2017), Benedict et al. (2018), Collins (2018); Hammond and Galizi (2018); Marshall and Akbari (2018), Meghani and Kuzma (2018), Scott et al. (2018), Thompson (2018), Brossard et al. (2019), Redford et al. (2019), Rode et al. (2019), Singh (2019), and von Gleich and Schröder (2020).

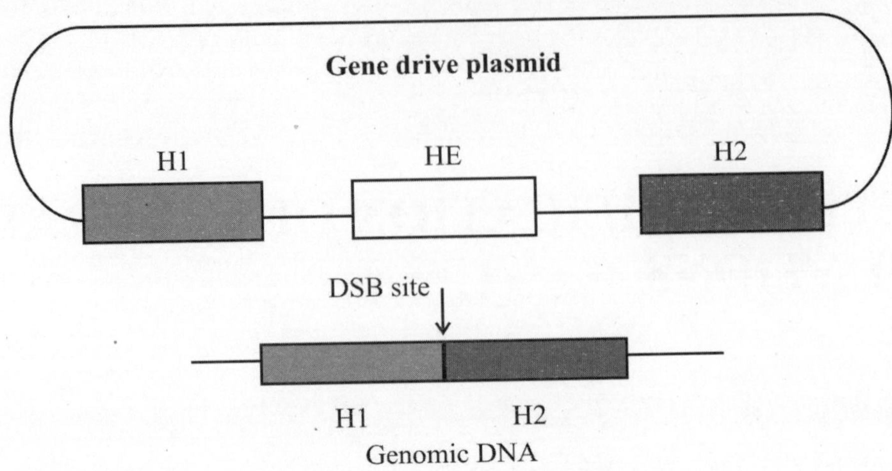

Fig. 11.1. A simplified representation of a gene drive based on homing endonuclease (HE). The drive has the HE expression casette flanked by the homology arms H1 and H2. H1 and H2 are genomic sequences flanking the DSB induced by the HE. H1 and H2 serve as templates for HDR (homology-directed repair) of the DSB.

11.2. GENE DRIVE SYSTEMS

Several different gene drive systems based on diverse mechanisms have been either developed or proposed to be developed for genetic modification of natural populations. These systems include homing-based gene drives, X-shredder (Y-linked meiotic drives), MEDEA (maternal effect dominant embryonic arrest), toxin-antidote underdominance (heterozygote inferiority), and heritable microorganism, *e.g.*, *Wolbachia*-based drives (Table 11.1). Although these gene drive systems offer considerable potential benefits, most of them have not been studied in sufficient detail to permit decision about their suitability for specific applications. The typical homing-based gene drives can efficiently spread in populations so that they can either eradicate, suppress, modify and replace or rescue the target populations. However, they are the least stable of the various drive systems since mutations in the sgRNA and, particularly, Cas9 genes may inactivate the drive. Such mutations may accumulate in eradication and suppression gene drives (but perhaps not in rescue gene drives) since there will be selection for non-functional gene drives. Further, the homing-based drives (Section 11.3) are the most likely to induce drive resistant alleles due to repair of the double-strand breaks (DSBs) by nonhomologous end joining (NHEJ). But these gene drives are the most widely applicable since CRISPR-Cas9 can be readily redesigned to target any genomic sequence of interest (Sinkins and Gould, 2006; Champer et al., 2016; Rode et al., 2019).

The meiotic drives are like homing-based drives in that they can suppress or eliminate populations, and there would be strong selection for resistance alleles. The MEDEA system is more stable and controllable than the homing-based drives, but it remains a challenge to engineer this system in the species of interest. The underdominance gene drive systems are highly stable over time, more likely to remain confined within the target population and can be completely removed from a population if so required. But this system requires release of large proportion of gene drive individuals, and it may be difficult to implement this drive system in species other than *Drosophila*. The heritable microorganism-based system exists naturally as does the translocation-based underdominance system. But modification/creation of these systems for the desired populations and/or applications does not appear to be feasible. All the gene drive systems induce biased inheritance patterns in one of the following two ways: (i) homing, *i.e.*, copying themselves in the wild type homologue at the concerned locus, and (ii) reducing the viability of gametes carrying the wild type allele of gene drive. Biased inheritance pattern refers to increased proportion of the gene drive individuals in the progeny as compared to that expected on the basis of Mendelian inheritance (Sinkins and Gould, 2006; Champer et al., 2016; Rode et al., 2019).

Table 11.1. Salient features of different types of gene drives

Type of gene drive	Architecture	Salient features
Homing-based drives	Based on HEG or CRISPR-Cas9	Nuclease induces DSB in the wild type allele in heterozygotes and HDR copies the drive in the place of wild type allele; CRISPR-Cas9-based drives the most widely used since any genomic site can be targeted
X-shredder	Y-linked meiotic drives	Prevents maturation of gametes lacking the meiotic driver; known in several species; X chromosome in sperms is broken; only Y containing sperms survive; can function only in species having X chromosome
MEDEA	microRNA (toxin) and recoded version of the gene targeted by the microRNA (antidote)	*MEDEA:* maternal effect dominant embryonic arrest system; based on microRNA toxin (disrupts an essential gene in all embryos from MEDEA females) and a tightly linked antidote (a recoded version of the target gene not affected by the microRNA) that is expressed in the zygotes having the MEDEA element; confers frequency-dependent fitness advantage; robust gene drives in *Drosophila*
Toxin-antidote underdominance (heterozygote inferiority)	Combination of a toxin and its antidote	High release threshold necessary for spreading through a population; may remain confined to a local population; demonstrated in *Drosophila*; system cannot carry a payload gene
Wolbachia	Heritable microorganisms, *e.g.*, *Wolbachia*	Cytoplasmically inherited microorganisms; estimated to reside in roughly half of all insects; highly invasive; known to induce male-killing, feminization, parthenogenesis and cytoplasmic incompa-tibility; proposed to be used as gene drive

The homing-based gene drives utilize a sequence-specific nuclease, and drives based on HEGs and CRISPR-Cas9 have been designed and demonstrated. The Cas9-based gene drives have been the most widely studied, and many different types of gene drives have been developed for achieving specific objectives. Gene drives have been demonstrated to work in many species, including yeast, *Drosophila*, at least two species of mosquitoes and mice.

11.3. HOMING-BASED GENE DRIVES

Homing-based gene drives, synthetic gene drives or, simply, gene drives are CRISPR-based genetic devices that are capable of self-propagation by homing. Several different types of gene drives have been designed for achieving various objectives (Table 11.2). The simplest of these drives was designed to induce mutations typically by inactivation of the targeted genes and rapidly propagating these mutations in random mating populations of the concerned species. This strategy consists of expression cassettes for Cas9 and sgRNA specific for the targeted locus flanked by homology arms, *i.e.*, the sequences on either side of the DSB induced by this CRISPR-Cas9 system (Fig. 11.2A). This design yields a simple CRISPR-Cas9 gene drive that has been called 'copy-cat'. The CRISPR-Cas9 encoded by the gene drive will induce a DSB at the targeted genomic site. The homology-directed repair (HDR) of this DSB using the gene drive as template will insert the gene drive at the DSB site, and lead to inactivation of the targeted gene. Once this gene drive becomes integrated in one chromosome of a cell, it will be copied into the homologous chromosome by the same process and yield homozygous gene drive individual. The gene drive will be similarly copied into the wild type homologous chromosomes of the heterozygous progeny produced by mating of gene drive individuals with wild type individuals (Fig. 11.3). This will happen in each generation in all the progeny produced by such mating. As a result, the mutant allele will spread to all the progeny of mutant individuals, and it will eventually replace the wild type allele in the population (Gantz and Bier, 2015; Ledford, 2015a). This method has been referred to as mutagenic chain reaction (MCR) by Gantz and Bier (2015).

Table 11.2. Salient features of different types of homing-based gene drives. Only CRISPR-Cas9 - based drives are described

Type of gene drive	Architecture*	Salient features	Remarks
A. *Invasive or global gene drives: drives that propagate indefinitely in both time and space*			
a. Simple gene drive	Cas9 and sgRNA	Cas9 and sgRNA integrated at the same site	Invasive gene drive; spreads in the population; inactivates the endogenous gene within which it is inserted; basis for mutagenic chain reaction (MCR) method
b. Standard/ conventional gene drive	Cas9, sgRNA and cargo gene	Cas9, sgRNA and cargo genes integrated at the same genomic site	This generic architecture is used for eradication/suppression/genetic modification, etc. gene drives; endogenous gene, within which the drive is integrated is inactivated; the cargo gene generates additional effects
i. Eradication gene drive	Cas9, sgRNA and cargo gene	Generates strong deleterious effect	Designed to eradicate a natural population; cargo gene selected for its deleterious effects; drive individuals are selected against
ii. Suppre-ssion gene drive	Cas9, sgRNA and cargo gene	Exerts mild deleterious effects	Designed to suppress a natural population; cargo gene selected for mild deleterious effects; drive individuals selected against
iii. Rescue gene drive	Cas9, sgRNA and cargo gene	Enhances fitness of individuals	The cargo gene modifies an endogenous gene or produces beneficial effects by itself; drive individuals selected for
iv. Precision gene drive	Cas9, sgRNA and cargo gene	The drive is specific for a sequence unique to the targeted population	The gene drive will be limited to the targeted population; the unique sequence may be created by first releasing an immunizing gene drive in the population
v. Immun-izing gene drive	Cas9, sgRNA and cargo gene	Makes the target gene immune to another drive specific for the wild type sequence of this gene	Cargo gene is a recoded (base sequence changed without affecting the amino acid sequence of the encoded polypeptide) version of the target gene; can be used to protect a population from another drive that recognizes the wild type sequence

Type of gene drive	Architecture*	Salient features	Remarks
vi. Immunizing reversal gene drive	Cas9, sgRNA and cargo gene	sgRNA specific for the sequence changes induced by an earlier drive	Cargo gene is a recoded version of the target gene; it reverses the changes induced by an earlier drive and simultaneously makes the gene resistant to the earlier drive; the target gene function, but not the base sequence, is restored to the wild type
vii. Precision reversal drive	Cas9, sgRNA and cargo gene	sgRNA specific for the sequence changes induced by an earlier drive	Cargo gene is the wild type target gene; reverses the altered sequence to the wild type; best used after an immunizing reversal drive
viii. Integral gene drive	Cas9 and sgRNA (element B) + cargo gene and sgRNA (element A)	Elements A and B integrated in-frame and close to the initiation codons of two different endogenous genes	Endogenous gene function remains unaffected; drive effect only due to the cargo gene; likely to facilitate field testing of gene drives using only element A, which cannot propagate in the absence of element B
B. Local gene drives: drives that propagate for a limited period of time after release			
i. Split gene drive	Cas9 and sgRNA	Cas9 and sgRNA integrated at unlinked genomic locations	Self-limiting since homing of sgRNA requires Cas9 function; Cas9 itself is incapable of homing so that it will be diluted out of the population; basis for daisy chain, integral and allelic gene drives
ii. Daisy chain gene drives	sgRNAs (element C)**; Ca9 and sgRNAs (element B); cargo gene (element A)	Elements A, B and C integrated in different chromosomes so that they are unlinked	A self-limiting gene drive; propagation of gene drive requires all the three elements; element C is diluted out first leading to fading out of element B, and A will be ultimately lost from the population
Daisy-field gene drives	Cas9 and cargo gene (element A), four unlinked sgRNA genes (element B) that target the locus with element A	Element A is driven by any one of the four copies of element B, which itself is nondriving; the two elements are unlinked	The number of sgRNA genes per nucleus is reduced by a factor of 1/2 after each generation so that it is lost after some generations; in few more generations, cargo gene is also diluted out
Daisy-quorum gene drives	Three-element daisy chain drive combined with underdominance, *i.e.,* heterozygote inferiority	The elements organised as in daisy chain drive; in addition, a sequence generating under-dominance is included	After the elements C and B are lost, fitness becomes frequency-dependent so that beneficial cargo genes would become fixed, while detrimental ones will be lost from the population
C. Allelic gene drives	Two sgRNAs in the chromosome with the target gene; Cas9 located elsewhere in the genome	One sgRNA propagates the drive; the other targets the undesirable allele of the selected gene	Cas9 activity is provided in *trans*; it induces a DSB in the undesirable allele; HDR copies the desirable allele in the place on undesirable one; can be used for accumulating desirable alleles of several genes in one line; for use in domesticated species

Type of gene drive	Architecture*	Salient features	Remarks
D. Cas9-based gene drive 'brake'	sgRNA targeting a sequence within the Cas9 gene	sgRNA integrates within and disrupts the Cas9 gene	Prevents spread of all Cas9-based gene drives; it requires Cas9 function in *trans* for homing so that it will spread so long as functional Cas9 is available; of theoretical significance so far
Excision gene drives#	Two sgRNA genes targeting the sequences flanking the gene drive to be excised	The two sgRNAs comprise the excision gene drive	The two sgRNAs generate one DSB on each side of the targeted gene drive, which is deleted; HDR copies the excision drive in its place; can function only when a preexisting gene drive provides Cas9 activity

* All gene drive constructs include at their ends homology arms, *i.e.*, sequences homologous to the genomic regions flanking the double-strand breaks (DSBs) induced by the gene drive encoded nucleases. The homology arms are required for homology-directed repair (HDR) of the DSBs leading to homing (copying into the homologous chromosome) of the gene drive.

** Element C is non-driving, *i.e.*, is not able to propagate itself, since there is no sgRNA specific for its integration site. Therefore, it is rapidly diluted out from the population due to selection and random drift.

This drive can be based on a site-specific recombinase instead of the two sgRNAs. However, the gene drive to be excised must then be flanked by the sequences recognized by the concerned recombinase.

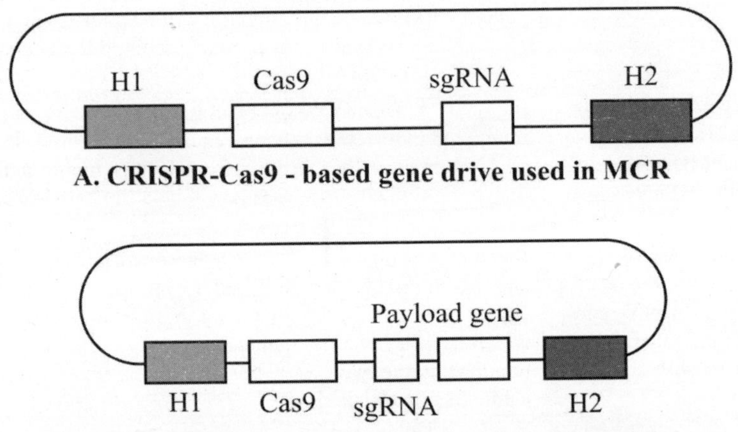

Fig. 11.2. A generalized schematic representation of a CRISPR-Cas9-based gene drive. **A.** Gene drive used for MCR (mutagenic chain reaction). **B.** Gene drive used to spread a payload gene. H1 and H2 sequences homologous to the genomic regions flanking the DSB induced by Cas9 nuclease; Cas9 and sgRNA, gene constructs encoding Cas9 nuclease and sgRNA, respectively; payload gene, the gene to be spread in the population for achieving the desired objective; DSB, double-strand break; MCR, mutagenic chain reaction.

Two designs of gene drives based on CRISPR-Cas9 system are depicted in Fig. 11.2. The gene drive construct may have only Cas9 and sgRNA genes (Fig. 11.2A) or it may have a payload or cargo gene as well (Fig. 11.2B). The cargo gene product performs some specific function, and it is selected for achieving a specific objective by spreading it in the concerned natural population. The gene drive construct present in a suitable plasmid is introduced into early embryos of animal species or in cells/protoplasts of plant species. Inside the cells, Cas9 and sgRNA will be produced and the Cas9-sgRNA complex will induce DSB at the targeted site in the genome. HDR of the DSB using the gene drive construct as template will lead to its integration at the DSB site (Fig. 11.3). subsequently, the gene drive will be copied into the homologous chromosome lacking the gene drive. Thus, a transgenic individual homozygous for the gene drive will be recovered (Fig. 11.4).

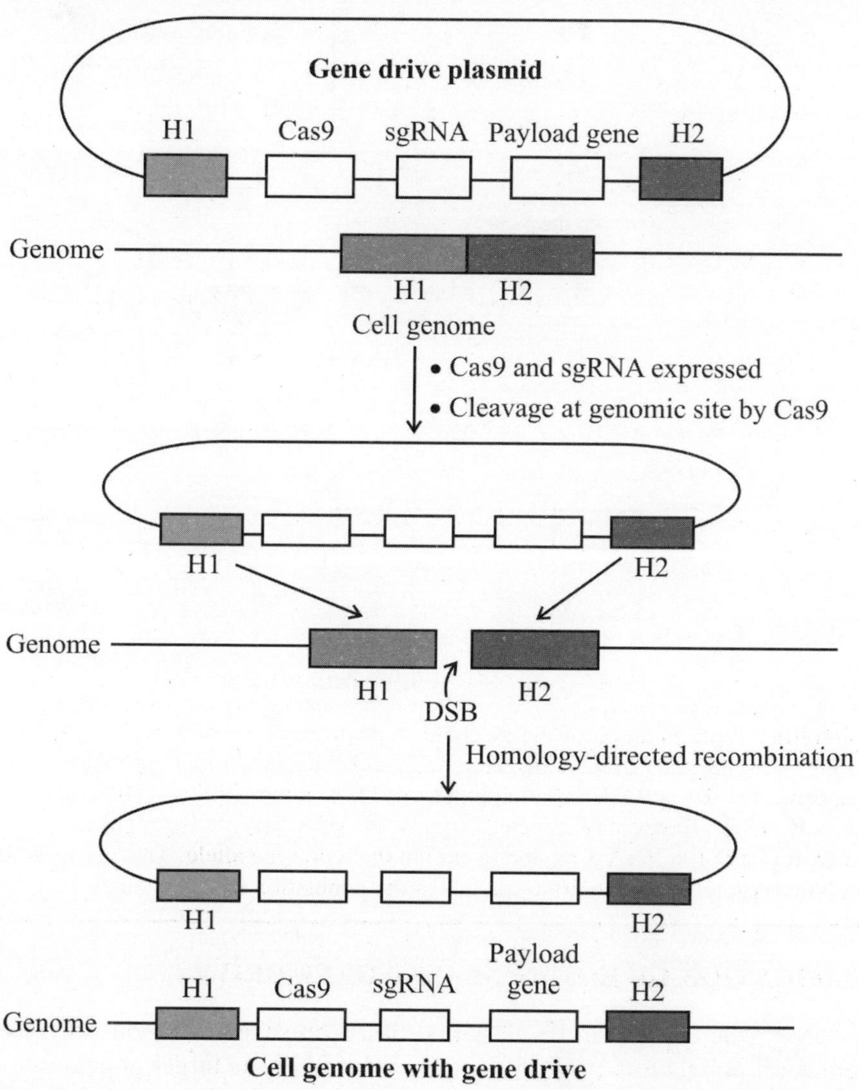

Fig. 11.3. A simplified representation of creation of a transgenic individual with a gene drive. H1, H2, sequences homologous to cell genome DSB site, Cas9, sgRNA, payload gene, expression casettes for Cas9, sgRNA and the gene for achieving the desired objective; DSB, double-strand break.

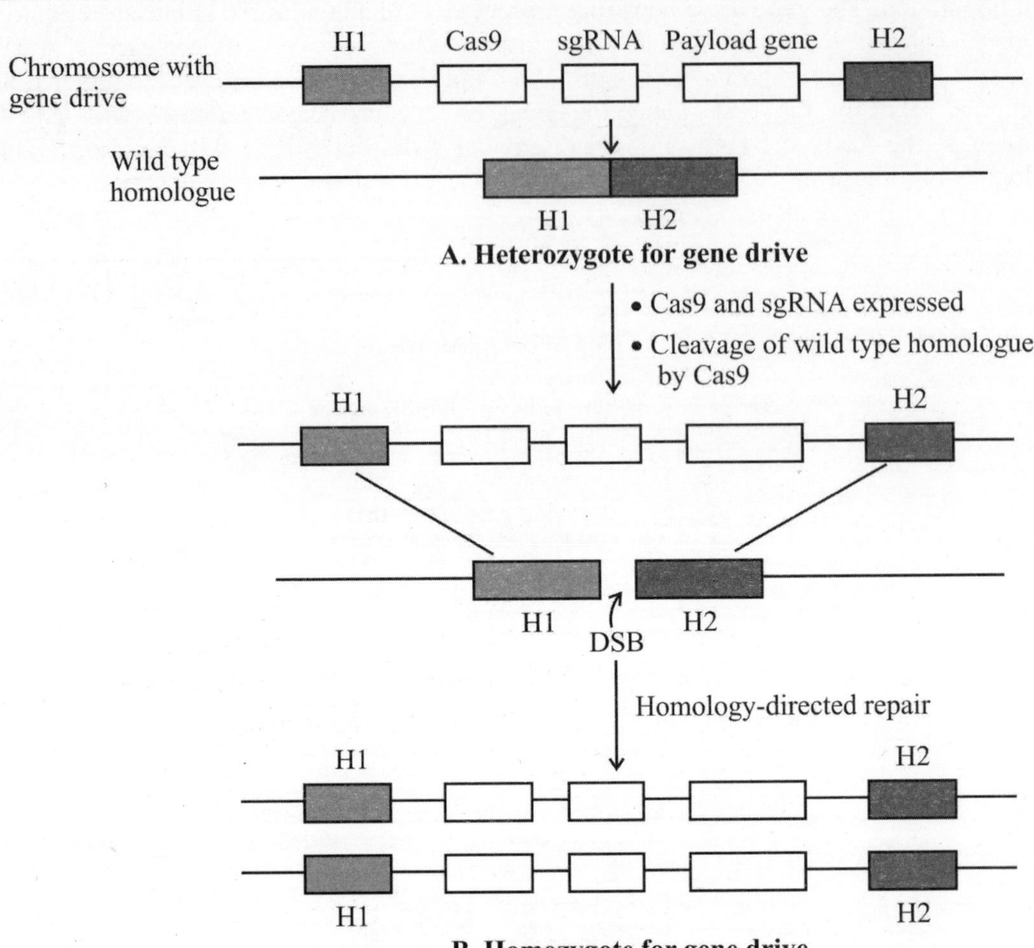

Fig. 11.4. A simplified representation of the mechanism of propagation of a gene drive in a population. **A.** Mating between gene drive transgenic and wild type individuals produces heterozyogotes for gene drive. Cas9 + sgRNA expression induces DSB in the wild type allele of the locus with gene drive. **B.** HDR (homology-directed repair) of DSB leads to integration of the gene drive construct (Cas9 + sgRNA + payload gene) in the wild type allele. This converts the heterozygote into homozygote. The gene drive spreads in the population in this manner.

11.4. CLASSIFICATION OF HOMING-BASED GENE DRIVES

The homing-based gene drives can be classified based on several criteria, such as invasiveness, effect on the natural population, type of change produced in the target population, etc. The gene drives are classified into two groups, *viz.*, global and local, depending on their ability to spread to other populations of the target species or to related species. Global gene drives are also referred to as standard drives or invasive gene drives as they can continue to spread indefinitely over time and space. Each such gene drive has all the elements (Cas9, sgRNA and, where needed, cargo genes) linked to each other and integrated at the same genomic locus (Fig. 11.2). The typical standard gene drives are global drives, but the precision drives and synthetic sequence drives have the standard design although they have been created to remain confined to the targeted population; thus, these gene drives essentially function as local gene drives (Section 11.14.5, 11.14.6). The local drives are designed in such a way that they typically remain confined in both space and time since they are designed to get diluted out of the population over time. Generally, the different elements of such drives are separated and integrated at selected sites in different chromosomes; these drives are described in some detail in Section 11.14.8.

From the viewpoint of conservation biology, gene drives can be classified as eradication drives, suppression drives, and rescue drives based on their effects on the fitness of the gene drive individuals and, ultimately, the targeted population. The eradication and suppression gene drives exert strong and mild deleterious effects, respectively, on fitness either by inactivation of carefully selected endogenous genes or by introducing appropriate cargo genes. For example, such gene drives may target genes necessary for fertility or viability to achieve population suppression. These drives can be used to protect human health, control agricultural pests and eradicate invasive species that threaten ecosystems. Gene drive deployment is expected to be cheaper than traditional methods for population management. The rescue drives, on the other hand, enhance the fitness of individuals and populations by introducing beneficial cargo genes/mutations in endogenous genes or sequence corrections of exiting deleterious gene mutations. Therefore, rescue drives can help save endangered populations and facilitate introduction of individuals from other regions into an endangered population. These drives could increase tolerance to an abiotic stress, generate resistance to a pathogen or pest, decrease susceptibility of useful insects like honeybees to certain chemical insecticides, etc. (Rode et al., 2019).

Based on their effects on genomes of the target populations, gene drives can be classified as modification drives, reversal drives, and immunizing drives. The modification gene drives introduce genetic changes in the genomes of target species. The eradication, suppression and rescue drives are all modification drives. But some modification gene drives can modify certain characteristics of the target species without affecting its fitness so that it becomes less harmful (or more useful) to humans, *e.g.*, making mosquitoes incapable of transmitting the malarial parasite to humans. Reversal drives, on the other hand, undo the genetic changes induced by the modification drives that were used earlier and restore the wild type function and even sequence of the concerned gene. The immunizing gene drives replace the wild type sequence of the target gene with recoded sequence so that the gene becomes immune to gene drives targeting the wild type sequence (Section 11.14.1). In addition, inactivation and excision gene drives inactivate (Section 11.14.3) and excise or remove (Section 11.14.4), respectively, the existing gene drives introduced earlier in the population.

11.5. ARCHITECTURE OF HOMING-BASED GENE DRIVES

In the simplest form, a gene drive comprises the expression cassette for a site-directed nuclease (SDN) flanked by homology arms required for HDR; initially the SDN was a homing endonuclease (Fig. 11.1) but now CRISPR-Cas9 seems to be exclusively used. The homology arms flanking the SDN genes consist of the genomic sequences located on either side of the DSB induced by the SDN encoded by the gene drive. A cargo gene may be included in the gene drive if and where needed. The cargo gene is carefully selected to bring about the desired change in the characteristics of the targeted population. The cargo gene may increase (rescue drives) or decrease (eradication and suppression drives) fitness of the population, reduce its harmfulness to humans, etc. In addition, a marker gene like *DsRed* (red fluorescent protein gene from *Discosoma* sp.) may be added to facilitate phenotypic tracking of the gene drive in the target population (Fig. 11.5).

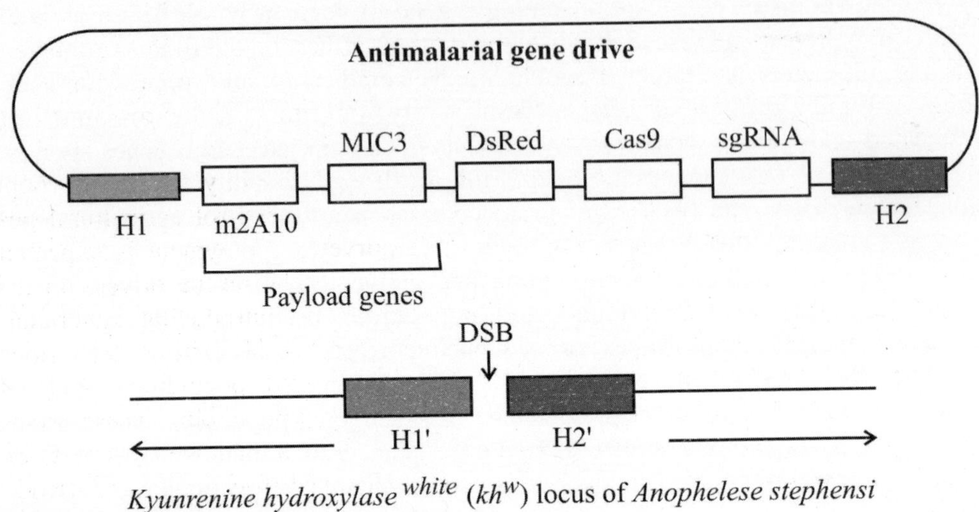

Fig. 11.5. A simplified schematic representation of an antimalarial gene drive based on CRISPR-Cas9. H1, H2, sequences homologous to the regions H1' and H2', respectively, flanking the DSB induced by CRISPR-Cas9 in the *Kh*^{w} locus of *A. stephensi,* m1C3 and m2A10, th single chain antibody gene constructs for resistance to *Plasmodium falciparum;* DsRed, eye colour marker gene construct. Cas9, Cas9 gene construct driven by *vasa* promoter; sgRNA, sgRNA gene construct with *U6A* promoter. The gene constructs *m1C3* and *m2A10* represent the payload genes (based on Gantz et al., 2015).

Generally, gene drives are designed to target a single locus, and all the elements of the drive, *viz*., Cas9, sgRNA and cargo gene, are linked together and located at the same locus (Fig. 11.2B). Multiple loci can be simultaneously targeted by using as many gene drives, each of which comprises Cas9, sgRNA and, if needed, cargo genes; in this case, each drive independently targets one genomic locus. Alternatively, Cas9 gene may be included in a single gene drive, while the remaining gene drives have only sgRNA and, where needed, cargo gene. In this case, the drive with Cas9 gene provides Cas9 activity in *trans* to the other drives. The second scheme (one drive provides Cas9 activity to all the drives) offers certain advantages, including inactivation of all the gene drives by excision/inactivation of the single gene drive with the Cas9 gene. The allelic gene drive is a variant of this scheme. This drive has Cas9 gene and two sgRNA genes; one sgRNA propagates the drive, while the other targets a nearby locus to induce conversion of an undesirable allele into a desirable one (Section 11.10.5). The gene drive architecture has been further modified by separating the Cas9, sgRNA and cargo (when included) genes into unlinked elements to create different daisy drives (split, daisy chain, daisyfield, and daisy quorum gene drives) for achieving specific objectives.

11.6. SELECTION OF GENOMIC SITES FOR GENE DRIVE INTEGRATION

It is desirable to integrate gene drives at such genomic sites that do not adversely affect the gene drive function. In theory, the integration of gene drive in the genome should not reduce the fitness of gene drive individuals, especially in the case of rescue drives. But in case of eradication and suppression drives, the gene drive can be integrated within a gene, inactivation of which imposes mild to strong fitness cost, but the gene drive heterozygotes should show either no reduction or only a small reduction in fitness so as to allow gene drive spread in the population. For this reason, the expression of nuclease genes should be confined to the germ cells only (Section 11.15).

It may be pointed out that neutral genomic sites are likely to show sequence variation, which could affect CRISPR-Cas9 action and give rise to gene drive resistance in the population. It has been suggested that the target population should be surveyed for sequence variation in the proposed genomic regions for gene drive integration and a site with little or no variation should be selected. However, the conserved sequences may be expected to be functionally important and, therefore, could generate detrimental effects after their disruption. A proposed global gene drive, called integral gene drive (IGD), has the design that would remove this dilemma. In IGD, the gene drive components are integrated within the coding sequences of genes in such a way that the functions of the targeted genes are not affected. Thus, IGD could be targeted into highly conserved sequences without creating an adverse effect (Nash et al., 2019).

11.7. SPREADING OF GENE DRIVES IN NATURAL POPULATIONS

When gene drive individuals, *i.e.*, individuals homozygous for a gene drive (*GD GD*), are released in a natural population, the gene drive rapidly spreads in the population. Eventually, all the individuals of the population acquire the gene drive and the gene drive is said to have reached fixation (Fig. 11.6). Since the proportion of released individuals will be much smaller than the wild type individuals (- -) present in the population, most of the mating will be between gene drive and wild type individuals, *i.e.*, *GD GD* x - -. The genotype of wild type individuals is depicted as - - since they will have no allele for the gene drive (*GD*) at the concerned locus. All the progeny zygotes from such mating will be initially heterozygous for the gene drive, *i.e.*, *GD* -. But during embryonic development, the nuclease will be expressed especially in the germ cells and will induce DSBs in the wild type homologous chromosomes. Ultimately, the gene drive construct will be copied into this chromosomes by HDR of the DSBs so that these individuals would become homozygous for the gene drive (*GD GD*).

In the case of normal nuclear genes, the progeny from *GD GD* x - - mating will remain heterozygous (*GD* -) and will transmit equal frequencies of the two alleles to their progeny. Thus, in the case of a gene drive, the segregation becomes biased toward the gene drive allele, and this abnormal transmission pattern is often referred to as super-Mendelian inheritance. In fact, the wild type allele may have one of the following three fates in gene drive heterozygotes: (i) it may not be cleaved by the nuclease and remain unchanged, (ii) it may be cleaved and repaired by

HDR so that it is converted to the gene drive allele, and (iii) it may be cleaved but repaired by NHEJ, which would generate gene drive resistant allele (Section 11.15). The frequency of conversion of wild type allele to the gene drive allele varies from 14 to nearly 100% depending on several factors, including the biological species, the genetic background and sex of heterozygotes, the number of sgRNAs included in the gene drive construct, the tissue and the timing of expression of nuclease, the genomic location of the targeted site, etc. (Rode et al., 2019).

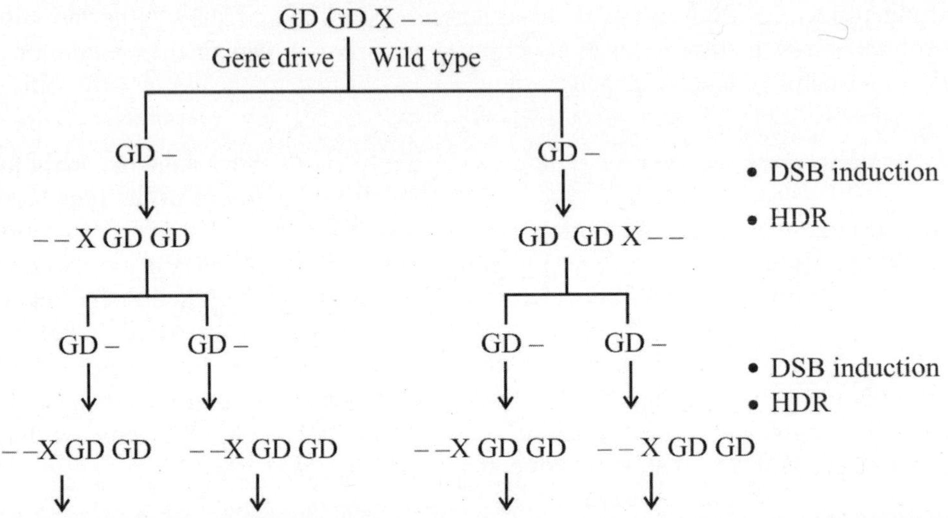

Fig. 11.6. A simplified representation of the mechanism of gene drive invasion in a population. Mating between individuals with gene drive and wild type individuals will generate progeny heterozygous for the gene drive. This progeny will become homozygous for the gene drive following DSB induction and HDR. Thus, in each generation, gene drive will spread in the population till some wild type individuals are present. There will, of course, be other types of matings. *viz.*, GD GD x GD GD and --x-- but they will not affect the ultimate consequence. GD, gene drive, -, gene drive absent; DSB, double-strand break; HDR, homology-directed repair.

In the next and the subsequent generations, mating between gene drive and wild type individuals will generate gene drive heterozygotes (*GD* -), which will eventually become homozygous for the gene drive. In this manner, the gene drive will rapidly spread in the populations and eventually all the members of the population will acquire the gene drive. In fact, there will be mating among gene drive individuals (*GD GD* x *GD GD*) as well as among the wild type individuals (- - x - -). But these mating will have little impact on the ultimate consequence of gene drive spread. It is assumed here that the gene drive allele imposes no fitness cost, the target natural population is random mating, *i.e.*, there is no discrimination against the gene drive individuals during mating, and there is 100% conversion of the – allele to *GD* allele. But one or more of these assumptions may fail in most of the cases. For example, the rate of conversion of wild type to gene drive allele shows considerable variation and the gene drive alleles generally impose some fitness cost. It has been estimated that with a conversion rate of ~90% and a 30% reduction in fitness, the gene drive allele would eventually become fixed, *i.e.*, attain the frequency of 1.0. This situation contrasts with the normal nuclear genes, which can reach fixation only if they increase the fitness of individuals carrying and expressing them.

The above can be explained in simple population genetics terms as follows. In random mating populations, gene and genotype frequencies follow the Hardy-Wienberg law. Accordingly, a gene with two alleles (*A*, *a*) produces the genotypes *AA*, *Aa* and *aa* in frequencies p^2, $2pq$ and q^2, respectively, where p and q are the frequencies of alleles *A* and *a* in the population. The above equilibrium frequencies remain constant over generations unless evolutionary forces like mutation, selection, etc. disturb them. In a population in which a small number of gene drive individuals are released (generation, g_0), *GD GD* x *GD GD*, *GD GD* x - - and - - x - - mating will take place. These will produce *GD GD*, *GD* – and - - progeny in the next generation (g_1 generation) in the proportion p^2, $2pq$ and q^2, where p and q are the frequencies of *GD* and – alleles in the g_0 generation. However, gene drive will convert the *GD* – heterozygotes into *GD GD* homozygotes. This will reduce the frequency of – allele to q^2 in generation g_1 as compared to q in g_0 generation. In the same way, the frequency of – allele will decline to q^4 in g_2, q^8 in g_3, and so on, leading to a rapid drop in the value of q. For example, if the frequency of – allele were 0.9 (the proportion of drive individual was 10% of the natural population) in g_0, it will be reduced to 0.81 in g_1, 0.66 in g_2, 0.43 in g_3 and so on; it will be reduced to 0.000002 (= 0.0002%) in merely seven generations. Even if the proportion of released individuals were only 0.1% of the natural population, *i.e.*, value of q was 0.999 in g_0, the value of q will drop to 0.003 in only 14 generations.

11.8. FACTORS AFFECTING THE RATE OF SPREAD OF GENE DRIVES

The rate of gene drive spread is affected by several factors, including the timing of gene drive expression, genetic features of gene drive, *e.g.*, the effect of *GD* allele on fitness, the frequency of conversion of the wild type (–) allele to *GD* allele by HDR, and the various features like mating system, generation time, social structure, etc. of the target population. In case of eradication and suppression drives, expression of the nuclease gene should occur in gonads or late in the life cycle since these drives reduce fitness. In contrast, the expression of rescue drive nuclease should occur early so that there is an increase in the fitness of gene drive individuals and the drive spread is promoted. The timing of gene drive expression can be regulated by using appropriate promoters for the gene drive genes. In theory, random genetic drift is generally not important in large populations, but it plays an important role in small fragmented populations. Therefore, release of higher frequencies of gene drive individuals in small populations increases the likelihood of eventual fixation of gene drives. However, rescue gene drives ultimately reach fixation irrespective of population size (Rode et al., 2019).

The drive spread would also be affected by target sequence polymorphism, level of inbreeding and the extent of genetic background variation present in target population. Target sequence variation will be particularly important if it involves the PAM sequence or the seed region (five bases immediately upstream of the PAM sequence) since this would abolish cleavage by Cas9. Such sequence variation is known to exist in natural populations of at least some species. In addition, the genetic background is known to exert a notable influence on the rate of gene drive propagation. Another relevant factor is the level of inbreeding in the targeted natural population. For example, natural populations of mosquitoes and mice show substantial levels of inbreeding, which will significantly reduce the rate of gene drive spread. The negative impact of inbreeding can be reduced by repeated release of gene drive individuals in high volumes to saturate the targeted natural population. Further, gene drive might adversely affect the mating fitness of gene drive individuals. This would limit the rate of gene drive spread and might require releases of larger size than would be necessary otherwise. In addition, the gene drive may be only partially copied during the HDR of DSBs; this would impose significant fitness costs on the affected individuals and would reduce the spread of the gene drive (Drury et al., 2016; Zentner and Wade, 2017; Rode et al., 2019).

In case of eradication and suppression drives, the fitness of gene drive heterozygotes is superior to that of gene drive homozygotes. Therefore, gene conversion only in gonads promotes the spread of these gene drives, while conversion in zygotes would favour the spread of rescue drives. Gene drives are much less effective in species with mixed modes of reproduction, *e.g.*, both outcrossing and self-fertilization, and sexual as well as asexual reproduction, which are often found in invasive species. Gene drive spread is much faster on time scale in species with short life cycles than in those with long life cycles, although the number of generations involved remains the same. When populations are fragmented either in space or over time, the speed of gene drive spread is likely to be reduced. Social structures tend to limit random mating in populations, which may adversely affect gene drive spread (Rode et al., 2019).

11.9. EFFICACY OF GENE DRIVES

Efficacy of a gene drive may be described as its homing efficiency, *i.e.*, the frequency of drive being copied into the wild type chromosomes in gene drive heterozygotes. Homing efficiency of gene drives varies widely since it is affected by several factors like the tissue and timing of expression of Cas9 nuclease, the genomic location of the targeted site, evolution of resistance to the gene drive, etc. Resistance to a gene drive may be due to the preexisting sequence polymorphism or it may arise due to NHEJ-mediated repair of the DSBs induced by the drive (Section 11.15). In addition, behavioral resistance to gene drives may be generated by discrimination against gene drive individuals during mating. The frequency of NHEJ varies among species, among and even within populations of a single species; it may also differ between gene drive heterozygotes and the wild type homozygotes. Further, gene conversion in gonads only would promote the spread of eradication and suppression gene drives, while conversion in zygotes would favour the spread of rescue drives (Rode et al., 2019).

11.10. APPLICATIONS OF GENE DRIVE

Gene drives are powerful genetic devices for implementing desired genetic alterations in random mating natural populations. Several different types of gene drives have been designed for achieving diverse objectives by either introducing an appropriate transgene into a population, inactivating selected endogenous genes, inducing desired mutations in the target genes or eliminating undesired mutations from endogenous genes. Gene drives can be used to achieve many different objectives, some of which are briefly discussed here (Table 11.3).

Table 11.3. Various applications of gene drives

Application	Type of gene drive	Remarks
Control of disease vectors:		
i. Eradication	Eradication gene drive	May be more effective over time than gene drives for parasite/pathogen resistance; drive targeting female fertility gene being field-tested in Brazil
ii. Resistance to parasite/ pathogen	Modification gene drive	Drive with single-chain antibodies makes mosquitoes incapable of transmitting malarial parasite and degue-2 virus to humans; tested in laboratory
Control of invasive species	Eradication/suppression gene drives	Gene drives proposed to be developed for black rats in New Zealand and some other invasive species in other countries; might affect food web and ecosystem; suppression drive developed for spotted-wing *Drosophila* in U.S.A.
Rescue of endangered populations	Rescue gene drives	Drives carry beneficial cargo genes like those for resistance to biotic and abiotic stresses; might affect food web and ecosystem
Promotion of agricultural production	Typically, modification drives	Replacing genes for resistance to pesticides or herbicides with those for sensitivity in pests or weeds; protected areas will be needed for the development of inoculums for repeated releases of the sensitive individuals
Development of desirable strains/lines	Allelic gene drives	For replacing undesirable alleles with desirable ones at one or more loci; desirable animal and crop strains/lines may be developed rapidly and cheaply
Basic research	Standard gene drives with the desired gene as cargo	Strains/lines can be created with specific combinations of many genes in a short period of time (1-2 years); very helpful in evolutionary studies; this application has been termed as 'active genetics'

11.10.1. Controlling disease vector populations

Many diseases, such as malaria, dengue, yellow fever, chikungunya, trypanosomiasis, leishmaniasis, etc. are transmitted by insect vectors. The incidence of such diseases can be minimized by suppressing the vector populations by chemical, biological or other measures. For example, mosquito populations were successfully suppressed by release of sterile mosquitoes produced by chemical or radiation treatments generally in numbers that are substantially larger than the targeted insect populations. Gene drives provide a means to minimize/eliminate the risk of disease transmission by making appropriate genetic changes in the insect vectors. In case of many diseases, transgenes that can be introduced into vectors or the vector genes that can be inactivated for interfering with disease transmission have been identified. Alternatively, the concerned insect vector population might be eradicated using an eradication gene drive. Once the disease is eliminated from the human population, pathogen-free vector population could be reintroduced either from laboratory or an island (Esvelt et al., 2014).

Gantz et al. (2015) designed a CRISPR-Cas9-based gene drive with cargo genes for the single-chain antibodies (scFvs) m1C3 and m2A10 (Fig. 11.5). These antibodies are specific for the ookinete protein Chitinase 1 (miC3) and the circumsporozoite protein (m2A10) of

Plasmodium falciparum, the human malarial parasite. *Anophelese stephensi* is a vector of *P. falciparum* in India (mainly in urban areas) and Africa. The salivary glands of transgenic *An. stephensi* expressing the scFvs m1C3 and m2A10 were free from sporozoites of the malarial parasite. As a result, the transgenic mosquitoes cannot transmit the malarial parasite to humans. This antimalarial gene drive has the expression cassettes for m1C3 and m2A10 scFvs, DsRed, Cas9 and sgRNA, and is flanked by sequences homologous to the DSB site in *kynurenine hydroxylase*^{white} (*kh*^{w}) gene of *An. stephensi*. The DsRed is a dominant marker that allows detection of transgenic larvae by fluorescence monitoring. The use of *vasa* promoter for the Cas9 gene limits its expression to germline in both males and females. A major limitation of this gene drive system was that progeny from mating between gene drive females and wild type males showed almost Mendelian transmission of the gene drive. It was suggested that this limitation might be resolved by using a male germ-line specific promoter for the Cas9 gene.

The sterile insect technologies for mosquito population control require releases at inundative levels, *i.e.*, in numbers substantially larger than the targeted populations. But in case of gene drive-based strategies, the number of released individuals would be much smaller, and in some cases, it can be even one 0.1% of the population. This would greatly reduce the time, effort and the cost involved in the vector population management (Gantz et al., 2015). In addition, Hammond et al. (2016) have identified a gene whose disruption leads to recessive female sterility. A gene drive designed to inactivate this gene showed transmission rates of 91.4 to 99.6% in caged populations; this drive could be used to suppress mosquito populations.

11.10.2. Agricultural productivity and sustainability

Gene drives can be useful tools for increasing agricultural productivity, safety and sustainability. The agricultural insect pest and weed species have become resistant to common insecticides and herbicides, respectively. For example, in U.S.A., western corn rootworm has developed resistance to Bt toxins, and horseweed and pigweed populations have become resistant to the eco-friendly herbicide glyphosate. Gene drives can be used to replace the resistant alleles with the respective susceptibility alleles in the pest and weed populations, and thereby facilitate their control. In addition, gene drives could introduce into these species such genes/genetic changes that would make them susceptible to novel pesticides and herbicides that are more specific, safer to use and/or readily biodegradable.

It may be pointed out that the gene drive individuals will have to be released in suitably large numbers and pest control measures cannot be used during the period of gene drive spread in the population. Therefore, the gene drive individuals will have to be raised in isolated populations protected from pesticides/herbicides from where they could spread into the surrounding fields. In addition, gene drives could be used to target genes related to reproduction or general fitness of storage grain pests which are generally confined to food storage facilities (Esvelt et al., 2014; Zentner and Wade, 2017). But repeated releases of susceptibility gene drive individuals will be necessary and this may be a challenging task in many situations. An alternative strategy could be to deliver a lethal cargo gene in the pest/weed population through a gene drive integrated at a neutral genomic site. The cargo gene would be driven by a suitable chemically inducible promoter. This chemical would be sprayed in the field after sufficient number of generations to allow the gene drive to spread to saturation in the target population. An alternative strategy could be to deliver a lethal cargo gene in the pest/weed population through a gene drive integrated at a neutral genomic site. The cargo gene would be driven by a suitable chemically inducible promoter. This chemical would be sprayed in the field after suffici-ent number of generations to allow the gene drive to spread to saturation in the target population.

11.10.3. Control of invasive species

Global economic activities often lead to introduction of invasive animal and/or plant species into different areas of the world. Such species often cause suppression or even annihilation of some native animal/plant species especially in isolated ecosystems like small islands. In addition, invasive species inflict considerable economic losses. For example, brown rats cause annual loss of US $19 billion to United States of America, while top 10 invasive species cause total loss of about US $42 billion annually. Eradication or suppression drives could be designed to eradicate or suppress populations of such invasive species (Esvelt et al., 2014). A gene drive has been developed to control the invasive spotted-wing *Drosophila*, a fruit fly species native to Asia; it damages cherry fruits in California's farms çausing annual loss of US $700 million (Table 11.4).

Table 11.4. Some examples of gene drives being developed for eradication of invasive species and rescue of endangered species (based on Rode et al., 2019)

Country	Target species	Gene drive	Remarks
New Zealand	Black rats	Eradication drive	It is a predator of many animal species and seeds of plant species; gene drives developed in house mice; risk, gene drive may spread in other rodent species and countries
U.S.A. (Hawaii)	*Culex quinquefasciatus*	Rescue drive; proposed to be developed	Vector of avian malarial parasite (*Plasmodium relictum*), to which endemic birds are susceptible
South America, Europe, Africa, Asia, Australia	Common ragweed (*Ambrosia artemisiifolia*)	Suppression drive; proposed to be developed	Causes allergies in humans; gene drives targeting orthologous *Rht* genes are proposed with a view to suppress this invasive weed species
Many	Amphibian species to be rescued	Rescue drives; proposed to be developed	An MHC allele increases chances of survival of individuals infected by the chytrid fungus, *Batrachochytrium dendrobatidis*; drive proposed to be designed to spread this MHC allele, preferably, without affecting the endogenous MHC alleles
U.S.A. (California)	Spotted-wing *Drosophila*	Suppression gene drive developed	It damages cherry fruits; causes annual loss of US $700 million

The potential risks due to eradication/suppression gene drives are spread of the drive into related species due to rare interspecies mating, and into the original source populations of the invasive species. The risk to related species might be minimized by developing precision drives, while the native source population could be protected by an immunizing drive (Section 11.14.1). However, care will have to be taken to preclude the inadvertent spread of immunizing drives into the invasive populations themselves. In addition, some improved gene drives like one of the daisy drives could be used. But implementation of any of these approaches should be preceded by extensive discussions among experts from various relevant fields, concerned governmental and nongovernmental organizations, and the public at large (Esvelt et al., 2014).

11.10.4. Rescue of natural populations

Some natural populations become endangered due to various reasons, including presence of some deleterious gene mutations, susceptibility to some new pathogens or pathogen races, etc. Gene drives can be created to rescue such populations and save them from extinction (Section 11.4; Table 11.4). It is feared that rescue gene drives could have a destabilizing effect on the food web as they might help some endangered species to emerge as invasive species.

11.10.5. Animal and plant breeding

The use of gene drives for improvement of domestic animals and field crops has but limited scope since these species have long been subjected to systematic breeding leading to the development elite breeds/lines. These breeds/lines are subjected to controlled mating and random mating is seldom followed. Recently, allelic gene drives were designed to facilitate controlled conversion of undesirable alleles into desirable alleles at one or more loci. The Cas9 expression cassette of a CRISPR-Cas9 gene drive (Fig. 11.2A) can be separated from the sgRNA cassette and placed at an unlinked genomic location to create a split gene drive (Fig 11.7A). In this case, the Cas9 activity is provided in *trans* to propagate the sgRNA cassette into the wild type homologous chromosomes. The split gene drive is refined as an allelic gene drive by adding a second sgRNA expression cassette; this sgRNA is specific for a genomic region that has a gene with one desirable and another undesirable allele. In case, the base(s) responsible for the allelic difference are such that they make the desirable allele resistant to CRISPR-Cas9 cleavage, *e.g.*, by altering a PAM sequence, the sgRNA is designed to target this sequence; this yields a specific allelic gene drive. As a result, the undesirable allele in the homologous chromosome would be cleaved and DSB repair by HDR would copy the desirable allele sequence in place of the undesirable allele (Fig. 11.7B; Guichard et al., 2019).

But in many cases, the bases critical for allelic differences may not confer on the desirable allele resistance to Cas9 cleavage. In such cases, a genomic site that is cleaved in the undesirable allele but not in the desirable allele is identified in the vicinity of the sequence involved in the allelic difference. Therefore, the second sgRNA of the allelic drive is specific for this neighbouring site; this drive design has been referred to as general allelic gene drive. The cleavage of the undesirable allele would be followed by HDR and conversion of the undesirable allele to the desirable one (Fig. 11.7C). The allelic gene drives have been shown to be efficient in conversion of Notch wing alleles in *Drosophila*. These gene drives may be useful in combining favourable alleles for different traits into a single line to improve complex traits like yield, stress resistance, etc. of crop plants, and to reverse pesticide resistance of agricultural pests. In case of animals, the allelic gene drives would facilitate the creation of complex genotypes useful in biomedical and basic research (Guichard et al., 2019).

11.10.6. Basic studies

Gene drives can be used to facilitate basic research like studies in evolutionary biology. These genetic devices will greatly facilitate introduction of desired genes from related species into the genome of selected species either in addition to or in the place of selected endogenous genes. This could help find answers to specific questions pertaining to evolutionary biology. Conventional breeding methods would require many years for development of strains with specific combinations of many genes in species like mice. The use of gene drives would achieve the same objective in about one or two years. Some researchers are already using gene drives to develop lines/strains with specific gene combinations for use in various fields of basic research. This application of gene drives has been termed as 'active genetics.'

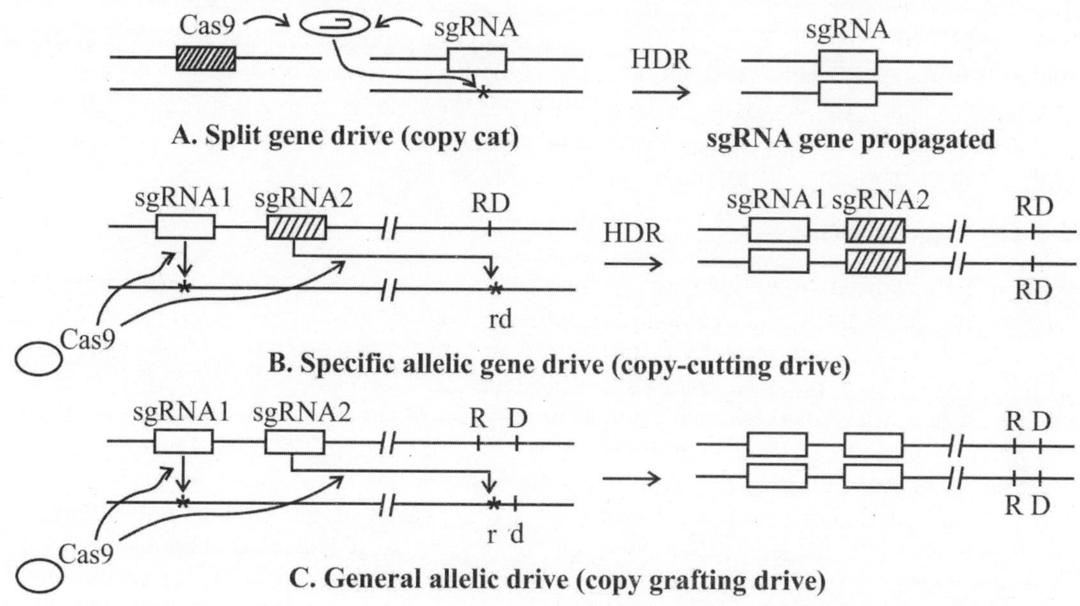

Fig. 11.7. Split and allelic gene drives. **A.** In split gene drive; the Cas9 and sgRNA expression casettes of a MCR drive are located at separate genomic locations. Cas9 supplied in *trans* allows sgRNA propagation. **B.** Specific allelic gene drive; a second sgRNA is included in gene drive. The second sgRNA targets a sequence of target gene that distinguishes the desirable and undesirable alleles of the gene; it also makes the desirable allele cleavage resistant. **C.** General allelic drive : the second sgRNA targets a sequence in the desired gene, which is cleaved in the undesirable allele but not in the desirable allele. This target sequence is close to the site responsible for the allelic differences, but is cleaved in both the alleles. RD, desirable sequence resistant to cleavage; rd, undesirable sequence susceptible to cleavage; R, cleavage resistant sequence, r, cleavage susceptible sequence; D, desirable allele sequence (cleavage resistant); d, undesirable allele sequence (cleavage sensitive). In all the cases (A, B, C), Cas9 activity is provided in *trans*.

11.11. GENE DRIVE MONITORING

The presence of gene drives should be monitored in both laboratory and natural populations following their release for either field testing or population control/replacement. The monitoring is necessary to generate data on spread and persistence of the gene drives in natural populations, and their unintended introduction into other populations or related species. The presence of gene drives can be detected either by a fluorescent marker included in the gene drive architecture (Fig. 11.5) or by genome sequencing. The marker gene generates a phenotype that is not present in the wild type individuals and enables rapid and inexpensive detection of gene drive individuals in a population. But marker genes can become separated form gene drives, *e.g.*, by incomplete copying of the drive, and lead to errors in detection of gene drive individuals. Therefore, whole genome sequencing is preferable to fluorescent markers for a reliable detection of gene drives. It may be pointed out that genome sequencing is becoming faster and cheaper and may soon become a routine diagnostic tool (Nash et al., 2019).

11.12. UNINTENDED EFFECTS OF GENE DRIVES (BIOSAFETY ISSUES)

The deployment of gene drives in natural populations may pose potential biosafety hazards at molecular to ecosystem levels. At the molecular level, there is risk of off-target effects and loss of heterozygosity. The effects at the ecosystem level include movement of gene drives into non-target populations, unexpected negative effects on the ecosystem, etc. Some of the important biosafety issues are briefly considered in the following sections.

11.12.1. Off-target activity

The CRISPR-Cas9 used in gene drives is known to induce DSBs at off-target sites. Protein engineering has been used to develop Cas9 mutants with improved specificity but even these mutants exhibit some off-target activity (Section 5.9). This risk is particularly relevant since the CRISPR-Cas9 will be expressed generation after generation at least in the case of standard gene drives. In addition, there is a risk that a random mutation in the protospacer sequence of sgRNA could alter its specificity leading to an increase in off-target activity; this has been called sgRNA "retargeting". This would ultimately lead to induction of indels at off-target sites and is likely to inactivate the affected genes, which could impose a fitness cost on the affected individuals. If the off-target activity were frequent, there will be an increase in the level of mutation load in the targeted population. The off-target activity might enhance the effectiveness of eradication and suppression drives, but it would reduce the efficacy of rescue drives (Rode et al., 2019).

11.12.2. Loss of heterozygosity

The repair of both on-target and off-target DSBs by HDR will use the homologous chromosomes as template. This process may sometimes include the genomic regions extending several kilobases beyond the gene drive. When this happens, HDR will convert few to several heterozygous loci into homozygous ones and, thereby, reduce heterozygosity. In case of yeast, loss of heterozygosity over entire chromosome arms has been frequently observed. A reduction in heterozygosity will reduce the effectiveness of rescue drives as the positive effects of the drive could be overcome by the increased inbreeding depression. But increase in homozygosity would enhance the effectiveness of eradication and suppression gene drives (Rode et al., 2019).

11.12.3. Spread of gene drives into non-target populations

The global gene drives can spread to other populations of the same species even when there is sporadic gene flow among them. The gene drive spread would be facilitated by several natural processes including migration as well as human activities. Therefore, several strategies like deployment of precision and immunizing gene drives have been suggested for its mitigation (Section 11.14), but this will take several generations. Further, in case the final drive used in this strategy were a rescue drive, the individuals carrying the drive will be favoured and they may hybridize with individuals in non-target populations. A more attractive strategy is to use a self-limiting gene drive, *e.g.*, one of the daisy gene drives (Section 11.14.8). These drives can spread only for a limited number of generations after release. Although they can move into nontarget populations, their spread will be trivial since only few gene drive individuals will enter new populations and the drive will be eventually lost from the population. But in case of rescue daisy drives, the selective advantage conferred by the cargo gene could help it survive and spread in the new population (Noble et al., 2017; Rode et al., 2019).

11.12.4. Movement of gene drives into non-target species

Gene drives can move from the targeted population into the population of a related species if there is even infrequent and low level of gene flow between them. Once a gene drive enters the gene pool of a related species, it will spread in the population due to its invasive nature. In addition to the normal sexual route, gene drives might be horizontally transferred to even unrelated species. Horizontal gene transfers are mediated by pathogens, endosymbionts, parasites and insect pests. For example, homing endonuclease genes have been transferred to distantly related species by horizontal transfer: the *omega* gene of yeast has been transferred among different yeast species at 15 different times (Rode et al., 2019). It has been concluded that the horizontal 'transfer of functional genetic sequences between animals is extremely rare' (Royal Society, 2018). However, NASEM (2016) has recommended risk assessment for horizontal gene transfer prior to the release of gene drive individuals in the environment.

11.12.5. Unanticipated negative consequences for the ecosystems

Eradication of invasive species might have some unforeseen negative effects on the ecosystem mainly through its impact on the food chain. In case the targeted invasive species is a prey, its elimination might result in increased predation of some endemic species by the predators of the invasive species. But when the exterminated invasive species is a predator/herbivore, its removal might allow the species it predated/fed on to emerge as an invasive species. When two invasive species compete with each other, targeting of only one of them may result in increased population of the other. One or more instances of the above unexpected consequences from control measures other than gene drives have already been reported. In addition, removal of an invasive species would disturb the ecosystem equilibrium and may even make it more prone to invasion by some other invasive species or reinvasion by the same invasive species (Rode et al., 2019).

11.12.6. Failure of ongoing gene drive control measures

One of the proposed measures to control the spread of eradication and suppression gene drives is to release in the affected population such individuals that are resistant to the concerned gene drives. Since eradication and suppression drives impose a fitness cost, individuals with gene drive resistance would be selected for. But this approach will not be effective against rescue drives. Other approaches are to deploy gene drive 'brakes' (Section 11.14.3), excision gene drives (Section 11.14.4) and immunizing reversal drives (Section 11.14.2). In *Drosophila*, the gene drive 'brake' was able to inactivate gene drives with high (>90%) efficiency. But the 'brake' cannot reverse the genetic changes, including increased homozygosity, resulting from the reduced population size or induced by the earlier drive in the population. The immunizing drive would carry a functional copy of the gene inactivated by the earlier drive and, thereby, undo the genetic effects (except for reduced heterozygosity) of the earlier drive. Thus, immunization reversal drives appear to be more desirable for controlling the actions of earlier gene drives. But immunizing reversal drives may fail due to random mutations in the concerned sequences, incomplete copying of the reversal drive and evolution of gene drive resistance in the population (Rode et al., 2019).

11.12.7. Use of gene drive organisms as bioweapons

A bioweapon comprises an organism that can be deployed as a weapon to cause overwhelming damage to humans, domestic animals, crops and even ecosystems. The gene drive organisms can serve as efficient bioweapons by targeting specific populations of random mating species. Their impact on humans, however, may not be significant in view of the long generation time of humans. It is noteworthy that the "Safe Genes Program" of the American Defense Advanced Research Programs Agency includes gene drive research projects.

11.13. RISK ASSESSMENT FOR GENE DRIVES

Risk from an action, in simple terms, is a function of the level of hazard from the action and the probability of this hazard being realized. A hazard is any imaginable, identifiable and measurable undesirable consequence of an activity. Thus, the degree of risk describes the likelihood and the level of hazards from a given activity. Risk assessment consists of identification of hazards from an activity, analysis of the identified hazards, determination of the consequences of the hazards, risk determination and risk evaluation. Thus, risk assessment is a subjective activity. Thus far, systematic efforts for risk assessments of gene drive laboratory research and field tests have not been made. It may be pointed out that safety guidelines for laboratory research on gene drives have been published (Akbari et al., 2015). However, there is no accepted procedure for risk assessment for the deployment of gene drives in natural populations.

Moro et al. (2018) have proposed a framework for environmental risk assessment for gene drive deployment on the basis of the National Academies of Sciences, Engineering, and Medicine report on gene drives (NASEM, 2016). They identified the following eight key features of the targeted species for risk assessment: (i) life history and fecundity, (ii) availability of reference genome (a critical requirement), (iii) extent of gene flow to predict gene drive spread through the target population, (iv) density dependant reproduction and mate selection (critical for risk assessment), (v) pathways for movement of the target species across national borders, (vi) community interactions between invasive species and resident pest and native species, (vii) invasiveness of the target species, and (viii) breeding behaviour and reproductive performance of the target species in response to fertility control measures.

Risk management strategies have been discussed in some detail by Oye et al. (2014). They proposed evaluation of the efficacy of reversal drives prior to release of a gene drive, assessment of long-term effects of a gene drive on genetic diversity of the target population, implementation of molecular containment strategies, conducting initial field tests in areas lacking native populations of the target species, testing gene drives in tandem with corresponding immunizing and reversal drives, monitoring the presence and persistence of gene drives in target populations, evaluation of gene drives on case-by-case basis, evaluation of deliberate misuse of gene drives by a multidisciplinary team of experts, etc.

11.14. GENE DRIVE SAFEGUARDS AND CONTROL STRATEGIES

Since standard gene drives can alter genetic makeup of the entire populations, it is essential to develop proper and efficient strategies for reversal of the harmful effects, if any. generated by them. Esvelt et al. (2014) proposed the following approaches for safeguarding against such eventualities: (i) reversal of genetic alterations induced by earlier drives, (ii) immunization against a gene drive, (iii) precise targeting of subpopulations, and (iv) limiting the suppression of targeted populations.

11.14.1. Immunization of a population

Immunization against an unwanted gene drive involves modification of the genomic sequence targeted by this drive in such a way that it is no longer recognized by the gene drive. At the same time, it is ensured that the amino acid sequence of the polypeptide encoded by the altered genomic sequence is not changed; this is called recoding of a sequence. An immunizing gene drive has the recoded target sequence and sgRNAs specific for the wild type target sequence. This drive will be copied at the targeted site and, thereby, spread the recoded genomic target in the population making it immune to the unwanted gene drive (Fig. 11.8). An immunizing gene drive will protect populations from invasion by the concerned unwanted gene drive released in the same or other populations of the same species. Immunising gene drives can also remove the undesirable changes induced by the earlier gene drives and make the genome sequences immune to the earlier drives; such drives are termed as immunizing reversal drives (Fig. 11.9A). Immunizing drives can be designed to target the genomes of DNA viruses for generating in the host plants resistance to the concerned viruses (Esvelt et al., 2014).

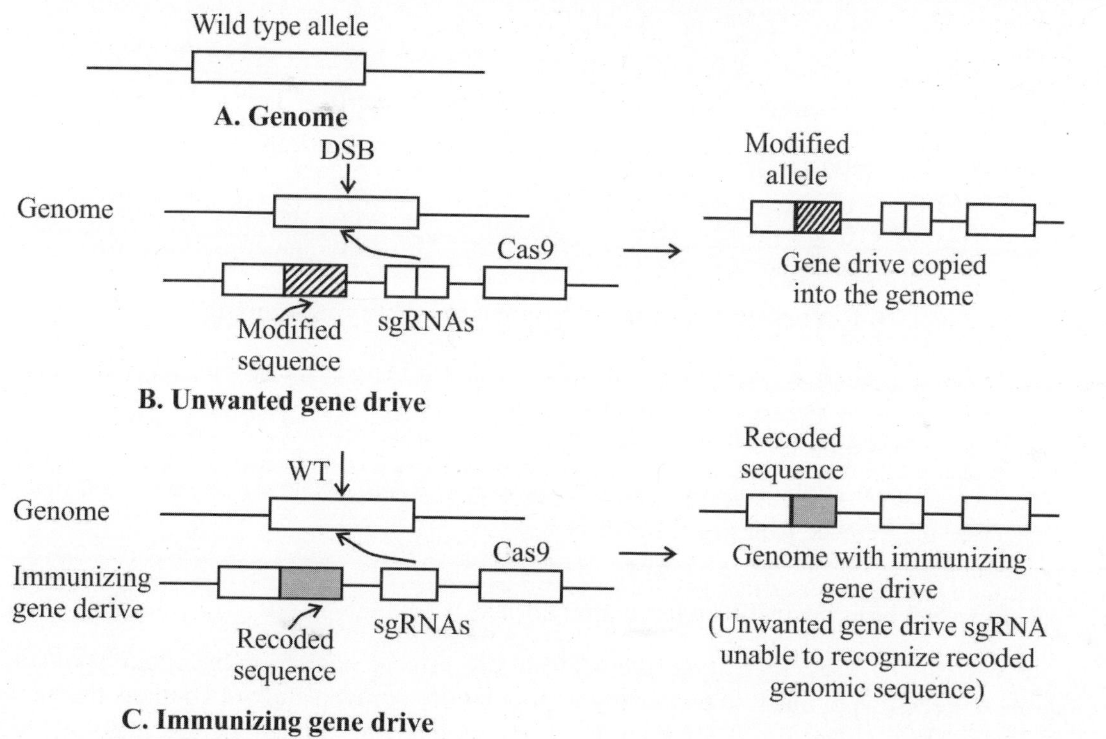

Fig. 11.8. Immunization of a population against an unwanted gene drive. The unwanted drive (B) sgRNA targets a sequence of the wild type allele *A* and becomes integrated into the genome. The immunizing drive (C) also targets the wild type allele and is copied into the genome. But the target sequence in this drive is recorded so that the base sequence is altered without affecting the amino acid sequence of the encoded polypeptide. As a result, the target sequence is no longer recognized by the unwanted gene drive and the population becomes immune to it. Immunization is carried out before the population is exposed to the unwanted gene drive.

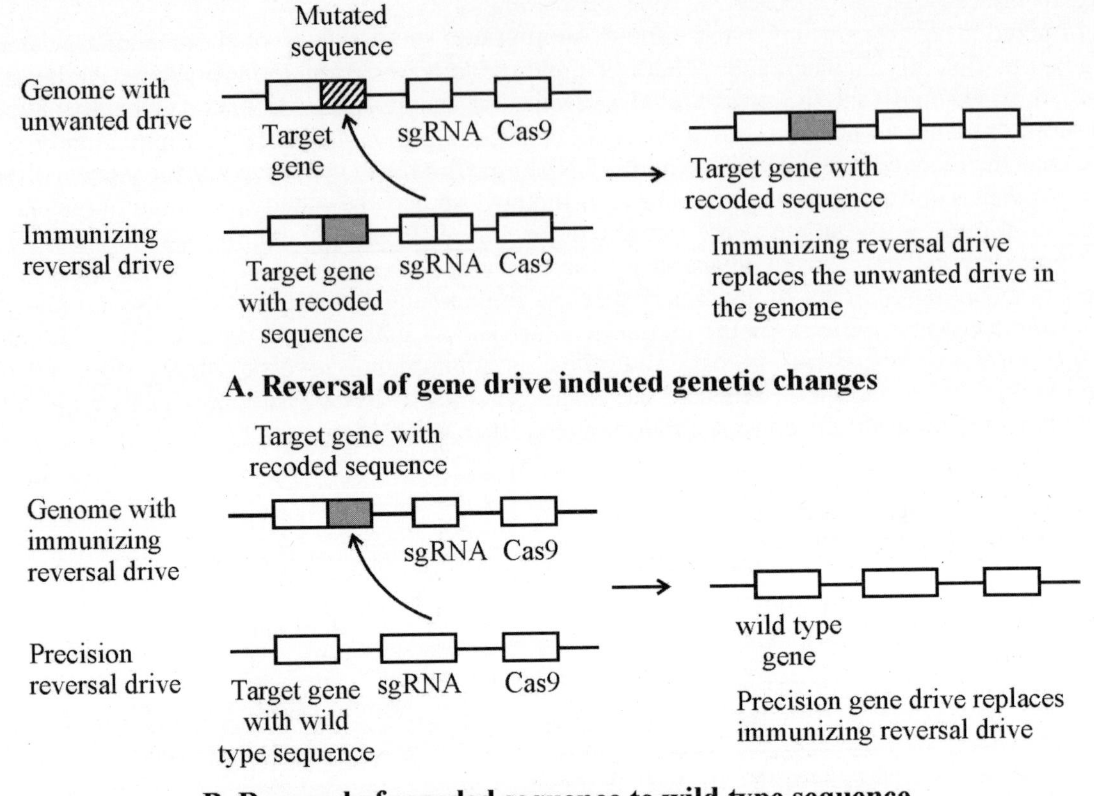

Fig. 11.9. Reversal gene drives. **A.** An immunizing reversal drive targets the genome sequence altered by the earlier drive and replaces it with the recorded sequence of the wild type allele of target gene. The recording protects it from the action of the earlier drive. **B.** A precision reversal drive replacces the mutated sequence with wild-type sequence. It is best used after an immunizing reversal drive has removed the earlier drive from the population. The sgRNA in this case will target the recorded sequence as shown here.

11.14.2. Reversal of undesirable genetic alterations

Gene drives can be used to reverse the undesirable effects generated by either spontaneous mutations or gene drives released earlier in the population. A reversal drive contains the genomic region that was modified by the earlier drive or spontaneous mutation; the genomic region included in the drive has the intended corrections and replaces the sequence altered by the earlier drive/mutation. When the genomic sequence included in a reversal drive is the same as the wild type sequence, the drive is termed as precision reversal drive (Fig. 11.9B). The earlier gene drive existing in the population will act on the precision reversal drive in the same way as it did on the wild type chromosome. Therefore, precision reversal drive will not eliminate the earlier drive from the population, and both will coexist in an equilibrium state. Therefore, an immunizing reversal drive (Section 11.14.1) should be used to remove an unwanted drive from the population and to reverse the genetic alterations induced by it; this drive will rapidly eliminate the earlier drive from the population. After this, the concerned genomic sequence can be reverted to the original wild type sequence by using an appropriate precision reversal drive (Fig.11.9A). Reversal drives can also be used to correct undesirable spontaneous mutations and to remove transgenes introduced into populations. The reversal drives provide a very powerful safeguard as they remove earlier gene drives as well as reverse the undesirable genetic changes induced by them, but they cannot reverse the ecological alterations, if any, generated by the earlier drives (Esvelt et al., 2014; Vella et al., 2017).

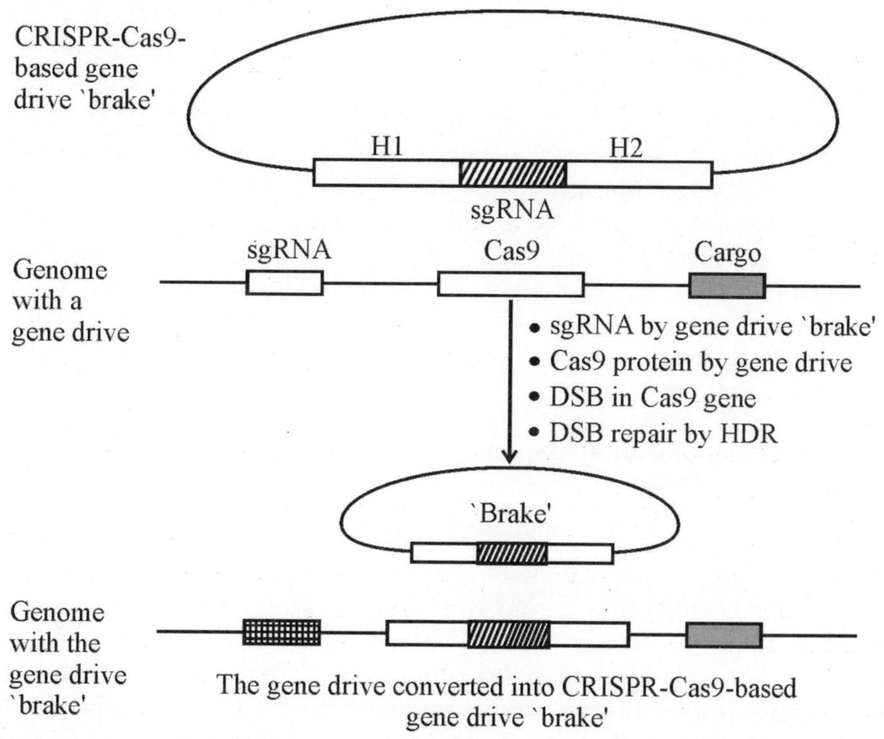

Fig. 11.10. The design and function of a Cas9-based gene drive `brake'. The `brake' comprises a sgRNA gene targeting a sequence within the Cas9; this sgRNA gene is flanked the Cas9 sequences located on either side of the site targeted by the `brake' sgRNA.

11.14.3. Inactivation of gene drives

A Cas9-based gene drive 'brake' has been developed for inactivation of Cas9-based genes drives and, thereby, convert them into gene drive 'brakes'. This 'brake' comprises a gene for sgRNA targeting a sequence within the Cas9 gene flanked by Cas9 gene sequences located on either side of the sgRNA target site. Cas9 function for this 'brake' is provided by the Cas9-based gene drive itself. When the 'brake' is present with a gene drive in the same nucleus, the 'brake' sgRNA would combine with Cas9 protein and induce a DSB in the Cas9 gene. HDR-mediated DSB will copy the 'brake' in the Cas9 gene: this will disrupt the Cas9 gene and convert the gene drive into a gene drive 'brake' (Fig. 11.10). A single 'brake' can target different Cas9-based gene drives. This 'brake' was highly efficient in *Drosophila.* The function of gene drive 'brakes' needs to be investigated in detail and their design improved before their deployment (Wu et al., 2016). The 'brake' has not aroused much interest possibly because no Cas9-based gene drive can be deployed in a population that has a 'brake' in its gene pool unless the Cas9 sequence targeted by the 'brake' is recoded.

11.14.4. Removal of gene drives

A gene drive can be excised from the genome by either CRISPR-Cas9 or recombinase action. In the first strategy, called elements to reverse the autocatalytic chain reaction (ERACR), two sgRNAs targeting the sequences flanking the drive and homology arms constitute the excision drive. The two sgRNAs combine with the Cas9 protein encoded by the drive itself to produce DSBs on both the sides of the drive. The existing gene drive is deleted and the excision drive is copied in its place. In the second strategy, the drive to be deleted must be specifically designed to contain at its both ends recognition sites for a site-specific recombinase, *e.g.*, *loxP* sequence for Cre recombinase. Individuals expressing the appropriate recombinase (in this case Cre recombinase) are released in the population having the gene drive. The progeny from the crosses between the released and the gene drive individuals will produce heterozygous progeny expressing the Cre recombinase. The Cre protein will catalyse recombination between the *LoxP* sequences flanking the gene drive. As a result, the gene drive will be excised leaving a *LoxP* sequence at the gene drive site (Zentner and Wade, 2017).

11.14.5. Precise targeting of subpopulations

A gene drive can be confined to the desired species by targeting the drive to a gene/genomic sequence unique to the species; such a drive is called precision gene drive. The drive could be restricted to even a specific population of a species provided this population has a unique genomic sequence that distinguishes it from other populations. In case a population does not have a unique sequence, it can be created by a suitably designed gene drive provided there is only sporadic gene flow between this and other populations. It has been suggested that a unique sequence should differ from similar genomic sequences for the PAM sequence and at least five base pairs in the spacer region (the region that pairs with sgRNA) immediately upstream of the PAM. For example, an immunizing gene drive may be first used to suitably alter the base sequence of the targeted genomic region in the concerned population. Then the precision drive, targeting the modified sequence is released in the population (Fig. 11.11; Esvelt et al., 2014). But the immunizing gene drive itself might spread to other populations/species creating an opportunity for their invasion by the precision drive.

11.14.6. Confining gene drives to laboratory strains

There is always risk of accidental release of gene drive individuals in the environment. Synthetic target site gene drives and split gene drives have been designed to confine the drives to laboratory strains of organisms and to virtually eliminate the risk of their spreading in natural populations following accidental release. The synthetic target site gene drives target a synthetic sequence like enhanced green fluorescent protein (eGFP) gene that has been introduced into the genomes of experimental strains. Such gene drives cannot spread to other laboratory strains and natural populations of the species since the latter will not have the concerned synthetic sequence in their genomes. The split gene drives comprise sgRNA and Cas9 expression cassettes integrated at unlinked genomic locations (Fig. 11.9A). Therefore, only the sgRNA cassette is propagated and Cas9 cassette follows Mendelian inheritance. In case of accidental release of few split gene drive individuals, the Cas9 cassette will be rapidly lost and the sgRNA cassette will also disappear soon thereafter. The conventional drives, synthetic target site drives and split drives show comparable homing efficiency and development of gene drive resistance (Champer et al., 2019).

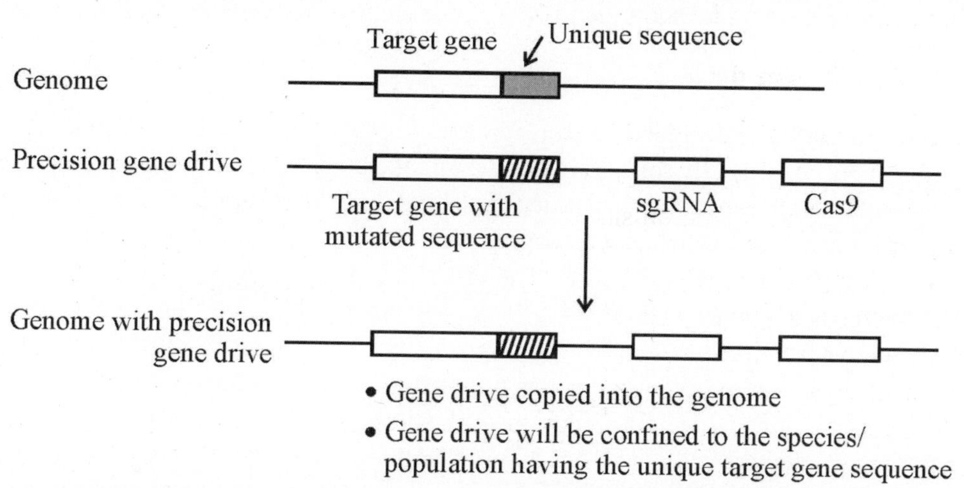

Fig. 11.11. A precision gene drive is confined to a specific species/population. Its sgRNA targets a gene/ sequence unique to the species/population of the species. Therefore, it can affect only those individuals that have this unique sequence. If the target species population lacks a unique sequence, the same can be created by first using an immunizing drive, which recodes the target sequence.

Various groups of experts have proposed guidelines for best laboratory practices for gene drive research. For example, the National Academies of U.S.A. have issued guidelines that recommend the use of a combination of confinement and containment strategies to minimize the risks from accidental releases. The proposed measures include (i) gene drive research in countries devoid of natural populations of the target species or where the species cannot establish in the wild (ecological confinement), (ii) use of strains that cannot produce progeny on mating with wild individuals (reproductive confinement), (iii) using self-limiting gene drives (molecular confinement), and (iv) inclusion of dominant visual markers in the gene drive architecture for rapid and reliable identification of gene drive individuals (molecular identification). But there is no guideline covering transport of gene drive organisms, including their sharing between laboratories (Rode et al., 2019).

11.14.7. Limiting population suppression

In many situations, it may be desirable to merely suppress rather than eradicate a natural population. Several strategies can be employed to achieve this objective. For example, a sensitizing drive can make a population susceptible to a specific molecule by one of the following alterations: (i) reversal of a mutation for resistance to the molecule to the wild type (sensitive) sequence/allele, (ii) conversion of an otherwise nontoxic molecule into a toxic one, and (iii) replacement of a gene whose product is insensitive to the molecule with a version of this gene that generates highly sensitive product. Such drives would allow an efficient control on the extent of suppression of a population simply by regulating its exposure to the concerned molecule. Another approach is to design two or more interacting drives in such a way that suppression is generated only when all the drives are present in the same nucleus. Finally, the daisy chain drives (Section 11.14.8; Fig. 11.12) are self-exhausting and have limited duration of spread, and suppression daisy drives will be soon lost from the population (Noble et al., 2019).

11.14.8. Daisy chain gene drives

A daisy chain gene drive is designed to stop spreading after a period and be ultimately lost from the population (Fig. 11.12). In this drive, the components of CRISPR-Cas9 gene drive, *viz*., sgRNA, Cas9 and cargo genes, are separated and integrated into different chromosomes of the genome so that they show independent segregation. In the simplest form, daisy chain drive comprises two unlinked elements A and B. The element A has the payload gene and is driven by element B, which has the genes encoding Cas9 and sgRNA for driving (= homing) the element A. However, the element B itself is nondriving so that it cannot propagate itself, shows Mendelian inheritance and, as a result, is lost from the population after some generations. After this, the element A ceases to spread in the population (DiCarlo et al., 2015; Smidler et al., 2017). The daisy chain drive depicted in Fig. 11.12A comprises three components, *viz*., A (cargo gene), B (genes for Cas9 and sgRNA-a specific for wild type allele for locus with element A), and C (gene for sgRNA-b specific for wild type allele for locus with element B). Thus, the element C (provides sgRNA-b) is necessary for driving or propagating element B, and element B is essential for spread of element A. Since there is no sgRNA specific for the locus with element C, it is inherited in a Mendelian mode so that it is rapidly diluted out of the population (Noble et al., 2019).

When daisy drive individuals are released in a population, initially the heterozygotes will have all the three (C, B and A) elements of the drive and elements B and A will be copied in the wild type chromosomes. But in few generations, the element C will be diluted out and, as a result, element B will cease to spread and show Mendelian inheritance, but element A will be able to spread in the presence of element B. In few more generations, element B will also be lost so that element A will stop spreading. Eventually, element A will also be lost from the population; this loss will be much faster if the cargo gene had a fitness cost. A daisy drive with three elements and homing efficiency of 95% will drive the cargo to near fixation if the release level was 2% of the natural population. It is assumed that the cargo itself has fitness cost and it is integrated in an essential gene so that the homing resistant allele resulting from NHEJ-mediated repair of DSBs will generally be dominant lethal. The potency of daisy drives can be greatly improved by increasing the number of elements to 6 or 7. However, a recombination between certain elements may convert a daisy drive into a self-propagating "daisy necklace" gene drive. This can be circumvented by ensuring that the daisy elements, including their promoters, do not have regions with sequence homology (Noble et al., 2019).

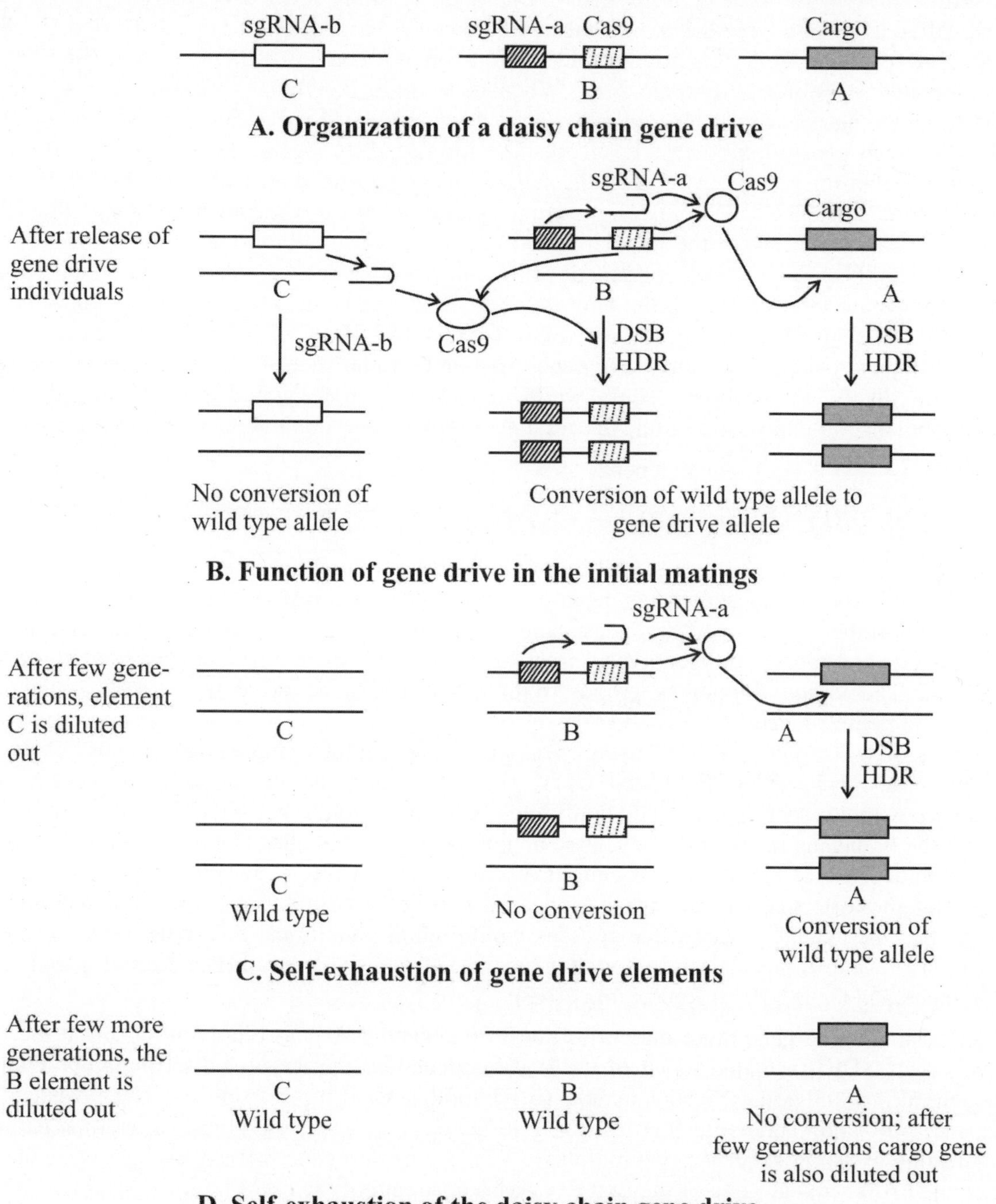

Fig. 11.12. The organization and function of daisy gene drive (see text for details). sgRNA-a and sgRNA-b, sgRNAs specific for the wild type homologues of the chromosomes with elements A and B, respectively; DSB, double-strand break; HDR, homology-directed repair. Each element having sgRNA gene would have more than one sgRNA gene (based on Noble et al., 2019).

Min et al. (2017a, b) have proposed two improved versions of the daisy chain drive. One of them, called daisyfield gene drive, has only two elements, *viz.*, sgRNA (element B) and Cas9 + cargo gene (element A). In this drive, genes for four sgRNAs targeting the locus with element A are integrated at unlinked genomic sites. When gene drive individuals are released in a wild population, the number of sgRNA genes per gene drive individual will decline, on an average, by 50% after each generation so that it will decline to one sgRNA gene per individual only two generations after the release. After this, the sgRNA genes will be rapidly lost and the cargo gene will stop spreading. The daisyfield design, therefore, puts a limit on the number of generations the gene drive can propagate in the population. In case of daisy quorum design, the typical daisy chain drive (Fig. 11.12) is combined with underdominance, a phenomenon in which heterozygotes are inferior to both the homozygotes. In this case, the cargo gene will continue to spread till the elements C and B are lost. But after this, the fitness of individuals is expected to become frequency-dependent due to the gene drive underdominance. Therefore, beneficial cargo genes are likely to become fixed, while deleterious ones would be eliminated. The underdominance is expected to limit the mixing of gene drive population with the surrounding wild type populations.

11.15. GENE DRIVE RESISTANCE

Gene drive resistance refers to the inability of a gene drive to propagate itself in the homologous locus of wild type chromosome; this generally results from inability of Cas9 to cleave the wild type chromosome. The rate of drive resistance development varies considerably depending on drive system, and species and genotype (target site polymorphism and genetic background) of the organism. The various strategies proposed for combating gene drive resistance include, (i) combined release of gene drives targeting different genomic sites, (ii) use of gene drives with multiple sgRNAs targeting different sequences preferably within a single gene, (iii) use of novel endonucleases, *e.g.*, dCas9-FokI and Cpf1, (iv) targeting gene drives to highly conserved sequences within essential genes, (iv) confining Cas9 expression to germline and other strategies to limit the tissue and the period of nuclease action, (v) novel strategies like daisyfield drives, and (vi) combined use of homing-based gene drives with a killer-rescue system. However, each of these strategies offers certain advantages and suffers from some limitations. Mathematical models predict that the available gene drive systems would require additional measures to alleviate the evolution of gene drive resistance for their continued invasive spread in a natural population (Hammond and Galizi, 2018; Rode et al., 2019).

Resistance to a gene drive may arise due to preexisting sequence polymorphism at the the drive target, NHEJ-mediated repair of the DSBs induced by the drive or discrimination against the gene drive individuals during mating (behavioral gene drive resistance). The presence of sequence polymorphism conferring resistance to gene drives has been described in some natural populations. Existing sequence polymorphism would generate gene drive resistance particularly when the PAM sequence and/or the first five bases upstream of the PAM are involved. This issue can be resolved by sgRNA multiplexing, *i.e.*, including in the gene drive sgRNAs specific for four different sites within the target gene. There is evidence for existence of genetic variation for NHEJ-mediated DSB repair among different organisms, among populations of a single species, and even within a single population of a given species. Further, NHEJ seems to be more frequent in the heterozygotes for gene drives than in the wild type homozygotes of at least *Drosophila*. In certain organisms like plants, NHEJ is the predominant mode of DSB repair. Since NHEJ leads to indels, the sequence at the DSB site will be altered making the concerned wild type chromosomes resistant to the gene (Zentner and Wade, 2017; Oberhofer et al., 2018; Champer et al, 2019; Rode et al., 2019).

The use of a suitable promoter for Cas9 gene may help reduce the development of gene drive resistance. For example, use of germline specific *nanos* promoter reduced gene drive resistance development as compared to the *vasa* promoter. In mice, there was a high rate of NHEJ-mediated DSB repair and gene drive resistance when the Cas9 expression was limited to male germ line, while gene drive showed high efficacy when Cas9 expression was confined to female germ line (Grunwald et al., 2019). In contrast, the development of resistance was faster with Cas9 expression in female germ line than in the male germ line of mosquitoes. Behavioral gene drive resistance would arise when there is discrimination against the gene drive individuals during mating. In any case, resistance to gene drives is an important risk for eradication and suppression drives since it would not only slow down the spread of gene drives but also reduce the extent of their spread since gene drive resistance will be selected for. However, gene drive resistance will not be a limitation for rescue drives since the resistance allele will be selected against in view of the benefits conferred by the rescue drives (Zentner and Wade, 2017; Oberhofer et al., 2018; Champer et al, 2019; Rode et al., 2019).

Gene drive resistance alleles will be selected for if they impose a lower fitness cost than the gene drive allele. Conversely, they will be selected against or even eliminated if the gene drive was integrated within an essential gene and contained a functional copy of this gene. In such cases, this essential gene will be inactivated in majority of the products of NHEJ-mediated DSB repair. But in about one-third or so of NHEJ products, the indels would be in-frame and a proportion of such products would be fully functional. These chromosomes, therefore, would incur no fitness cost and would lead to evolution of gene drive resistance especially in case of eradication and suppression gene drives. The predominance of NHEJ and the emergence of gene drive resistance in an early generation will retard the spread of eradication and suppression gene drives; this is regarded as the one of chief reasons for the failure of many gene drive-based experimental studies on evolution (Rode et al., 2019).

11.16. ETHICAL ISSUES RELEVANT TO GENE DRIVES

The word ethics is derived from a Greek word meaning ‘relating to one’s character.’ Ethics is a branch of philosophy, called moral philosophy, that attempts to systemize, defend and recommend the concept of right and wrong behaviour. In case of gene drives, a variety of ethical issues will emerge since they can be used to eradicate, suppress, rescue or simply change the genetic makeup of wild species, and they can also affect non-target populations as well as related species. The ethical issues will vary to some extent depending on the objective of a gene drive application, and the objections would be the strongest to eradication drives. It is argued that ‘all life is sacred’ and that humans do not have the moral right to eradicate a species and ‘play God’. Pugh (2016) has examined these objections in detail and rejected them. The first objection that ‘all life is sacred’ is inconsistent with human survival as it would deprive us of our daily food, and measures for prevention and treatment of infectious diseases. Similarly, the ‘playing God’ objection is untenable since it is tantamount to prohibiting the development and use of new biotechnologies. He further argued that in cases like mosquitoes, eradication may provide long-term protection from the mosquito-transmitted diseases than other measures like making them resistant to the concerned parasites. It was, therefore, concluded that the decision to use gene drive technology should be based on relevant experimental data on the risks posed and the benefits offered by the technology.

Ethics and science should work together to promote human welfare and to protect all life forms and the ecosystems. The four main ethical considerations relevant to new technologies like gene drives are, (i) the science behind the technology should be sound and comprehensive, (ii) benefits to human health and welfare must outweigh risks, (iii) the effects on biodiversity and ecosystems must be clearly understood, and the reversal of these changes should be feasible, and (iv) the likelihood that the technology can be misused must be determined and guarded against. It has been recommended that gene drive technology should be considered with alternative technologies, and it should be used only when its side effects, and societal and ecological effects are reliably determined (Neves and Druml, 2017).

The NASEM report (NASEM, 2016) 'called for a responsible science approach to developing and applying gene drive technologies… Responsible science ultimately rests on values–deeply held, complicated, sometimes evolving beliefs about what kinds of things, in human lives and the world at large, should be fostered, protected, or avoided, and therefore about what people should and should not do.' The decision regarding the use of gene drives can be taken based on the risks and benefits from them as compared to the potential alternative measures for achieving the same objectives. Further, the possible ecological consequences of gene drives should be assessed 'on a species-by-species basis in each location' and should include assessment of the possibility of its spreading to other locations from the area of its release (Royal Society, 2018).

11.17. REGULATION OF GENE DRIVES

The gene drive technology is very attractive for resolving many difficult issues related to ecosystem, agriculture and public health. But it poses many potential threats to the ecosystems that must be evaluated, and the experiences from classical biological control strategies should be used to develop safer and effective gene drive technologies. In this context, suitable guidelines should be developed for deployment of gene drives in natural populations. In addition, the inclusion of gene drives in the conservation toolkit for protection of endangered species or ecosystems should be done only when they provide added value in relation to the available alternative strategies. The following three types of information are needed for decisions on using a gene drive strategy. (i) *Genetic and technical information* pertaining to rearing, breeding and experimental use of the target species should be available. In addition, genome editing technology for and an annotated reference genome of the target or related species is needed to facilitate gene drive research. (ii) The important *ecological information* include behavioural and demographic aspects of the target species, and system of mating and gene flow between different populations; these would be useful in modelling of gene drive dynamics. (iii) Finally, *ecological and evolutionary* data required about the likely non-target species relate to the frequency of gene flow (including horizontal gene transfers) to and from the target species, presence of putative target sites for the proposed gene drives in their genomes, and proper modelling of food web structure (Rode et al., 2019).

Gene drives have the potential to spread across national boundaries once they are released in a local population. Therefore, field release even for safety evaluation would require public support and regulatory approval, often international regulatory approval. But regulatory procedures for field release of gene drives can only be developed when relevant experimental data, including those from field tests become available. This dilemma is likely to delay the deployment of gene drives. Since the ecosystems are shared by communities and often nations, the development of regulatory framework should involve all the stakeholders who will use and/or be affected by the gene drives as well as the experts from various relevant fields. But engaging the stakeholders and evolving an agreement among them becomes increasingly difficult as the size of the relevant region increases. In this context, one group of researchers is 'scaling up cage experiments and building ecological models' to generate data on risks from release of gene drives in the field. Another approach is to design self-limiting gene drives and integral gene drives (IGD) that might greatly simplify experimental field testing and thereby accelerate the development of regulatory procedures for field release (Noble et al., 2019; Scudellari, 2019).

The IGD was proposed to facilitate field testing of cargo genes without the risk of their spreading to non-target populations. An IGD comprises two elements, say, A and B. Element A has the cargo gene with an intronic sgRNA sequence. The element A is integrated in-frame within a somatic gene (expressed in the target tissue) close to the initiation codon, and the intronic sgRNA is specific for this site of the gene. The element B has the Cas9 gene with intronic sgRNA sequence; this element is integrated within a germline gene in the same manner as element A. The Cas9 and cargo genes are transcribed using the regulatory sequences of the endogenous genes within which they are integrated to yield compound transcripts. The translation of these compound mRNAs yields separate Cas9 and cargo polypeptides since their gene sequences are separated from those of the endogenous genes by 2A ribosome-skipping signal sequences. It is expected that integration of A and B elements will not affect the functions of the endogenous genes. Therefore, the gene drive effects will be generated only by the cargo gene product. The element B is capable of autonomous homing, while element A can undergo homing only in the presence of element B. Therefore, cargo genes alone can be field-tested using inundative releases without the risk of their invasive spread in the population. In the next step, element A can be combined with a non-homing element B to provide Cas9 activity. Since this element B will be rapidly lost from the population, IGD performance and perhaps its impact on, say, disease transmission can be evaluated in the local population. Finally, when the gene drive deployment is approved, the complete IGD, *i.e.*, both elements A and B capable of homing, can be deployed in the target population (Nash et al., 2019).

Since gene drive individuals contain recombinant foreign DNA, they are subject to the national and international regulations applicable to GMOs. At the international level, gene drives would be regulated by the Cartagena Protocol on Biosafety (SCBD, 2003) and the two directives of EU. However, it would be necessary to modify some of the provisions of GMO regulatory framework to make it applicable to gene drive organisms; these modifications will have to take into account the specific features of the rapidly evolving gene drive technology. A call for global moratorium on research and development of gene drives and their release in the environment was issued by 160 civil society organizations in 2016 (Rode et al., 2019). However, UN decision in 2018 permits research on gene drives, but it requires "free, prior and informed consent" of people living in the areas affected by the proposed field test/experiment.

The NASEM report (NASEM, 2016) concluded that the evidence to support field release of gene drive organisms was insufficient. But 'laboratory research and controlled field trials should continue', and the research should progress in five phases as follows:

Phase 0: Research preparation

Phase 1: Laboratory-based research

Phase 2: Field-based research

Phase 3: Staged environmental release

Phase 4: Post-release surveillance

Table 11.5. Gene drives designed for molecular containment

Gene drive	Feature relevant to containment	Containment
A. Gene drive spread limited to laboratory strains :		
Split gene drive	Cas9 and sgRNA genes integrated at unlinked loci	Cas9 gene dilutes out of the population after which the sgRNA gene is eventually lost
Synthetic site gene drive	Standard gene drive targets a synthetic sequence introduced in the genome	Gene drive cannot invade any strain/population other than that having the concerned synthetic sequence in its genome
B. Gene drive spread limited to the targeted population:		
Precision gene drive	Gene drive targets a unique sequence present only in the genome of target population*	Since other populations/species will not have this sequence, gene drive cannot invade them
C. Gene drive spread and persistence limited to a small number of generations:		
Daisy chain drive	Elements C, B and A with sgRNA, Cas9 and sgRNA, and cargo genes, respectively, at unlinked loci; C needed for driving B, B needed for driving A, and C itself is nondriving	First C is diluted out of the population, then B is lost and ultimately A is lost; C is lost since it follows Mendelian transmission; A is amplified the most compared to B and C due to super-Mendelian transmission
Daisy-field drive	Two unlinked elements: element A has Cas9 and cargo gene, nondriving element B comprises four unlinked sgRNA genes targeting the locus with element A	The number of sgRNA genes per nucleus reduced by a factor of 1/2 after each generation so that it is lost after some generations; in few more generations, cargo gene is also diluted out
Daisy-quorum drive	The daisy chain drive combined with underdominance, *i.e.*, heterozygote inferiority	After the elements C and A are lost, fitness become frequency-dependent so that beneficial cargo genes would become fixed, while detrimental ones will be lost from the population

* The unique sequence can be created by an immunizing drive prior to the deployment of the precision drive. But the immunizing drive itself might spread into other populations/species creating an opportunity for their invasion by the precision drive

In each phase, safety should be given high priority and it should use both confinement (based on ecological conditions) and (ii) containment (due to natural physical barriers or restrictions created by humans) to prevent accidental release of gene drive organisms and gene drive spread in natural populations (Collins, 2018). Gene drives have been specifically designed to restrict them to laboratory strains and to minimise the risk of their spread into natural populations due to accidental release of few gene drive individuals. Other designs of gene drives either limit their spread to the targeted population or limit their invasiveness to a small number of generations and, thereby, eliminate the risk of their spread to other populations and related species, and persistence in the target population except when the cargo gene is selected for. These approaches of gene drive containment have been termed as molecular containment (Table 11.5).

11.18. ADVANTAGES OF GENE DRIVES

Gene drives provide a potent means for introduction of desired genetic changes in natural crossbreeding populations. These self-propagating tools can spread in and, thereby, modify the genetic architecture of entire populations. Gene drives offer several advantages as listed below.

1. Gene drives enable genetic modification of entire natural populations for tackling many otherwise intractable and pressing problems like human health, agricultural and ecological issues, etc. (Section 11.10).
2. Since gene drives are self-propagating, the number of gene drive individuals needed for release in wild populations is very small as compared to those in the conventional technologies. Gene drives can spread their cargo genes in the populations even when the release size is merely a small fraction of the targeted population and the cargo genes impose a wide range of fitness costs.
3. Gene drive deployment promises to be much cheaper than the alternative biocontrol methods. At the same time, their scale of action is expected to be much larger. Many authors consider these tools as 'multi-purpose silver bullets in conservation biology, agriculture and public health' (Rode et al. 2019).
4. Reversal gene drives can reverse the genetic changes induced by earlier gene drives, and immunizing gene drives can protect a population from invasion by unwanted gene drives (Section 11.14.1).
5. Gene drives have been designed to inactivate or remove previous gene drives already present in the gene pool of a population (Section 11.14.3, 11.14.4).
6. Allelic gene drives enable accumulation of favourable alleles of multiple genes in a single line/strain. These gene drives are being used in basic biological investigations, and they might be useful in plant and animal breeding (Section 11.10.5).
7. Self-limiting gene drives (the daisy gene drives) minimize or even eliminate the risk of gene drive spread to non-target populations and to populations of related species (Section 11.14.8).
8. The IGDs have been designed to facilitate field tests of gene drives and cargo genes without the risk of invasion of other populations.

11.19. LIMITATIONS OF GENE DRIVES

The gene drive systems suffer from several important limitations listed below. Efforts are being made to overcome one or more of these limitations by suitable modifications in gene drive designs and evolving effective procedures for their field evaluation and eventual deployment.

1. The first and foremost limitation of all gene drives is that they can affect only sexually reproducing and cross-fertilizing species. They cannot be used in species that are either self-fertilizing or reproduce asexually. In addition, their effectiveness is doubtful even in species with mixed modes of reproduction (Esvelt et al., 2014).
2. Gene drives require many generations to spread through a population. Therefore, the time required for a drive to approach fixation would vary from around one year to centuries depending on the reproductive cycle of target species. Thus, this technology would be the most useful in species with short reproductive cycles (Esvelt et al., 2014).
3. The changes introduced in natural populations by gene drives are not everlasting on an evolutionary timescale. Although the cargo genes can spread in the target population, deleterious cargo genes will ultimately be displaced by more fit alleles that may arise due to spontaneous mutations especially after the drive has reached fixation (Esvelt et al., 2014).
4. The information available on the risks associated with release of gene drives in natural populations, their ecological consequences, and the strategies for management of these risks is mainly based on studies with laboratory populations of mosquitoes. It is, therefore, imperative that such studies are carried out with populations of other species as well to evolve comprehensive risk assessment procedures, effective management strategies and regulatory guidelines for gene drive deployment in natural populations (Esvelt et al., 2014).
5. Gene drives are based on CRISPR-Cas9 system, which can induce off-target cleavage and gene editing (Section 5.9). This has forced the researchers to devise a variety of diverse strategies, including restricted period of Cas9 expression and activity to minimize off-target effects. In case of gene drives, the Cas9 expression can be confined to the germline cells, but prolonged activity would be preferable to ensure high allele conversion rates for faster spread of the drives. This would also increase the risk of off-target activity, which will be aggravated by continued expression over generations as the drives will persist in the targeted populations. This may not be important in case of eradication and suppression drives, but it will be of considerable significance for rescue drives. There is virtually no information on off-target effects of gene drives; this information is essential for informed decision on their use in natural populations.
6. Gene drives released in a local population might invade all the populations of the target species that may be distributed across national boundaries. This would be assisted by both natural forces and human activities. In some cases, this might lead to extinction of the targeted species. In addition, it complicates the development of procedures for and the actual regulation of gene drive deployment in natural populations (Esvelt et al., 2014).
7. Gene drives might invade related species if there is occasional gene flow between them and the target species. This may generate unpredictable consequences to the related species and the ecosystem. It may be pointed out that even if the direct impact of the traits being modified by the gene drive were well known for the target species, its unintended side effects on the ecosystem cannot be predicted.

8. Mutations in the gene drive sequences might alter their properties that might have unpredictable effects on the target populations. Significantly, these effects will be propagated by the gene drives.
9. There is evidence that a proportion of the wild type homologues of the genomic loci having gene drives can be resistant to cleavage by Cas9 and, as a result, conversion to the gene drive allele. The resistance may already be present in the population or it may arise due to NHEJ mode of DSB repair. This limitation can be overcome/mitigated by using multiple sgRNAs in the gene drive and some other strategies like proper selection of target site, use of suitable promoters, etc. (Section 11.15). Gene drive resistance is an important risk for eradication and suppression drives, but not for rescue drives (Zentner and Wade, 2017; Oberhofer et al., 2018; Champer et al, 2019; Rode et al., 2019).
10. Homing efficiency of gene drives is affected by several factors and shows wide range of variation. This variation makes it very difficult to reliably predict the various consequences of introduction of a gene drive in a natural population and to devise safe release protocols.
11. Gene drives may reduce heterozygosity in the targeted populations, which would tend to reduce fitness and interfere with efficacy of rescue drives (Section 11.12.2).
12. Gene drives can be inactivated by spontaneous mutations in Cas9 gene, sgRNA gene and the targeted genomic sequence (especially, the PAM sequence).
13. Gene drives cannot be used for altering traits of human populations, say for therapeutic applications, in view of long human reproductive cycle (Oye et al., 2014).

11.20. ACHIEVEMENTS

CRISPR-Cas9-based homing gene drives have been designed and successfully tested in laboratory populations of several random mating species. The standard gene drive scheme has been modified in a variety of ways to make gene drives safer for deployment in natural populations and to achieve specific objectives. Some modifications of gene drives are self-limiting (split drives, daisy gene drives) and/or safer for field-testing (integral gene drives). In addition, gene drives have been developed for reversing the undesirable effects generated by earlier gene drives (reversal drives), immunization of populations against specified gene drives and to confine gene drives within specific populations of a species. Cas9-based gene drive 'brake' for gene drive inactivation and gene drives designed to excise or remove existing gene drives have also been developed. The split gene drives and synthetic site drives were designed to minimize the effects of gene drives on natural populations due to accidental release of few gene drive individuals from laboratories. Allelic gene drives are designed for genetic enhancement of lines/strains of animals and plants by conversion of undesirable alleles of one to several genes into desirable ones.

Gene drives for eradication/genetic modification of mosquitoes for control of mosquito-borne diseases have been developed and tested in laboratory populations. Gene drives have been used to make mosquitoes incapable of transmission of malarial parasite and dengue-2 pathogen (Section 11.10.1). In addition, gene drive has been used to induce female sterility and thereby eradicate the targeted mosquito populations. The gene drive causing female sterility is being field-tested by the British start-up firm Oxitec in Brazil for suppression of *Aedes aegypti* mosquitoes. Oxitec calls this technology as Friendly™ Mosquitoes Platform.

11.21. FUTURE PERSPECTIVES

Several different designs of gene drives have been either proposed or tested in laboratory. But gene drive behaviour after their release in natural populations and the effects of large-scale genetic alterations of natural populations on the ecosystem are not yet known. Therefore, emphasis of gene drive designs has been to minimize or eliminate their spread to non-target populations and related species after they are released in the field. Some of these gene drives have been evaluated in laboratory but their field tests are yet to be done. At present field tests cannot be done since the regulations and procedures for field testing of gene drives are yet to be developed. The main difficulty in evolving regulatory guidelines is the lack of relevant data concerning gene drive performance and spread in natural populations. At the same time, the lack of regulatory guidelines precludes field tests for collecting the data needed for guideline development. In addition, the risk assessment procedures relevant for gene drives are yet to be developed. Thus, resolving the predicament concerning field testing, and the development of appropriate risk assessment and regulatory procedures should be one of the priority areas of future work on gene drives (Rode et al., 2019).

Another area that needs to be tackled is the phenomenon of gene drive resistance; this is one of the main areas of current gene drive research. Efforts are being made to unravel the nature of resistance, and to identify the factors contributing to its evolution as well as those that would mitigate its spread in the population. Further, such gene drives should be designed that have reduced invasiveness and are likely to be safer for use in natural populations. The significant activities in this research area include, (i) use of suitable regulatory sequences for limiting the expression of drive genes to specific tissues, (ii) identification of genomic sites and drive designs that minimize the risk of gene drive resistance, and (iii) enhancing the proportion of HDR in comparison to NHEJ. Further, gene drive integration into the genomes of target species may lead to unexpected effects on fitness, reduced homing efficiency and incomplete gene drive copying during homing. These issues need to be investigated and resolved to facilitate safe and effective use of gene drives for management of natural populations (Nash et al., 2019).

Nearly all the homing-based gene drives are CRISPR-Cas9-based, and this SDN system is known to induce off-target gene editing. However, there is virtually no data on off-target effects of gene drives. Further, data on the risk of horizontal transfer of gene drives should be accumulated before their deployment in natural populations (NASEM, 2016).

11.22. CONCLUSIONS

The development of appropriate technologies for introduction of large-scale genetic alterations in natural populations has been an important objective of researchers in various fields, including human health. In 2003, Burt proposed the development of self-propagating gene drives based on homing endonuclease genes. About a decade later, in 2015, the CRISPR-Cas9-based gene drives were designed and demonstrated to work in yeast and *Drosophila*. A simple gene drive has expression cassettes for Cas9 and sgRNA specific for the targeted genomic site; these gene constructs are flanked by homology arms required for HDR. Where needed, cargo and/or fluorescent marker genes may be included in the gene drive design. The CRISPR-Cas9 induces DSBs in the wild type chromosomes, and DSB repair by HDR leads to copying of the gene drive at the DSB site; this is called homing. Thus, in heterozygotes the gene drive is copied into the wild type chromosome so that they become homozygous for the drive. This leads to super-Mendelian inheritance pattern, in which the frequency of gene drive allele continues to increase in the population. Thus, the cargo gene included in a gene drive keeps on spreading till it reaches fixation. The rate of spread of gene drives is affected by several factors, including the species, and the strain/genotype of the target organism. The gene drives work effectively in laboratory populations, but their behaviour in natural populations remains unknown in the absence of field testing.

The homing-based gene drives may serve as potent tools for handling of many intractable problems in several fields. They can be used to control human/animal disease vectors like mosquitoes, promote agricultural production and sustainability, control invasive species, rescue endangered natural populations, develop desirable animal and crop strains/lines and facilitate many basic investigations in biology. For example, gene drives have been developed or eradication of mosquitoes and for making them incapable of transmitting malarial parasite and dengue-2 pathogen. The first field testing of gene drives for eradication of *Aedes aegypti* mosquitoes is being done in Brazil by the British firm Oxitec even though the procedures for regulation of such tests as well the transport and import/export of gene drive organisms are yet to be developed.

In view of the risk of gene drives persisting indefinitely in a population, self-limiting gene drives, *e.g.*, three types of daisy chain gene drives, have been developed. The invasion of a population by an unwanted gene drive can be prevented by deployment of an immunizing gene drive. In addition, the unwanted gene drive can be inactivated by Cas9-based gene 'brake', or it can be removed from the population by excision gene drives. Reversal gene drives have been designed to undo the undesirable genetic alterations induced by an earlier gene drive. Precision gene drives are proposed to target sequences unique to specific populations and, thereby, confine them to the respective populations. Finally, split gene drives and synthetic site gene drives are designed to confine the gene drives to laboratory strains. In addition, integral gene drives have been proposed to facilitate field testing, and allelic gene drives were developed to enable improvement of animal and crop plants.

Gene drives offer several advantages, such as rapid spread in the entire population, very small release size, management of important and intractable problems related to heath, agriculture and ecosystem, etc. The risks from gene drives include off-target activity, spreading of gene drives into non-target populations of the same species and even into related species, reduction in heterozygosity, and unanticipated negative consequences to ecosystems. Gene drive designs and deployment procedures are being refined to minimize risks and enhance benefits from them. The presence of gene drives in natural populations can be monitored by whole genome sequencing of the concerned individuals, and by detection of fluorescent markers included in the gene drive design. The procedure for risk assessment for gene drives is yet to be developed as are the regulations and procedures for field testing and the eventual deployment of gene drives in natural populations.

■■■■

12

BIOSAFETY AND ETHICAL CONCERNS

12.1. INTRODUCTION

The site-directed nucleases (SDNs) were developed to facilitate integration of transgenes at specific genomic sites by homology-directed repair (HDR) of the double-strand breaks (DSBs) induced by them. But DSBs are predominantly repaired by nonhomologous end-joining (NHEJ) leading to indel (insertion/deletion) formation at the DSB sites. Thus, SDNs can be used for gene inactivation by NHEJ-mediated indel formation, mutation correction/induction by HDR-mediated sequence replacement, and whole gene (transgene) integration at the targeted genomic sites. The term 'gene editing' or 'genome editing' is used to describe these targeted genome modifications. Gene editing technology has been used in a wide range of organisms for furthering both basic as well as applied researches. Several gene-edited lines of crops and some animal strains either have been or are in the process of being commercialized. More notably, gene editing has emerged as a promising tool for gaining insights into the origin and progression of human diseases. It has also been successfully used for the development of therapies and even preventive procedures for several genetic diseases and disabilities, many of which lead to devastating consequences (Cribbs and Perera, 2017; NASEM, 2017).

In human subjects, gene editing may be limited to somatic cells and, as a result, to the concerned individuals and, more particularly, the generation. But when germ cells and/or embryos are edited, the edited cells/embryos can be used to initiate pregnancies to produce gene-edited babies; this has already been done in China (Section 8.5). In addition, SDN technology has been used to create gene drives that can eliminate, suppress or rescue natural populations of crossbreeding species. These and other applications of gene editing technology raise issues of safety, ethics, and societal and religious values. The safety and ethical issues relevant to gene editing of plants, animals and management of natural populations have been discussed in some detail in Chapters 7, 8 and 11, respectively, and they are recapped here as well. In addition, the science, ethics and other issues relevant to gene drives have been deliberated in considerable detail in NASEM (2016). In this chapter, the various issues associated with gene editing in humans are considered in some detail.

There are several publications devoted to the safety, regulation, ethics and other issues related to gene editing, some of which are as follows: Baltimore et al. (2015); Ledford (2015b, 2019a,b,c); Whelan and Lema (2015); Charo (2016), NASEM (2016, 2017); Cribbs and Perera (2017); Ishii and Araki (2017); Shim et al. (2017); Smyth (2017); de Wert et al., (2018), Cavaliere (2019), and Tuerlings (2019).

12.2. THE ECOSYSTEM FOR INNOVATION IN BIOTECHNOLOGY

Innovations in biotechnology have generated a variety of products for human use and welfare; these products include improved genotypes of organisms with valuable/novel traits, novel diagnostic tools and therapeutic agents, genetic tools for management of natural populations, etc. The biotechnological innovations are based on basic science and applied researches conducted in various universities and institutions. The innovations occur and thrive in an 'ecosystem' comprising the government, public institutions, private industry, and the general public. The innovations need to progress from laboratories of institutions/companies to the consumers for distributing the benefits arising out of them to all the stakeholders. In this journey, the research and development activities leading to the generation of innovative products and those for their commercialization must conform to the relevant legislations and regulations of the countries concerned and ultimately fulfil the consumer's needs and gain their acceptance. Therefore, the prevailing 'ecology of this ecosystem' determines the types of innovations produced in a country as well the ease and speed of their commercialization (Charo, 2016).

Consumer demand and acceptance has a profound impact on the types of innovative products developed and commercialized by the industry. It may be pointed out that even when the basic and applied research necessary for the development and commercialization of a product are available, the industry is not very likely to invest in a product that it visualizes to have little demand and/or acceptance by the consumers. In addition, the preferred areas of research in the universities and public institutes influence the areas in which innovations are likely to be made. This, in turn, is affected by the availability of research grants in the given discipline and the conditions attached to them. For example, there is no legal ban on gene editing of human embryos for implantation in the United States of America, but federal funds cannot be used for supporting such work. This condition coupled with certain other requirements make it extremely difficult for a scientist to undertake heritable germline gene editing in U.S.A. (Charo, 2016).

An important component of the innovation ecosystem consists of intellectual property rights regime, which ranges from patent policy of the concerned country to the prevalent international trade laws (Table 12.1). This component has as strong impact on creation, production and marketing of biotechnology products in different countries, which together determine the economic benefits obtainable from various biotechnology innovations. Another component is the regulatory framework governing innovations in the concerned country. This component determines the ease and speed of commercialization of an innovation and is reflected in such activities as risk assessment approach, emphasis on public information, *e.g.*, product labelling, etc. The regulatory framework in a country may be promotional, neutral or precautionary. The promotional regulatory system follows a permissive approach to risk assessment, which assumes a product to be safe until it is proven dangerous. As a result, this system encourages innovation since the approval of new products is relatively easier, quicker and less expensive so that the product reaches the market in a relatively much shorter time. In the precautionary system of regulation, on the other hand, a precautionary approach to risk assessment is followed; this approach presumes that a product is dangerous until it is proven safe. Therefore, this system tends to discourage innovation as it either does not fund the new technology or it may even make the use of an innovative product a criminal offence. U.S.A. follows a promotional approach of regulation, while the approach adopted by EU (European Union) is precautionary. The neutral regulatory approach follows the middle path of least regulation by the government and, as a result, has the minimum impact on the development of innovations (Charo, 2016).

Table 12.1. A list of the various policy issues that make up the 'ecosystem' for biotechnology innovations (based on Charo, 2016)

Policy issue	Effect on biotechnology innovation	Remarks
Intellectual property rights (from patent policy to international trade laws)	Huge impact on production and marketing of biotechnology products in different countries, *i.e.*, deriving economic benefits from biotechnology innovations	Strong patent laws viewed to promote innovation, but many consider them to hinder basic research by reducing the opportunity to collaborate or take advantage of one another's work by free exchange of information
Regulatory framework in a country	Determines the ease and speed of commercialization of an innovation; reflected in IPR regime, risk assessment approach, and emphasis on public information, *e.g.*, product labelling	In a single country, different areas of biotechnology may be treated differently, *e.g.*, the approach to gene editing in plants in USA is promotional, while that to human germline gene editing is precautionary
i. Promotional	The regulations promote innovation	A permissive approach to risk assessment; it is presumed that a product is safe until proven dangerous; therefore, labelling of a product from modern biotechnology is required only when there is evidence for a higher level of risk than the conventional product
Ii. Neutral	Minimum government direction	It has minimum impact on the development of innovations
iii. Precautionary	A prohibitive system: it either does not fund the technology or even makes it a criminal offence	A precautionary approach to risk assessment; it is presumed that a new product is dangerous until it is proven safe; therefore, labelling of all products from modern biotechnology is mandatory irrespective of the level of risk from them
Consumer demand and acceptance	Profound impact on the types of products developed	Information dispersal, public discussions/debates, etc. modify the public opinion/demand
Preferred areas of public research and conditions on research grants	Affects the preferred areas of innovation	Serves as a strong regulator; easier to adapt to the changing experience and knowledge

12.3. REGULATION OF INNOVATIONS IN BIOTECHNOLOGY

The innovations in biotechnology are regulated by a variety of different mechanisms ranging from voluntary self-regulation to legislations enacted by competent authorities (Table 12.2). These mechanisms differ markedly in their features like flexibility, enforceability, etc. Voluntary self-regulation involves development of and adherence to a set of rules and procedures by the concerned community, *e.g.*, the scientific community for research activities. This mechanism can be highly effective yet quite flexible, and it can accommodate subtle changes without sacrificing credibility. However, voluntary self-regulation is technically not enforceable, except by segregation of defaulters by the concerned community. Some of the rules devised for voluntary self-regulation may later become part of national guidelines or regulations. For example, during 1970s, the research community responded to the potential risks from recombinant DNA technology by voluntarily developing and adhering to a set of rules and precautionary measures to minimize these risks (Charo, 2016).

Table 12.2. Various mechanism for implementation of the biotechnology regulatory framework (based on Charo, 2016)

Mechanism	Salient features	Remarks
Voluntary self-regulation	Set of rules developed and adhered to by the concerned community, *e.g.*, scientific community; some of the rules may later become part of national guidelines or regulation	Can be flexible and able to accommodate subtle changes without sacrificing reliability
Government guidelines	Set of rules and practices expected to be followed by the concerned persons; developed by the concerned government agency	Technically not enforceable but strongly persuasive since their compliance tends to protect from violation of actual regulations/laws
International organizations/ conventions	Provide global standards for regulation; nations can adopt and implement with necessary changes; occasionally have enforcement-level value	Example, Council of Europe's Oviedo Convention specifically calls for a prohibition of genetic engineering of human germline
National laws	They are binding, politically more credible especially in democracies	Legislation is extremely rigid and difficult to change or remove; takes several years to enact/amend a law
Regulations	Detailed administrative rules adopted in accordance with the relevant law	Much more responsive to expert opinion; tends to retrogress into a rule by the expert and become detached from the public opinion
Public consultation	The government policy is allowed public scrutiny; public opinion helps policy modification	The least enforceable approach; considered as an alternative to a centralized directive form of governance

Another regulatory mechanism comprises guidelines containing a set of rules and practices developed by an academic body or the concerned government agency. The guidelines are expected to be followed by the concerned persons/institutions/industries, but they are technically not enforceable. However, guidelines are strongly persuasive since their compliance tends to protect the persons/organizations from violation of other existing relevant regulations/laws. International organizations and/or conventions provide global standards for regulation of research and development activities in the concerned fields. For example, the 1997 Convention on Human Rights and Biomedicine (also called the Oviedo Convention) of the European Council specifically calls for a prohibition of genetic modification of human germline. Individual nations can adopt and implement these regulations with some modification where necessary. Occasionally, the international conventions and/or guidelines may be enforceable as well (Charo, 2016).

The most rigid and enforceable regulation takes the form of a national law enacted by the competent authority of the concerned country. Legislations are extremely rigid and difficult to modify and/or remove. But they are binding on the concerned parties, and are politically more dependable, especially in democratic countries. The laws are implemented through regulations comprising detailed administrative rules devised in accordance with the relevant laws. The regulations are more responsive to expert advice and public opinion, but they tend to regress into a 'rule by expert' and become detached from the public opinion. In countries like United States, extensive public consultations take place during the development of regulations and procedures for the governance of novel technologies. In this process the government policy is allowed public scrutiny, and the public opinion is used to introduce appropriate policy modifications. The public consultation is the least enforceable approach of regulation, but it can lead to adjustments in the relevant regulations and procedures. It can also create pressure on the industry leading to modifications in the direction and speed of innovations. 'Public consultation is considered as an alternative to a centralized directive form of governance' (Charo, 2016).

A fundamental issue in regulation of biotechnology innovations is whether to regulate the product or the process of innovation. The EU follows the approach of regulating the process/technology; this approach is more easily understood and appreciated by the public. It also allows one to 'focus on key aspects of the science that create key questions about the effects of a particular kind of innovation'. 'It also has the problem of needing much more specific legislation to focus on individual products.' The USA, on the other hand, regulates the product, which considers the technology to be 'neither inherently dangerous nor safe.' It recognizes that some products of a technology may be safe, some others may generate detrimental effects, while in some other cases it may not be easy to either predict or determine the consequences. Similarly, different products may have different impacts on the environment. Therefore, product-based regulatory system can determine the risks from specific products, and take advantage of the accumulated scientific knowledge in devising regulatory procedures appropriate for different categories of products, *e.g.*, therapeutic agents, foods, etc. However, this type of regulation tends to confuse the public since there are too many different regulations applicable to different biotechnology products (Charo, 2016). The consequences of process-based and product-based regulations are highlighted by the decision by USA to exempt gene-edited crop varieties from regulation, while EU has decided to regulate them in the same way as GM (genetically modified) crops.

The regulatory requirements should be evidence-based and proportionate to the risk assessed for the concerned innovations. Further, the process should be transparent, globally harmonized and able to accommodate new developments. Those innovations for which no major hazards are identified, may be exempted from pre-market regulation and they may be subjected to a post-market monitoring system. Ambiguous and/or inconsistent regulations (*e.g.*, differentiating between mutagen-induced and SDN-induced varieties having indels) should be avoided. Where necessary, the existing legislations may be amended in the light of available scientific evidence and other relevant information. Finally, the risk assessment should be of the actual product; the process used to generate the product may also be considered, but it should not be the primary basis for the regulatory procedure used for the product (Jones, 2015).

In summary, efficient pre-market regulation can minimize adverse events later in the commercialization process. But a too rigid pre-market regulation would slowdown the pace of innovation and discourage smaller companies to engage in biotechnology innovation. For example, a Canadian company trying to develop a pig strain that would cause less environmental damage decided to abandon the project in view of the long, slow and expensive process of gaining approval for commercialization of the new strain. In addition, the conditions attached to research grants from the government or charitable organizations function as strong regulators of research and, thereby, development activities. Finally, the regulatory procedures need to be harmonized among countries since this would allow scientists to share research materials with confidence and without the fear of ethical and/or legal violations. This would also help maintain uniformly high standards of products/therapies and reduce conflicts and redundancies. It has been suggested that innovation and precaution should not be viewed in conflict with each other; in fact, they are complementary strategies since precaution will nurture confidence in new technologies and, thereby, facilitate innovation (Charo, 2016).

12.4. BIOSAFETY CONCERNS WITH GENE-EDITED CROPS AND ANIMALS

The main limitation of CRISPR-Cas9 and other SDN systems is their off-target action. The frequency of off-target action depends on the SDN system, the organism and the cell type used for gene editing, the targeted gene, and some other factors. Although high specificity variants of Cas9 have been developed and computer programmes for designing of sgRNAs with enhanced specificity greatly reduce the risk of off-target action, an SDN system free from off-target editing remains elusive. In addition, CRISPR-Cas9 is known to induce large deletions and other unintended complex genomic rearrangements at the targeted sites. Another biosafety issue relates to the efficiency and the mode of SDN delivery. Delivery of SDNs as DNA poses the risk of integration of plasmid DNA into the cell genome, particularly in the case of non-viral delivery methods. Therefore, SDNs are preferably delivered as protein, mRNA or ribonucleoprotein. But even with such modes of delivery, the donor DNA plasmid delivered for HDR (SDN-2 and SDN-3 technologies) may integrate at the target site (Section 8.6; Cribbs and Perera, 2017). Therefore, the delivery systems need to be refined to increase their efficiency and, especially, safety. Further, due care needs to be taken to ensure that the gene-edited cells/zygotes/ individuals used for the succeeding steps of the gene editing protocol are free from unintended genetic changes, including integration of foreign DNA.

It has been argued that the current methods of detection and monitoring of off-target and on-target genomic alterations are not adequate and even whole genome sequencing may not identify unintended sequence changes unless the genome sequence data are appropriately analysed (Section 8.6; Cribbs and Perera, 2017). It has been pointed out that the negative public perception toward GMOs (genetically modified organisms) is partly due to the distrust in the regulatory procedures and the motives of the organisations involved in the development of GMOs. Therefore, it is important that gene-edited lines/varieties are thoroughly analysed for unintended sequence changes before they are submitted for regulatory clearance. In addition, the regulatory authorities should carry out thorough risk assessment leading to timely approvals to avoid suspicions in the minds of the public at large (Ishii and Araki, 2017; Smyth, 2017).

It has been observed that some off-target edits might lead to gain of function mutations. These mutations may or may not cause safety concerns depending on the nature of activated genes. In addition, some trait modifications by gene-editing may raise environmental issues. For example, gene editing has been used to generate herbicide resistant mutants of many crop species, and an ODM (oligonucleotide-directed mutagenesis)-induced sulfonylurea resistant canola mutant has been commercialized as SU Canola. Herbicide resistance can be transferred to wild relatives of the crop leading to production of herbicide resistant hybrids (Ishii and Araki, 2017). Finally, the cultivation of herbicide resistant varieties would lead to emergence of herbicide resistant variants of weed species and this would ultimately reduce the usefulness of the herbicide resistant variety. This problem can be overcome by growing another variety resistant to a different herbicide after every few years so that the resistant mutants/hybrids are eliminated by the application of a different herbicide to which they are susceptible.

12.5. ETHICAL ISSUES RELEVANT TO GENE EDITING IN PLANTS AND ANIMALS

Genetic manipulations including gene editing of plants are well accepted by the society and there is virtually no ethical objection to these activities. However, gene editing of animals raises several ethical concerns. Gene editing of livestock for basic research is generally well accepted by the biologists and, to a large extent, public at large. The bioethical concerns related to the use of gene-edited animals as disease models for various biomedical researches depend on the moral status assigned to the concerned animal species. For example, there is very little objection to the use of fruit fly, mouse and rat models of human diseases, while many people object to the use of nonhuman primates like monkeys since they are evolutionarily far closer to humans than fruit fly, mouse or rat. But scientists argue that NHPs are matchless as animal models for brain research and biomedical studies on neurological and cardiac disorders (Section 8.7). This situation creates the following paradox: the closer an animal is to humans in terms of evolutionary relationship, the better it would serve as a disease model and the more ethically objectionable would be its use as a model. One of the possible reasons for objections to the development and use of novel tools for genetic modification in animals like NHPs is the fear that sooner or later the same tools might be used in humans. Another objection is that such uses of animal species tends to reduce their status to mere commodities, which may lead to inappropriate treatments being meted out to them (Ogden, 2019).

12.6. REGULATION OF GENE-EDITED PLANTS

Regulation of biotechnology products relies heavily on two international instruments, *viz.*, the Cartagena Protocol on Biosafety (SCBD, 2000) and the updated version of Risk Analysis Principles for Foods Derived from Biotechnology by the Codex Alimentarius Commission (CAC, 2014). Many national regulations have adopted the definition of Cartagena Protocol for 'living modified organisms' (LMOs). The Protocol defines an LMO as 'any living organism that possesses a novel combination of genetic material obtained through the use of modern biotechnology', *i.e.*, application of '*in vitro* nucleic acid techniques, including recombinant DNA' or 'fusion of cells beyond the taxonomic family.' The various national regulations have been classified as product-based or process-based depending on their definition of GMOs. When an organism is classified as GMO if it has a specific genetic status, *e.g.*, 'a novel combination of genetic material obtained through the use of modern biotechnology', the regulation is treated as product-based. But if the definition of GMO focuses on the process used for the genetic modifications, *e.g.*, 'the genetic material has been altered in a way that does not occur naturally by mating and/or natural recombination', the regulation is treated as process-based (Ishii and Araki, 2017).

Nearly equal proportion of countries that cultivate GM crops follow product-based and process-based regulations, while in case of few countries like Paraguay, Myanmar, Chile and Vietnam the position is not clear. Five main countries that follow product-based regulation are USA, Argentina, Canada and Uruguay, while EU, Brazil, India, China, Pakistan and South Africa follow process-based regulation. It seems that process-based regulations discourage GM crop cultivation, whereas product-based procedures encourage commercialization of GM crops (Table 12.3), although Japan, the Republic of Korea and the Russian Federation follow a product-based regulation, GM crops are not commercially cultivated in Japan and Korea, while Russia has banned GM crop cultivation (Ishii and Araki, 2017). A product-based regulatory system would consider gene-edited organisms produced by SDN-1 and some SDN-2 technologies as non-GM, while a process-based system would tend to classify them as GM (Table 12.4). SDN-3 products, however, would be treated as GM under both the systems.

Table 12.3. Salient features of regulation of GM crop varieties in selected countries

Country	Regulation based on	Trigger for regulation	Remarks
EU	Process	Alteration of genetic material 'in a way that does not occur naturally by mating and/or natural recombination'	The most demanding regime; only one variety approved so far
USA	Product	Presence of foreign DNA, including plant pest DNA and/or the submitted line being or likely to become a plant pest	Regulated under 7 CFR Part 340 and Part 360
Canada	Product	'Plants with novel traits'	It does not consider the breeding method used
Argentina	Product	Presence of 'a new combination of genetic material'	One of the oldest and the most articulate regulation
India	Process	Insertion of foreign genetic material, presence of new combinations of genetic material, deletion and removal of parts of the heritable material	GM crops other than Bt cotton are banned since 2019

The crop genotypes developed by SDN-1 and SDN-2 technologies are expected to be transgene-free unless a novel DNA sequence has been used for SDN-2 technology (Section 7.13). Further, these products will be modified by deletion, addition or substitution of few base pairs as are the mutants generated by conventional mutation breeding. Significantly, gene editing generally produces fewer off-target sequence changes than conventional mutagenesis, and the lines derived from the latter are never subjected to any molecular analyies to assess unintended genetic alterations. It was therefore, expected that crop varieties developed by gene editing without any transgene integration will be exempt from the regulatory requirements applicable to GM crop varieties. The position adopted by the APHIS (Animal and Plant Health Inspection Service) of USDA (United States Department of Agriculture) is in conformity with this expectation, while the European Union has adopted the opposite view.

The APHIS considers genome editing as an expansion of the existing plant breeding tools and recognizes it to be quicker and more precise than the latter. It regulates plant varieties based on their features and not on the basis of tools and techniques used for their development. There was no specific policy formulated for gene-edited plant varieties, which were regulated on case-by-case basis under the US Plant Protection Act. APHIS does not consider the gene-edited varieties as GM crops and exempts them from regulation provided 'they are themselves not plant pests, and they do not contain any plant pest sequences or any other inserted genetic material.' In addition, the genetic changes induced by gene editing procedure should be such that can be induced by cell's own DNA repair machinery. In case of sequence replacement, the introduced sequences should be from the same species and preferably from an allele known to occur naturally in the crop species. Thus, the gene-edited line should be free from foreign DNA. Finally, the genetic changes generated by gene editing should not be such that they might increase the persistence and weediness of the edited lines (https://www.aphis.usda.gov/aphis/ourfocus/biotechnology/am-i-regulated). In March 2018, the USDA excluded genome-edited plants from regulatory oversight. Similarly, Australia has decided not to regulate crop varieties (as well as animal and human cell lines) developed by SDN-1 technology (Whelan and Lema, 2015; Mallapaty, 2019).

Table 12.4. The salient features of products from different SDN technologies relevant for regulations of the products (based on Jones, 2015)

SDN technology	Process	Genetic consequences	Remarks
SDN-1*	SDN induces DSB at the targeted site; DSB repaired by NHEJ	Indels produced at the targeted site leading to inactivation of the targeted gene(s)	Transgenes and plasmid sequences removed by backcrossing; mutant lines comparable to those produced by conventional mutagenesis; should be exempted from regulation
SDN-2	SDN and a donor DNA used; the DSB is repaired by HDR using the donor DNA as repair template	The sequence of donor DNA replaces the homologous sequence of the targeted gene; generally, used for mutation correction or induction of predecided mutations	Most difficult to decide the mode of regulation; likely action: sequence change to wild type or a known allele may be treated as SDN-1, while that to a novel allele may be considered as SDN-3
SDN-3**	SDN and donor DNA containing the complete transgene construct; the DSB is repaired by HDR using the donor DNA leading to transgene integration at the DSB site	The transgene is generally integrated within a 'safe' harbor site of genome; transgene expression will not be subject to position effects associated with the site of integration	Multiple transgenes can be stacked in the same genomic region; resulting lines treated and regulated as GM; biosafety evaluation would be less demanding, *e.g.*, less event-specific data may be required, since the unpredictable effects due to integration at random sites are avoided

* DSB, double-strand break; HDR, homology-directed repair; NHEJ, non-homologous end joining

**`Safe' harbor, a genomic site within which transgene integration does not lead to disruption of an essential endogenous gene, activation of a detrimental native gene, and other position effects. If necessary, a 'trait landing pad' (TLP) site can be created in a suitable genomic region for trait stacking.

The regulatory system followed in Canada is product-based and the regulatory trigger is 'Plants with Novel Traits' (PNT). The phrase 'novel trait' means a trait that is not present in the cultivated varieties of the same species irrespective of the breeding procedure used for development of the line/variety under consideration. Canadian plant breeders generally consult the regulators if the targeted trait expressions in their new varieties are 20-30% higher or lower than those in the conventional varieties; generally, varieties with changes within the above range have been exempted from regulation (Whelan and Lema, 2015; Smyth, 2017). But the regulatory system in India is process-based and is derived from the "Rules for Manufacture, Use/Import/ Export & Storage of Hazardous Microorganisms/Genetically Engineered Organisms or Cells, 1989". This rule uses a very broad definition of genetic engineering that includes insertion of recombinant DNA and "formation of new combinations of genetic material… by deletion and removal of parts of heritable material". 'This implies that all new technologies will be subject to regulation under the provision of Rules, 1989' (Ahuja, 2018). The broad-based and somewhat ambiguous definition of the regulated articles is likely to stifle research and development efforts in genome editing since their products will be classified as GMOs. It is noteworthy that commercial cultivation of only a single GM crop, Bt-cotton, has been approved in India although five more Bt-events were approved for cultivation after the first approval of Bollgard-1 event of Bt-cotton in 2002. India is in the process of revising its regulatory guidelines for gene-edited organisms; the draft guideline seems to exempt SDN-1 products from biosafety evaluation.

The EU follows precautionary approach to risk assessment and regulation is process-based. As a result, the regulatory procedure focuses on the tools and techniques used for the development of new crop varieties, and not on the features of varieties themselves. Further, GM crop varieties are regulated according to the European Council Directive 2001/18/EC, which defines GMO as 'an organism, with the exception of human beings, in which the genetic material has been altered in a way that does not occur naturally by mating and/or natural recombination'. However, the directive excludes plant varieties and other organisms developed by conventional mutagenesis and cell/protoplast fusion provided recombinant DNA molecules were not used in the process. Further, the directive is silent about gene-edited lines/varieties since at the time it was formulated, the development of gene editing tools had barely begun. In view of the above, a gene-edited variety will be treated at par with GM varieties and would be regulated as per the above directive. This situation is anomalous since a gene-edited variety is to be regulated differently from those developed by conventional mutation breeding and somatic hybridization even when the two groups of varieties would have identical genetic changes and other relevant features (Georges and Ray, 2017).

Therefore, a clarification was sought from the European Court of Justice (ECJ) regarding the application of this directive to gene-edited crop varieties. On 25 July 2018, the ECJ decided that gene-edited crop varieties will be treated at par with GM crop varieties and will be subject to the same regulations. It is noteworthy that the EU has one of the most rigorous and expensive regulatory mechanisms in the world. There is stringent requirement for a diverse set of data, and standards of risk assessment are high. In addition, the process of approval is slow, and outcome of the process is unpredictable. The discouraging impact of this demanding regulatory procedure is evident from the fact that till 2017 the insect resistant maize variety MON810 was the only GM variety approved by EU for commercial cultivation. It may be pointed out that this GM variety was allowed commercialization in 1998 under the earlier approval system. Since then, several requests for approval of GM varieties have not been acceded to despite positive reports on risk assessment from the European Food Safety Authority (EFSA; Georges and Ray, 2017). The reason for this situation seems to be that the approvals for commercial cultivation of crop varieties are granted by the Standing Committee on Food Chain and Animal Health of the

European Commission, and this committee 'has become co-opted by the European environmental movement' (Smyth, 2017).

There have been representations that the gene editing be treated as an extension of the conventional plant breeding tools and techniques and the gene-edited crop varieties be relieved from regulatory requirements. It has been pointed out that genome editing tools act as benign and highly specific mutagens, and they produce sequence alterations, including the occasional changes due to editing at potential off-target sites, that are analysed and characterized in the mutant lines. In contrast, wheat plants subjected to conventional mutagenesis are likely to contain in their genomes, on an average, 415,000 mutations (= base sequence changes) per plant. Further, the actual number of genetic changes and their nature are never determined in the varieties derived by conventional mutagenesis before they are commercialized. Most scientists cannot understand why the law should treat mutational changes like single nucleotide polymorphisms (SNPs) and small indels produced by spontaneous errors or physical/chemical mutagen treatments differently from those induced by treatments with SDNs. At the same time, the enforcement of above ECJ ruling will pose tough technological and legal challenges since the regulatory authorities will find it extremely difficult to reliably distinguish between the DNA sequence alterations produced by SDNs (SDN-1 and some SDN-2 events) from those generated spontaneously or by conventional mutagenesis. The detection of gene-edited sequences may become easier provided the companies commercializing these varieties voluntarily provide the data necessary for unequivocal identification of their gene-edited varieties (Jones, 2015; Georges and Ray, 2017; Ledford, 2019b).

In view of the above discrimination, it is argued that the EU Directive 2001/18 should be revised and updated taking into account the major advances in the field of targeted genome editing. It is encouraging that some member states like Germany, Belgium, Holland and Sweden have decided not to consider varieties developed by ODM as GMOs. It may be pointed out that the application of GM regulatory procedures to gene-edited crop varieties is stifling innovative research and development activities, and it is encouraging their increasing confinement to few very big multinational companies. The application of GM regulatory procedure to gene-edited varieties is not logical, and it will further confuse the already complicated trade and labelling of bulk commodity shipments into EU. It has been suggested that if 'the standards applied to GM crops were applied with equal rigour to the products of classical plant breeding, few new cultivars would survive the process' (Jones, 2015; Georges and Ray, 2017).

The Argentinian regulatory system for GMOs is one of the oldest, and Argentine Biosafety Commission CONABIA was recently recognized by the Food and Agriculture Organization (FAO) as Centre of Reference for the Biosafety of GMOs. In addition, the Argentinian experience contributed to the formulation of CPB and the existing Argentinian regulatory system is fully compatible with the latter. The regulatory system for gene-edited crop varieties in Argentina is described in detail in their published guidelines. A plant is considered 'as a GMO when it possesses a new combination of genetic material created by the insertion of one or more genes or DNA sequences that are a part of a genetic construct have been inserted into the plant genome.' Therefore, the lines developed by the SDN-1 technology will be treated as non-GM and exempt from regulation. In addition, Argentina also provides preliminary advice, during the formulation stage of projects, concerning the manner of regulation of new crop varieties likely to be developed by these projects. This is very thoughtful, logical and supportive approach to regulation of products of new biotechnologies. Argentina considers deregulation on case-by-case basis and uses the presence of "novel combination of genetic material" as the key to decide if a product is GMO (Jones, 2015; Whelan and Lema, 2015).

In summary, no country has enacted laws specifically for the regulation of gene-edited crop varieties. The regulatory decisions concerning gene-edited varieties of crops are based on the existing laws for regulation of GM crops. Countries like Argentina and New Zealand have taken decisions within their existing laws to treat SDN-1 products as non-GM and GM, respectively. The Swedish authorities have also indicated to deregulate transgene-free plant varieties. It is expected that gene-edited crop varieties developes by SDN-1 and SDN-2 technologies will soon reach commercial cultivation and might be approved for use as food by the health regulators in USA and Argentina (Ishii and Araki, 2017). In fact, the FDA, USA approved in 2019 the gene-edited high oleic acid soybean for human consumption.

12.7. REGULATION OF GENE-EDITED ANIMALS

It has been argued that gene editing produces comparable types of genetic changes irrespective of the biological kingdom of the concerned organism. Therefore, gene-edited animals may be expected to be subjected to the same regulations as are the gene-edited crop varieties. Countries like Argentina and Brazil regulate gene-edited animals in conformity with this expectation, but the situation in USA is very different. According to the draft proposal of the United States Food and Drug Administration (FDA; "Regulation of Intentionally Altered Genomic DNA in Animals"), gene-edited food animals would be treated as drugs and regulated accordingly. This proposal has been criticized on the grounds that it treats a change in DNA base sequence as drug even when the same change occurs naturally and is accepted without a hint of any concern. The proposal seeks to apply the same criteria for regulation of gene-edited animals that have been applied to transgenic animals. This is justified for animals produced by SDN-3 technology but appears to be too harsh for the products of SDN-1 and SDN-2 technologies (Van Eenennaam, 2018; Van Eenennaam et al., 2019). In EU also, gene-edited animals and their products are considered as regulated articles and require approval by regulatory authorities.

It is feared that such a harsh regulatory regime will stifle research and development in the area, encourage monopoly of large multinationals, and drive research, development and commercialization activities to other countries with more amicable regulatory regimes. Therefore, the 2019 Plant and Animal Genome meeting has petitioned that plant and animal lines developed by modern methods like gene editing, which can be developed by traditional breeding methods as well, should not be subjected to regulation (Van Eenennaam, 2018; Van Eenennaam et al., 2019).

12.8. REGULATION OF GENE DRIVES

Gene drives are SDN-based self-propagating genetic devices that spread in the targeted populations and reduce, do not affect or enhance their fitness either by inactivating an endogenous gene or introducing a novel gene into the genome of concerned crossbreeding populations. Gene drives can be used to eradicate harmful/parasitic populations, rescue endangered populations or modify the characteristics of a harmful population to render it harmless without affecting its survival/fitness. Gene drives are extremely powerful tools for managing natural populations by altering their genetic makeup, which can have important implications for human health; they offer feasible solutions to agricultural and ecological issues that may otherwise be extremely difficult to manage. The deployment of gene drives promises to be inexpensive and require much less effort than that of conventional biocontrol measures (Section 11.18).

The gene drives, however, suffer from several limitations, including invasion of other populations of the concerned species in the same and even neighbouring countries. They can also spread to populations of related species that show even sporadic mating with the target species. Once a gene drive is introduced into a population, it can be inactivated or eliminated only by introducing of another gene drive. Therefore, gene drives that protect the population from invasion of unwanted gene drives, or remain confined to the targeted populations have been developed. In addition, gene drives like various daisy drives have been developed that are self-limiting and are eventually lost from the population. Since gene drives can lead to human-driven extinction of targeted species, numerous ethical issues have been raised, which range from 'playing God' to unforeseen ecological consequences and the risk of this technology being misused, *e.g.*, by extremist groups (Sections 11.16, 11.19).

There have been demands for a moratorium on gene drive research, but the current view seems to be to continue 'to evaluate the potential adverse effects of gene drive applications, and to formulate effective measures to prevent the hijacking of this technology by extremists and to avoid misleading public expectations with overinflated promises' (Chneiweiss et al., 2017). The NASEM report (NASEM, 2016) 'called for a responsible science approach to developing and applying gene drive technologies' and recommended that data on the risk of horizontal transfer of gene drives should be accumulated before their deployment in natural populations. The behaviour of gene drives has been studied only in laboratory populations, and this information might be only of limited value for extrapolation to natural populations. The information on gene drive behaviour in natural populations cannot be obtained without field tests, which cannot be undertaken in the absence of appropriate guidelines for such tests. But the guidelines and regulatory procedures themselves cannot be developed unless dependable information on potential risks and benefits from gene drives become available from suitably designed field experiments. This leads to the dilemma that field tests cannot be conducted without guidelines, which cannot be developed in the absence of relevant data from field tests.

12.9. HUMAN SOMATIC CELL GENOME EDITING

Human genome editing can be used to genetically alter somatic cells, germ cells, gametes, zygotes and embryos for a variety of purposes, including biomedical research. It is broadly classified into two groups, *viz.*, somatic cell and germline gene editing. Both these groups are further classified into basic, preclinical and clinical researches. The existing regulations to address ethical issues are considered adequate for oversight of gene editing for basic research. Clinical applications involve editing of somatic cells and ultimately introduction of the edited cells into ailing individuals for various therapies. The genetic changes generated by gene editing remain confined to the individuals receiving the treatments. These applications are subject to the same ethical norms and regulatory oversight as the conventional gene therapy procedures (Section 12.9.1), and they must undergo clinical trials like any other therapeutic agent/procedure. However, in case of gene editing for *in vivo* gene therapies, concerns relate to efficient editing of the targeted cell types and unintended editing of germline cells. In spite of these limitations, clinical trials of *in vivo* gene editing for haemophilia B and mucopolysaccharidosis I are already under way (NASEM, 2017).

12.9.1. Ethical issues in clinical somatic cell genome editing

There are two main groups of views on the ethical issues related to clinical human somatic cell genome editing. The first group considers somatic cell editing as an extension of the existing gene therapy procedures. It argues that there are no novel challenges posed by genome editing provided the standards of safety and efficiency are ensured. The relevant ethical issues are assessment of safety of the technology, evaluation of risks/benefits from their application, protection of vulnerable subjects, informed consent of the patients, monitoring of the patients, etc. It may be pointed out that these issues are the same as those for existing clinical research for the development of novel human gene therapies. The second group, however, argues that CRISPR-based genome editing is an extraordinary technology that would require reassessment of the existing 'acceptable thresholds for harms and benefits', ethical issues involved, the regulatory mechanisms and revision of 'clinical trials pathways'. However, there is general public acceptance of human somatic cell genome editing and the use of edited cells for clinical applications (Cavaliere, 2019).

12.9.2. Regulation of gene therapies

The regulation of gene therapies varies to considerable extent among different countries (Table 12.5). In USA, gene therapies are considered as biological drugs and are regulated accordingly. They are subject to comprehensive regulation by FDA under 'multiple laws focusing on infection control, efficacy, and safety.' Each gene therapy product and procedure must undergo properly conducted clinical trials and ultimately be approved by FDA for clinical application. However, once a therapy is approved, the regulation of its clinical use is much weaker than the premarket regulation. South Korea follows a similar regulatory mechanism, but it allows extended access to the new drugs. It is also evolving a system of 'conditional approval' that would permit some use of the therapy before its formal approval for clinical use. The EU has special provisions for medicinal products used for advanced therapy, including a centralized review procedure for quality control. The UK has, in addition to the EU directives, its own very strong premarket review process coupled with an equally strong post-market regulation of procedures involving human embryos/fertilization (Charo, 2016).

In Japan, the mechanism of regulation is dependent on the risk assessment of individual gene therapies. The therapies are prospectively classified into high, medium and low risk categories, and they are regulated accordingly. Singapore also follows a similar approach, but the risk assessment procedure is more comprehensive than that in Japan. But the regulatory process of USA considers all new drugs to be equally dangerous and pose comparable risks so that they all undergo the same evaluation process for safety and efficacy. In Brazil, the regulations relevant to gene therapies are layered over earlier and more general rules, *viz.*, sale of human tissues and patenting of human biological materials, which makes the actual regulatory regime quite complex. Thus, the Brazilian regulation is 'an example of regulation and governance by accretion', which creates confusion (Charo, 2016).

Table 12.5. The salient features of regulation of human gene therapy in selected countries

Country	Salient features of regulation
USA	Treated as biological drug and approved by FDA after properly conducted clinical trials; clinical applications not regulated; somatic cell genome editing permitted for treatment or prevention of disease or disability
European Union	EU has a centralized review procedure for quality control; regulations for gene therapy harmonized across member states and applied to somatic genome editing for clinical applications
United Kingdom	Operates within the context of the EU and its directives; also has its own very strong pre-market review process and very strong post-market regulation of procedures involving human embryos/fertilization
South Korea	Pathway similar to that of USA; but it favours expanded access to investigational drugs; also developing a system of conditional approval*
Japan	Initial determination of risk level and classification of therapies as high, medium, or low risk and regulates them accordingly; also added a system conditional approval*
Singapore	Risk-based approach similar to that of Japan; follows a detailed evaluation of the risk levels of new therapies
Brazil	The regulations relevant to gene therapies are layered over earlier, more general rules, which creates confusion

* It is a challenge 'to balance the desire to move forward as quickly as possible while avoiding the kinds of adverse outcomes that not only injure individuals but could slow progress' in the field.

In brief, the existing regulatory frameworks involve assessment of the balance of anticipated risks and benefits to the patients. It was noted that the off-target effects vary with the gene editing technology, cell type, targeted gene, and some other factors. As a result, a single standard for the efficiency and/or off-target activity for somatic cell gene editing cannot be prescribed yet. When conducted carefully and with proper oversight, gene therapy research has enjoyed support from many stakeholder groups (NASEM, 2017).

12.10. HUMAN GERMLINE GENOME EDITING

Germline genome/gene editing may be defined as editing of germ cells, gametes, zygotes or embryos. It may be classified as basic, preclinical and clinical gene editing depending on the use of edited cells/embryos. Germline editing for basic research aims to answer questions related to human embryology, DNA repair pathways in germline cells, the efficacy and safety of gene editing platforms, etc. But pre-clinical editing is designed to evaluate the 'effectiveness and safety of possible future clinical/reproductive' uses of genome editing procedures. Finally, clinical germline editing aims to use genome editing protocols for effective treatment of specific serious diseases or conditions in the babies produced from the edited cells/embryos. The embryos used for germline editing may be spare embryos from IVF (*in vitro* fertilization) procedures or they may be specifically created for research from donated oocytes fertilized *in vitro*; the latter are called research embryos. Since this involves collection and use of germline cells, the ethical and regulatory issues related to the collection of cells and the ultimate use of gene-edited cells must be adequately addressed. The basic research generates information that would facilitate preclinical research, while preclinical research would yield information necessary for planning and execution of clinical germline genome editing (de Wert et al., 2018).

The rumour that Chinese scientists had used CRISPR-Cas9 for gene editing of preimplantation stage "nonviable" human embryos (this work was later published in Protein & Cell; Liang et al., 2015) triggered two publications in Nature and Science. The paper in Nature called for a 'voluntary moratorium on germline genome editing', while the publication in Science recommended a 'prudent path forward'. In addition to the above publications, several articles in scientific journals and popular press, and many professional bodies, international organizations, and national academies from all over the world expressed concern about human genome editing, particularly heritable genome editing, and for regulation of its appropriate uses. In December 2015, scientists of major world academies called for a moratorium on heritable human germline editing, but supported continuance of basic research and somatic cell gene editing. In February 2016, British scientists were given permission by regulators to genetically modify human embryos on the condition that the embryos must be destroyed in seven days (Baltimore et al., 2015; Cavaliere, 2019).

In 2015, the U.S. National Academies of Sciences and Medicine announced a human genome-editing initiative to cover the science, ethics, and regulation of this upcoming discipline. This initiative convened the 'International Summit on Human Gene Editing: A Global Discussion' jointly with the Chinese Academy of Sciences and The Royal Society of U.K. The discussions during and the conclusion reached by the summit provided important inputs for the Committee on Human Gene Editing: Scientific, Medical, and Ethical Considerations. The committee comprised an international group of scientists, clinicians, ethicists, patient advocates, and government officials; it was convened to continue the dialogue initiated by the International Summit. The report of this committee was published by NASEM in 2017; this report outlines the principles for supervision and regulation of research on human genome editing. It also proposes a possible pathway for its ultimate use for therapeutic applications. At present, government funds cannot be used for human germline editing research in USA and several other countries, and FDA (USA) does not approve clinical trials based on germline genome editing (Dunbar et al., 2018; NASEM, 2017).

12.10.1. Safety concerns in human germline genome editing

The objections to human germline genome editing are based on both biosafety and ethical considerations (Table 12.6). The biosafety issues relate to off-target effects, detrimental pleiotropic effects of edited genes, unplanned on-target genetic changes, incomplete editing of embryos leading to mosaicism, and undesirable epigenetic effects. In addition, low efficiency of on-target editing and the lack of sufficient information on safety of gene editing agents and procedures are other relevant concerns. Despite remarkable refinements, the SDNs used for gene editing do produce a variable frequency of off-target effects and unintended on-target genomic alterations. Further, the frequencies of these effects are affected by a variety of factors, including the cell type and the targeted gene. The bulk of data on various safety aspects of genome editing have been collected from experiments with somatic cells, and there is paucity of information from studies with germ cells, gametes and embryos. Therefore, it is important that basic research on gene editing in all cell types is fostered to establish the feasibility, efficacy and safety of the technology. In addition, studies should be undertaken to collect adequate information to facilitate a decision on the acceptable levels of off-target and other undesirable effects associated with gene editing (Chneiweiss et al., 2017).

Table 12.6. A summary of the different types of potential risks due to gene editing in humans using CRISPR-Cas9 (based on Cribbs and Perera, 2017)

Gene editing application	Unintended editing*	Random integration of the delivery vector	Toxicity due to immunogenicity of delivery vector	Aggravation of social inequalities
Somatic cell editing (*ex vivo* gene therapy)	High (edited cells can be cloned and clones free from unintended edits can be selected and used)	Low (clones with integrated vector can be identified and removed)	Low	Low
Somatic cell editing (*in vivo* gene therapy)	High (selection against cells with unintended edits not feasible; nontarget cells and even germline cells might be edited)	High (monitoring of vector integration not feasible; consequence will depend on the integration site)	High (vector components can initiate immune or inflammatory response)	Low
Germline editing	High (the embryos could be screened and only those that are free from unintended edits are used for implantation)	Medium (the embryos could be screened and only those that are free from vector sequences are used for implantation)	Low	High (gene editing can be used for enhancement)

* Off-target editing (indels) and unintended on-target editing

One of the challenges for policy makers is presented by the fact that the type and the genomic location of off-target effects associated with gene editing cannot be predicted. It would, therefore, be critical to establish reliable protocols for detection and characterisation of the unplanned off-target and on-target changes. At the same time, a method of risk assessment needs to be developed that would consider the likely consequences of these alterations. It may be pointed out that the consequences of off-target edits will be greatly affected by the genomic sequence affected and the type of sequence change induced by gene editing. Undesirable effects of gene editing may also arise from pleiotropic action of the edited genes. For example, gene *CCR5* encodes the receptor for AIDS virus. Inactivation of this gene generates resistance to HIV infection, but it may also increase susceptibility to certain other viruses. Further, it is very difficult to assess the long-term effects of germline genome editing since these will extend over several generations.

According to one view, germline editing is exceptional, and one should exercise extreme care in proceeding with its application. But the second view considers it to be an extension of the assisted reproductive technologies and objects to it being labelled as exceptionally risky. A combination of precautions has been suggested to reduce the health risks for children produced by germline gene editing; these measures include, (i) adequate preclinical research, (ii) restriction of clinical applications to such genes that are the direct cause of the concerned diseases, (iii) the mutant alleles planned to be induced are known to occur in relatively high frequencies in the population, (iv) comprehensive genome analyses to exclude off-target effects, (v) coupling clinical application with long-term follow-ups, etc. (de Wert et al., 2018; Cavaliere, 2019; Tuerlings, 2019).

The general view seems to be that further pre-clinical studies are a prerequisite for clinical application of germline editing. At the same time, a 'zero risk' criterion seems unjustified since it would prevent any clinical innovation. Therefore, a 'coherent and transparent policy' should be developed on the safety standards of clinical germline editing (de Wert et al., 2018). Further, a mechanism needs to be developed for fixing the liability in cases where undesirable consequences arise from gene editing and/or the unintended changes associated with it. This will discourage any tendency for inadequate safety evaluation to speed up commercialization of a new technology. Further, such measures would also contribute to the confidence of potential users of innovative therapies (Cribbs and Perera, 2017).

Finally, biosecurity concerns have been expressed mainly in USA and to some extent in countries like France, Singapore, Malaysia, and UK. It is feared that gene editing could be used to create pathogen strains capable of causing deadly disease epidemics. Further, genome-edited pathogen strains may complicate the identification of sources of disease outbreaks.

12.10.2. Ethical issues in human germline genome editing

The genetic alterations produced for basic and preclinical researches are confined to the edited germline cells, while those generated for clinical research/application are transmitted to the future generations. Therefore, the safety and ethical issues relevant to the three types of researches vary considerably; these issues are the most controversial and the most widely and hotly debated. The chief concerns relate to safety of the procedures and real need of germline genome editing; these and some other relevant issues are briefly discussed in simple terms in the following sections. Readers interested in more detailed treatment of the relevant issues may consult NASEM (2017), de Wert et al. (2018), Cavaliere (2019), Tuerlings (2019) and other publications on the subject.

12.10.2.1. *Ethical issues in germline gene editing for basic research*

The basic research on human embryos, including germline gene editing, is criticised for reducing human embryos from 'person-like' moral status to mere instruments of scientific research. But the general view appears to be that human embryos have a much lower moral status than children or even foetuses, and many countries tend to permit some research with human embryos. Many researchers have used parthenotic embryos and 3PN embryos to avoid criticism since these embryos cannot develop into foetuses. But for many types of researches, it is preferable to use normal embryos. The use of normal leftover or supernumerary embryos created during IVF in fertility clinics, *i.e.*, 'spare' embryos, may be more ethically acceptable than creating normal research embryos. At present, the embryos can be maintained and studied for up to 14 days after fertilization. Many argue for an extension in this time limit on the grounds that this would enable a better understanding of certain aspects of human embryo development, such as the genes involved in gastrulation, which occurs on 16th or 17th day after fertilization. Any decision on the time limit issue should be based on 'further ethical and societal' debate and deliberations (de Wert et al., 2018; Cavaliere, 2019).

It is important to protect the oocyte donors from 'disproportionate medical risks and exploitation' by creating 'pressure to donate'. In addition, their right to know or not to know about their own genetic status should be respected. This information would be generated from comprehensive analysis of the whole genome sequence (WGS) data of edited embryos, which would be done to ensure the absence of off-target edits and unintended on-target edits in them. It has been suggested that WGS analysis may be limited to detection of possible off-target and on-target actions of the genome editing protocols. Another concern is popularly called the 'slippery slope' argument. This argument is based on the presupposition that basic research would ultimately lead to clinical application of germline genome editing, and that such application is 'ethically unjustified'. It may be pointed out that the slippery slope objection is more applicable to preclinical research than to basic research since the latter must be followed by pre-clinical research prior to clinical applications (Section 12.10.2.2; de Wert et al., 2018).

The basic research on germline genome editing is not as widely acceptable to public as somatic cell gene editing. This is evident from the fact that several countries forbid human embryo research and many countries impose various restrictions on such research. But most scientific and academic bodies support basic research on germline gene editing 'under appropriate oversight mechanism' (NASEM (2017; Cavaliere, 2019; Tuerlings, 2019).

12.10.2.2. *Ethical issues in germline gene editing for preclinical research*

The ethical issues related to germline gene editing for preclinical research are the same as those for basic research, *viz*., 'instrumentalization' of human embryos, protection of oocyte donors, and the risk of 'slippery slope'. The ethical aspect of 'slippery slope' argument is that heritable germline gene editing is ethically unjustified. It may be pointed out that this premise is opposed to the majority view of various groups, including the public at large. Most people favour clinical use of germline gene editing, and for them the only relevant concern is the safety of the technology. Further, the primary objective of preclinical germline gene editing research is to generate information on effectiveness and safety necessary for clinical application of this procedure. Therefore, if the 'ethically unjustified' view were accepted, there would be no need for preclinical research on germline gene editing. In view of the above, the preclinical research may be kept on hold till a consensus is reached on the moral acceptability of heritable germline gene editing even when it was demonstrated to be 'acceptably' safe (Section 12.10.1). But if this procedure were ethically acceptable provided it were shown to be effective and 'acceptably' safe, preclinical research should continue to generate the data necessary for a decision on its clinical application (de Wert et al., 2018).

12.10.2.3. *Ethical issues in germline gene editing for clinical application*

Clinical germline gene editing would be the most relevant for traits governed by single genes that produce large phenotypic effects, and the trait heritability is near 100%. At present, it is not feasible to modify complex traits that are affected by multiple genes and show large environmental effects on their phenotypes. It is important that both intended and unintended genetic changes generated by clinical genome editing will be transmitted to the future generations. These changes, therefore, will affect the human gene pool, which may lead to irreversible and unforeseen consequences. Therefore, it is highly desirable that the various pros and cons of any proposed heritable gene editing should be both extensively and intensively deliberated. In this context, the questions that need to be addressed relate to the diseases that should be the targets of gene editing, the consequences including the pleiotropic effects of the proposed genetic changes, the risks of off-target as well as unintended on-target changes caused by the SDNs, the acceptance of the proposed action by the society, the availability of alternative methods like preimplantation genetic diagnosis (PGD) for achieving the same objective, the acceptability of gene editing being used for trait enhancement or non-therapeutic cosmetic purposes, and the mechanisms for limiting the application of this technology only to the agreed upon uses. Many scientists believe that genetic modification of human germline should not be permitted at present. But in USA, 65% of people approve of therapeutic germ line gene editing, which is slightly higher than those (64%) supporting therapeutic gene editing of somatic cells. Further, mitochondrial replacement therapy for diseases caused by defective mitochondria has been approved in UK. The mitochondrial transplantation does lead to heritable changes in the plasmagenes, but it has been argued that this involves only a small number of genes located in and transmitted via cytoplasm (NASEM, 2017).

The ethical objections to heritable gene editing range from its in-principle rejection to the fear that it will promote inequality in the society by enabling 'enhancement'. The in-principle rejection is based on the premise that human germline is 'inviolable'; therefore, the genetic makeup of human germline must not be deliberately altered. A similar criticism is that genetic change in germline would 'threaten human dignity' and 'violate the autonomy of future children'. It is extremely difficult to define 'human dignity', and one fails to understand how genetic changes leading to correction of serious diseases/conditions would either reduce the 'dignity' or 'violate the autonomy' of affected children. Another concern relates to undermining of the reproductive autonomy of affected parents since they may come under social pressure either not to give birth to a child with a harmful mutation or to use gene editing for correction of this defect. In this context, the overwhelming majority view is that prospective parents are morally bound to attempt to protect their children form 'at least high risks of serious suffering'. Therefore, germline genome editing should be considered as a potential option for treatment of serious diseases/conditions in future children (de Wert et al., 2018; Cavaliere, 2019).

One of the main concerns is that germline gene editing may be used for 'enhancement'. Enhancement may be described as elimination of a trait, introduction of a novel trait, or increasing or decreasing the expression level of a trait beyond the normal range. This creation of "designer babies" or "super humans" would raise many ethical questions like 'fairness, social norms, personal autonomy, and the role of government'. 'Catholic-based think tanks' consider enhancement to be equivalent to "playing God", and that this would "herald a dystopian future." At the same time, some applications of germline genome editing like increasing resistance to a disease are not easy to classify as therapeutic intervention since many feel they represent enhancement. Enhancement applications would intensify social inequalities and could create pressure for using otherwise avoidable technologies. It may lead to 'genetic classism' as the expensive procedures would be available to only the wealthy people. It may be pointed out that somatic cell gene therapy can also be used for similar purposes, *e.g.*, for enhanced muscle development, improved intelligence, etc.; these applications would be open to the same ethical objections of increasing inequality and promoting discrimination. It is comforting that the public support in many countries, including UK, USA, Austria, Denmark, Germany, Netherlands, etc., is much lower for germline gene editing for enhancement than that for therapy, and in some countries enhancement finds little or no support. Further, since enhancement would invariably involve complex traits, this application is not likely soon. But companies and professional gene editing enthusiasts may attempt to take advantage of the prospective parents by making exaggerated promises for promoting enhancement (NASEM, 2017; de Wert et al., 2018; Tuerlings, 2019).

The circumstances and the types of cases that qualify for germline gene editing should be decided by the society based on the viewpoints of diverse groups, and in the light of well accepted alternative methods, *e.g.*, PGD and prenatal diagnosis to terminate foetuses having genetic diseases, available for the purpose. PGD involves implantation of only such IVF derived embryos that are free from genetic diseases. Each of these measures prevents birth of babies with genetic diseases, but these measures would also inadvertently select for the genetic backgrounds being transmitted to the future generations. PGD is accepted and practised in many countries, but it is useless when both the parents are homozygous for an autosomal recessive disease or at least one of the parents is homozygous for a dominant autosomal genetic disease; in such cases, germline editing will be the only option for producing disease-free children. The same will be the case for Y chromosome-determined defects. But the decision whether to go for therapeutic germline gene editing will be ultimately taken by the potential parents in accordance with their religious and ethical beliefs, and possibly the norms of their respective societies (Porteus, 2016; Cribbs and Perera, 2017; Cavaliere, 2019).

The more relevant question is whether therapeutic germline gene editing is in any way ethically more objectionable than the alternative measures listed above? It may be pointed out that germline gene editing is a proactive measure to achieve the same end that is achieved by the above alternative passive approaches. Modern medical interventions allow many persons afflicted by genetic diseases to survive and reproduce, which tends to increase the frequenies of disease-causing mutant alleles in the population. In contrast, therapeutic germline gene editing would help reduce the frequencies of disease causing mutant alleles, many of which generate no benefit but cause tremendous suffering to the affected persons, their relatives and even future generations (when the affected persons reproduce; Porteus, 2016).

It is feared that gene editing may transform the society 'in terms of societal values, its economy status, individuality, injustices, and accessibility', which may influence 'the overall moral and ethical texture of the society'. Further, the unequal access to gene editing may be influenced by such factors as gender, sexuality, ethnicity, disability, class, and location. It is important that further public discussion precedes the formulation of a policy for governance of heritable germline editing. This discussion should include a wide range of groups, such as scientists, industry experts, government representatives, medical practitioners, religious scholars, insurers, lawyers, ethicists, politicians, and ordinary citizens, etc. (NASEM, 2017; Cribbs and Perera, 2017). The unequal access due to wealth could be remedied by implementing plans like 'Médicine France Génomique 2025' designed to enable sick people gain an easy and affordable access to expensive therapies like gene editing.

The consequentialist ethical valuation of medical interventions considers the potential risks and benefits from the respective interventions. This viewpoint recognizes some moral obligation to relieve human suffering from disease and promote wellbeing without imposing a significant level of risk. Gene editing of somatic cells is ethically accepted since it is associated with low risk-to-benefit ratio in view of the induced changes being confined to the concerned individuals (Table 12.7) and the availability of informed consent from them. But in the case of germline editing, the risk-to-benefit ratio is high in view of the genetic changes being heritable, and the debatable nature of informed consent from the prospective parents on behalf of their furture child. In the cases of interventions like IVF and PGD, the informed consent from the would-be parents is accepted as ethical possibly because in these procedures there is little risk of 'unforeseen consequences' to the resulting child. In contrast, such risk is present in germline gene editing, which makes the informed consent from the would-be parents questionable (Cribbs and Perera, 2017).

Table 12.7. Oversight and use of human gene editing: a summary of recommendations by the Committee on Human Gene Editing: Scientific, Medical, and Ethical Considerations (NASEM, 2017)*

Objective	Features	Recommendation
Basic laboratory research	i. Somatic cells and even germ cells used for gene editing ii. Gene-edited cells used for various basic investigations iii. Human subjects not involved	'Use existing regulatory processes to oversee human genome-editing laboratory research'
Somatic genome editing	i. Gene editing of human somatic cells for treatment or prevention of disease or disability ii. Changes confined to the concerned individuals	'Use existing regulatory processes for human gene therapy to oversee somatic human genome-editing research and uses. Limit clinical trials or therapies to treatment and prevention of disease or disability at this time. Evaluate safety and efficacy in the context of risks and benefits of intended use. Require broad public input prior to extending its use for purposes 'other than treatment or prevention of disease and disability'
Germline (heritable) genome editing	i. Gene editing of germ cells and embryos ii. Edited cells/embryos used to establish pregnancy and birth of babies iii. Changes transmitted to future generations	'Permit clinical research trials only for compelling purposes of treating or preventing serious disease or disability, and only if there is a stringent oversight system able to limit uses to specified criteria'
Enhancement	Improvement in a condition or trait beyond the normal level for the condition or trait	'Do not proceed at this time with human genome editing for purposes other than treatment or prevention of disease and disability'

* 1. 'Consider and apply the global principles in governance of human genome editing, *viz.*, promoting well-being, transparency, due care, responsible science, respect for persons, fairness, transnational cooperation'.

2. 'Public input should precede any clinical trials for an extension of human genome editing beyond disease treatment and prevention'

3. 'Ongoing reassessment and public participation should precede any clinical trials of heritable germline editing'

4. 'Incorporate public participation into the human genome editing policy process concerning "enhancement"'

The biosafety concerns can be alleviated by technical refinements in the gene editing technology. The concern related to social inequalities can be resolved by effective regulation of gene editing applications coupled with public discussions, consultations and awareness programmes (Chneiweiss et al., 2017). But the in-principle rejection of heritable genome editing is much more difficult to overcome. It is encouraging that some ethics groups like German Ethics Council now agree that human germline 'is not inviolable' and that there are no convincing philosophical arguments against human germline editing. But the Council concluded that germline editing will not be ethical unless the risks are reduced to an acceptable level, a view shared by many other groups. Finally, a US-based study has revealed that a large majority of couples undergoing IVF (the persons most likely to be the first users of germline gene editing) do not share the objections raised by many ethicists and scientists.

In summary, the decision for permitting germline gene editing for therapeutic applications will depend on the following factors: (i) the safety of the gene editing platforms, (ii) the balance of risk-benefit assessment of the proposed intervention, (iii) the religious and ethical beliefs of the concerned patients and their societies and (iv) the existence of efficient and effective mechanisms for oversight and governance to prevent inappropriate applications. An international group comprising ethicists and researchers have gone to the extent of calling 'for a moratorium on clinical use of human germline editing' until its safety is adequately assayed, and its acceptable uses are agreed upon. It is relevant that this call has been supported by the US National Institutes of Health (Editor, 2019; Tuerlings, 2019).

12.10.3. The NASEM (2017) Report

The NASEM (2017) report concluded that genome editing for basic research and somatic cell gene therapy for treatment and prevention of disease is a valuable tool, and at present its application is adequately regulated (Table 12.8). Further, considerable research as well as public engagement are necessary before a cautious application of germline gene editing may be considered for treatment and prevention of serious human diseases according to strict criteria and under rigorous oversight. Further, this tool should not be used for enhancement till the matter is debated in public and some consensus is reached. Since the definition of enhancement depends on the concept of 'normal level', public discussion to define the normal range for various conditions/traits would be necessary. The report sets forth criteria that must be met before clinical trials of heritable germline editing are permitted, and highlights the critical need for public education and engagement.

The NASEM (2017) proposes seven general principles for the governance of human genome editing as follows: (i) promotion of well-being (providing benefit and preventing harm to those affected), (ii) transparency (openness in sharing of information in a manner that is understandable to the stakeholders), (iii) due care (proceeding carefully and deliberately, and only when supported by sufficient and strong evidence), (iv) responsible science (adherence to the highest standards of research and professional norms), (v) respect for persons (recognition of personal dignity, personal choice, and respect for individual decisions), (vi) fairness (like cases should be treated alike, and risks and benefits should be equitably distributed), and (vii) transnational cooperation (commitment to collaborative approaches to research and governance, while respecting different cultural contexts) (Coller, 2019).

Table 12.8. The legal status of human embryo gene editing in some selected countries (based on Ledford, 2015b; Tuerlings, 2019)

Country	Status of gene editing of human embryo
United States of America	Federal funds not available for review of research with human embryos; NIH 'will not fund any use of gene editing technologies in human embryos'; probably regulatory approval would be required for human gene editing; clinical use of the techniques is not officially banned
Canada	Human germline editing is banned; may result in fine of up to 500,000 Canadian dollars and imprisonment of up to 10 yrs
United Kingdom*	Prior approval is required for research involving human embryos; research is permitted provided the embryos are destroyed after 14 days; clinical use not allowed; creation of research embryos allowed; mitochondrial replacement therapy is permitted since 2015
France*	Embryo editing for research is permitted provided it does not lead to live birth; prior approval is required for work on human embryos and germ cells; production of research embryos is not allowed
Germany*	Artificial genetic alteration in human germ cells and embryos limited to research; modified germ cells or embryos cannot be used for establishment of pregnancy and eventual birth of babies; violation could result in criminal charges
China	Guidelines restrict editing of human embryos: approval required for clinical use; editing for research permitted and perhaps requires approval
Chile	Research on human embryos is prohibited
Japan	Draft guidelines restrict editing of human embryos for reproduction, but they are not legally binding; creation of research embryos and editing of embryos for research permitted, but perhaps requires approval
India	Guidelines prohibit germ line therapy, enhancement of genetic characteristics, eugenic genetic engineering and reproductive cloning, but they are not legally binding; editing for research permitted and perhaps requires approval
Australia	Clinical use is banned; approval is required for embryo editing for research, which is permitted provided it does not lead to live birth
New Zealand	Implantation of a genetically modified gamete, human embryo, or hybrid embryo into a human being is an offence
Italy*	Prohibits research on embryos and production of research embryos, but permits the use of imported embryonic cell lines for research
Ireland*	Guidelines restrict editing of human embryos, but they are not legally binding; editing for research permitted and perhaps requires approval

* In EU, the regulatory regimes applicable to 'genome editing on human germline cells' 'differ significantly among Member States and are sometimes ambiguous'. Further, 'many Member States have not signed nor ratified the Oviedo Convention'. EU funds research and clinical editing of somatic cells, but it does not fund creation of research embryos and heritable germline editing.

Germline genome editing would allow persons suffering from any of the thousands of genetic diseases caused by a single gene mutation to produce children free from these diseases. It was concluded that 'heritable genome-editing research trials might be permitted, but only after much more research aimed at meeting existing risk/benefit standards for authorizing clinical trials and even then, only for compelling reasons and under strict oversight.' The regulatory criteria must include, absence of reasonable alternatives, edited gene definitely known to cause a serious disease/condition, the allele produced by gene editing is prevalent in the population and known to have no deleterious effects, credible data on potential risks and health benefits are available, and rigorous oversight, long-term follow-up, continued reassessment of health, maximum transparency, and use of the technique only for prevention of serious diseases/conditions are ensured. The expressions, such as 'reasonable alternatives' and 'serious disease or condition' will be interpreted differently by different societies as well as individuals. In addition, issues like balancing the potential benefits against the risk of unplanned harms, establishing effective mechanisms for governing the use of gene editing technologies, incorporation of societal values into relevant clinical and policy considerations, and respecting the inevitable differences arising out of national cultures and the viewpoints on whether and how to use these technologies would have to be satisfactorily resolved (NASEM, 2017).

12.10.4. Regulation of human germline gene editing

The legal status of human germline gene editing ranges from legally enforceable ban to a ban that is difficult to enforce (Table 12.8). There have been two key waves of legislations related to human germ cells, gametes and embryos: one was sparked by the production of induced human pluripotent stem cells (ihPSCs) and the other was triggered by reproductive cloning. The creation and use of ihPSCs did raise various concerns but it is now largely accepted, while reproductive cloning is generally considered unsafe and is banned in most countries, The research on human embryos in Germany is subject to restrictions, and artificial genetic alteration in human germ cells and embryos is limited to research. The genetically altered germ cells cannot be used for fertilization and to establish pregnancy; violation of this limitation could result in criminal charges. The situation is similar in France, UK and Australia where research on human embryos is permitted, but prior approval is required, the embryos must be destroyed after the specified period, and clinical use of germline gene editing is not allowed.

On the other hand, human germline gene editing is not formally banned in USA, but the prevalent rules and regulations make it extremely difficult to carry out such work. For example, the use of federal funds for research with human embryos is not allowed, and regulatory approval will be required for human gene editing. The gene-edited germ cells and embryos will be treated as new drugs that must be approved by FDA before they can be used, say, for implantation. Therefore, heritable germline gene editing without FDA approval will be illegal and lead to criminal charges. These and other restrictions are strictly followed and they in effect are like a ban on germline gene editing. The regulatory situations in countries like China, India and Japan are somewhat unclear. These countries have developed guidelines that restrict gene editing of human embryos and ban germline gene editing. However, these guidelines are not legally binding, and the degree of their enforceability is unclear. It may be pointed out that gene editing of human germ cells and embryos is permitted for basic research but it may require prior approval (Table 12.8).

12.11. INTERNATIONAL HARMONIZATION OF REGULATORY REGIMES

The views on clinical gene editing, especially clinical germline gene editing are known to vary immensely among different countries. At present, there is no international regulatory framework for gene editing and its applications, including in human beings. Individual countries are trying to develop their own regulations for gene editing research and its various applications. It will be highly desirable to develop a consensus globally acceptable regulatory framework that takes into account both differences and similarities among the viewpoints of different countries, achieving which would be quite a challenge. Various suggestions range from 'soft instruments (*e.g.*, agreed principles) to legal instruments/treaties or even a global governance body'. In 2015, the International Bioethics Committee (IBC) of UNESCO (United Nations Educational, Scientific and Cultural Organization) proposed that 'states and governments' should decide not to allow germline gene editing without an international consensus (Whelan and Lema, 2015; Tuerlings, 2019).

It has been advocated that the policies/regulations should include the following: (i) clear demarcation of germline gene editing for research from its probable clinical use, (ii) preference for product-based regulation over process-based regulation, (iii) a flexible and evidence-based system that considers the potential risk/benefit ratio and applies regulatory procedures proportionate to this ratio, and (iv) 'the best possible balance between free scientific inquiry and social values'. However, many stakeholders, including the IBC of UNESCO have issued a call for moratorium on human germline editing. The IBC and some other organisations have recommended the development of the least controversial rules, procedures, etc. and continued public engagement, including a global debate. Different models of public engagement, have been proposed, but it is not clear whether the participants in these discussions should have some knowledge about the technology being discussed. A consensus set of 'guiding principles' has been evolved by a group of funding organizations for research on gene drives; similar set of principles may also be developed for human germline gene editing. It has also been proposed to set up a global registry 'by funding bodies or governments to record preclinical research associated with gene editing in human embryos' (Cavaliere, 2019; Tuerlings, 2019).

The global regulatory framework may follow the pattern of some existing frameworks like that followed by the UK Human Fertilisation and Embryology Authority. The proposals for embryo gene editing could be submitted to such an agency for approval. The agency could ask the researchers to 'adhere to a set of principles by signing a code of conduct'. At the same time, research institutions and funding agencies 'should define and monitor clear protocols for germline genome-editing research.' The institutions should carefully review research projects on germline gene editing both at the start and during operation. The journals should develop a consensus set of 'minimum standards for ethical conduct and reporting' of human gene editing research, and evaluate all manuscripts in the light of these standards (Whelan and Lema, 2015; Editor, 2019).

The World Health organization (WHO) of the United Nations has set up an Expert Advisory Committee on Developing Global Standards for Governance and Oversight of Human Genome Editing. The first meeting of this committee was held during 18-19 March, 2019 (Cavaliere, 2019; Tuerlings, 2019). The committee 'has agreed to work towards a strong international governance framework' after discussions with 'all those conducting human genome editing research'. The committee proposes to 'develop essential tools and guidance' 'to ensure maximum benefit and minimal risk to human health' from gene editing. The committee concluded 'that it is irresponsible at this time for anyone to proceed with clinical applications of human germline genome editing'. The committee has asked WHO to initiate the establishment of 'a central registry on human genome editing research … in order to create an open and transparent database of ongoing work' (WHO, 2019).

12.12. SOCIO-ECONOMIC ASPECTS OF REGULATORY REGIMES

The innovations in biotechnology occur and thrive in an 'ecosystem' comprising the government, public institutions, private industry and the general public, and they are regulated by several different mechanisms. The regulatory environment determines the countries where research and innovation flourish, the type of products developed, the speed of their commercialization, and the benefits that will accrue to the innovators and the society. A scientific, evidence-based, efficient, predictable and transparent regulatory regime responsive to the level of risk/benefit ratio of the products has a promontory effect on innovation. In addition, it boosts consumer confidence in products of innovation and encourages investment in innovation activities. In contrast, a rigid, inflexible, inefficient and unpredictable regulatory regime has a discouraging effect on innovation, discourages investment in research and development, and undermine consumer acceptance of innovative products. Delays in approvals add to the cost of product development and commercialization and reduce the profit margin of the developers. For example, a delay of one year in approval of a product can reduce the return on the investment by a public institution by up to 20% (Smyth, 2017).

The differences in regulatory procedures may lead to either acceptance of a technology or its almost total rejection. For example, USA and EU follow different approaches to regulate GM crops leading to the acceptance of GM technology by USA and its rejection by EU. This has cost EU billions of dollars in terms of research funding: during mid-1990s, the EU's share in global investment in agricultural research and development was around 33%, which had dropped below 10% by 2017. A factor that can lead to delays in approvals is the basis of safety assessment of a product. For example, the herbicide (sulfonylurea) tolerant canola developed by ODM was granted non-regulated status in 2004 by APHIS, USA since it used the characteristics of the product for deciding its regulatory status. But Canada approved this variety in 2013 because it had to carry out biosafety evaluation applicable to the GM crops since it follows the 'Plants with Novel Trait' criterion for determining the regulatory status of crop varieties (Whelan and Lema, 2015; Smyth, 2017).

In some countries, the approval process is extremely sensitive to the public opinion and pressure groups. In spite of several applications with favourable biosafety assessments, the European Commission has not been able to approve a single GM crop variety after 1998 mainly due to the European environmental movement (Smyth, 2017). Similarly, an insect resistant Bt-brinjal variety developed by a seed company in India was not approved for commercialization despite favourable biosafety assessment and recommendation for commercialization primarily due to vociferous opposition from several interest groups. This brinjal variety has been commercialized in the neighbouring Bangladesh from where some of the harvest might find its way into the Indian markets. In any case, the cultivation of GM crops other than Bt-cotton is now banned in India. This along with a case pending in the Indian Supreme Court have had a strong dampening effect on GM crop research and development activities in India.

The policymakers consider the opinions from experts in the fields of regulation, academics, industry, agriculture and trade sector, non-governmental organizations (NGOs) and even public at large. They consider the views of the academic community and attempt to balance it against the ever-evolving social attitudes, perceptions and claims. Therefore, social studies relevant to the various biotechnology innovations, including the gene-edited crop varieties and animal strains would provide the policymakers inputs necessary for development of appropriate regulations. Such studies constitute the field of 'social studies of science and technology', which is an established but rapidly developing interdisciplinary field. 'It covers how social, political, and cultural factors interplay with scientific research and technological innovation'. There are several reports on the socioeconomic impacts of GM crops as well as on the effects of public opinion and regulatory regimes on their development and adoption, but such studies on genome-edited crops are negligible (Whelan and Lema, 2015).

12.13. FUTURE PERSPECTIVES

Gene/genome editing has been used to develop several useful strains of crops and farm animals. The gene-edited plants are being regulated under the existing laws, and specific regulatory policies and procedures for their regulation have not been developed by any country. USA, Argentina, Brazil and Australia have exempted the products of SDN-1 and some products of SDN-2 technologies from regulation, while EU treats all of them as GM. Gene-edited animals are proposed to be regulated as drugs by FDA, USA, while Argentina and Brazil intend to treat them at par with gene-edited plants. Clearly, policies and procedures for specific regulation of gene-edited plants and animals need to be developed and the regulatory policies and procedures of different countries need to be harmonized to foster research and development in this highly promising technology. Gene editing of plants and farm animals is generally well accepted, but the use of large animals, especially non-human primates for biomedical research raises some ethical issues. Further, large animals are promising sources of tissues and organs for human patients. Public engagement may be necessary to arrive at a consensus about ethical issues related to these uses. Gene drives are potent genetic devices for management of natural crossbreeding populations, but they remain confined to laboratory research. It is important that appropriate international guideline are developed to facilitate their field-tests to enable a rational decision on their deployment in the field. Finally, effective mechanisms need to be developed for preventing the misuse of gene editing and gene drive technologies by extremist groups.

Human genome editing for basic research and somatic cell gene editing for clinical uses is widely accepted and some gene therapies have already reached the clinics. The most controversial issue relates to human germline gene editing. Germline gene editing for basic research is generally acceptable with some restrictions, but its clinical use is considered 'irresponsible' at present. Both basic and preclinical studies need to be continued to make the gene editing technologies more efficient and safer, and to generate adequate data on these aspects for reaching a sound decision on the clinical use of germline gene editing. Further, consensus needs to be developed on several important issues like the diseases/conditions that would warrant germline gene editing, the levels of acceptable off-target and unintended on-target effects of gene editing platforms, clear demarcation of therapeutic and cosmetic or enhancement applications, etc. In addition, efficient and effective mechanisms for oversight and governance need to be developed to prevent inappropriate uses of the technology. It is noteworthy that a strong international regulatory framework for germline editing is being developed by the WHO of UNO. WHO is also to develop 'a central registry on human genome editing research' that would serve as transparent database on the subject.

12.14. CONCLUSIONS

The direction and the pace of biotechnology innovations are affected by government policies, researches pursued in public institutions, and acceptance of the innovative products by the general public. The regulation of innovations takes several different forms ranging from voluntary self-regulation to legislations enacted by competent authorities. The regulations may be either product-based (seems to promote innovation) or process-based (tends to discourage innovation). In addition, the approach to risk assessment may be either permissive (assumes a product to be safe until it is proven dangerous) or precautionary (presumes that a product is dangerous until it is proven safe). The regulation in USA is product-based with permissive risk assessment; this regulatory regime promotes innovation and encourages consumer acceptance. In contrast, the regulation in EU is process-based coupled with precautionary risk assessment, which tends to discourage innovations as well as consumer acceptance.

Genome editing of plants has been well-accepted and there have been little or no ethical concerns. The main limitation of the CRISPR-Cas9 and other SDN systems is their off-target action, and the induction of large deletions and other unintended genomic rearrangements. Delivery of SDNs as DNA poses the risk of integration of plasmid DNA into the cell genome. Therefore, SDNs should preferably be delivered as protein, mRNA or ribonucleoprotein. Further, the donor DNA plasmid delivered for HDR (SDN-2 and SDN-3 technologies) may integrate at the target site; this should be avoided by comprehensive WGS analyses. So far, no country has developed specific policies and procedures for the regulation of gene-edited crop varieties. But many gene-edited crop varieties have been approved for commercial cultivation in U.S.A., Canada and Brazil under the existing regulations that were developed for GM crops. The countries following product-based approach to regulate GM crops tend to consider SDN-1 and some SDN-2 products as non-GM, while process-based regulatory systems treat them as GM. Thus, there is a need for international harmonization of the regulatory procedures for gene-edited crop varieties.

The regulation of gene-edited food animals is beingly hotly debated since FDA, USA proposes to treat them as drugs, which seems unreasonable. Argentina and Brazil, on the other hand, would apply the same regulatory process to gene-edited animals that is applied to gene-edited plants. Various scientists have been demanding that authorities in USA follow the examples of Argentina and Brazil. Gene editing of larger animals for use as human disease models and as organ donors raises ethical questions especially in relation to nonhuman primates. This situation creates the following paradox: the closer an animal is to humans in terms of evolutionary relationship, the better it would serve as a disease model but the more ethically objectionable would be its use as model. Another objection is that such use of animals tends to reduce their status to mere commodities, which may result in their inappropriate treatment.

Gene drives are SDN-based self-propagating genetic devices that spread in the targeted crossbreeding populations and either reduce, do not affect or enhance their fitness either by inactivating an endogenous gene or introducing a novel gene into them. Gene drives are potent tools for genetic modification, rescue, suppression or eradication of target populations. They can be used for cost-effective and efficient management of natural populations, but they cannot be used until they are field-tested. But field-tests cannot be carried out as there are no guidelines for the same. Since gene drives can move across national boundaries, it would be necessary to develop a regulatory framework that is acceptable to various countries. At the same time, effective measures must be devised to prevent the misuse of gene drives by extremists.

Human genome editing can be broadly divided into two categories, *viz,* somatic and germline gene editing. Both somatic and germline gene editing are of three types, *viz.,* gene editing for basic, preclinical and clinical researches. Somatic cell gene editing is generally accepted by the public and several gene therapies are either in clinical use or in different stages of clinical trials. But germline gene editing, especially heritable germline editing, is one of the most controversial subjects and is faced with many ethical and regulatory issues. The ethical positions range from in-principle rejection to concerns about safety and societal inequalities, and potential exploitation of the would-be parents by gene editing enthusiasts and commercial companies. The regulation of germline gene editing ranges from an enforceable and explicit ban on heritable germline editing through implicit but effective ban to explicit ban with doubtful enforceability. There is general agreement that it is 'irresponsible' to undertake heritable germline gene editing at present since more information on efficiency and safety needs to be accumulated.

Different countries are trying to develop their own regulations for human gene editing. There is a strong need for the development of a consensus international regulatory framework and WHO of the United Nations Organization has already taken the initiative in this direction. It has appointed an Expert Advisory Committee on Developing Global Standards for Governance and Oversight of Human Genome Editing, which held its first meeting during March 18-19, 2019. The committee intends to develop 'a strong international governance framework' 'to ensure maximum benefit and minimal risk to human health' from the gene editing technology. The committee has recommended the establishment of 'a central registry on human genome editing research' to enable the creation of 'an open and transparent database of ongoing work' (WHO, 2019).

■■■■

GLOSSARY

'Active' state of chromatin. The genes present in 'active' chromatin are available for transcription initiation.

'Central 4' base pairs. The four base pairs at positions -2, -1, +1, +2 of the target site of a MN or LHE; recognition by MN is critical for cleavage.

'Copy-cat' gene drive. See, *Simple gene drive*.

'Humanized' animal models. Animal models having genomes that mimic the human genome.

'Off-the-shelf' T cells. See, *'Universal' T cells.*

'Prime editing'. Use of 'prime editor' to bring about desired sequence replacements at the targeted sites.

'Prime editor'. It comprises a long guide RNA, called pegRNA and a Cas9 nickase fused to an engineered reverse transcriptase.

'Safe-harbour' site. A genomic locus that is not likely to generate undesirable effects because of integration of a transgene expression cassette within this locus.

'Third-party' T cells. See, *'Universal' T cells.*

'Universal' T cells. T cells, from which the gene encoding either α- or β-chain of the T cell receptor is deleted by a SDN; much less likely to lead to graft-versus-host disease.

"Gene scissor". See, Site-directed nuclease.

Active genetics. Use of gene drives for rapid creation of model organism strains with specified genotypes useful for basic research.

Agricultural breeding of animals. For traits relevant to commercial raising of herds/flocks and utilization of animal products, including product quantity and quality.

Allelic gene drive. An allelic gene drive has a Cas9 gene and two sgRNA genes; one sgRNA propagates the drive, while the other targets a nearby locus to induce the conversion of an undesirable allele of this locus into a desirable one.

Allergy. An abnormal immune response to common environmental substances that are harmless to most people.

Animal bioreactors. Transgenic animals used for production of human proteins used as therapeutic agents.

Antirepressor proteins. Proteins that bind to enhancers and recruit chromatin modifying factors that change the chromatin from inactive to active state.

Antisense gene construct. A gene construct, in which the coding region of the gene is inverted in relation to its promoter so that its sense strand is transcribed to yield antisense RNA.

Antisense RNA technology. Suppression of an endogenous gene mediated by its antisense RNA transcribed from the antisense construct of this gene.

Antisense RNA. RNA transcribed from the sense strand of a gene; complementary to mRNA.

Aptamer. Single-stranded DNA (ssDNA) or single-stranded RNA (ssRNA) that binds specific ligands with high affinity and specificity.

Aptazyme. A self-cleaving ribozyme with an integrated aptamer domain; the ribozyme is inactive till the aptamer interacts with the appropriate ligand molecule.

Atopy. A person being genetically predisposed to develop allergic diseases.

Autologous. Derived from the person/patient who will receive the cells/tissues.

Backcross. Crossing of a mutant line or progeny from a cross to its parental line.

Base editing product distribution. The frequencies of various types of genetic changes generated by the base editing procedure.

Base editing. A direct and irreversible conversion of one base into another at the targeted genomic site without induction of a DSB; based on deamination of cytidine and adenosine.

Base editor. It comprises DNA binding module of a SDN and a specific base deaminase, *e.g.*, cytidine deaminase or adenosine deaminase.

Behavioral gene drive resistance. Discrimination against gene drive individuals during mating.

Biased inheritance pattern of gene drive. Increased proportion of the gene drive individuals in progeny as compared to that expected based on Mendelian inheritance.

Biohacking. The activity of exploiting genetic material(s) experimentally without regards to accepted standards and/or outside of standard traditional research institutions

Bioweapon. An organism that can be used as a weapon to cause devastating damage to humans, domestic animals, crops and/or ecosystems.

Blastocyst complementation. Production of chimeric embryos, foetuses and animals by introducing ESCs or iPSCs of one species into blastocyst stage embryos of another species, the blastocysts being genetically deficient in development of a specific organ, *e.g.*, pancreas.

Blocking sequence (sgRNA). A sequence that is complementary to and base-pairs with the sgRNA spacer sequence so that it blocks the sgRNA function.

Breeding by design. See, In vitro *breeding*.

Bystander mutation. A change in a base located within the editing window but other than the one targeted for editing.

Cancer. An abnormal or uncontrolled cell growth with the potential for invasion of other body tissues and organs.

Cargo gene. A gene included in a gene drive that performs some specific function; it is selected for achieving a specific objective by spreading it in the concerned natural population,

CAR-T cells. T cells expressing a chimeric antigen receptor.

Cas9. The protein component of a Type II CRISPR-Cas system that functions as an RNA-guided sequence-specific endonuclease when associated with crRNA–tracrRNA or sgRNA.

Cas9 nickase (nCas9). A Cas9 mutant, in which one of the nuclease domains is inactivated

CD19 CAR-T cells. T cells expressing the CD19-specific CAR.

Cellular disease models. Models of certain human diseases at cultured cell level.

Chemometrics. Application of mathematical and statistical procedures to infer the relationships between quality or physical parameters and the data obtained from analytical instruments.

Chimeraplasty. See, Oligonucleotide-directed mutagenesis.

Chimeric oligonucleotide. Oligonucleotide comprising a single, generally, 68 nt long DNA strand that has two short (10 nt long) RNA sequences at strategic locations and is present in double-stranded form due to internal base pairing.

Chromatin remodelling complexes. Protein complexes that mediate changes in chromatin organization; they are recruited to chromatin by transcription activator/repressor proteins.

Chromothripsis. A mutational process by which up to thousands of clustered chromosomal rearrangements occur in a single catastrophic event in localised and confined genomic regions in one or a few chromosomes. It is known to be involved in both cancer and congenital diseases. Occasionally, it may delete a disease-causing mutant gene leading to cure of the disease; this is sometimes termed as 'natural gene editing'.

Circular RNAs (circRNAs). These are covalently closed mammalian RNAs produced by back-splicing of exons from a single pre-mRNA. They have tissue- and cell type-specific expression patterns; their biogenesis is tightly regulated by specific cis-acting elements and trans-acting factors (splicing factors); they act as decoys for microRNA or RNA-binding proteins; involved in regulation of gene expression and implicated in diseases, such as diabetes mellitus, neurological disorders, cardiovascular diseases and cancer.

Clinical trial. Research on human volunteers designed to compare the efficacy of 'a new medical approach to a standard one that is already available, to a placebo that contains no active ingredients, or to no intervention'

Cognate base pair. The base pair that is the target of action by a base editor.

Complementation test. Introduction of the normal allele of an inactivated gene into the mutant line to determine if it can restore the normal phenotype.

Conversion tract. The sequence of a gene involved in a gene conversion event.

Crispr babies. Babies produced from young embryos subjected to gene editing by CRISPR-Cas9; first reported by a Chinese scientist in November 2018.

CRISPR barcode. A DNA sequence that acquires sequence variation due to the action of CRISPR-Cas9; it can be used to follow cell lineage during development and differentiation.

CRISPR-Cas systems: CRISPR loci contain multiple short direct repeats separated by variable sequences; these systems provide acquired immunity to bacteria and archaea against invading foreign DNA.

CRISPR-activation. Transcription activation using CRISPR-dCas9-based activators.

CRISPR-interference. Transcription repression using CRISPR-dCas9-based repressors.

crRNA. CRISPR RNA base pairs with tracrRNA to form a structure that associates with Cas9 and guides it to specific DNA sequences and activates its endonuclease activity.

Daisy chain gene drive. The components of CRISPR-Cas9 gene drive, *viz.*, sgRNA, Cas9 and cargo gene, are separated and integrated into different chromosomes of the genome so that they show independent segregation.

Daisy quorum gene drive. The typical daisy chain gene drive is combined with underdominance.

Daisyfield gene drive. It has element B (sgRNA targeting the locus with element A) and element A (Cas9 + cargo gene) segregating independently; four copies of the sgRNA gene are integrated at unlinked genomic sites.

Dead Cas9 (dCas9). A Cas9 mutant, in which both the nuclease domains are inactivated.

Democratic deliberations. A form of collective dialogue and decision-making that emphasizes open, reasoned dialogue and aims to improve policy-making by broadening the policy conversation to include as many perspectives as possible.

Directed evolution of proteins. Creation of numerous mutants of a proteins in the laboratory and selection of the protein(s) with desired functions from among them.

Diversifying base editing. Base editing for generating local sequence diversity, *i.e.*, generating mutations in several different C or A residues located in the targeted localized genomic region

DNA barcode. A short DNA sequence that enables unambiguous identification of cells, individuals or species from other similar entities.

Donor DNA. A DNA fragment that is used as template for HDR of DSBs; it is flanked by sequences homologous to those flanking the DSB to allow homologous recombination between the donor DNA and the genomic sequences flanking the DSB.

Editing window. A short, ~5 bp long region of single-stranded non-target strand of the DNA duplex in which base editing takes place.

Effector-target proteins. Host proteins, with which pathogen effectors specifically interact; this interaction leads to the suppression of host resistance to the concerned pathogens.

Efficiency of a site-directed nuclease. The percent of on-target sites edited by the site-directed nuclease.

Efficiency of base editing. The proportion of all the targeted genomic sites exposed to the base editor that have undergone the desired base editing.

Endogenous gene suppression. Preventing the expression of specific endogenous genes, *e.g.*, by gene inactivation.

Endonuclease. A nuclease that generates internal cuts in DNA.

Engineered nuclease. See, *Site-directed nuclease.*

Enhancement. Elimination of a trait, introduction of a novel trait, or increasing or decreasing the expression level of a trait beyond the normal range.

Enzyme engineering. Production of enzymes with altered properties and/or functions

Epigenetic effector. A programmable effector that carries out epigenetic modification of chromatin

Epigenetic marks of chromatin. These marks comprise DNA methylation and histone modifications like acetylation, methylation, and phosphorylation of specific amino acid residues of certain histones; they together indicate the active/inactive state of chromatin.

Epigenetic modifications. Changes in DNA and/or histones by methylation, demethylation, acetylation and/or deacetylation.

Epigenome editing tools. Tools used for epigenome editing.

Epigenome editing. Transcription regulation using programmable transcription regulators.

Epigenome editors. Programmable transcription activators and repressors used for generating epigenetic modifications for targeted gene regulation.

Epigenome engineering. See, *Epigenome editing*.

Epigenome. The collection of various chemical modifications of the genomic DNA and the histone proteins of an organism.

Epistasis. See, *Gene interaction.*

Eradication gene drive. It exerts a strong deleterious effect on fitness either by inactivation of a carefully selected endogenous gene or by introducing an appropriate cargo gene.

Ethics. A branch of philosophy that attempts to systemize, defend and recommend the concept of right and wrong behaviour.

Eugenics. A range of social and reproductive practices aimed at improving the quality of human gene pool based on fitness, physical beauty or other traits valued by a society.

Evolving CRISPR barcode. The barcode comprises a homing gRNA encoding locus; the sequence of this locus continues to change due to CRISPR-Cas9 action.

Excision gene drive. It excises existing gene drives introduced earlier in the population; comprises gene constructs for two sgRNAs targeting the sequences flanking the drive and homology arms for HDR; Cas9 activity is provided by the gene drive that is to be excised.

Fluorescence resonance energy transfer. A donor fluorophore is excited by radiation; it transfers its excitation energy to an acceptor chromophore located nearby, as a result of which the latter generates fluorescence.

FokI. A Type IIS endonuclease; recognition site 5′-GGATG-3′.

FRET imaging. Imaging based on specific pairs of donor and acceptor fluorophores.

Gene 'knockin'. Integration of a novel gene into the genome.

Gene 'knockout'. Inactivation of an endogenous gene.

Gene construct. The coding region of a gene along with the appropriate regulatory sequences needed for its expression in the desired host organism.

Gene disruption. Integration by homologous recombination of a DNA sequence within a gene that prevents expression of the concerned gene.

Gene drive confinement. Spread of gene drives is limited by ecological conditions.

Gene drive containment. Spread of gene drives is limited by natural physical barriers or restrictions created by humans.

Gene drive efficacy. The frequency of a gene drive being copied into the wild type chromosomes of gene drive heterozygotes.

Gene drive individual. An individual homozygous for a gene drive.

Gene drive. A SDN-based self-propagating genetic device designed for introducing genetic alterations in cross-breeding natural populations.

Gene editing. Alteration in the base sequence of target gene as specified by the experimenter; typically achieved by DSB induction by a SDN followed by its repair with or without donor DNA.

Gene inactivation. Elimination of gene expression.

Gene interaction. The functions of alleles of one interacting gene are affected by those of the other interacting gene and *vice-versa*.

Gene knockdown. Suppression of the expression of a gene, *e.g.*, by RNAi.

Gene therapy, *ex vivo*. Appropriate cells are isolated from the body of concerned patient, subjected to the desired genetic modification, including gene editing, *in vitro* and then introduced back into the patient.

Gene therapy, *in vivo*. The therapeutic reagents, including a therapeutic vector carrying the necessary gene constructs, are either delivered locally into the concerned tissue or introduced into the blood stream.

Gene therapy. Favourably changing the base sequence or expression pattern of an endogenous gene so that the disease phenotype is reversed to the normal or near normal phenotypic state.

Gene/allele replacement. Induction of a new mutation in the targeted gene by sequence replacement.

Gene-targeting. See, *Gene editing*.

Geneticization. A term used to describe how science of genetics is influencing society in general and medicine in particular, especially the increasing use of genetic explanations to describe differences between individuals and groups.

Genome editing. See, *Gene editing*.

Genome engineering. See, *Gene editing.*

Global gene drive. The gene drive can continue to spread indefinitely over time and space; all the components of gene drive are linked together and present at the same locus in the genome.

Guidelines. A set of rules and practices developed by an academic body or the concerned government agency; they are expected to be followed by the concerned persons, institutions and/or industries.

Hazard. An imaginable, identifiable and measurable undesirable consequence of an activity.

hgRNA. A type of sgRNA that can target its own locus for Cas9 action.

Histone code. The epigenetic marks of chromatin together constitute the histone code.

Homing. Integration of a copy of a self-splicing intron or intein in an allele of the concerned gene, which lacks the intron/intein.

Homing-based gene drive. See, *Gene drive.*

Homing efficiency of a gene drive. See, *Gene drive efficacy.*

Homing endonuclease. A nuclease encoded by a mobile intron or intein encoding sequence; recognizes typically 20 to 30 bp long DNA sequence.

Homology arms. In case of donor DNA; the sequences flanking the DSB induced by a SDN and included as arms of the donor molecule to enable homology-directed repair of the DSB.

Homology-directed repair (HDR). This pathway for DSB repair uses a homologous donor DNA as template and leads to insertion of the donor DNA at the targeted site.

Homology-independent targeted integration. Insertion of a DNA fragment/transgene at the target site by DSB induction by CRISPR-Cas9, followed by NHEJ-mediated repair; the target site and the donor DNA plasmid are simultaneously cleaved.

Horizontal gene transfer. Gene transfer mediated by pathogens, endosymbionts, parasites and insect pests.

Hybrid SDN. A SDN created by linking the DNA binding domain from one SDN system with the nuclease domain from another SDN system.

Hyperacute rejection. Rejection of pig to human xenotransplants immediately after transplantation.

Immunising reversal gene drives. These drives remove the undesirable changes induced by an earlier gene drive and, at the same time, make the genome sequence immune to gene drives, including the earlier drive, that target the wild type sequence.

Immunization against a gene drive. Modification of the genomic sequence targeted by a gene drive in such a way that it is no longer recognized by this gene drive, but the polypeptide encoded by the genomic sequence remains unchanged.

Immunizing drives. They replace the wild type sequences targeted by the unwanted gene drives with recoded sequences so that the gene drives specific for the wild type sequences are unable to target the recoded sequences.

***In vitro* breeding**. Breeding schemes incorporating gene or genome editing procedures.

Inactivation gene drives. They inactivate existing gene drives introduced earlier in the population.

Induced pluripotent stem cells. Pluripotent stem cells induced by introduction of stem cell transcription factors into somatic cells.

Inducible CRISPR. Inducible promoters used to regulate CRISPR-Cas9 SDN expression.

Insertional mutagenesis. Disruption of gene function by integration of a DNA sequence within the coding region of the gene.

Instrumentalization of human embryos. Reduction of human embryos from 'person-like' moral status to mere instruments.

Integral gene drive. In this drive, the gene drive components are integrated within the coding sequences of targeted genes in such a way that the functions of these endogenous genes are not affected.

Intron targeting. Replacement of complete exon by targeting sequences in the two flanking introns; the donor DNA has the sequence for replacing the exon and its arms contain the same sequence, for which the two sgRNAs targeting the intron sequences are specific; simultaneous cleavage of introns and the donor DNA leads to insertion of the donor DNA sequence into the genome by NHEJ resulting in exon replacement.

Invasive gene drive. See, *Global gene drive.*

Ku-dependent pathway of NHEJ repair. The Ku complex first binds to the two DNA ends at the DSBs; Ku complex is then joined by MRX complex (MRN complex in mammals), Artemis, etc. proteins, which complete the repair process.

Ku-independent pathway of NHEJ repair. The Ku complex is not involved in repair of DSBs; leads to longer indels than the Ku-dependent pathway; several variants of the pathway exist.

LAGLIDADG domain. The nuclease domain of LAGLIDADG homing endonucleases or meganucleases.

Local gene drive. A gene drive designed to typically remain confined in both space and time since it exhausts itself over time; generally, the different elements of such a drive are separated and integrated at separate loci in different chromosomes.

Maturases. Proteins participating in post-transcriptional splicing of introns.

Meganickases. MNs that induce single strand breaks, *i.e.*, nicks, at their target sites; generated by inactivation of one of the two nuclease sites of the concerned MNs.

Meganuclease. Homing endonuclease with LAGLIDADG nuclease domain.

MegaTAL. SDN created by fusing a meganuclease (functions as DNA cleavage domain) with a TALE protein (serves as DNA binding domain).

Metabolic engineering. Modification of metabolic pathways of an organism by either silencing or overexpressing endogenous genes or introducing novel transgenes that affect enzymatic, transport and/or regulatory functions in the cells.

Microhomologous sequences. Short homologous sequences of 5-25 bp.

Microhomology-mediated end-joining. A type of NHEJ; it uses short homologous sequences of 5-25 bp for error-prone joining of the broken ends. It leads to loss of the DNA segment located between the short homologous sequences on the two sides of DSBs.

Mitochondrial replacement therapy. Transplantation of normal mitochondria into young embryos for the treatment of diseases caused by defective mitochondria.

Modification gene drives. They introduce genetic changes in the genomes of target species.

Molecular containment. Limiting the spread of gene drives in time and/or space by molecular strategies like gene drive design.

Multiplexed gene editing. Simultaneous editing of more than one gene in the same cell.

Multiplexing. See, *Multiplexed gene editing.*

Mutagenic chain reaction. Use of a simple gene drive to inactivate a gene in a crossbreeding population and thereby propagate this mutation in the population due to self-propagation of the gene drive.

Mutant library. A large collection of mutant lines/strains of the selected organism; ideally, it should include mutants in all the genes of the organism.

Mutant. In the context of gene editing: an individual that contains the intended mutation in homozygous or heterozygous state.

Mutation breeding. Use of mutagenic agents to create genetic variation for development of new crop varieties.

Mutation. A sudden heritable change in the characteristics of an organism. In the context of gene editing: the intended on-target sequence change induced by the gene editing procedure.

Nick. Cleavage of a single strand of a DNA duplex.

Noncoding RNAs. RNAs that do not encode polypeptides; they participate in gene regulation by affecting chromatin remodelling, transcription and translation.

Nonhomologous end-joining (NHEJ). A DSB repair pathway that joins the two broken ends together; generally, leads to production of insertions or deletions of one to few bases at the DSB site.

Nontarget strand for CRISPR-Cas system. The DNA strand that does not pair with the crRNA, gRNA or sgRNA.

Off-target sites. Sites similar to the target sites in base sequence; may be cleaved by SDNs.

Oligonucleotide-directed gene targeting. See, *Oligonucleotide-directed mutagenesis.*

Oligonucleotide-directed mutagenesis. It uses chimeric oligonucleotides for induction of desired mutations.

On-target sites. Sites specifically targeted by SDNs or some other sequence-specific reagents.

Organoid disease models. Human disease models in form of organoid-like systems (*e.g.*, models of intestines, brain, kidney, etc.) developed in iPSC cultures.

Oviedo Convention. The 1997 Convention on Human Rights and Biomedicine of the European Council; it calls for, among other things, a prohibition of genetic modification of human germline.

PAMmer. An oligodeoxynucleotide with PAM sequence and a short 3' as well an 18 nt long 5' extension that base-pairs with the targeted sequence of ssDNA or ssRNA.

Parthenocarpy. Development of fruits without fertilization.

Payload gene. See, *Cargo gene.*

Permissive risk assessment. Risk assessment assumes a new product to be safe until it is proven dangerous.

PGMS line. A line that is male sterile under long-day, but male fertile under short-day conditions.

Photoperiod-sensitive genic male sterility. A genic male sterility system in which sterility and fertility are regulated by the prevailing photoperiod regime.

Plant breeding. The technology for the development of new crop cultivars better suited to human needs.

Pluripotent stem cells. Cells capable of continued cell division and giving rise to all the three primary germ layers, namely, ectoderm, mesoderm and endoderm.

Precautionary risk assessment. Risk assessment presumes a new product to be dangerous until proven safe.

Precise integration into target chromosome. MMEJ-based integration of recombinant DNA or transgene cassettes into the targeted chromosomes.

Precision base editing. Editing a single C or A at the targeted site either to correct a mutation or to induce a desired point mutation.

Precision drive. A drive targeting a gene/genomic sequence unique to the species so that it remains confined to the species.

Precision of base editing. The ratio of the number of desired C-to-T or A-to-G transitions to all the other types of base transitions.

Precision reversal drive. The genomic sequence included in the reversal drive is the same as the wild type genomic sequence.

Preimplantation genetic diagnosis. Screening of several embryos produced by *in vitro* fertilization for specific mutations to select a healthy embryo for implantation.

Primary cells. Cells obtained directly from normal body tissues.

'Prime editing' guide RNA. A long guide RNA that has the sequence required for the desired sequence replacement in addition to the sequences of the standard sgRNA.

Product purity (C-to-T base editors). The percentage of the targeted C-to-T transitions among all the different types of genetic changes induced by C-to-T base editors.

Programmable epigenome effectors. See, *Epigenome editing tools.*

Programmable hybrid effector. The desired effector domain attached to a programmable DNA binding component.

Programmable nuclease. See, *Site-directed nuclease.*

Programmable transcription activators. Chimeric proteins comprising DNA binding domain from a SDN and the activation domains from one or more transcription activators.

Programmable transcription repressors. Chimeric proteins comprising DNA binding domain from a SDN and the repressor domains from one or more transcription repressors.

Protein engineering. Production of proteins with altered properties and/or functions.

Protospacer adjacent motif. Short nucleotide sequence present in crRNA and the nontarget strand of the targeted DNA duplex, specifically recognized by Cas9; required for Cas9 action.

Recoded TALEN. A TALEN with TALE protein encoding DNA sequence optimized for facilitating assembly, enhancing expression and enabling safe delivery by lentiviral vectors.

Recoding of a DNA sequence. Modification of a DNA sequence in such a way the polypeptide encoded by the sequence is not altered.

Recombinant DNA technology. The suite of techniques used for creation, propagation and transfer of recombinant DNAs.

Recombinant DNA. A vector into which the gene or DNA fragment to be cloned (DNA insert) has been integrated.

Redesigning of a SDN. Making appropriate alterations in a SDN to enable it to specifically recognize and bind to a new target site.

Regulations. A set of detailed administrative rules devised to implement the relevant laws.

Rescue gene drive. It enhances the fitness of individuals and populations by introducing beneficial cargo genes/mutations in endogenous genes or sequence corrections of exiting deleterious gene mutations.

Research embryos. embryos specifically created for research from donated oocytes that are fertilized *in vitro*.

Resilient animals. The animals do not develop symptoms of the disease and maintain an acceptable level of productivity when they are infected by the concerned disease.

Restriction enzyme. An endonuclease that generates internal cuts in DNA at specific sites in response to a specific sequence recognized by it.

Restriction factors. Proteins that interfere with key steps in virus propagation.

Reversal gene drives. These drives undo the genetic changes induced by earlier modification drives or deleterious mutations and restore the wild type function and even sequence of the affected genes.

Reverse PGMS. A line that is male sterile under short-day, but male fertile under long-day conditions.

Risk assessment. Identification of hazards from an activity, analysis of the identified hazards, determination of the consequences of the hazards, and determination and evaluation of the risk.

Risk. Risk is a function of the level of hazard from, say, an activity and the probability of this hazard being realized.

RNA editing. Irreversible conversion of one base present in an RNA molecule into another base after it has been produced by transcription; typically A-to-I transition.

RNA interference (RNAi). The process, in which small double-stranded RNA molecules prevent or knockdown the expressions of genes that encode mRNAs with sequences complementary to these small RNAs; natural eukaryotic defence mechanism against RNA viruses.

RNA-targeted Cas9. An adaptation of CRISPR-Cas9 for cleavage of ssRNA.

SDN technology. The use of SDNs to induce targeted genetic changes.

SDN-1 technology. It uses only SDNs and relies on NHEJ-mediated DSB repair to inactivate the targeted genes.

SDN-2 technology. The selected SDN is used along with appropriate donor DNA to achieve pre-decided sequence replacement in the targeted gene.

SDN-3 technology. The selected SDN is used along with appropriate donor DNA to achieve integration of a complete gene construct at the specified genomic site by HDR-mediated DSB repair.

Self-consciousness. 'The ability to reason, think, remember, and know'.

Sentience. 'The ability to feel and to suffer'.

Sequence-specified chimeric effector. See, *Programmable hybrid effector.*

sgRNA "retargeting". A change in sgRNA specificity due to a random mutation in its protospacer region.

Simple gene drive. A gene drive comprising expression cassettes for Cas9 and sgRNA specific for the targeted locus flanked by homology arms.

Single-chain variable fragment (scFv) antibodies. Antibodies created by fusing the epitope-binding regions of light and heavy chains to yield a single polypeptide.

Site-directed chimeric effector. See, *Programmable hybrid effector.*

Site-directed mutagenesis. Introduction of specified changes at predetermined sites of the targeted genes and, thereby, their products.

Site-directed nuclease. Endonucleases designed to bind specific sequences of duplex DNA and to induce DSBs at the targeted genomic sites.

Site-specific mutagenesis. See, *Site-directed mutagenesis.*

Split gene drive. The Cas9 and sgRNA expression cassettes of a gene drive are separated and placed at unlinked genomic loci.

Standard gene drive. See, *Global gene drive.*

SunTag. An array of oligopeptide repeats that specifically interact with and bind to antibodies.

Super-Mendelian inheritance pattern. The inheritance pattern generated by conversion of heterozygous progeny into homozygotes leading to rapid increase in the frequency and, usually, fixation of the concerned allele in the population.

Suppression gene drive. It exerts a mild deleterious effect on fitness either by inactivation of a carefully selected endogenous gene or by introducing an appropriate cargo gene.

Susceptibility genes. Host genes activated by pathogens to facilitate invasion and pathogenesis.

Switchable guide RNA. A sgRNA with an aptamer sequence in its scaffold region that either enables or disables sgRNA function in response to trigRNA of trigDNA.

Synthetic gene drive. See, *Gene drive.*

Synthetic lethal. A synthetic lethal is based on a pair of interacting genes; inactivation of any one gene of this pair causes a disease, but cells die when both are inactivated. The expression of that gene of this pair, which is still active in the diseased cells is blocked by a drug or repressed by CRISPRr to kill the diseased cells.

Synthetic sequence gene drive. A gene drive targeting a synthetic sequence introduced into the genome of a laboratory strain; the gene drive cannot spread to other populations.

T_0 plants. Plants regenerated from cultured cells or tissues.

T_1 plants. Progeny of T_0 plants generally obtained by selfing.

TALE scaffold. The C- and N-terminal domains of TALE proteins flanking their TALE repeat domains, *i.e.*, the DNA binding domains.

TALE-recombinase. Fusion of recombinase with TALE DNA binding domain; it efficiently recombines DNA molecules in cells.

Target strand for CRISPR-Cas system. The strand of target DNA, which base-pairs with the crRNA, gRNA or sgRNA.

Targeted gene correction. Correction of a deleterious mutation in the targeted gene by sequence replacement.

Targeted gene editing. See, *Gene editing*.

Temperaturee-sensitive genic male sterility. A genic male sterility system in which sterility and fertility are regulated by the prevailing temperature regime.

Toehold-mediated strand displacement technique. Activation or inactivation of modified sgRNAs, called swigRNAs, by an interaction with trigger RNAs/DNAs that base-pair with the aptamers present in the swigRNAs.

Tolerant animals. See, *Resilient animals*.

Trait stacking. Integration of genes for two or more traits at the same locus.

***trans*-activating CRISPR RNA.** It promotes processing of crRNA, base-pairs with the 3' region of crRNA and they together associate with and activate Cas9.

Transcription activator proteins. Transcription factors that activate transcription initiation.

Transcription activator-like effector nuclease (TALEN). A TALE protein fused with the cleavage domain of FokI nuclease.

Transcription regulator proteins. Transcription factors that either activate or repress transcription initiation.

Transcription repressor proteins. Transcription factors that repress transcription initiation.

Transgene. A gene transferred by recombinant DNA technology.

Transgenic organism. An organism carrying and/or expressing a transgene.

Transgenic technology. Gene transfers into selected organisms using recombinant DNA technology.

Trigger DNA. Single-stranded DNA complementary to the aptamer sequence present in swigRNA.

Trigger RNA. Single-stranded RNA complementary to the aptamer sequence present in swigRNA.

Underdominance. The heterozygotes for a gene being inferior to both the homozygotes.

USER cassette. A PacI recognition site flanked by binding sites for Nt.BbCI, the binding sites for the two enzymes being separated by a variable nucleotide.

USER cloning. It employs a USER-compatible vector that has a USER cassette for cloning of PCR product amplified by using a pair of specially designed primers.

USER fusion. Utilization of the USER cloning technique for fusing two or more PCR products together.

Variable di-residues. The highly variable amino acid residues at positions 12 and 13 of TALE repeat units.

Voluntary self-regulation. Development of and adherence to a set of rules and procedures by the concerned community.

Welfare traits. Traits important for protection of both humans as well as animals from injury, *e.g.*, polled or hornless trait in cattle.

Xeno-generation. Production of humanized organs in gene-edited pigs or pigs produced by blastocyst complementation.

Xenotransplantation. Transplantation of organs, tissues and/or cells from one species into another species.

Zinc finger. A DNA binding motif that has about 30 amino acids; one type of ZF contains the invariant Cys2-His2 residues that together bind one zinc atom.

Zinc-finger nickases. A pair of ZFNs in which the FokI nuclease domain of one ZFN is inactivated by suitable amino acid substitutions.

Zinc-finger nuclease. A Zinc-finger protein fused with the DNA cleavage domain from FokI restriction endonuclease; a pair of ZFNs needed for inducing site-specific DSBs.

Zinc-finger protein. A eukaryotic transcription factor comprising zinc fingers and transcription activation domains; used to create ZFNs.

■■■■

REFERENCES

Abudayyeh, O.O., Gootenberg, J.S., Konermann, S., Joung, J. et al. 2016. C2c2 is a single-component programmable RNA-guided RNA-targeting CRISPR effector. Science 353: aaf5573. doi:10.1126/science.aaf5573.

Acharya, S., Mishra, A., Paula, D., Ansari, A.H. et al. 2019. *Francisella novicida* Cas9 interrogates genomic DNA with very high specificity and can be used for mammalian genome editing. Proc. Natl. Acad. Sci. USA 116: 20959–20968.

Ahuja, V. 2018. Regulation of emerging gene technologies in India. BMC Proc. 12 (Suppl 8):14. https://doi.org/10.1186/s12919-018-0106-0

Ain, Q.U., Chung, J.Y. and Kim, Y.-H. 2015. Current and future delivery systems for engineered nucleases: ZFN, TALEN and RGEN. J. Controlled Release 205: 120–127.

Akbari, O.S., Bellen, H.J., Bier, E., et al. 2015. Safeguarding gene drive experiments in the laboratory. Science 349: 927–929.

Alagoz, Y., Gurkok, T., Zhang, B., and Unver, T. 2016. Manipulating the biosynthesis of bioactive compound alkaloids for next generation metabolic engineering of opium poppy using CRISPR-Cas9 genome editing technology. Sci. Rep. 6: 30910. doi.10.1038/srep30910.

Ali, Z., Abul-Faraj, A., Li, L., Ghosh, N. et al. 2015. Efficient virus-mediated genome editing in plants using the CRISPR/Cas9 system. Mol. Plant. 8: 1288–1291. doi. 10.1016/j.molp.2015.02.011.

Andolfo, G., Iovieno, P., Frusciante, L. and Ercolano, M. R. 2016. Genome-editing technologies for enhancing plant disease resistance. Front. Plant Sci. doi. 10.3389/fpls.2016.01813.

Anonymous, Engineering LAGLIDADG homing endonucleases: notes, citations and suggestions. http://www.homingendonuclease.net/info/engineering_LHEs.pdf

Anzalone, A.V., Randolph, P.B., Davis, J.R., Sousa, A.A. et al. 2019. Search-and-replace genome editing without double-strand breaks or donor DNA. Nature, doi:10.1038/s41586-019-1711-4.

Anzalone, A.V., Koblan, L.W. and Liu, D.R. 2020. Genome editing with CRISPR–Cas nucleases, base editors, transposases and prime editors. Nat. Biotechnol. 38, 824–844. https://doi.org/10.1038/s41587-020-0561-9

Aquino-Jarquin, G. 2019. CRISPR-Cas14 is now part of the artillery for gene editing and molecular diagnostic. Nanomed. Nanotech. Bio. Med. 18: 428-431.

Arnould, S., Chames, P., Perez, C., Lacroix, E. et al. 2006. Engineering of large numbers of highly specific homing endonucleases that induce recombination on novel DNA targets. J. Mol. Biol. 355: 443–458.

Arora, L., and Narula, A. 2017. Gene editing and crop improvement using CRISPR-Cas9 system. Front. Plant Sci. 8: 1-21.

Asano, K., Hirano, K., Ueguchi-Tanaka, M., Angeles-Shim, R.B. et al. 2009. Isolation and characterization of dominant dwarf mutants, *Slr1-d*, in rice. Mol. Genet. Genomics 281: 223–231.

Auer, T.O., Duroure, K., De Cian, A., Concordet, J.-P. et al. 2014. Highly efficient CRISPR/Cas9-mediated knock-in in zebrafish by homology-independent DNA repair. Genome Res. 24: 142–153.

Bacman, S.R., Williams, S.L., Pinto, M., Peralta, S. et al. 2013. Specific elimination of mutant mitochondrial genomes in patient–derived cells by mitoTALENs. Nat Med. 19: 1111–1113. doi:10.1038/nm.3261.

Bailey-Serres, J., Parker, J.E., Ainsworth, E.A., Oldroyd, G.E.D. et al. 2019. Genetic strategies for improving crop yields. Nature 575:109-118.

Baltimore, D., Berg, P., Botchan, M., Carroll, D. et al. 2015. A prudent path forward for genomic engineering and germline gene modification. Science 348: 36–38. doi:10.1126/science.aab1028.

Batista, A.C. and Pacheco, L.G.C. 2018. Detecting pathogens with zinc-finger, TALE and CRISPR- based programmable nucleic acid binding proteins. J. Microbiol. Methods. 152: 98-104. doi: 10.1016/j.mimet.2018.07.024.

Beadle, G.W. and Tatum, E.L. 1941. Genetic control of biochemical reactions in *Neurospora*. Proc. Natl. Acad. Sci. U.S.A. 27: 499-506.

Beerli, R.R., Segal, D.J., Dreier, B., and Barbas, C.F., 3rd. 1998. Toward controlling gene expression at will: specific regulation of the *erbB-2/HER-2* promoter by using polydactyl zinc finger proteins constructed from modular building blocks. Proc. Natl. Acad. Sci. USA 95: 14628–14633.

Beetham, P.R., Kipp, P.B., Sawycky, X.L., Arntzen, et al. 1999. A tool for functional plant genomics: Chimeric RNA/DNA oligonucleotides cause *in vivo* gene-specific mutations. Proc. Natl. Acad. Sci. USA 96: 8774–8778.

Belfort, M. and Bonocora, R.P. 2014. Homing endonucleases: from genetic anomalies to programmable genomic clippers. Methods Mol Biol. 1123: 1–26. doi:10.1007/978-1-62703-968-0_1.

Belhaj, K., Chaparro-Garcia, A., Kamoun, S., Patron, N.J. et al. 2015. Editing plant genomes with CRISPR/Cas9. Curr. Opinion Biotechnol. 32: 76–84.

Benedict, M.Q., Burt, A., Capurro, M.L. et al. 2018. Recommendations for laboratory containment and management of gene drive systems in arthropods. Vector-Borne Zoonotic Dis. 18:2–13. https://doi.org/10.1089/vbz.2017.2121

Beumer, K.J., Trautman, J.K., Christian, M., Dahlem, T.J. et al. 2013. Comparing ZFNs and TALENs for gene targeting in *Drosophila*. G3 3: 1717–1725.

Bhowmik, P.. Ellison, E., Polley, B., Bollina, V. et. al., 2018. Targeted mutagenesis in wheat micrspores using CRISPR/Cas9. Sci. Rep. 8: 6502-6511.

Bibikova, M., Beumer, K., Trautman, J.K. and Carroll, D. 2003. Enhancing gene targeting using designed zinc finger nucleases. Science 300: 764.

Bibikova, M., Golic, M., Golic, K.G. and Carroll, D. 2002. Targeted chromosomal cleavage and mutagenesis in *Drosophila* using zinc-finger nucleases, Genetics 161: 1169–1175.

Billon, P., Bryant, E.E., Joseph, S.A., Nambiar, T.S. et al. 2017. CRISPR-mediated base editing enables efficient disruption of eukaryotic genes through induction of stop codons (iSTOP). Mol. Cell 67: 1068.

Bogdanove, A.J. and Booher, N.J. 2016. Online tools for TALEN design. *In* Ralf Kühn et al. (eds) *TALENs: Methods and Protocols*, *Methods in Molecular Biology*, vol. 1338, pp., 43-48. DOI 10.1007/978-1-4939-2932-0_1, ©Springer Science+ Business Media, New York.

Bogdanove, A.J., Bohm, A., Miller, J.C., Morgan, R.D. et al. 2018. Engineering altered protein–DNA recognition specificity. Nucleic Acids Res. doi: 10.1093/nar/gky289.

Boissel, S. and Scharenberg, A.M. 2015. Assembly and Characterization of megaTALs for hyperspecific genome engineering applications. In S.M. Pruett-Miller (ed.), Chromosomal Mutagenesis, Methods in

Molecular Biology, vol. 1239, DOI 10.1007/978-1-4939-1862-1_9, © Springer Science+Business Media, New York 2015

Boissel, S., Jarjour, J., Astrakhan, A., Adey, A. *et al.* 2014. megaTALs: a rare-cleaving nuclease architecture for therapeutic genome engineering. Nucleic Acids Res. 42: 2591-2601.

Bolukbasi, M.F., Gupta, A., Oikemus, S., Derr, A.G. et al. 2015. DNA-binding-domain fusions enhance the targeting range and precision of Cas9. Nat Methods 12: 1–9.

Bortesi, L. and Fischer, R. 2015. The CRISPR/Cas9 system for plant genome editing and beyond. Biotechnol. Adv. http://dx.doi.org/10.1016/j.biotechadv.2014.12.006

Broeders, M., Herrero-Hernandez, P., Ernst, M.P.T., van der Ploeg, A.T. et al. 2020. Sharpening the molecular scissors: advances in gene-editing technology. iScience 24; 23(1):100789. doi: 10.1016/j.isci.2019.100789.

Brooks, A.K. and Gaj, T. 2018. Innovations in CRISPR technology. Curr. Opin. Biotechnol. 52: 95–101.

Brossard, D., Belluck, P., Gould, F., Wirz, C.D. 2019. Promises and perils of gene drives: navigating the communication of complex, post-normal science. Proc Natl Acad Sci USA. https//doi.org/10.1073/pnas.18058 74115

Burt, A. 2003. Site-specific selfish genes as tools for the control and genetic engineering of natural populations. Proc. R. Soc. Lond. B 270: 921–928.

Butler, J.R., Paris, L.L. and Blankenship, R.L. et al. 2016. Silencing porcine CMAH and GGTA1 genes significantly reduces xenogeneic consumption of human platelets by porcine livers. Transplantation 100: 571–576.

CAC (Codex Alimentarius Commission). 2014. Procedural Manual. Joint FAO/WHO Food Standards Programme 22nd edition 2014. FAO/WHO http://www.codexalimentarius.org/

Carlson, D.F., Tan, W., Lillico, S.G., Stverakova, D. et al. 2012. Efficient TALEN-mediated gene knockout in livestock. Proc. Natil. Acad. Sci., 109: 17382-17387

Carroll, D. 2014. Genome engineering with targetable nucleases. Annu. Rev. Biochem. 83: 409–439.

Carroll, D. 2015. Genome editing by targeted chromosomal mutagenesis. S.M. Pruett-Miller (ed.), Chromosomal Mutagenesis, Methods in Molecular Biology, vol. 1239, p.1-13. DOI 10.1007/978-1-4939-1862-1_1, © Springer Science+Business Media New York.

Carroll, D. 2016. Genome editing: progress and challenges for medical applications. Genome Med. 8: 120. Doi. 10.1186/s13073-016-0378-9.

Cavaliere, G. 2019. WHO expert advisory committee on developing global standards for governance and oversight of human genome editing: background paper, the ethics of human genome editing. http://www.Who.int>ethics>human-genome-editing>

Cermak, T., Curtin S.J., Gil-Humanes, J., Cegan, R. et al. 2017. A multipurpose tool kit to enable advanced genome engineering in plants. Plant Cell 29: 1196-1217.

Cermak, T., Doyle, E.L., Christian, M., Wang, L. et al. 2011. Efficient design and assembly of custom TALEN and other TAL effector-based constructs for DNA targeting. Nucleic Acids Res 39: e82

Champer, J., Buchman, A. and Akbari, O.S. 2016. Cheating evolution: engineering gene drives to manipulate the fate of wild populations. Nat. Rev. Genet. 17:.146–159. https ://doi.org/10.1038/nrg.2015.34

Champer, J., Chung, J., Lee, Y.L., Liu, C. et al. 2019. Molecular safeguarding of CRISPR gene drive experiments. eLife 8: e41439. DOI: https://doi.org/10.7554/eLife.41439

Chan, Y.-S. 2011. Insect population control by homing endonuclease-based gene drive: an evaluation in *Drosophila melanogaster*. Genetics. 188: 33-44.

Chandrasegaran, S. 2017. Recent advances in the use of ZFN-mediated gene editing for human gene therapy. Cell Gene Ther. Insights. 3: 33–41. doi: 10.18609/cgti.2017.005

Chandrasegaran, S. and Carroll, D. 2016. Origins of programmable nucleases for genome engineering. J Mol Biol. 428: 963–989.

Chandrasegaran, S. and Smith, J. 1999. Chimeric restriction enzymes: what is next? Biol. Chem. 380: 841–848.

Chandrasekaran, J., Brumin, M., Wolf, D., Leibman, D. et al. 2016. Development of broad virus resistance in non-transgenic cucumber using CRISPR/Cas9 technology. Molec. Plant Pathol. 17: 1140–1153.

Chang, C.W., Lai, Y.S., Westin, E. et al. 2015. Modeling human severe combined immunodeficiency and correction by CRISPR/Cas9-enhanced gene targeting. Cell Rep. 12: 1668–1677.

Chari, R., Mali, P., Moosburner, M. and Church, G.M. 2015. Unravelling CRISPR-Cas9 genome engineering parameters via a library-on-library approach. Nature Methods. doi:10.1038/nmeth.3473

Chari, R., Yeo, N.C., Chavez, A. and Church, G.M. 2017. sgRNA scorer 2.0: a species independent model to predict CRISPR/Cas9 activity. ACS Synth Biol. 6: 902–904.

Charo, R.A. 2016. The legal and regulatory context for human gene editing. Issues Sci. Technol. 32: 39

Chaudhary, K., Pratap, D. and Sharma, P.K. 2016. Transcription activator - like effector nucleases (TALENs): An efficient tool for plant genome editing. Engg. Life Sci. 16: 330-337.

Chen, J.S., Dagdas, Y.S., Kleinstiver, B.P. and Welch, M.M. 2017. Enhanced proof- reading governs CRISPR–Cas9 targeting accuracy. Nature doi:10.1038/nature24268

Chen, S., Yu, X., and Guo, D. 2018. CRISPR-Cas targeting of host genes as an antiviral strategy. Viruses. 10: 40; doi:10.3390/v10010040

Chneiweiss, H., Hirsch, F., Montoliu, L., Muller, A.M. et al. 2017. Fostering responsible research with genome editing technologies: a European perspective. Transgenic Res. 26: 709–713, DOI 10.1007/s11248-017-0028-z

Choulika, A., Perrin, A. Dujon, B. and Nicolas, J-F. 1995. Induction of homologous recombination in mammalian chromosomes by using the I-*Sce*I system of *Saccharomyces cerevisiae*. Mol. Cell. Biol. 15: 1968–1973

Christian, M., Cermak, T., Doyle, E.L., Schmidt, C. et al. 2010. Targeting DNA double-strand breaks with TAL effector nucleases. Genetics 186: 757–761.

Chugunova, A.A., Dontsova, O.A. and Sergiev, P.V. 2016. Methods of genome engineering: a new era of molecular biology. Biochemistry (Moscow) 81: 662-677.

Clasen, B.M., Stoddard, T.J., Luo, S., Demorest, Z.L. et al. 2016. Improving cold storage and processing traits in potato through targeted gene knockout. Plant Biotechnol. J. 14:169-176. doi: 10.1111/pbi.12370.

Cole, J.B. 2017. Management of Mendelian traits in breeding programs by gene editing: a simulation study. bioRxiv, doi: http://dx.doi.org/10.1101/116459

Cole-Strauss, A, Yoon, K., Xiang, Y., Byrne, B. C. et al. 1996. Correction of the mutation responsible for sickle cell anemia by an RNA-DNA oligonucleotide. Science 273: 1386–1389.

Coller, B.S. 2019. Ethics of human genome editing. Ann. Rev. Medicine 70:289-305. https://doi.org/10.1146/annurev-med-112717-094629

Collins, J.P. 2018. Gene drives in our future: challenges of and opportunities for using a self-sustaining technology in pest and vector management. BMC Proce. 12(Suppl 8):9. https://doi.org/10.1186/s12919-018-0110-4

Cong, L., Ran, F.A., Cox, D., Lin, S. et al. 2013. Multiplex genome engineering using CRISPR/Cas systems, Science 339: 819–823.

Cox, D.B.T., Gootenberg, J.S., Abudayyeh, O.O., Franklin, B. et al. 2017. RNA editing with CRISPR-Cas13. Science 358: 1019–1027. doi: 10.1126/science.aaq0180

Cribbs, A.P. and Perera, S.M.W. 2017. Science and bioethics of CRISPR-Cas9 gene editing: an analysis towards separating facts and fiction. Yale J. Biol. Med. 90: 625-634.

Crispo, M., Mulet, A.P., Tesson, L., Barrera, N, et al. 2015. Efficient generation of myostatin knock-out sheep using CRISPR/Cas9 technology and microinjection into zygotes. PLoS One. 10:e0136690. https://doi.org/10.1371/journal.pone.0136690.

Csörgő, B., León, L.M., Chau-Ly, I.J. *et al.* 2020. A compact Cascade–Cas3 system for targeted genome engineering. *Nat Methods* (2020). https://doi.org/10.1038/s41592-020-00980-w.

Cui, X., Ji, D., Fisher, D.A., Wu, Y.et al. 2011. Targeted integration in rat and mouse embryos with zinc-finger nucleases. Nat. Biotechnol. 29: 64–67.

Curinga, G., Romano, G., Sather, B., Gwiazda, K. et al. 2015. MegaTAL nucleases outperform TALENS in promoting homology-directed-gene modification in primary human T cells. Gene Targeting Gene Correction I, S48. DOI: https://doi.org/10.1016/S1525-0016(16)33721-2

Curtin, S.J., Voytas, D.F. and Stupar, R.M. 2012. Genome engineering of crops with designer nucleases. Plant Genome 5: 42-50.

Cyranoski, D. 2016. CRISPR alternative doubted. Nature 536: 136-137

Cyranoski, D. 2019a. Gene-edited monkey clones stir excitement and debate. Nature 566: 15-16.

Cyranoski, D. 2019b. What's next for CRISPR babies? Nature 566: 440-443.

de Wert, G., Heindryck, B., Pennings, G., Clarke, A. et al. 2018. Responsible innovation in human germline gene editing. Background document to the recommendations of ESHG and ESHR. Human Reproduction Open 1–18. doi:10.1093/hropen/hox024

Delwart, E.L., Shpaer, E.G., Louwagie, J., McCutchan, F.E. et al. 1993. Genetic relationships determined by a DNA heteroduplex mobility assay: analysis of HIV-1 env genes, Science 262: 1257e1261.

Deng, D., Yan, C., Pan, X., Mahfouz, M. et al. 2012. Structural basis for sequence-specific recognition of DNA by TAL effectors. Science 335: 720–723

Dicarlo, J. E., Chavez, A., Dietz, S. L. Esvelt, K. M. et al. 2015. RNA-guided gene drives can efficiently and reversibly bias inheritance in wild yeast. bioRxiv 013896.

Dinesh-Kumar, S.P. and Voytas, D.F. 2020. "https://www-nature-com.ezp1.lib.umn.edu/articles/s41477-020-0716-1#citeas"Editing through infection. Nat. Plants 6: 738–739. https://doi.org/10.1038/s41477-020-0716-1

Ding, Q., Strong, A., Patel, K.M., Ng, S.L. et al. 2014. Permanent alteration of PCSK9 with in vivo CRISPR-Cas9 genome editing. Circ. Res. 115: 488-492. doi: 10.1161/CIRCRESAHA.115.304351.

Doman, J.L., Raguram, A., Newby, G.A. and Liu, D.A. 2020. Evaluation and minimization of Cas9-independent off-target DNA editing by cytosine base editors. Nat. Biotechnol. 38:620-628. https://doi.org/10.1038/s41587-020-0414-6.

Doudna, J.A. and Charpentier, E. 2014. The new frontier of genome engineering with CRISPR-Cas9. Science 346: 1077-1087. DOI: 10.1126/science.1258096.

Doudna, J.A. and Sternberg, S.H. 2017. A Crack in Creation, Gene Editing and the Unthinkable Power to Control Evolution. Houghton Mifflin Harcourt, Boston, N.Y., U.S.A.

Drury, D.W., Siniard, D.J., Zentner, G.E., and Wade, M.J. 2016. CRISPR/Cas9 gene drives in genetically variable and non-randomly mating wild populations. bioRxiv 2016; doi: http://dx.doi.org/10.1101/071670.

Du, D., Roguev, A., Gordon, D.E., Chen, M. et al. 2017. Genetic interaction mapping in mammalian cells using CRISPR interference. Nature Meth. doi:10.1038/ nmeth.4286

Dunbar, C.E., High, K.A., Joung, J.K., Kohn, D.B. et al. 2018. Gene therapy comes of age. Science eaan4672. DOI: 10.1126/science.aan4672

Durai, S., Mani, M., Kandavelou, K., Wu, J. et al. 2005. Zinc finger nucleases: custom-designed molecular scissors for genome engineering of plant and mammalian cells. Nucleic Acids Res. 33: 5978–5990.

Durr, R., Papareddy, R., Nakajima, K. and Gutierrez-Marcos, J. 2018. Highly efficient heritable targeted deletions of gene clusters and non-coding regulatory regions in *Arabidopsis* using CRISPR/Cas9. Sci. Rep. DOI:10.1038/s41598-018-22667

Editor, 2019. Set rules for germline gene editing. Nature 567: 145.

EFSA, Panel on Genetically Modified Organisms (GMO): Scientific opinion addressing the safety assessment of plants developed through cisgenesis and intragenesis, EFSA J., 10, 2561 [33 pp] (2012)

Esvelt, K.M. and Gemmell, N.J. 2017. Conservation demands safe gene drive. PLOS Biol 15: e2003850. https ://doi.org/10.1371/journal.pbio.20038 50

Esvelt, K.M., Smidler, A.L., Catteruccia, F. and Church, G.M. 2014. Concerning RNA-guided gene drives for the alteration of wild populations. *eLife* 3: e03401.

Feng, X., Peng, C., Chen, Y., Liu, X. et al. 2017. Discrimination of CRISPR/Cas9-induced mutants of rice seeds using near-infrared hyperspectral imaging. Sci. Reports 7: 15934. DOI:10.1038/s41598-017-16254-z

Feng, Y., Zhang, S. and Huang, X. 2014. A robust TALENs system for highly efficient mammalian genome editing. Sci. Rep. 4: 3632. DOI: 10.1038/srep03632

Fu, Y., Sander, J.D., Reyon, D., Cascio, V.M. et al. 2014. Improving CRISPR-Cas nuclease specificity using truncated guide RNAs. 32: 279–284. doi:10.1038/nbt.2808.

Gaj, T., Gersbach, C.A. and Barbas, C.F. 2013. ZFN, TALEN, and CRISPR/Cas-based methods for genome engineering. Trends Biotechnol. 31: 397-405.

Gaj, T., Shannon J. Sirk, S.J., Shui, S.-l. and Liu, J. 2019. Genome-Editing technologies: principles and applications. Additional Perspectives on Synthetic Biology. doi: 10.1101/cshperspect.a023754

Gamper, H.B., Parekh, H., Rice, M.C., Bruner, M. et al. 2000. The DNA strand of chimeric RNA/DNA oligonucleotides can direct gene repair/conversion in mammalian and plant cell-free extracts. Nucleic Acids Res. 28: 4332-4339.

Gantz, V.M. and Bier, E. 2015. The mutagenic chain reaction: a method for converting heterozygous to homozygous mutations. Science 348: 442–444. doi:10.1126/ science.aaa5945.

Gantz, V.M., Jasinskiene, N., Tatarenkova, O., et al. 2015. Highly efficient Cas9-mediated gene drive for population modification of the malaria vector mosquito *Anopheles stephensi*. Proc. Natl. Acad. Sci. U.S.A. 112: E6736–6743.

Gao, C. 2019. Precision plant breeding using genome editing technologies. Transgenic Res. 28: 53-55.

Gao, Y. and Zhao, Y. 2014. Self-processing of ribozyme-flanked RNAs into guide RNAs *in vitro* and *in vivo* for CRISPR-mediated genome editing. J. Integr. Plant Biol. 56: 343-349.

Gaudelli, N.M., Komor, A.C., Rees, H.A., Packer, M.S. et al. 2017. Programmable base editing of A-T to G-C in genomic DNA without DNA cleavage. Nature 551: 464-471. doi:10.1038/nature24644.

Gehrke, J.M., Cervantes, O., Clement, M.K., Pinello, L. et al. 2018. High-precision CRISPR-Cas9 base editors with minimized bystander and off-target mutations. bioRxiv. doi: http://dx.doi.org/10.1101/273938.

Georges, F. and Ray, H. 2017. Genome editing of crops: A renewed opportunity for food security. GM Crops Food 8: 1–12.

Geu-Flores, F., Nour-Eldin, H.H., Nielsen, M.T., and Halkier, B.A. 2007. USER fusion: a rapid and efficient method for simultaneous fusion and cloning of multiple PCR products. Nucleic Acids Res 35: e55

Gilbert, L.A., Horlbeck, M.A., Adamson, B., Villalta, J.E. et al. 2014. Genome-scale CRISPR-mediated control of gene repression and activation. Cell 159: 647–661.

Gilbert, L.A., Larson, M.H., Morsut, L., Liu, Z. et al. 2013. CRISPR-mediated modular RNA-guided regulation of transcription in eukaryotes. Cell 154: 442–451.

Gogolok, S., Garcia-Diaz, C. and Pollard, S.M. 2016. STAR: a simple TAL effector assembly reaction using isothermal assembly. Sci. Rep. 6: 33209. DOI: 10.1038/srep33209

Gomez, M.A., Lin, Z.D., Moll, T, Chauhan, R.D. et al. 2018. Simultaneous CRISPR/ Cas9-mediated editing of cassava eIF4E isoforms nCBP-1 and nCBP-2 reduces cassava brown streak disease symptom severity and incidence. Plant Biotechnol. J. doi: 10.1111/pbi.12987.

Goodman, M.A., Manesh, D.M., Malik, P. and Rothenberg, M.E. 2017. CRISPR/Cas9 in allergic and immunologic diseases. Expert Rev. Clinical Immunol. 13: 5–9.

Gootenberg, J., Abudayyeh, O.O., Lee, J.W., Essletzbichler, P. et al. 2017. Nucleic acid detection with CRISPR-Cas13a/C2c2. Science 356: 438–442

Grunwald, H.A., Gantz, V.M., Poplawski, G., Xu, X.S. 2019. Super-Mendelian inheritance mediated by CRISPR/Cas9 in the female mouse germline. Nature 566: 105–109. doi:10.1038/s41586-019-0875-2.

Guha, T.K. and Edgel, D.R. 2017. Applications of alternative nucleases in the age of CRISPR/Cas9. Int. J. Mol. Sci. 18: 2565. doi:10.3390/ijms18122565

Guichard, A., Haque, T., Bobik, M., Xu, X.-R.S. et al. 2019. Efficient allelic-drive in *Drosophila*. Nat. Commun. doi.org/10.1038/s41467-019-09694-w https://doi.org

Guilinger, J.P., Pattanayak, V., Reyon, D., Tsai, S.Q. et al. 2014. Broad Specificity Profiling of TALENs results in engineered nucleases with improved DNA cleavage specificity. Nat Methods. 11: 429–435. doi:10.1038/nmeth.2845.

Gutschner, T., Baas, M. and Diederichs, S. 2011. Noncoding RNA gene silencing through genomic integration of RNA destabilizing elements using zinc finger nucleases. Genome Res.21:1944–1954.

Haapaniemi, E., Botla, S., Persson, J., Schmierer, B. et al. 2018. CRISPR–Cas9 genome editing induces a p53-mediated DNA damage response. Nat. Med. 24: 927–930.

Halperin, S.O., Tou, C.J., Wong, E.B., Modavi, C. et al. 2018. CRISPR-guided DNA polymerases enable diversification of all nucleotides in a tunable window. Nature 560: 248 -252.

Hammond, A., Galizi, R., Kyrou, K., Simoni, A. et al. 2016. A CRISPR-Cas9 gene drive system targeting female reproduction in the malaria mosquito vector *Anopheles gambiae*. Nat. Biotechnol. 34:78-83

Hammond, A.M. and Galizi, R. 2018. Gene drives to fight malaria: current state and future directions, Pathogens Global Health 111: 412-423, DOI: 10.1080/ 20477724.2018.1438880

Hanna, J., Wernig, M., Markoulaki, S., Sun, C.W. et al. 2007. Treatment of sickle cell anemia mouse model with iPS cells generated from autologous skin. Science 318: 1920–1923.

Harrington, L.B. Doxzen, K.W., Ma, E. and Liu, J.-J. 2017. A Broad-spectrum inhibitor of CRISPR-Cas9. Cell 170: 1224–1233

Haun, W., Coffman, A., Clasen, B.M., Demorest, Z.L. et al. 2014. Improved soybean oil quality by targeted mutagenesis of the *fatty acid desaturase 2* gene family. Plant Biotechnol. J. 12: 934–940.

Hendel, A., Fine, E.J., Bao, G., and Porteus, M.H. 2015. Quantifying on- and off-target genome editing. Trends Biotechnol. 33: 132–140.

Hess, G.T., Fresard, L., Han, K., Lee, C.H., et al. 2016. Directed evolution using dCas9-targeted somatic hypermutation in mammalian cells. Nat. Methods 13, 1036–1042.

Hess, G.T., Tycko, J., Yao, D. and Bassik, M.C. 2017. Methods and applications of CRISPR-mediated base editing in eukaryotic genomes. Mol. Cell 68: 26-43. doi.org/10.1016/j.molcel.2017.09.029

Hsu, P.D., Lander, E.S. and Zhang, F. 2014. Development and applications of CRISPR-Cas9 for genome engineering. Cell 157: 1262–1278. doi: 10.1016/j.cell.2014.05.010

Hua, K., Tao, X., Yuan, F., Wang, D. et al. 2018. Precise A.T to G.C base editing in the rice genome. Mol. Plant 11: 627-630.

Hubbard, B.P., Badran, A.H., Zuris, J.A., Guilinger, L.P. et al. 2015. Continuous directed evolution of DNA-binding proteins to improve TALEN specificity. Nature Methods 12: 939-934. doi:10.1038/nmeth.3515

Igoucheva, O. and Yoon, K. 2000. Targeted single-base correction by RNA–DNA oligonucleotides. Human Gene Ther. 11: 2307–2312.

Ihry, R.J. Worringer, K.A., Salick, M.R., Frias, E. et al. 2018. p53 inhibits CRISPR–Cas9 engineering in human pluripotent stem cells. Nat. Med. 24: 939–946.

Ishii, T. and Araki, M. 2017. A future scenario of the global regulatory landscape regarding genome-edited crops,. GM Crops Food 8: 44-56, DOI: 10.1080/ 21645698.2016.1261787

Ito, Y., Nishizawa-Yokoi, A., Endo, M., Mikami, M. et al. 2015. CRISPR/Cas9-mediated mutagenesis of the *RIN* locus that regulates tomato fruit ripening. Biochem. Biophys. Res. Commun. 467: 76e8

Jakimo, N., Chatterjee, P. and Jacobson, J.M. 2018. ssRNA/DNA-sensors via embedded strand-displacement programs in CRISPR/Cas9 guides. bioRxiv doi: http://dx.doi.org/10.1101/264424

Jenko, J., Gorjanc, G., Cleveland, M.A. and Varshney, R.K. 2015. Potential of promotion of alleles by genome editing to improve quantitative traits in livestock breeding programs. Genet. Sel. Evol. 47: 55. doi: 10.1186/s12711-015-0135-3

Jinek, M., Chylinski, K., Fonfara, I., Hauer, M. et al. 2012. A programmable dual-RNA-guided DNA endonuclease in adaptive bacterial immunity. Science 337: 816–821

Jones, H.D. 2015. Future of breeding by genome editing is in the hands of regulators. GM Crops Food 6: 223–232.

Joung, J.K. and Sander, J.D. 2013. TALENs: a widely applicable technology for targeted genome editing. Nat. Rev. Mol. Cell Biol. 14: 49–55.

Juillerat, A., Pessereau, C., Dubois, G., Guyot, V. et al. 2015. Optimized tuning of TALEN specificity using non-conventional RVDs. Sci. Rep. 5: 8150. Doi: 10.1038/srep08150

Jung, C., Capistrano-Gossmann, G., Braatz, J., Niharika Sashidhar, N. et al. 2018. Recent developments in genome editing and applications in plant breeding. Plant Breeding 137: 1–9.

Kalhor, R., Mali, P. and Church, G.M. 2016. Rapidly evolving homing CRISPR barcodes. Nat. Methods. 2016; doi:10.1038/nmeth.4108

Kamburova, G.L., Nikitina, E.V., Shermatov, S.E., Buriev, Z.T. et al. 2017. Genome editing in plants: an overview of tools and applications. Int. J. Agron. https://doi.org/10.1155/2017/7315351.

Kasinathan, P., Wei, H., Xiang, T., Molina, J.A. et al. 2015. Acceleration of genetic gain in cattle by reduction of generation interval. Sci. Rep. 5: 8674. https://doi.org/ 10.1038/srep08674.

Katrekar, D., Hu, M. and Mali, P. 2017. Advances in CRISPR-Cas based genome engineering. Current Opinion Biomed. Engg. 1: 78–86

Kelliher, T., Starr, D., Richbourg, L., Chintamanani, S. et al. 2017. MATRILINEAL, a sperm-specific phospholipase, triggers maize haploid induction. Nature 542: 105–109.

Khatodia, S., Bhatotia, K. and Tuteja, N. 2017. Development of CRISPR/Cas9 mediated virus resistance in agriculturally important crops. Bioengineered 8: 274–279. https://doi.org/10.1080/21655979.2017.1297347

Kim, D., Lim, K., Kim, S.T., Yoon, S.H. et al. 2017. Genome-wide target specificities of CRISPR RNA-guided programmable deaminases. Nat. Biotechnol. 35, 475-480.

Kim, Y.B., Komor, A.C., Levy, J.M., Packer, M.S. et al. 2017. Increasing the genome-targeting scope and precision of base editing with engineered Cas9-cytidine deaminase fusions. Nat. Biotechnol. 35: 371–376.

Kim, Y.-G., Cha, J. and Chandrasegaran, S. 1996. Hybrid restriction enzymes: Zinc finger fusions to FokI cleavage domain. Proc. Natl. Acad. Sci. U.S.A. 93: 1156–1160.

Kim, Y-G. and Chandrasegaran, S. 1994. Chimeric restriction endonuclease. Proc. Natl. Acad. Sci. USA 91: 883–887.

Klap, C., Yeshayahou, E., Bolger, A. M., Arazi, T. et. al. 2016. Tomato facultative parthenocarpy results from SlAGAMOUS-LIKE 6 loss of function. Plant Biotechnol. J. 15: 634–647.

Kleinstiver, B.P., Wolfs, J.M., Kolaczyk, T., Roberts, A,K, et al. 2012. Monomeric site-specific nucleases for genome editing. Proc. Natl. Acad. Sci. U S A 109: 8061–8066.

Klimek-Chodacka, M., Oleszkiewicz, T., Lowder, L.G., Qi, Y. et al. 2018. Efficient CRISPR/Cas9-based genome editing in carrot cells. Plant Cell Rep. 37: 575–586. doi: 10.1007/s00299-018-2252-2.

Koike-Yusa, H., Li, Y., Tan, E,P., Velasco-Herreram, M.D.C. et al. 2014 Genome-wide recessive genetic screening in mammalian cells with a lentiviral CRISPR-guide RNA library. Nat. Biotech. 32: 267-73. doi: 10.1038/nbt.2800.

Komor, A. C., Kim, Y. B., Packer, M. S., Zuris, J. A. et al. 2016. Programmable editing of a target base in genomic DNA without double-stranded DNA cleavage. Nature 533, 420–424.

Komor, A.C., Badran, A.H. and Liu, D.R. 2017a. Editing the genome without double-stranded DNA breaks. ACS Chem. Biol. DOI: 10.1021/acschembio.7b00710.

Komor, A.C., Badran, A.H. and Liu, D.R. 2017b. CRISPR-based technologies for the manipulation of eukaryotic genomes. Cell 168: 20-36.

Komor, A.C., Zhao, K.T., Packer, M.S., and Gaudelli, N.M. 2017c Improved base excision repair inhibition and bacteriophage Mu Gam protein yields C:G to T:A base editors with higher efficiency and product purity. Sci. Adv. 3: eaao4774.

Konermann, S., Mark D. Brigham, M.D., Trevino, A.E. et al. 2015. Genome-scale transcriptional activation by an engineered CRISPR-Cas9 complex. Nature 517: 583–588. doi:10.1038/nature14136

Koo, T., Lee, J. and Kim, J.S. 2015. Measuring and reducing off-target activities of programmable nucleases including CRISPRCas9. Mol. Cell 38: 475–481.

Kosicki, M., Tomberg, K. and Bradley, A. 2018. Repair of double-strand breaks induced by CRISPR-Cas9 leads to large deletions and complex rearrangements. Nat. Biotechnol. 36: 765-771.

Krebs, J.E., Goldstein, E.S. and Kilpatrick, S.T. 2014. Lewin's Genes XI. Jones and Bartlett Learning, Burlington, Massachusetts, U.S.A.

Kühn, R., Wurst, W. and Wefers, B. (eds.) 2016. *TALENs: Methods and Protocols*, Methods in Molecular Biology, vol. 1338, DOI 10.1007/978-1-4939-2932-0_1, © Springer Science+Business Media, New York.

Kundert, K., Lucas, J.E., Watters, K.E., Fellmann, C. et al. 2018. Controlling CRISPR-Cas9 with ligand-activated and ligand-deactivated sgRNAs. bioRxiv doi: http://dx.doi.org/10.1101/323105/ncomms15939.

Kuscu, C., Parlak, M., Tufan, T., Yang, J. et al. 2017. CRISPR-STOP: gene silencing through base-editing-induced nonsense mutations. Nat. Methods 14: 710–712.

Lai, L. and Prather, R.S. 2003. Creating genetically modified pigs by using nuclear transfer. Reprod. Biol. Endocrin. 1: 82.

Lanigan, T.L., Kopera, H.C. and Saunders, T.L. 2020. Principles of Genetic Engineering. Genes 11: 291; doi:10.3390/genes11030291

Ledford, H. 2015a. CRISPR, the disrupter. Nature 522: 20-24.

Ledford, H. 2015b. The landscape for human genome editing. Nature 526: 310-311

Ledford, H. 2019a. CRISPR babies, when will the world be ready? Nature 570: 293-297.

Ledford, H. 2019b. EU regulators face CRISPR crop condundrum: Testing labs are struggling to enforce a court ruling on gene-edited foods. Nature 572: 15.

Ledford, H.2019c. Creators of gene-edited animals bypass US market. Nature 566: 433-434.

Lee, J., Chung, J.-H. Kim, H.M., Kim, D.-W. 2016. Designed nucleases for targeted genome editing. Plant Biotechnol. J. 14: 448–462.

Lei, L., Chen, H., Xue, W., Yang, B. et al. 2018. APOBEC3 induces mutations during repair of CRISPR–Cas9-generated DNA breaks. Nature Struct. Mol. Biol. 25: 45–52.

Li, C., Zong, Y., Wang, Y. Jin, S. et al. 2018. Expanded base editing in rice and wheat using a Cas9-adenosine deaminase fusion. Genome Biol. 19: 59-67 (https://doi.org/10.1186/s13059-018-1443-z)

Li, H.L., Fujimoto, N., Sasakawa, N., Shirai, S. et al. 2015. Precise correction of the *dystrophin* gene in Duchenne muscular dystrophy patient induced pluripotent stem cells by TALEN and CRISPR-Cas9. Stem Cell Reports 4: 143-154.

Li, J., Meng, X., Zong, Y., Chen, K. et al. 2016. Gene replacements and insertions in rice by intron targeting using CRISPR–Cas9. Nat. Plants 2: 16139.

Li, J., Sun, Y., Du, J., Zhao, Y. et al. 2017. Generation of targeted point mutations in rice by a modified CRISPR/Cas9 system. Mol. Plant 10, 526–529.

Li, L., Wu, L.P. and Chandrasegaran, S. 1992. Functional domains in FokI restriction endonuclease. Proc. Natl. Acad. Sci. U.S.A. 89: 4275–4279.

Li, M., Li, X., Zhou, Z., Wu, P. et al. 2016. Reassessment of the four yield-related genes *Gn1a, DEP1, GS3,* and *IPA1* in rice using a CRISPR/Cas9 system. Front. Plant Sci. 7: 377. doi: 10.3389/fpls.2016.00377

Li, Q., Zhang, D., Chen, M., Liang, W., Wei, J., Qi, Y., & Yuan, Z. 2016. Development of japonica photo-sensitive genic male sterile rice lines by editing *Carbon Starved Anther* using CRISPR/Cas9. Journal of Genetics and Genomics, 43, 415–419.

Li, T., Liu, B., Spalding, M.H., Weeks, D.P. et al. 2012. High- efficiency TALEN-based gene editing produces disease-resistant rice. Nat. Biotechnol. 30: 390–392. doi:10.1038/nbt.2199.

Li, X., Wang, Y., Liu, Y., Yang, B. et al. 2018. Base editing with a Cpf1–cytidine deaminase fusion. Nat Biotechnol. 36:324-7

Li, Y., Ma, S., Sun, L., Zhang, T. et al. 2018. Programmable single and multiplex base-editing in *Bombyx mori* using RNA-guided cytidine deaminases. Genes Genomes Genet. doi:10.1534/g3.118.200134

Liang, L, Liu, D-P. and Liang, C-C. 2002. Optimizing the delivery systems of chimeric RNA·DNA oligonucleotides, beyond general oligonucleotide transfer. Eur. J. Biochem. 269: 5753–5758. doi:10.1046/j.1432-1033.2002.03299.x.

Liang, P., Xu, Y., Zhang, X., Ding, C. et al. 2015. CRISPR/Cas9-mediated gene editing in human tripronuclear zygotes. Protein Cell. 6: 363–372.

Liang, Z., Chen, K., Li, T., Zhang, Y. et al. 2017. Efficient DNA-free genome editing of bread wheat using CRISPR/Cas9 ribonucleoprotein complexes. Nat. Commun, 8: 14261. doi: 10.1038/ncomms14261.

Liang, Z., Zhang, K, Chen, K., and Gao, C. 2014. Targeted mutagenesis in Zea mays using TALENs and the CRISPR/Cas system. J Genet Genomics 41: 63-68. doi: 10.1016/j.jgg.2013.12.001.

Lin, Q., Zong, Y., Xue, C. et al. 2020. Prime genome editing in rice and wheat. Nat Biotechnol. 38, 582–585. https://doi.org/10.1038/s41587-020-0455-x

Lin, S., Zhao, Y., Zhu, Y., Gosney, M. et al. 2016. An effective and inducible system of TAL effector-mediated transcriptional repression in *Arabidopsis*. Molec. Plant 9: 1546–1549.

Liu, H., Wei, Z., Dominguez, A., Li, Y. et al. 2015. CRISPRERA: a comprehensive design tool for CRISPR-mediated gene editing, repression and activation. Bioinformatics 31: 3676–3678.

Lloyd, A., Plaisier, C.L., Carroll, D. and Drews, G.N. 2005. Targeted mutagenesis using zinc-finger nucleases in *Arabidopsis*. Proc. Natl. Acad. Sci. U.S.A. 102: 2232–2237. doi:10.1073/pnas.0409339102

Lu, Y., and Zhu, J.K. 2017. Precise editing of a target base in the rice genome using a modified CRISPR/Cas9 system. Mol. Plant 10:523–525

Lusser, M., Parisi, C., Plan, D., and Rodriguez-Cerezo, E. 2012. Deployment of new biotechnologies in plant breeding. Nat. Biotechnol. 30: 231-239. DOI: 10.1038/nbt.2142

Lutz, A.J., Li, P., Estrada, J.L. et al. 2013. Double knockout pigs deficient in N-glycolylneuraminic acid and galactose alpha-1,3-galactose reduce the humoral barrier to xenotransplantation. Xenotransplantation 20: 27–35.

Ma, A.C., Lee, H.B., Clark, K.J. and Ekke, S.C. 2013. High efficiency *in vivo* genome engineering with a simplified 15-RVD GoldyTALEN design. PLoS ONE 8: e65259. doi:10.1371/journal.pone.0065259

Ma, H., Tu, L.-C., Naseri, A., Huisman, M, et al. 2016. Multiplexed labeling of genomic loci with dCas9 and engineered sgRNAs using CRISPRainbow. Nat. Biotechnol. 34: 528–530.

Ma, Y., Zhang, J., Yin, W., Zhang, Z. et al. 2016. Targeted AID-mediated mutagenesis (TAM) enables efficient genomic diversification in mammalian cells. Nat. Methods 13: 1029–1035

Maher, M.F., Nasti, R.A., Vollbrecht, M., Starker, C.G. et al. 2020. Plant gene editing through de novo induction of meristems. Nat. Biotehnol. 38:84-89. doi: 10.1038/s41587-019-0337-2.

Mali, P., Aach, J., Stranges, P.B., Esvelt, K.M. et al. 2013. Cas9 transcriptional activators for target specificity screening and paired nickases for cooperative genome engineering. Nat. Biotechnol. 31: 833–838. doi:10.1038/nbt.2675.

Mallapaty, S. 2019. Australian gene-editing rules adopt 'middle ground'. Nature News 23 April, 2019.

Malzahn, A., Lowder, L. and Qi, Y. 2017. Plant genome editing with TALEN and CRISPR. Cell Biosci. 7: 21-38. DOI 10.1186/s13578-017-0148-4

Mansour, S.L., Thomas, K.R., Capecchi, M.R. 1988. Disruption of the proto-oncogene *int-2* in mouse embryo-derived stem cells: a general strategy for targeting mutations to non-selectable genes. Nature 336: 348–52.

Marshall, J.M. and Akbari, O.S. 2018. Can CRISPR-based gene drive be confined in the wild? A question for molecular and population biology. ACS Chem Biol 13:424–430. https ://doi.org/10.1021/acsch embio .7b009 23

Mashimo, T. 2014. Gene targeting technologies in rats: Zinc finger nucleases, transcription activator-like effector nucleases, and clustered regularly interspaced short palindromic repeats Develop. Growth Differ. 56: 46–52

Matsoukas, I.G. 2018. Commentary: RNA editing with CRISPR-Cas13. Frontiers Genet. doi: 10.3389/fgene.2018.00134

McCreath, K.J., Howcroft, J., Campbell, K.H.S., Colman, A. et al. 2000. Production of gene-targeted sheep by nuclear transfer from cultured somatic cells. Nature 405: 1066-1069.

Meghani, Z. and Kuzma, J. 2018. Regulating animals with gene drive systems: lessons from the regulatory assessment of a genetically engineered mosquito. J. Responsible Innov. 5:S203–S222. https://doi.org/10.1080/23299 460.2017.14079 12

Miki, D., Zhang, W., Zeng, W., Feng, Z. et al. 2018. CRISPR/Cas9-mediated gene targeting in *Arabidopsis* using sequential transformation. Nature Commun. DOI: 10.1038/s41467-018-04416-0

Min, J., Noble, C., Najjar, D. and Esvelt, K. 2017a. Daisyfield gene drive systems harness repeated genomic elements as a generational clock to limit spread. BioRχiv. https://doi.org/10.1101/104877.

Min, J., Noble, C., Najjar, D. and Esvelt, K. 2017b. Daisy quorum drives for the genetic restoration of wild populations. BioRχiv.; https://doi.org/10.1101/115618.

Mok, B.Y., de Moraes, M.H., Zeng, J. *et al.* 2020. A bacterial cytidine deaminase toxin enables CRISPR-free mitochondrial base editing. Nature. https://doi.org/10.1038/s41586-020-2477-4

Montague, T.G., Cruz, J.M., Gagnon, J.A., Church, G.M. et al. 2014. CHOPCHOP: a CRISPR/Cas9 and TALEN web tool for genome editing. Nucleic Acids Res. 42, Web Server issue W401–W407. doi: 10.1093/nar/gku410

Montecillo, J.A.V., Chu, L.L. and Bae, H. 2020. CRISPR-Cas9 system for plant genome editing: current approaches and emerging developments. Agronomy 10: 1033; doi:10.3390/agronomy10071033.

Montiel-González, M.F., Vallecillo-Viejo, I.C., and Rosenthal, J.J. 2016. An efficient system for selectively altering genetic information within mRNAs. Nucleic Acids Res. 44: e157. doi: 10.1093/nar/gkw738

Moreno, A.M. and Mali, P. 2017. Therapeutic genome engineering via CRISPR-Cas systems. WIREs Syst. Biol. Med. 2017, e1380. doi: 10.1002/wsbm.1380

Moro, D., Byrne, M., Kennedy, M., Campbell, S. et al. 2018. Identifying knowledge gaps for gene drive research to control invasive animal species: The next CRISPR step. Global Ecol. Conserv. 13: e00363

Mueller, M.L., Cole, J.B., Sonstegard, T.S. and Van Eenennaam, A.L. 2018. Simulation of introgression of the *POLLED* allele into the Jersey breed via conventional breeding vs. gene editing. Transl. Anim. Sci. 2: S57–S60. doi: 10.1093/tas/txy054

Nakade, S., Tsubota, T., Sakane, Y., Kume, S. et al. 2014. Microhomology-mediated end-joining-dependent integration of donor DNA in cells and animals using TALENs and CRISPR/Cas9. Nature Commu. DOI: 10.1038/ncomms6560.

NASEM (National Academies of Sciences, Engineering, and Medicine). 2017. Human Genome Editing: Science, Ethics and Governance. National Academies Press, Washington, DC.

NASEM, 2016. Gene Drives on the Horizon: Advancing Science, Navigating Uncertainty, and Aligning Research with Public Values. National Academies Press, Washington, D.C.

Nash, A., Urdaneta, G.M., Beaghton, A.K., Hoermann, A. et al. 2019. Integral gene drives for population replacement. Biol. Open 8: bio037762. doi:10.1242/bio.037762

Nelles, D.A., Fang, M.Y., O'Connell, M.R., Xu, J.L. et al. 2016. Programmable RNA tracking in live cells with CRISPR/Cas9. Cell 165: 488–496.

Neves, M.P. and Druml, C. 2017. Ethical implications of fighting malaria with CRISPR/Cas9. BMJ Glob Health **2**:e000396. doi:10.1136/bmjgh-2017-000396

Nihongaki, Y., Yamamoto, S., Kawano, F. Suzuki, H. et al. 2015. CRISPR-Cas9-based photoactivatable transcription system. Chem. Biol. 22:169-74. doi: 10.1016/j.chembiol.2014.12.011.

Nishida, K., Arazoe, T., Yachie, N., Banno, S. et al. 2016. Targeted nucleotide editing using hybrid prokaryotic and vertebrate adaptive immune systems. Science 353: 1248-1261.

Niu, D., Wei, H.J., Lin, L., George, H. et al. 2017. Inactivation of porcine endogenous retrovirus in pigs using CRISPR-Cas9. Science 357: 1303–1307.

Niu, Y., Tenney, K., Li, H. and Gimble, F.S. 2008. Engineering variants of the I-SceI homing endonuclease with strand-specific and site-specific DNA-nicking activity. J. Mol. Biol. 382: 188–202.

Noble, C., Adlam, B., Church, G.M., Esvelt, K.M. et al. 2017. Current CRISPR gene drive systems are likely to be highly invasive in wild populations. bioRxiv; doi: http://dx.doi.org/10.1101/219022.

Noble, C., Min, J., Olejarz, J., Buchthal, J. et al. 2019. Daisy-chain gene drives for the alteration of local populations. Proc. Natl. Acad. Sci. U.S.A. 116: 8275–8282

Norris, A. L., Lee, S.S., Greenlees, K.J., Tadesse, D.A. et al. 2019. Template plasmid integration in germline genome-edited cattle. doi: https://doi.org/10.1101/715482

Nour - Eldin, H.H., Geu - Flores, F. and. Halkier, B.A. 2010. USER cloning and USER fusion: the ideal cloning techniques for small and big laboratories. Methods in molecular biology (Clifton, N.J.). DOI: 10.1007/978-1-60761-723-5_13.

O'Connell, M.R., Oakes, B.L., Sternberg, S.H., East-Seletsky, A. et al. 2014. Programmable RNA recognition and cleavage by CRISPR/Cas9. Nature 516: 263–266. doi:10.1038/nature13769.

Oberhofer, G., Ivy, T. and Hay, B.A. 2018. Behavior of homing endonuclease gene drives targeting genes required for viability or female fertility with multiplexed guide RNAs. Proc. Nat. Acad. Sci. U.S.A. 115: E9343–E9352

Ogden, L.E. 2019. Cloning gene-edited monkeys: a bioethical dilemma? BioScience 69: 580. doi:10.1093/biosci/biz048

Old, R.W. and Primrose, S.B. 2001. Principles of Gene manipulations: An introduction to Genetic Engineering, 6th edn. Blackwell Science Ltd., London.

Osakabe, Y. and Osakabe, K. 2014. Genome editing with engineered nucleases in plants. Plant Cell Physiol. 56: 389-400.

Ousterout, D.G. and Gersbach, C.A. 2016. The development of TALE nucleases for biotechnology. *In* Ralf Kühn et al. (eds) *TALENs: Methods and Protocols*, Methods in Molecular Biology, vol. 1338, pp. 27-42. DOI 10.1007/978-1-4939-2932-0_1, © Springer Science+Business Media, New York.

Ousterout, D.G., Kabadi, A.M., Thakore, P.I., Majoros, W.H. et al. 2015. Multiplex CRISPR/Cas9-based genome editing for correction of dystrophin mutations that cause Duchenne muscular dystrophy. Nat. Commun. 6: 6244

Oye, 1., K.A., Esvelt, K., Appleton, E., Catteruccia, F. et al. 2014. Regulating gene drives. Science 345: 626-628.

Parekh, U., Yusupova, M. and Mali, P. 2017). Genome engineering in human pluripotent stem cells. Curr. Opin. Chem. Engg. 15: 56–67.

Pattanayak, V., Ramirez, C.L., Joung, J.K. and Liu, D.R. 2011. Revealing off-target cleavage specificities of zinc-finger nucleases by *in vitro* selection. Nature Methods doi:10.1038/nmeth.1670.

Paul, M. and Ma, J.K. 2011. Plant-made pharmaceuticals: leading products and production platforms. Biotechnol Appl. Biochem. 58: 58-67.

Pausch, P., Al-Shayeb, B. et al. 2020. CRISPR-CasΦ from huge phages is a hypercompact genome editor. Science 369: 333-337

Peng, J., Wang, Y., Jiang, J. et al. 2015. Production of human albumin in pigs through CRISPR/Cas9-mediated knockin of human cDNA into swine albumin locus in the zygotes. Sci. Rep. 5: 16705.

Peng, Y., Clark, K.J., Campbell, J.M., Panetta, M. et al. 2014. Making designer mutants in model organisms. Development 142: 4042–4054.

Petek, L.M., Russell, D.W. and Miller, D.G. 2010. Frequent endonuclease cleavage at off-target locations *in vivo*. Mol. Ther. 18: 983–986.

Podevin, N., Davies, H.V., Hartung, F., Nogue. F. et al. 2013. Site-directed nucleases: a paradigm shift in predictable, knowledge-based plant breeding. Trends Biotechnol. 31: 376-383.

Porter, S.N., Levine, R.M. and Pruett-Miller, S.M. 2019. A practical guide to genome editing using targeted nuclease technologies. Comprehensive Physiol. 9: 665-714.

Porteus, M. 2016. Genome editing: a new approach to human therapeutics. Annu. Rev. Pharmacol. Toxicol. 56: 29.1–29.28

Porteus, M.H. and Baltimore, D. 2003. Chimeric nucleases stimulate gene targeting in human cells. Science 300: 763.

Proudfoot, C., Carlson, D.F., Huddart, R., Long, C.R. et al. 2015. Genome edited sheep and cattle. Transgenic Res. 24: 147–153.

Puchta, H. 2003. Towards the ideal GMP: homologous recombination and marker gene excision. J. Plant Physiol. 160: 743–754.

Puchta, H. 2017. Applying CRISPR/Cas for genome engineering in plants: the best is yet to come. Current Opinion Plant Biol. 36: 1–8.

Pugh, J. 2016. Driven to extinction? The ethics of eradicating mosquitoes with gene-drive technologies. J. Medical Ethics. 42: 578–581. http://dx.doi.org/10.1136/medethics-2016-103462

Pulecio, J. Verma, N., Mejıa-Ramırez, E., Huangfu, D. et al. 2017. CRISPR/Cas9-based engineering of the epigenome. Cell Stem Cell 21: 431-447.

Pyott, D.E., Sheehan, E. and Molnar, A. 2016. Engineering of CRISPR/Cas9-mediated potyvirus resistance in transgene-free Arabidopsis plants. Mol. Plant Pathol. 17: 1276–1288.

Qi, L.S., Larson, M.H., Gilbert, L.A., Doudna, J.A. et al. 2013. Repurposing CRISPR as an RNA-guided platform for sequence-specific control of gene expression. Cell 152, 1173–1183.

Ramakrishna, S., Dad, A.-B.K., Beloor, J., Gopalappa, R. et al. 2014. Gene disruption by cell-penetrating peptide-mediated delivery of Cas9 protein and guide RNA. Genome Res. 24: 1020-1027.

Ramalingam, S., Kandavelou, K., Rajendran, R. and Chandrasegaran, S. 2011. Creating designed zincfinger nucleases with minimal cytotoxicity. J. Mol. Biol. 405: 630-641

Ramirez, C.L., Certo, M.T., Mussolino, C., Goodwin, M.J. et al. 2012. Engineered zinc finger nickases induce homology-directed repair with reduced mutagenic effects. Nucleic Acids Res. 40: 5560–5568.

Redford, K., Brooks, T., Macfarlane, N. and Adams, J. (eds.) 2019. Genetic Frontiers for Conservation: An Assessment of Synthetic Biology and Biodiversity Conservation. Gland, Switzerland: IUCN.

Rees, H.A., Komor, A.C., Yeh, W.-H., Caetano-Lopes, J. et al. 2017. Improving the DNA specificity and applicability of base editing through protein engineering and protein delivery. Nat. Commun. 8, 15790 (2017)

Ren, B., Yan, F., Kuang, Y., Li, N. et al. 2017. A CRISPR/Cas9 toolkit for efficient targeted base editing to induce genetic variations in rice. Sci. China Life Sci. 33: 516–519.

Ren, B., Yan, F., Kuang, Y., Li, N. et al. 2018. Improved base editor for efficiently inducing genetic variations in rice with CRISPR/Cas9-guided hyperactive hAID mutant. Mol. Plant 2;11(4):623-626. doi: 10.1016/j.molp.2018.01.005.

Reynolds, M. P. 2010. Climate Change and Crop Production. CABI, London.

Reyon, D., Tsai, S.Q., Khayter, C., Foden, J.A. et al. 2012. FLASH assembly of TALENs enables high-throughput genome editing. Nat. Biotechnol. 30: 460–465

Richter, A., Streubel, J. and Boch, J. 2016. TAL effector DNA-binding principles and specificity. *In* Ralf Kühn et al. (eds) *TALENs: Methods and Protocols*, Methods in Molecular Biology, vol. 1338, pp. 9-26. DOI 10.1007/978-1-4939-2932-0_1, © Springer Science+Business Media, New York.

Rode, N.O., Estoup, A., Bourguet, D., Courtier-Orgogozo, V. et al. 2019. Population management using gene drive: molecular design, models of spread dynamics and assessment of ecological risks. Conserv. Genet. 20:671–690. https://doi.org/ 10.1007/s10592-019-01165-5.

Royal Society, 2018. The gene drive research, shy if matters, pp. 1-24. Issued: November 2018 DES5819.

Ryan, O.W., Skerker, J., Maurer, M.J., Li, X. et al. 2014, Selection of chromosomal DNA libraries using a multiplex CRISPR system. eLife 3, 03703. doi: 10.7554/eLife.03703.

Sadelain, M., Papapetrou, E.P. and Bushman, F.D. 2012. Safe harbours for the integration of new DNA in the human genome. Nat. Rev. Cancer 12: 51–58.

Sadhu, M.J., Bloom, J.S., Day, L. and Kruglyak, L. 2016. CRISPR-directed mitotic recombination enables genetic mapping without crosses. Science 352: 1113-1116 (http://dx.doi.org/10.1126/science.aaf5124).

Sadowski, I., Ma, J., Triezenberg, S., and Ptashne, M. 1988. GAL4-VP16 is an unusually potent transcriptional activator. Nature 335: 563–564.

Sainsbury, F. and Lomonossoff, G.P. 2014. Transient expressions of synthetic biology in plants. Curr. Opin. Plant Biol. 19: 1–7.

Sakuma, T. and Yamamoto, T. 2016. Engineering customized TALENs using the Platinum Gate TALEN kit. . *In* Ralf Kühn et al. (eds) *TALENs: Methods and Protocols*, Methods in Molecular Biology, vol. 1338, pp., 61-70. DOI 10.1007/978-1-4939-2932-0_1, © Springer Science+Business Media, New York.

Sander, J.D. and Joung, J.K. 2014. CRISPR-Cas systems for genome editing, regulation and targeting. Nat. Biotechnol. 32: 347–355. doi:10.1038/nbt.2842.

Sander, J.D., Dahlborg, E.J., Goodwin, M.J., Cade, L. et al. 2011. Selection-free zinc-finger-nuclease engineering by context-dependent assembly (CoDA), Nat. Methods 8: 67–69.

Sauer, N.J., Narvaez-Vasquez, J., Mozoruk, J., Miller, R. B. et al. 2016. Oligonucleotide-mediated genome editing provides precision and function to engineered nucleases and antibiotics in plants. Plant Physiol. 170:1917–1928.

Savic, N. and Schwank, G. 2016. Advances in therapeutic CRISPR/Cas9 genome editing. Translational Res. 168: 15–21

SCBD (Secretariat of the Convention on Biological Diversity) 2000. Cartagena Protocol on Biosafety to the Convention on Biological Diversity: Text and annexes. Secretariat of the Convention on Biological Diversity, Montreal, Canada.

Schiml, S. and Puchta, H. 2016. Revolutionizing plant biology: multiple ways of genome engineering by CRISPR-Cas. Plant Methods 12: 8-16.

Schindele, P., Wolter, F. and Puchta, H. 2018. Transforming plant biology and breeding with CRISPR/Cas9, Cas12 and Cas13. FEBS Letters . doi:10.1002/1873-3468.13073

Schmidt, T., Schmid-Burgk, J.L. and Hornung, V. 2017. Synthesis of an arrayed sgRNA library targeting the human genome. Sci. Rep. 5: 14987, DOI: 10.1038/srep14987

Scott, M.J., Gould, F,, Lorenzen, M. et al. 2018. Agricultural production: assessment of the potential use of Cas9-mediated gene drive systems for agricultural pest control. J Responsible Innov 5: S98–S120. https ://doi.org/10.1080/23299 460.2017.14103 43

Scudellari, M. 2019. Hijacking evolution. Narure 571: 160-162.

Segal, D.J. and Meckler, J.F. 2013. Genome engineering at the dawn of the golden age. Annu. Rev. Genomics Hum. Genet. 14: 135–158.

Senturk, S., Shirole, N.H., Nowak, D.G., Corbo, V. et al. 2017. Rapid and tunable method to temporally control gene editing based on conditional Cas9 stabilization. Nature Commun. DOI: 10.1038/ncomms14370

Shalem, O., Sanjana, N.E., Hartenian, E., Shi, X., Scott, D.A., et al. 2014. Genome-scale CRISPR–Cas9 knockout screening in human cells. Science 343: 84–87

Shen, J.P., Zhao, D., Sasik5, R., Luebeck, J. et al. 2017. Combinatorial CRISPR–Cas9 screens for *de novo* mapping of genetic interactions. Nature Meth. doi:10.1038/nmeth.4225

Shi, J., Gao, H., Wang, H., Lafitte, H.R. et al. 2017. ARGOS8 variants generated by CRISPR-Cas9 increase maize grain yield under field drought stress conditions. Plant Biotechnol. J. 15: 207-216.

Shim, G., Kim, D., Park, G.T., Jin, H. et al. 2017. Therapeutic gene editing: delivery and regulatory perspectives. Acta Pharmacologica Sinica 38: 738–753.

Shimatani, Z., Kashojiya, S., Takayama, M., Terada, R. et al. 2017. Targeted base editing in rice and tomato using a CRISPR-Cas9 cytidine deaminase fusion. Nat. Biotechnol. 35: 441–443.

Shin, H.Y., Wang, C., Lee, H.K., Yoo, K.H. et al. 2017. CRISPR/Cas9 targeting events cause complex deletions and 116 insertions at 17 sites in the mouse genome. Nat Commun 8: 15464.

Shin, J., Jiang, F., Liu, J-J., Bray, N.L. et al. 2017. Disabling Cas9 by an anti-CRISPR DNA mimic. Sci. Adv. 3: e1701620.

Sims, M. and First, N.L. 1994. Production of calves by transfer of nuclei from cultured inner cell mass cells. Proc. Natl. Acad. Sci. U.S.A. 91: 6143-6147.

Singh, B.D. 2012. Biotechnology, Expanding Horizons. Kalyani Publishers, Ludhiana, India

Singh, B.D. 2018. Plant Breeding, Principles and Practices, 11th edn. Kalyani Publishers, Ludhiana, New Delhi.

Singh, B.D. 2019. Genetics, 3rd edn. Kalyani Publishers, Ludhiana, New Delhi.

Singh, B.D. and Singh, A.K. 2015 Marker-Assisted Plant Breeding, Principles and Practices. Springer, New Delhi, India.

Singh, J. A., 2019. Informed consent and community engagement in open field research: lessons for gene drive science. BMC Medical Ethics 20 (54).

Sinkins, S.P. and Gould, F. 2006. Gene drive systems for insect disease vectors. Nat. Rev. Genet. 7:427-435. doi:10.1038/nrg1870

Smidler, A., Noble, C., Olejarz, J., Buchthal, J. et al. 2017. Daisy-chain gene drives for the alteration of local populations. https://www.responsivescience.org/ pub/daisydrives

Smith, A.M., Takeuchi, R., Pellenz, S., Davis, L. et al. 2009. Generation of a nicking enzyme that stimulates site-specific gene conversion from the I-AniI LAGLIDADG homing endonuclease. Proc. Natl. Acad. Sci. 106: 5099–5104.

Smith, J., Grizot, S., Arnould, S., Duclert, A. et al. 2006. A combinatorial approach to create artificial homing endonucleases cleaving chosen sequences. Nucleic Acids Res. 34: 22 e149. doi:10.1093/nar/gkl720

Smithies, O., Gregg, R,G,, Boggs, S.S., Koralewski, M.A. et al. 1985. Insertion of DNA sequences into the human chromosomal β-globin locus by homologous recombination. Nature 317: 230–234.

Smyth, S.J. 2017. Canadian regulatory perspectives on genome engineered crops. GM Crops Food 8: 35-43, DOI: 10.1080/21645698.2016.1257468

Songstad, D.D., Petolino, J.F., Voytas, D.F. and Reichert, N.A. 2017. Genome editing of plants. Critical Rev. Plant Sci. 36: 1-23.

Soyk, S., Muller, N. A., Park, S. J., Schmalenbach, I., Jiang, K. et al. 2017. Variation in the flowering gene *SELF PRUNING 5G* promotes day-neutrality and early yield in tomato. Nat. Genet. 49:162–168.

Stafforst, T. and Schneider, M.F. 2012. An RNA–deaminase conjugate selectively repairs point mutations. Angew. Chem. Int. Ed. Engl. 51: 11166–11169. doi: 10.1002/anie.201206489

Stoddard, B.L. 2011. Homing endonucleases: from microbial genetic invaders to reagents for targeted DNA modifications. Structure 19: 7–15.

Stoddard, B.L. 2014. Homing endonucleases from mobile group I introns: discovery to genome engineering. Mobile DNA 5: 7-21. doi:10.1186/1759-8753-5-7.

Storici, F., Lewis, L.K., and Resnick, M.A. 2001. *In vivo* site-directed mutagenesis using oligonucleotides. Nat. Biotechnol. 19: 773–776.

Sturtevant, A.H. 1965. A History of Genetics. Harper and Row, New York.

Sun, N. and Zhao, H. 2013. Transcription activator-like effector nucleases (TALENs): a highly efficient and versatile tool for genome editing. Biotechnol. Bioeng. 110: 1811-1821.

Suzuki, K., Tsunekawa, Y., Hernandez-Benitez, R., Wu, J. et al. 2016. *In vivo* genome editing via CRISPR/Cas9 mediated homology-independent targeted integration. Nature 540: 144–149. doi:10.1038/nature20565.

Swarts, D.C., van der Oost, J. and Jinek, M. 2017. Structural basis for guide RNA processing and seed-dependent DNA targeting by CRISPR-Cas12a. Mol. Cell 66: 221–233.

Tait-Burkard, C., Doeschl-Wilson, A., McGrew, M.J., Archibald, A.L. et al. 2018. Livestock 2.0 – genome editing for fitter, healthier, and more productive farmed animals. Genome Biol. 19: 204, https://doi.org/10.1186/s13059-018-1583-1

Takeuchi, R., Choi, M. and Stoddard, B.L. 2015. Engineering of customized meganucleases via *in vitro* compartmentalization and *in cellulo* optimization. Methods Mol. Biol. 1239: 105–132. doi:10.1007/978-1-4939-1862-1_6.

Tanenbaum, M.E., Gilbert, L.A., Qi, L.S., Weissman, J.S., Vale, R.D. 2014. A protein-tagging system for signal amplification in gene expression and fluorescence imaging. Cell. 159: 635-646. DOI: 10.1016/j.cell.2014.09.039.

Tang, W., Hu, J.H. and Liu, D.R. 2017. Aptazyme-embedded guide RNAs enable ligand-responsive genome editing and transcriptional activation. Nat. Commun. 8: 15939. doi: 10.1038/ncomms15939 .

Tasan, I. and Zhao, H. 2017. Targeting specificity of the CRISPR/Cas9 system. ACS Synth. Biol. 6: 1609–1613

Taylor, G.K., Petrucci, L.H., Lambert, A.R., Baxter, S.K. et al. 2012. LAHEDES: the LAGLIDADG homing endonuclease database and engineering server. Nucleic Acids Res. 40 (W1): W110-W116.

Tebas, P., Stein, D., Tang, W.W., Frank, I. et al. 2014. Gene editing of *CCR5* in autologous CD4 T cells of persons infected with HIV. N. Engl. J. Med. 370: 901–910.

Thakore, P.I. and Gersbach, C.A. 2016. Design, assembly, and characterization of TALE-based transcriptional activators and repressors. *In* Ralf Kühn et al. (eds) *TALENs: Methods and Protocols*, Methods in Molecular Biology, vol. 1338, pp., 71-88. DOI 10.1007/978-1-4939-2932-0_1, © Springer Science+Business Media, New York

Thakore, P.I., Black, J.B., Hilton, I.B. and Gersbach, C.A. 2016. Editing the epigenome: technologies for programmable transcription and epigenetic modulation. Nat Methods. 13:127-137. doi: 10.1038/nmeth.3733.

Thomas, K,R., Folger, K.R. and Capecchi, M.R. 1986. High frequency targeting of genes to specific sites in the mammalian genome. Cell 44: 419–428.

Thompson, P.B. 2018. The roles of ethics in gene drive research and governance. J. Responsible Innov. 5: S159–S179. https ://doi.org/10.1080/23299 460.2017.14155 87

Tovkach, A., Zeevi, V. and Tzfira, T. 2009. A toolbox and procedural notes for characterizing novel zinc finger nucleases for genome editing in plant cells. Plant J. 57: 747–757.

Truong, D-J. J., Kuhner, K., Kuhn, R., Werfel, S. et al. 2015. Development of an intein-mediated split–Cas9 system for gene therapy. Nucleic Acids Res. 43: 6450–6458. doi: 10.1093/nar/gkv601

Tsai, C.-J. and Xue, L.J. 2015. CRISPRing into the woods. GM Crops Food 6: 206-215, DOI: 10.1080/21645698.2015.1091553.

Tsai, S.Q. and Joung, J.K. 2016. Defining and improving the genome-wide specificities of CRISPR–Cas9 nucleases. Nat. Rev. Genet. 17: 300-312.

Tseng, Y.L., Kuwaki, K., Dor, F.J. et al. 2005. Alpha1,3-Galactosyltransferase gene knockout pig heart transplantation in baboons with survival approaching 6 months. Transplantation 80: 1493–1500.

Tuerlings, E. 2019. WHO Expert advisory committee on developing global standards for governance and oversight of human genome editing, background paper governance 1, human genome editing.

Tycko, J., Myer, V.E. and Hsu, P.D. 2016. Methods for optimizing CRISPR-Cas9 genome editing specificity. Mol. Cell 63: 355-370.

Tzfira, T., Weinthal, D., Marton, I., Zeevi, V. et al. 2012. Genome modifications in plant cells by custom-made restriction enzymes. Plant Bioteccnol. J. 10: 373–389.

Ueta, R., Abe, C., Watanabe, T., Sugano, S.S. et al. 2017. Rapid breeding of parthenocarpic tomato plants using CRISPR/Cas9. Sci. Reports 7: 507.

Urnov, F.D., Miller, J.C., Lee, Y.L. Beausejour, C.M. et al. 2005. Highly efficient endogenous human gene correction using designed zinc-finger nucleases. Nature 435: 646–651.

Urnov, FD, Rebar EJ, Holmes MC, Zhang HS, Gregory PD. 2010. Genome editing with engineered zinc finger nucleases. Nat. Rev. Genet. 11: 636–646.

Van Eenennaam, A.L. 2017. Genetic modification of food animals. Curr. Opin. Biotechnol. 44: 27–34

Van Eenennaam, A.L. 2018. The importance of a novel product risk-based trigger for gene-editing regulation in food animal species. CRISPR J 1:101–106. https://doi.org/10.1089/crispr.2017.0023

Van Eenennaam, A.L. 2019. Application of genome editing in farm animals: cattle. Transgenic Res. 28: 93–100.

Van Eenennaam, A.L., Wells, K.D. and Murray, J.D. 2019. Proposed U.S. regulation of gene-edited food animals is not fit for purpose. npj Sci. Food 3:3 ; https://doi.org/10.1038/s41538-019-0035-y.

Vella, M.R., Gunning, C.E., Lloyd. A.L. and Gould, F. 2017. Evaluating strategies for reversing CRISPR-Cas9 gene drives. Sci. Reports 7: 11038. DOI:10.1038/s41598-017-10633-2

von Gleich, A. and Schröder, W. (Eds) 2020. Gene Drives at Tipping Points: Precautionary Technology Assessment and Governance of New Approaches to Genetically Modify Animal and Plant Populations. Springer Open, Springer Nature, Switzerland AG

Vrebalov, J., Ruezinsky, D., Padmanabhan, V., White, R. et al. 2002. A MADS-box gene necessary for fruit ripening at the tomato ripening-inhibitor (*Rin*) locus. Science 296: 343–346.

Wah, D.A., Bitinaite, J., Schildkraut, I. and Aggarwal, A.K. 1998. Structure of FokI has implications for DNA cleavage, Proc. Natl. Acad. Sci. U.S.A. 95: 10564–10569.

Wang, B. and Zhou, J. 2003. Specific genetic modifications of domestic animals by gene targeting and animal cloning. Reprod. Biol. Endocrinol. 1:103-110. doi:10.1186/ 1477-7827-1-103

Wang, C.-H., Lundh, M., Fu, A., Kriszt, R. et al. 2020. CRISPR-engineered human brown-like adipocytes prevent diet-induced obesity and ameliorate metabolic syndrome in mice. Sci. Transl. Med. 12, eaaz8664. DOI: 10.1126/scitranslmed.aaz8664

Wang, M., Liu, Y., Zhang, C., Liu, J. et al. 2015. Gene editing by co-transformation of TALEN and chimeric RNA/DNA oligonucleotides on the rice *OsEPSPS* gene and the inheritance of mutations. PLoS ONE 10: e0122755. doi:10.1371/journal. pone.0122755

Wang, T., Wei, J.J., Sabatini, D.M. and Lander, E.S. 2014. Genetic screens in human cells using the CRISPR/Cas9 system. Science 343: 80–84

Wang, Y., Cheng, X., Shan, Q., Zhang, Y. et al. 2014a. Simultaneous editing of three homoeoalleles in hexaploidy bread wheat confers heritable resistance to powdery mildew. Nature Biotechnol. 32: 947–951.

Wang, Y., Khan, I.F., Boissel, S., Jarjour, J. et al. 2014b. Progressive engineering of a homing endonuclease genome editing reagent for the murine X-linked immunodeficiency locus. Nucleic Acids Res. 42: 6463–6475. doi: 10.1093/ nar/gku224.

Wang, Y., Liu, K.I., Sutrisnoh, N.-A.B., Srinivasan, H. *et al.* 2018. Systematic evaluation of CRISPR-Cas systems reveals design principles for genome editing in human cells. Genome Biol 19**:** 62. doi:10.1186/s13059-018-1445-x

Wang, Y., Liu, K.I., Sutrisnoh, N.B., Srinivasan, H. et al. 2016. Enhanced rice blast resistance by CRISPR/Cas9-targeted mutagenesis of the ERF transcription factor gene OsERF922. PLoS ONE, 11, e0154027.

Whelan, A.I. and Lema, M.A. 2015. Regulatory framework for gene editing and other new breeding techniques (NBTs) in Argentina, GM Crops Food 6: 253-265, DOI: 10.1080/21645698.2015.1114698

WHO, 2019. WHO expert panel paves way for strong international governance on human genome editing. https://www.who.int/news-room/detail/19-03-2019-who-expert-panel

Windbichler, N., Menichelli, M., Papathanos, P. A., Thyme, S. B. et al. 2011. A synthetic homing endonuclease-based gene drive system in the human malaria mosquito. Nature. 473: 212–215.

Wolt, J.D., Wang, K., Sashital, D. and Lawrence-Dill, C.J. 2016. Achieving plant CRISPR targeting that limits off-target effects. Plant Genome 9: doi: 10.3835/plantgenome2016.05.0047

Wolter, F., Klemm, J. and Puchta, H. 2018. Efficient *in planta* gene targeting in *Arabidopsis* using egg cell-specific expression of the Cas9 nuclease of *Staphylococcus aureus*. Plant J. 94: 735–746.

Wright, A.V., Sternberg, S.H., Taylor, T.W., Staahl, B.T. et al. 2015. Rational design of a split-Cas9 enzyme complex. Proc. Natl. Acad. Sci. U.S.A. 112: 2984–2989.

Wright, D.A., Li, T., Yang, B. And Spalding, M.H. 2014. TALEN-mediated genome editing: prospects and perspectives. Biochem. J. 462: 15–24. doi:10.1042/BJ20140295

Wright, D.A., Townsend, J.A., Winfrey, Jr., R.J. Irwin, P.A. et al. 2005. High-frequency homologous recombination in plants mediated by zinc-finger nucleases. Plant J. 44: 693–705. doi:10.1111/j.1365-313X.2005.02551.x

Wu, B., Luo, L. and Gao, X.J. 2016. Cas9-triggered chain ablation of Cas9 as a gene drive brake. Nat Biotechnol 34:137–138.

Wu, J., Platero-Luengo, A. Sakurai, M., Sugawara, A. et al. 2017. Interspecies chimerism with mammalian pluripotent stem cells. Cell 168: 473–486.e15.

Xia, E., Zhang, Y., Cao, H., Li, J. et al. 2019. TALEN-mediated gene targeting for cystic fibrosis-gene therapy. Genes 10: 39. doi:10.3390/genes10010039

Xie, K. and Yang, Y. 2013. RNA-guided genome editing in plants using a CRISPR-Cas system. Mol. Plant 6: 1975-1983.

Xie, K., Minkenberg, B, and Yang, Y. 2015. Boosting CRISPR/Cas9 multiplex editing capability with the endogenous tRNA-processing system. Proc. Natl. Acad. Sci. (USA). 112: 3570-3575.

Xu, S., Cao, S., Zou, B., Yue, Y. et al. 2016. An alternative novel tool for DNA editing without target sequence limitation: the structure-guided nuclease. Genome Biol. 17:186. DOI 10.1186/s13059-016-1038-5

Yan, F., Kuang, Y., Ren, B., Wang, J. et al. 2018. Highly efficient A.T to G.C base editing by Cas9n-guided tRNA adenosine deaminase in rice. Mol. Plant 11: 631-634.

Yang, B., Li, X., Lei, L., Chen, J. 2017. APOBEC: From mutator to editor. J. Genet. Genomics 44: 423–437. doi: 10.1016/j.jgg.2017.04.009

Yang, H. and Wu, Z. 2018. Genome editing of pigs for agriculture and biomedicine. Front. Genet. 9: 360. doi: 10.3389/fgene.2018.00360

Yang, J., Guo, S., Yuan, P. and Wei, W. 2016. Assembly of customized TAL effectors through advanced ULtiMATE system. *In* Ralf Kühn et al. (eds) *TALENs: Methods and Protocols*, Methods in Molecular Biology, vol. 1338, pp., 49-60. DOI 10.1007/978-1-4939-2932-0_1, © Springer Science+Business Media, New York

Yang, L., Briggs, A.W., Chew, W.L., Mali, P. et al. 2016. Engineering and optimising deaminase fusions for genome editing. Nat. Commun. 7: 13330. DOI: 10.1038/ncomms13330

Yang, L., Guell, M, Byrne, S., Yang, J.L. et al. 2013. Optimization of scarless human stem cell genome editing. Nucleic Acids Res. 41: 9049–9061.

Yang, L., Guell, M. and Niu, D. et al. 2015. Genome-wide inactivation of porcine endogenous retroviruses (PERVs). Science 350: 1101–1104.

Ye, L., Wang, J., Beyer, A.I., Teque, F. et al. 2014. Seamless modification of wild-type induced pluripotent stem cells to the natural CCR5Delta32 mutation confers resistance to HIV infection. Proc. Natl. Acad. Sci. USA 111: 9591–9596.

Yeh, W-H., Chiang, H., Rees, H.A., Edge, A.S.B. et al. 2018. *In vivo* base editing of post-mitotic sensory cells. Nature Comm. DOI: 10.1038/s41467-018-04580-3

Yin, P., Deng, D., Yan, C., Pan, X. et al. 2012. Specific DNA-RNA hybrid recognition by TAL effectors. Cell Reports 2: 707–713

Yoon, K., Cole-Strauss, A. and Kmiec, E.B. 1996. Targeted gene correction of episomal DNA in mammalian cells mediated by a chimeric RNA.DNA oligonucleotide. Proc. Natl. Acad. Sci. USA. 93: 2071-2076.

Youds, J.L. and Boulton, S.J. 2011. The choice in meiosis—defining the factors that influence crossover or non-crossover formation. J. Cell Sci. 124: 501–513.

Yudin, N.S., Lukyanov, K.I., Voevoda, M.I. and Kolchanov, N.A. 2016. Application of reproductive technologies to improve dairy cattle, genomic selection. Russian J. Genet.: Appl. Res. 6: 321–329.

Yum, S.-Y., Youn, K-Y., Choi, W.J. and Jang, G. 2018. Development of genome engineering technologies in cattle: from random to specific. J. Animal Sci. Biotechnol. 9: 16. DOI 10.1186/s40104-018-0232-6

Zaka, A., Grande, G., Coronejo, T., Quibod, I.L. et al. 2018. Variations in the promoter of *OsSWEET13* and *OsSWEET14* expand the range of resistance against *Xanthomonas oryzae pv. oryzae*. PLoS ONE 13: e0203711. https://doi.org/10.1371/journal.pone.020371

Zentner, Z.E. and Wade, M.J. 2017. The promise and peril of CRISPR gene drives. Bioessays 39: 1700109

Zetsche, B., Gootenberg, J.S., Abudayyeh, O.O., Slaymaker, I.M. et al. 2015a. Cpf1 is a single RNA-guided endonuclease of a class 2 CRISPR-Cas system. Cell 163: 759–771.

Zetsche, B., Volz, S.E., and Zhang, F. 2015b. A split-Cas9 architecture for inducible genome editing and transcription modulation. Nat. Biotechnol. 33: 139–142.

Zhang F., Cong, L., Lodato, S., Kosuri, S., et al. 2011. Efficient construction of sequence-specific TAL effectors for modulating mammalian transcription. Nat. Biotechnol. 29: 149–153. [PubMed: 21248753]

Zhang, F., Maeder. M.L., Unger-Wallace, E., Hoshaw, J.P. et al. 2010. High frequency targeted mutagenesis in *Arabidopsis thaliana* using zinc finger nucleases. Proc. Natl. Acad. Sci. USA 107:12028-12033. doi: 10.1073/pnas.0914991107

Zhang, Y., Arango, G., Li, F., Xiao, X. et al. 2018. Comprehensive off-target analysis of dCas9-SAM-mediated HIV reactivation via long noncoding RNA and mRNA profiling. BMC Medical Genomics 11: 78. https://doi.org/10.1186/s12920-018-0394-2

Zhang, Y., Liang, Z., Zong, Y., Wang, Y. et al. 2016. Efficient and transgene-free genome editing in wheat through transient expression of CRISPR/Cas9 DNA or RNA. Nat. Comm. 7: 12617. DOI: 10.1038/ncomms12617.

Zhang, Y., Yin, C., Zhang, T., Li, F. et al. 2015. CRISPR/gRNA-directed synergistic activation mediator (SAM) induces specific, persistent and robust reactivation of the HIV-1 latent reservoirs. Sci. Reports 5:16277. DOI: 10.1038/srep16277.

Zhao, J., Lai, L., Ji, W. and Zhou, Q. 2019. Genome editing in large animals: current status and future prospects. Natl. Sci. Rev. 6: 402–420, 2019. doi: 10.1093/nsr/nwz013

Zhao, J., Lin, Q. Song, Y. and Liu, D. 2018. Universal CARs, universal T cells, and universal CAR T cells. J. Hematol. Oncol. 11:132-240. https://doi.org/10.1186/ s13045-018-0677-2

Zheng, Q., Lin, J., Huang, J., Zhang, H. et al. 2017. Reconstitution of UCP1 using CRISPR/Cas9 in the white adipose tissue of pigs decreases fat deposition and improves thermogenic capacity. Proc. Natl. Acad. Sci. U.S.A. 114: E9474–E9482. doi: 10.1073/pnas.1707853114

Zheng-yin, X., Li-fang, Z., Wen-xiu, M., Lu-lu, C. et al. 2017. Action modes of transcription activator-like effectors (TALEs) of *Xanthomonas* in plants. J. Integrative Agric. 16: 2736–2745

Zhou, H., He, M., Li, J., Chen, L. et al. 2016. Development of commercial thermo-sensitive genic male sterile rice accelerates hybrid rice breeding using the CRISPR/Cas9-mediated TMS5 editing system. Sci. Rep. 6: 37395. DOI: 10.1038/srep37395.

Zhu, Y., Zhang, F. and Huan, Z. 2018. Structural insights into the inactivation of CRISPR-Cas systems by diverse anti-CRISPR proteins. BMC Biol. 16:32. doi.org/10.1186/s12915-018-0504-9

Zong, Y., Wang, Y., Li, C., Zhang, R. et al. 2017. Precise base editing in rice, wheat and maize with a Cas9-cytidine deaminase fusion. Nat. Biotechnol. 35: 438–440.

■■■■

SUGGESTED READING

Baylis, F. 2019. Altered Inheritance-CRISPR and the Ethics of Human Genome Editing. Harvard University Press, London, UK

Coen, E. 1998/1999. The Art of Genes: How Organisms Make Themselves. Oxford University Press. Oxford, UK.

Doudna, J.A. and Sternberg, S.H. 2017. A Crack in Creation: Gene Editing and Unthinkable Power to Control Evolution. Houghton Mifflin Harcourt, Boston, USA

Judson, H.. 1996/2013. The Eighth Day of Creation: Makers of the Revolution in Biology. Simon and Schuster, New York, USA

Kozubek, J. 2016/2018. Modern Prometheus-Editing the Human Genome with CRISPR-Cas9. Cambridge University Press, Cambridge, UK (Kindle Edition)

Metzl, J. 2019. Hacking Darwin- Genetic Engineering and the Future of Humanity. Sourcebooks, Inc., Naperville, Illinois, USA. (Kindle edition)

Miglani, G.S. 2019. Genome Editing: A Comprehensive Treatise. Narosa Publishing House Pvt. Ltd., New Delhi, India.

Mukherjee, S. 2016. The Gene: An Intimate History. Allen Lane/Penguin Random House India Pvt. Ltd., Gurugram, Haryana, India.

National Academy of Medicine, National Academy of Sciences, and the Royal Society. 2020. Heritable Human Genome Editing. The National Academies Press, Washington, USA.

Parrington, J. 2016/2020. Redesigning Life - How Genome Editing Will Transform the World. Oxford University Press, London/U.K.

Ramakrishnan, V. 2018. Gene Machine: The Race to Decipher the Secrets of the Ribosome. Harper Collins Publishers, Noida, U.P, India.

Ridley, Matt. 2000. Genome-The Autography of a Species in 23 Chromosomes. Harper Collins Publishers Inc., New York, USA

Schrödinger, E. 1944. What is Life? With Mind and Matter and Autobiographical Sketches (Canto Classics). Cambridge University Press India Pvt.Ltd., New Delhi, India

Venter, J.C. 2014. Life at the Speed of Light-from the Double Helix to the Dawn of Digital Life. Penguin Books, New York, USA.

Watson, J.W. 1968. The Double Helix. Atheneum Press (US), Weidenfeld & Nicolson (UK)

Watson, J.W. 2000. A Passion for DNA: Genes, Genomes, and Society. Cold Spring Harbour Laboratory Press, New York, USA.

Watson, J.W. 2001. Genes, Girls, and Gamow: After the Double Helix. Vintage Books/Random House, Inc., New York, USA

Watson, J.W. 2003. DNA: The Secret of Life. Random House Group Ltd., London, UK

Watson, J.W. and Tooze, J. 1981. The DNA Story-A Documentary History of Gene Cloning. W.H. Freeman and Company, San Francisco, USA

Watson, J.D., Tooze, J. and Kurtz, D.T. 1983. Recombinant DNA – A Short Course. Scientific American Books, W.H. Freeman & Co., New York, USA.

Williams, G. 2019. Unravelling the Double Helix: The Lost Heroes of DNA. Weidenfeld & Nicolson, the Orion Publishing Group, Ltd., London, UK.

■■■■

AUTHOR INDEX

SUBJECT INDEX

I-CreI 25-28, 32, 33, 35